MESEMBS
OF THE WORLD

Supported by

NATIONAL
BOTANICAL INSTITUTE
OF SOUTH AFRICA

MESEMBS
OF THE WORLD

Authors:
Gideon F. Smith, Pascale Chesselet, Ernst J. van Jaarsveld,
Heidi Hartmann, Steven Hammer, Ben-Erik van Wyk,
Priscilla Burgoyne, Cornelia Klak and Hubert Kurzweil

Published by

BRIZA
PUBLICATIONS

Published by

Briza Publications

CD 90/11690/23

P.O. Box 56569
Arcadia
0007
Pretoria
South Africa

First Edition 1998

Text © National Botanical Institute of South Africa
Photographs © various photographers
Drawings by Marietjie Steyn, Pascale Chesselet and Gillian Condy
Cover design by Roger de Andrade
Language edited by Janine Smit
Typesetting by Ilse van Oudtshoorn
Reproduction by Unifoto, Cape Town
Printed and bound by Tien Wah Press, Singapore

All rights reserved. No part of this publication may be reproduced or transmitted in any form or by any means without written permission of the copyright holder.

ISBN 1 875093 13 3

Contents

ACKNOWLEDGEMENTS	6
INTRODUCTION	7
MAGIC OF MESEMBS	8
DISTRIBUTION AND ECOLOGY	10
CONSERVATION BY CRAIG HILTON-TAYLOR	12
USES OF MESEMBS	14
CULTIVATION	16
HOW TO IDENTIFY THE MESEMB GROUPS	20
GUIDE TO THE GROUPS	24
GENERA OF MESEMBS	
GROUP 1. WEEDY MESEMBS	27
GROUP 2. FLAT-LEAVED MESEMBS	57
GROUP 3. FLOWERING STONES	81
GROUP 4. TONGUE-LEAVED MESEMBS	133
GROUP 5. ROUGH-LEAVED MESEMBS	155
GROUP 6. TOOTH-LEAVED MESEMBS	177
GROUP 7. TUFTED MESEMBS	201
GROUP 8. BEAD-LEAVED MESEMBS	233
GROUP 9. MAT-FORMING MESEMBS	245
GROUP 10. DWARF SHRUBBY MESEMBS	275
GROUP 11. GLITTERING SHRUBBY MESEMBS	291
GROUP 12. *LAMPRANTHUS*-LIKE SHRUBBY MESEMBS	315
GROUP 13. *RUSCHIA*-LIKE SHRUBBY MESEMBS	345
GROUP 14. *LEIPOLDTIA*-LIKE SHRUBBY MESEMBS	369
FURTHER READING	380
KEY TO GENERA	381
INDEX	395

Acknowledgements

Many people contributed to making this book a reality. In particular, the publisher and authors would like to thank the following institutions and individuals for the use of information and/or photographic material:

The National Botanical Institute of South Africa and the Bolus Herbarium of the University of Cape Town.

Otto Leistner, Deon Viljoen, James Townsend, Emsie du Plessis, Hugh Glen, Bob Chinnock, Derek Clark, John Trager, Chris Barnhill, Graham Williamson, Adela Romanowski, Hellmut Toelken, Bruce Bayer, Piet Vorster, Gerhard Marx, David Lambie, John Braggins, Peter Linder, Ulrich Meve, Antje Burke, Jeanette Loedolff.

From the authors: Gideon F. Smith, Pascale Chesselet, Ernst J. van Jaarsveld, Heidi Hartmann, Steven Hammer, Ben-Erik van Wyk, Priscilla Burgoyne, Cornelia Klak and Hubert Kurzweil.

Introduction

"Mesembs" is a popular term used for succulent members of the family Aizoaceae, sometimes placed in a separate family of their own, the Mesembryanthemaceae. The succulent Aizoaceae are also known as "vygies", "fig-marigolds", "flowering stones", "ice plants" or "midday flowers" and they display remarkable variation in leaf architecture, flower colour and form and fruit structure. As a result, these plants fascinate many plant lovers and have become popular amongst collectors of succulent plants. The aim of this book is to provide an illustrated, easy-to-read overview of all 123 mesemb genera that are currently recognised. Illustrations are provided for a representative selection of species.

The genus rank (as opposed to the species or other taxonomic ranks) is used to refer to a group of species that are more closely related to one another than to other groups of species. In general, the species in a genus exhibit a number of similar characters, i.e. they show an overall resemblance, and ideally there should be discontinuities between the species included in one genus and those included in other genera. It is often easy to group species in a genus together on the basis of their external appearance. However, this is not always possible and such an oversimplification of the genus rank as a taxonomic category is by no means foolproof. Essentially, genera are man-made groups, but in cases where they have been well studied they can also be regarded as natural groupings to which recognition has been given to facilitate the classification of biological entities such as species. In contrast to animals, in which hybrids between species of different genera are almost unknown, plants do not observe this barrier to reproduction.

Topics that are covered in the generic treatments in this book include the derivation of the genus name, common names, a brief description, distinguishing characters, flowering time (mostly time of day and season), geographical distribution and ecology, and cultivation. Interesting anecdotes about a genus are given in a separate paragraph. Maps indicating the natural distribution are provided as a further aid to assist with identification at the generic rank. Full lists of all the species and infraspecific taxa that are recognised, are given for each genus, along with an indication of often used generic synonyms (other scientific names by which the genera are or have been known) and the conservation status of the species. Some of these topics are also discussed in supplementary chapters. Most importantly, a key (identification aid) is provided to all the mesemb genera.

To facilitate easy identification, the genera are divided into groups exhibiting more or less similar external characters, for example branching pattern, growth form, leaf shape and capsule type. These groups may or may not reflect a close relationship amongst the genera included. The purpose was not necessarily to attempt a reflection of affinity, but this work will hopefully fill a long-existing void in the literature of this important group of succulent plants.

Magic of Mesembs

"Magic" is not too strong a term for the appeal of this unique group of plants. Highly diverse in form and stature, with plants as small as wheat grains or as large as rhinos, radiant when in flower, and possessing fruits as complex as any in the floral kingdom or as simple as salt-shakers, mesembs have fascinated botanists, horticulturalists, and travellers to southern Africa since the 17th century.

In the present century the family has received a great deal of attention, in the field and in collections of preserved plant material (in herbaria). Many of the relevant treatments have been published in obscure and expensive journals, and Hans Herre's principal South African work on the mesembs is now out of print. The time therefore seems right for a new publication which should bring this wonderful group of plants to a wider public attention, giving recent insights into its character and demonstrating some of its matchless beauty.

It is an important group of plants in several senses. They stabilise the soil; their blossoms cater year-round (and around the clock) to various insects; their fruits nourish rodents; their leaves are vital as fodder. Namaqualand without mesembs would be poor and barren indeed. The different genera grow in many habitats, from rocky crevices to silty flats and saline wastelands. Apart from their use in grazing, plants have been used in many domestic contexts: soap-making, poultices, preserves, and even as a kind of stimulant.

The family has had a wide impact in horticulture, especially in the northern hemisphere, to which, with very few exceptions, it is not native. Americans know mesembs as ground covers and rockery plants; *Delosperma nubigenum* is an important ground cover in much of the west, and a hybrid of *Aptenia cordifolia* dominates landscapes in arid southern California, having displaced the heavier, soil-tugging mesembs like *Carpobrotus* species. In many areas with Mediterranean mild-winter climates, the shrubbier mesembs have for a long time been popular as landscape subjects; the easy cultivation and brilliant flowers of these plants have guaranteed their immortality. Northern Europeans know mesembs as treasured pot plants and brilliant garden annuals; the Bok Bay vygie is common in English and Dutch gardens. Since the late 18th century, the dwarf species, the so-called sphaeroids or pebble plants, have been cultivated in glasshouses. These adaptable plants are so highly ornamental, and so curious in their annual cycle of behaviour – they shed their patterned skins like a snake – that they have attracted a large group of devoted horticulturists, who eagerly await the latest news from South Africa. And there is always news. Since 1900 there has not been a year in which several significant new species were not discovered. This reflects an increase in fieldwork and stresses the fact that many species are extremely limited in their distribution. The smallest, most cryptic plants are easily overlooked, and many species resemble each other greatly, differing only in floral behaviour or in subtleties of fruit structure.

Those fortunate enough to have seen mesembs in their natural habitats will be able to testify that their visual appeal is increased by an extraordinarily fit match between habitat and plant. This is particularly true for the mimicry species, which resemble the rocks amongst which they grow. But once a year these plants give themselves away in a mass floral display of wonderful luminosity; the light dances with purple, gold, and more rarely, scarlet. Flowers can open at any hour from early morning to noon (hence the name midday flowers), late afternoon, twilight, and late evening, each with its characteristic scent and, probably, its favoured pollinator. Most of those are still unstudied, and that is true for so many aspects of this vast group of succulent plants. This book is an invitation to learn, to observe, and to contribute to our knowledge of southern Africa's most diverse plant family.

Drosanthemum speciosum

Pleiospilos compactus

Lithops steineckeana

Mesembryanthemum sp. showing water cells

Titanopsis calcarea

Conophytum sp.

Distribution and ecology

Mesembs occur mainly in the southwestern parts of the African continent, from Angola down to the Cape, extending well into the east over the central plateau of South Africa and reaching into Zimbabwe and Botswana. The highest number of species and genera is found in the mountainous area of the Richtersveld in the Northern Cape Province: about 30 genera are counted in that small region. The number of genera decreases towards the east, and hitherto only seven genera have been recorded from Lesotho.

Outside this core area, several species of *Delosperma* are found from northern Tanzania up to the Yemen and also in Arabia and Madagascar. A few species of *Mesembryanthemum* occur around the Mediterranean Sea, and most species of the genus *Disphyma* are found in Australia, New Zealand and the adjacent islands. The genus *Sarcozona* and some species of *Carpobrotus* are restricted to Australia, but species of *Ruschia* and *Lampranthus* may not be indigenous to Australasia. It is also not clear whether species of *Malephora* and *Carpobrotus* are recent introductions to South America, but the populations of *Carpobrotus* found along the west coast of North America were certainly introduced by man.

Winter rains dominate in the greatest part of all the distribution areas mentioned above, and plants react by forming their leaves and flowers mostly during that period. Fruits are generally developed towards or in the dry summer season, the seeds being ready for dispersal at the beginning of the next rainy period. Only a limited number of genera, such as *Delosperma* and *Khadia*, is well adapted to summer rains, but *Trichodiadema* and spiny members of *Ruschia* rapidly grow shoots after rains, regardless of the season.

Although the soils on which mesembs are found differ greatly, for most genera they can be described as rather sandy-gravel and well drained. Mesembs often dominate on gravelly plains, for example the Knersvlakte, but they may also be found in broken rocky places. In these habitats, shrubs like *Stoeberia* may settle in crevices amongst the rocks. Only a few genera, for example *Khadia* and *Ebracteola,* prefer humus, and equally few settle in flood plains, for example *Nananthus* and *Mestoklema*. Only a restricted number of species from some genera can be classified as pioneers or weeds, for example some in the genera *Psilocaulon* and *Malephora*, and a few in the genera *Cephalophyllum* and *Leipoldtia*. In the core areas, temperatures are generally rather low during winter, meaning that the main growing season corresponds with moderate, wet conditions. The dry, hot summer months are survived in a resting state. *Mitrophyllum* and many other genera are so strictly adapted to seasonal changes of heat and moisture that they grow themselves to death when permanently kept in moist conditions.

As a rule, pollination is performed by insects, which are attracted by the mostly brightly coloured petals. The flowers open either during the day, in the afternoon or at night. Bees collect nectar from day-flowering species, for example those of *Psilocaulon*. Dispersal of seeds in most genera is closely related to the opening of the capsule through moisture: expanding keels lift the valve when moistened, and in most cases raindrops expel the seeds from the locules. This process is triggered by rain, when conditions for germination and the establishment of seedlings are favourable. In a number of genera the capsules break off from their stalks and are dispersed by wind as tumble fruits. The dispersal of fruits by animals has only been observed in a few species. Nuts and dry fruits are rare.

Ruschia sp.

Mesembryanthemum sp., presently known as *Eurystigma clavatum*

Braunsia apiculata

Various mesembs growing in rock crevices

Drosanthemum speciosum

CONSERVATION

by Craig Hilton-Taylor

As is the case with most groups of succulent plants, the mesembs are also popular amongst collectors. Unfortunately, the removal of plants from the wild together with habitat destruction or modification and various other threats are having a marked impact. Of the approximately 1 800 mesembs (species and infraspecific taxa) listed in this book, 275 have had their conservation status investigated and of these, 242 have been assigned a Red Data conservation status. The IUCN Red Data categories, as defined below, are used for the purposes of this book. For each species or infraspecific taxon evaluated, one of the symbols below appears in brackets after its name. If no symbol is given, the plant has not yet been evaluated, even though it may not be threatened. More detailed definitions are given in the Red Data List of southern African Plants.

Extinct [Ex]: Plants (species, subspecies, varieties or forms) which are no longer known to exist in the wild after repeated searches of type localities and other known or likely places.

Endangered [E]: Plants in immediate danger of extinction if the factors causing their decline continue to operate.

Vulnerable [V]: Plants believed likely to move into the Endangered category in the near future if the factors causing their decline continue to operate.

Rare [R]: Plants with small world populations that are not at present Endangered or Vulnerable, but are at risk as some unexpected threat could easily cause a critical decline.

Indeterminate [I]: Plants known to be Extinct, Endangered, Vulnerable, or Rare but where there is not enough information to say which of the four categories is appropriate.

Rare/Vulnerable [R/V]: In a few instances where there was doubt as to the appropriate category to use, this hybrid category was used instead of Indeterminate.

Insufficiently Known [K]: Plants that are suspected but not definitely known to belong to any of the above categories, because of a lack of information.

Not threatened [nt]: This category is used for plants which are no longer in one of the above categories due to an increase in population sizes or to subsequent discoveries of more individuals or populations. Taxonomic changes, particularly the lumping of species, subspecies or varieties, may also result in them losing their threatened status.

A brief analysis of the main threats to mesembs showed that the vast majority (70%) of those threatened have restricted distributions, sometimes with only very small numbers of plants. As a result, the illegal removal of plants from the wild by collectors (42% of cases) and habitat destruction by ploughing, overgrazing, afforestation, urbanisation and mining can very quickly result in a species becoming highly threatened and possibly even extinct. Habitat destruction was almost certainly the cause of extinction in the cases of *Cephalophyllum parvulum*, *Circandra serrata*, *Erepsia promontorii*, *Lampranthus schlechteri* and *L. vanzijliae*. Other threats include habitat transformation by the invasion of alien plant species and frequent fires, and the building of dams and roads.

In terms of all the conservation legislation, collectors are required to have permits to collect seed or plants of mesemb species from the wild. Listing of species on the appendices of the Convention on International Trade in Endangered Species of Fauna and Flora (CITES) is a possible way to control illegal trade. At present no mesembs are listed.

Aloinopsis setifera – rare

Conophytum burgeri – vulnerable

Pleiospilos compactus subsp. *minor* – rare

Uses of mesembs

The main use of mesembs is as garden plants, because they produce a wonderful display in spring and early summer. They are also commonly grown as curiosity plants. In landscaping, species of *Carpobrotus* and *Aptenia* are used all over the world. They are particularly useful to stabilise slopes in dry areas because they do not require frequent watering and will survive dry periods. The horticultural value of mesembs is only briefly mentioned here, because details are given in the next chapter.

Mesembs have several interesting non-horticultural uses. *Carpobrotus* species are commonly utilised in the Western Cape Province of South Africa for their delicious edible fruits. These can be eaten in a fresh (fleshy) or dried state by simply biting off the bottom end of the leathery capsule (just above the stalk) and sucking out the slimy fruit pulp with its numerous tiny seeds. The fruit capsules are harvested commercially and are a common sight at informal markets in and around Cape Town. A delicious preserve is prepared from the fruits and in South Africa they are highly sought after as an ingredient of Eastern cooking. *Carpobrotus edulis* is the main commercial species, but *C. acinaciformis* and *C. deliciosus* are also used to some extent.

Medicinal uses of mesembs are relatively well known. The leaf juice of *Carpobrotus* species is highly astringent and is a traditional remedy for a sore throat (used as a gargle) and also for fungal infections. Species of *Sceletium* have been widely used as a stimulant known as *kougoed* (chewing stuff). These plants contain alkaloids of the mesembrine type that have a mild stimulating and hypnotic effect, not unlike that of nicotine. Unlike the latter, however, no physical or psychological dependency seems to develop with habitual use.

Species of *Khadia*, *Mestoklema* and *Trichodiadema* were formerly important ingredients of traditional beer. The common names *moerwortel* or *moerwortelvygie* for species of these genera testify to this traditional use. The combination *moer* (yeast) and *wortel* (root) refers to the value of the roots to enhance the fermentation process in brewing traditional sorghum beer and honey beer.

The traditional use of mesembs in soap-making is still sometimes encountered in rural areas in South Africa. Species of *Psilocaulon* (mainly *P. junceum* and *P. coriarium*) were once a highly valued source of ash, hence the common name *asbos* (ash bush). On incineration, these plants yield an ash rich in alkali, which was used to prepare the lye for soap-making. *Psilocaulon* species are well known in the deserts and semi-deserts of the northwestern parts of southern Africa, where they are harvested to construct the traditional *asbosskerm* or *kookskerm* (cooking shelter) around the fire. The plants are merely stacked into a circular wall of about 2,5 m in height, to keep out the cold desert wind and unwanted animals. It is not difficult to appreciate the value of these plants to the rural people in a desolate, treeless landscape. Other members of the WEEDY MESEMBS (particularly *Mesembryanthemum* species) have been used in a process to remove hair from animals' skins. In the desert, the fleshy leaves are also a valuable source of moisture in an emergency.

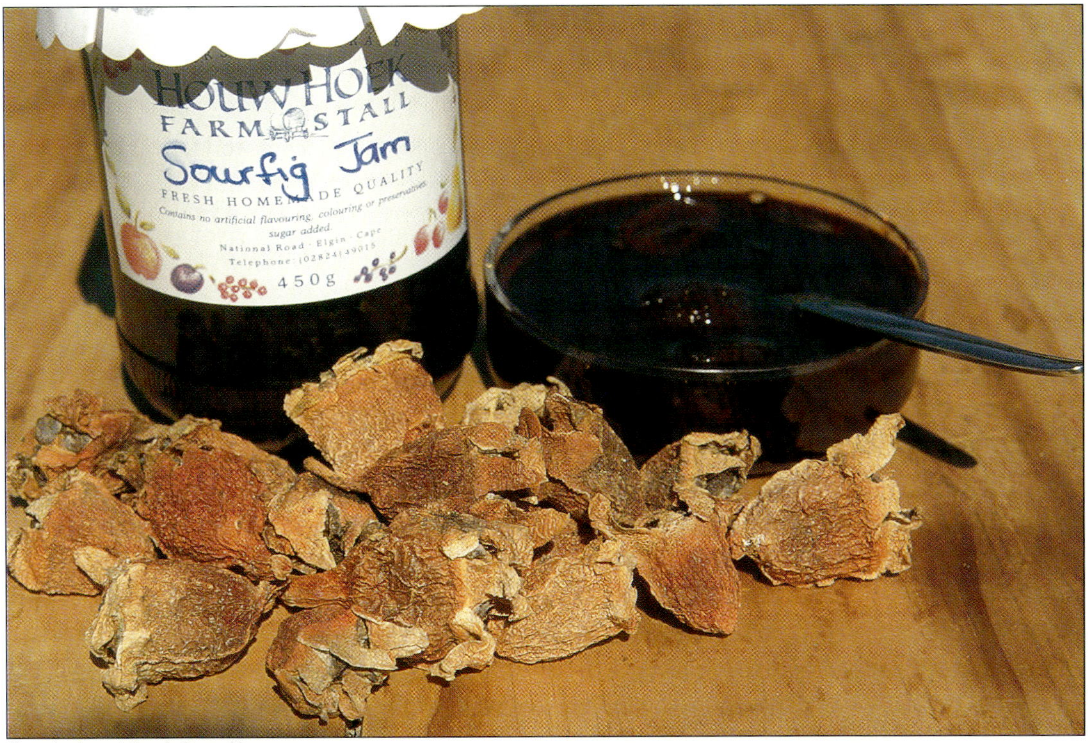

Ben-Erik van Wyk

Carpobrotus edulis – fruits and jam

Ben-Erik van Wyk

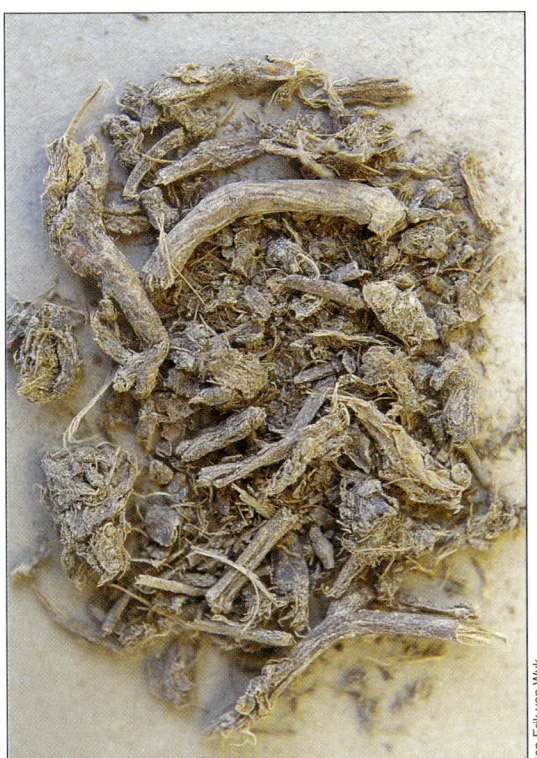

"Kougoed", prepared from *Sceletium* species

Ben-Erik van Wyk

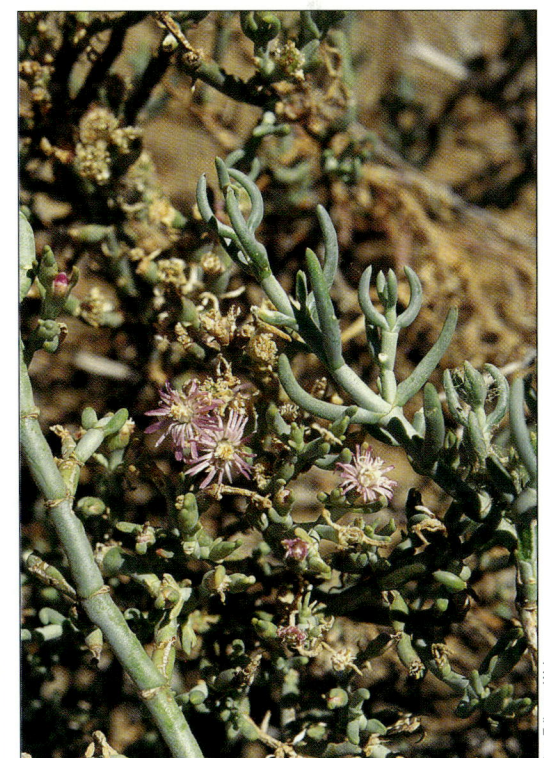

Psilocaulon junceum, source of ash used in soap-making

Cultivation

To deal adequately with the cultivation of this large and complex family would take – and has taken – a book. A few notes are useful, however. The shrubby species of whatever genus are easy to grow, if given porous but rich soil, deep pots and ample watering. When to water is an important question, since many shrubs, for example half of *Lampranthus*, come from winter rainfall areas. Even these are summer responsive, however, and can be adjusted to a summer moist regime.

It is the specialised plants, those which exhibit dormancy structures (whitish leaf sheaths, tubers, etc.) which require careful watering. It is sometimes difficult to tell if a plant is coming or going, so to speak. Most specialists, for example *Conophytum* species, are winter-rain responsive. That means that they start to grow actively in the late summer to early autumn, spend winter in a green and steady state, and form new but internally concealed leaves in spring when they should be watered only lightly. In summer they require only light misting and a shaded position will prevent stress. *Lithops* species behave in the opposite manner. They go dormant in autumn (i.e. right after flowering) and winter, when they should be kept quite dry. When they emerge in spring, they can use a good drink once a week. The fact that *Lithops* and *Conophytum* often grow together in nature tells us that the plants are flexible and opportunistic to some extent. The strictest species belong to the BEAD-LEAVED MESEMBS and to those *Phyllobolus* species with tuberous roots and deciduous stems. Both of these abstain from any summer-water response (unless we call rotting a response!) In the right season a plant with a healthy root system can and should accept a lot of water. Root systems are built up gradually, like muscles. It is the perpetually starved anorexic plant which succumbs to a sudden excess. No plant should be watered so heavily that it ruptures or produces excess growth, however. For the compact sphaeroid species one can adopt the maxim from Orwell's *Animal Farm*: two leaves good, four leaves bad!

Other important considerations include fresh air (mesembs rot in stagnant situations), well-drained soil with plenty of grit, good light (most welcome in the morning), light feeding (tomato fertilizer works well at one third of the recommended strength), and a keen eye for pests. Check your plants regularly for signs of rot, pests, and beauty.

Outdoor cultivation

Plants are grown for a colourful spring display, either as ground covers, small shrubs or annuals. Their requirements are a sunny position and well-drained soil. The ground cover, soil retention groups such as *Carpobrotus*, *Jordaaniella*, *Delosperma*, *Drosanthemum*, *Aptenia*, *Disphyma* and others can be established from cuttings planted *in situ* like lawn grass. It is always best to plant cuttings at the beginning of the growing season in order to save water. However, they are adaptable and can actually be planted at any time of the year as long as sufficent moisture is provided for successful rooting. Cuttings 100 to 150 mm in length are suitable. Cuttings are usually made during mid-summer, after the capsules have hardened. Rooting takes place within a month and the plants may provide a show as soon as the next spring. The cuttings can either be planted directly into the beds or first rooted in sand and planted out in autumn. Because mesembs are often rapid, opportunistic, relatively short-lived growers, some tend to die back after two or three years and need repropagation. During this time the soil could be improved by adding compost, ample bonemeal or fertilizers. The annuals such as *Dorotheanthus* and *Carpanthea* are also winter growers and seed

Mesemb display at Kirstenbosch National Botanical Garden, South Africa

Mass display of *Drosanthemum*

The commonly cultivated annual, *Dorotheanthus bellidiformis*

Drosanthemum bicolor (top) and *D. micans* (bottom)

Mass display of various mesembs

could first be sown into beds or containers and planted out once they are large enough. Seeds could also be sown by broadcasting them into the required beds and thinning them out later if necessary.

Indoor cultivation

Most mesembs need ample light and sufficient air movement when in active growth. It is important to bear this in mind if plants are to be grown indoors or in a small greenhouse. Where frost is severe, mesembs could be grown on sunny window sills or open patios and taken in during the mornings. Because most mesembs go into a resting phase during summer, care should be taken to provide some shade during this period, especially when the glass is closer than 1 m to the plants. Plants tend to overheat, leading to fatalities. The smaller collectors' items thrive in small containers but can also be grown socially in larger containers. Mesembs flourish in plastic or clay containers. Soil is the medium from which they obtain nourishment and is therefore of the utmost importance. Plants should do well in two parts of sand, one part clay-loam and one part sieved and well-decomposed leaf-mould or compost. It is always better to make use of sterilised soil. Feeding could be done with a mild solution during the growing season. Watering is also important and plants should only be watered during the growing season. Some plants are active in spring and autumn, and become dormant in the other seasons (as mentioned under the cultivation notes for each genus). As a result, water should often be withheld during winter and summer (the one being too cold, the other too hot for growth). *Dinteranthus* and *Titanopsis* are mainly autumn growers and should be adequately watered during this period, but kept dry at other times of the year. A fungicide could be added when necessary and should be applied by hand according to the needs and conditions of the plants. Many caterpillars thrive on mesembs. Watch the plants carefully and take precautionary measures before damage is done.

Harvesting seeds

Seeds can be harvested as soon as fruit capsules have hardened and dried. Seeds are easily obtained from the soft fruit types by breaking the capsules open by hand. However, with many robust capsule types, such as *Lampranthus*, the seed is best removed by beating them with a wooden hammer between two newspapers. The seed can then simply be rubbed out or removed by hand. Mesemb seeds usually have a long life span and can be stored for prolonged periods before planting.

Propagation from seeds

Mesemb seeds are best sown in shallow trays in sandy, well-drained soil. First water the tray with a fine rose before sowing. The seed should be sown evenly on the medium and covered with a thin layer of sand of about 1 mm deep. Germination is usually rapid and so is the subsequent growth. In some cases, the first leaves (cotyledons) look like miniature plants of the genus *Lithops* before they are replaced by the permanent leaves. Seedlings can be planted out after a season, or sooner depending on their performance. They need more care than mature plants and mild feeding can be beneficial.

Seedling pots

Vegetative propagation of *Conophytum flavum*

Seedlings of *Conophytum minimum*

Lithops julii 'chapped lips'

Mesemb enthusiast (Steven Hammer) in his greenhouse

How to identify the mesemb groups

Identifying mesembs is a notoriously difficult task. This book provides guidelines to identify the genera, as well as the groups into which they have been placed. Fruit capsules are fundamental to assigning mesembs to the correct genus. However, capsules are complex structures, often requiring a hand lens to see all the details, and are not always available on the plants. The use of capsule characters in identification has therefore been reduced to a minimum. As many clearly visible characters as possible have been used instead. As a consequence, a certain artificiality has been brought into the groups of genera as the genera included in a group are not necessarily closely related.

Before attempting to name any mesemb it is necessary to establish which of the two subfamilies it belongs to, the Mesembryanthemoideae (GROUP 1. WEEDY MESEMBS) or the Ruschioideae (ALL OTHER GROUPS). There is a fundamental distinction between species with axile placentation (ovules or seeds attached to the central axis of the ovary or fruit – Mesembryanthemoideae) and those with parietal placentation (ovules or seeds attached to the outer walls of the ovary or fruit – Ruschioideae). For this, one needs to look at an open capsule or make a cross-section of the base of the flower or ovary. With experience one can get a feeling for the WEEDY MESEMBS, members of which are usually floppy-looking, normally have prominent water cells on their leaves and light brown spongy fruit capsules. Only using these superficial characters might cause one to confuse them with species of *Delosperma*.

The non-weedy mesembs (the Ruschioideae) comprise far more genera and species, including well-known collectors' genera such as *Lithops*. A diversity of growth forms occurs in these groups. Leaves are usually opposite and fused to a greater or lesser degree. They may be flat and not very succulent or fused into cylindrical, highly succulent bodies, toothed, textured, crowded, erect and tufted or of different forms on the same plant (heterophyllous plants). The plants can appear stemless and reduced to a few leaf pairs, creeping, mat-forming or upright and shrub-like. One genus, *Stoeberia*, can grow up to 3,5 m tall and has been accorded tree status.

The non-weedy mesemb genera have been assigned to 13 groups based on leaf types and growth habit. The shrubby mesembs (groups 11, 12, 13 and 14) are distinguished primarily by their fruit type. The basic fruit types are described and illustrated here.

Mesemb fruit capsules

The mesembs have characteristic fruit capsules with locules (chambers or cavities) containing the seed. With few exceptions they open and close repeatedly in response to moisture (hygrochastic). The capsules are closed when dry and open when wet, limiting the dispersal of seeds to wet conditions. This repetitious dehiscence mechanism, whereby the valves that form the lid of the capsule are opened and fold back, relies on the response of expanding keels to moisture. Seeds are liberated from the open capsules. This unique dispersal mechanism is particularly suited to dessemination and germination in habitats where rainfall is erratic. Dispersal mechanisms are usually more complex in the Ruschioideae (non-weedy mesembs) than in the Mesembryanthemoideae (weedy mesembs). Various structures such as closing bodies, closing rodlets, and covering membranes over the seeds control and direct seed dispersal in the Ruschioideae.

1. *MESEMBRYANTHEMUM*-TYPE

Fruit capsules of the WEEDY MESEMBS are pale in colour with a corky rather than woody texture. They have an axile placenta, i.e. seeds are attached to the central axis of the fruit. The large seeds are easily visible when the capsules are open and may become lodged under the broad wings which are fused to the valves. Fruit of this type usually have four, sometimes five locules (rarely six, as in the case of some *Sceletium* species).

2. *DROSANTHEMUM*-TYPE

The genus *Drosanthemum* is included in the GLITTERING SHRUBBY MESEMBS. Its capsules typically have broad valve wings, covering membranes that are translucent yet persistent in their shape, and no closing bodies at the exits of the locules. This sort of capsule is also found in *Mestoklema*. (Some genera with *Drosanthemum*-type capsules have deliberately not been included in this group, for example *Chasmatophyllum* and *Gibbaeum*). These genera have other distinctive characters.

3. *DELOSPERMA*-TYPE

The *Delosperma*-type of fruit capsule is aptly named: *delos* means naked and *sperma* means seed. The name refers to the easy visibility of the seeds in the capsules of the genus. This type of fruit does not characteristically have covering membranes over its locules and seeds are easily washed out by raindrops when capsules are open. These capsules have broad valve wings and do not have closing bodies. In the GLITTERING SHRUBBY MESEMBS, *Delosperma*-type fruit occur in *Delosperma* and *Trichodiadema*. Numerous other mesembs from various different groups have this type of fruit, which are usually five or six-locular, for example *Conophytum*.

4. *LAMPRANTHUS*-TYPE

The most striking feature of the *Lampranthus*-type of fruit is the bunches of sterile funicular hairs at the exits of the locules, i.e. in lieu of closing bodies. Covering membranes are quite rigid and persistently convex in shape. They have a distinct recurved rim above and closing ledge on the distal undersurface of the membrane. Valve wings may be present or absent in this capsule type. This type of fruit capsule, which usually has five but may have up to 10 locules, characterises the *LAMPRANTHUS*-LIKE SHRUBBY MESEMBS and also several genera in the DWARF SHRUBBY MESEMBS, i.e. *Braunsia*, *Esterhuysenia* and *Zeuktophyllum*.

5. *RUSCHIA*-TYPE

Ruschia-type fruit are easily recognised by the consistent absence of valve wings (except in the case of *Eberlanzia*) in combination with small, rod-shaped closing bodies and stout, convex covering membranes over the locules. These usually have upright rims and closing rodlets at their distal extremities. This fruit type, which is usually five-locular, characterises the *RUSCHIA*-LIKE SHRUBBY MESEMBS.

6. *LEIPOLDTIA*-TYPE

The fruits of the *LEIPOLDTIA*-LIKE SHRUBBY MESEMBS are easily recognised by the large round, often whitish, stalked closing bodies at the exits of the locules. Valve wings are broad and covering membranes concave, unlike the previous types, and quite stiff. They are distinctly recurved at their distal margins. Fruit capsules in this group have five to many locules and rounded tops.

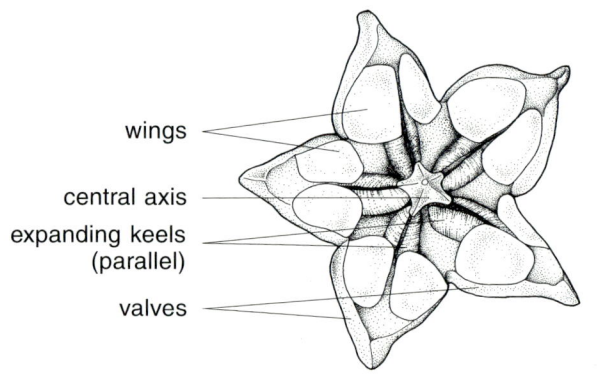

***Mesembryanthemum*-type**

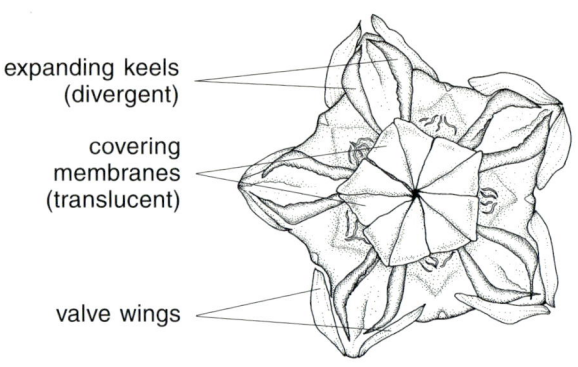

***Drosanthemum*-type**

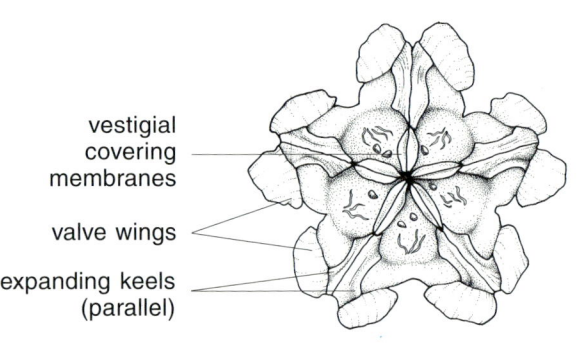

***Delosperma*-type**

- valve wings (not always present)
- rigid covering membranes
- bunches of funicular hairs
- expanding keels (divergent)

LAMPRANTHUS-TYPE

 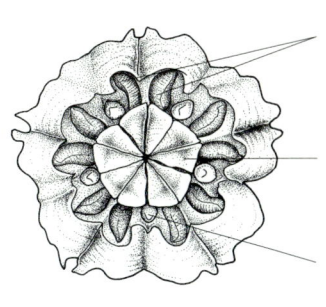

- expanding keels (divergent)
- covering membranes with upright rims and closing rodlets
- small rod-shaped closing bodies

RUSCHIA-TYPE

- large, whitish closing bodies
- covering membranes
- expanding keels (divergent at tips)
- valves (curved back, valve wings not visible)

LEIPOLDTIA-TYPE

Guide to the groups

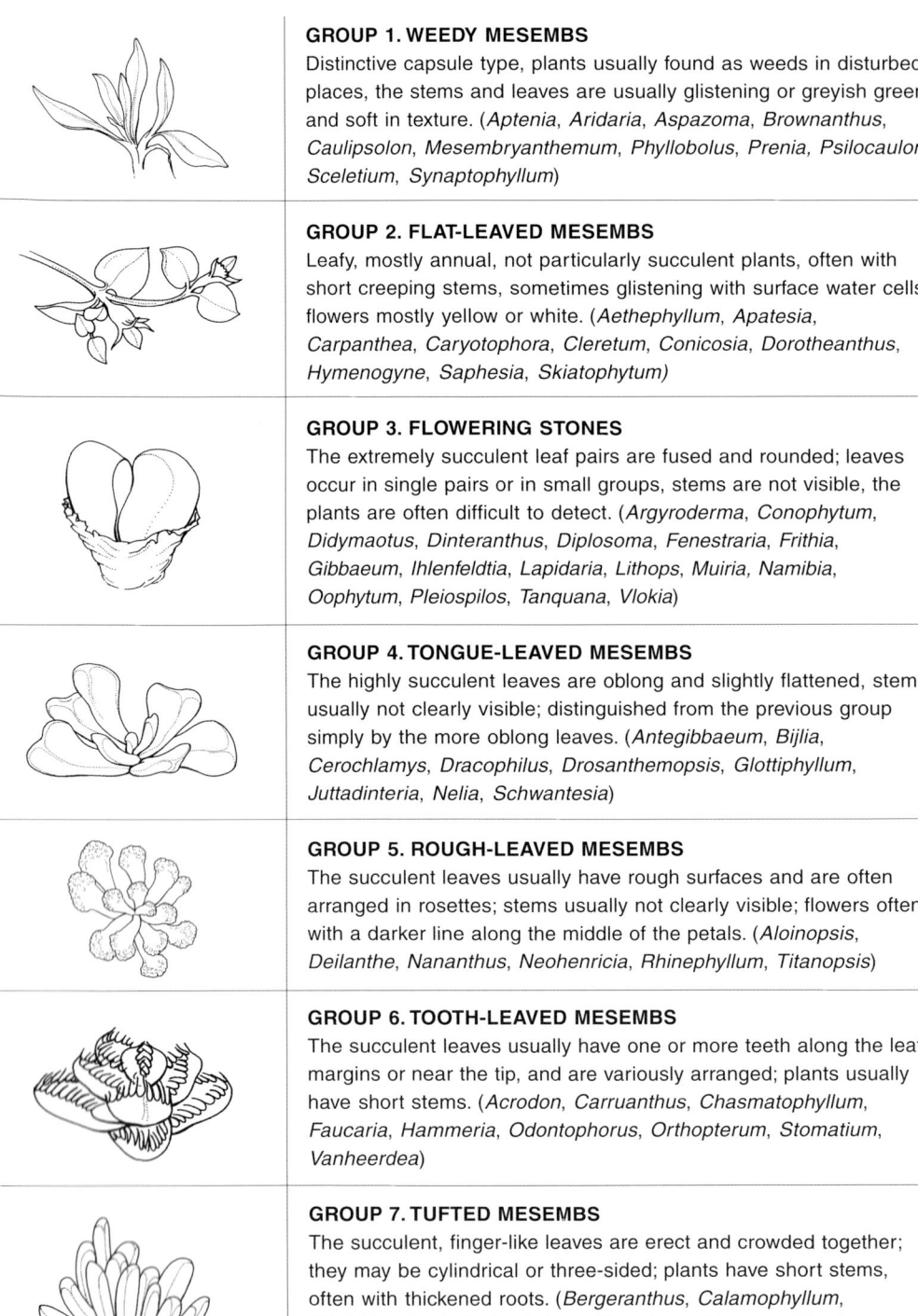

GROUP 1. WEEDY MESEMBS
Distinctive capsule type, plants usually found as weeds in disturbed places, the stems and leaves are usually glistening or greyish green and soft in texture. (*Aptenia, Aridaria, Aspazoma, Brownanthus, Caulipsolon, Mesembryanthemum, Phyllobolus, Prenia, Psilocaulon, Sceletium, Synaptophyllum*)

GROUP 2. FLAT-LEAVED MESEMBS
Leafy, mostly annual, not particularly succulent plants, often with short creeping stems, sometimes glistening with surface water cells, flowers mostly yellow or white. (*Aethephyllum, Apatesia, Carpanthea, Caryotophora, Cleretum, Conicosia, Dorotheanthus, Hymenogyne, Saphesia, Skiatophytum*)

GROUP 3. FLOWERING STONES
The extremely succulent leaf pairs are fused and rounded; leaves occur in single pairs or in small groups, stems are not visible, the plants are often difficult to detect. (*Argyroderma, Conophytum, Didymaotus, Dinteranthus, Diplosoma, Fenestraria, Frithia, Gibbaeum, Ihlenfeldtia, Lapidaria, Lithops, Muiria, Namibia, Oophytum, Pleiospilos, Tanquana, Vlokia*)

GROUP 4. TONGUE-LEAVED MESEMBS
The highly succulent leaves are oblong and slightly flattened, stems usually not clearly visible; distinguished from the previous group simply by the more oblong leaves. (*Antegibbaeum, Bijlia, Cerochlamys, Dracophilus, Drosanthemopsis, Glottiphyllum, Juttadinteria, Nelia, Schwantesia*)

GROUP 5. ROUGH-LEAVED MESEMBS
The succulent leaves usually have rough surfaces and are often arranged in rosettes; stems usually not clearly visible; flowers often with a darker line along the middle of the petals. (*Aloinopsis, Deilanthe, Nananthus, Neohenricia, Rhinephyllum, Titanopsis*)

GROUP 6. TOOTH-LEAVED MESEMBS
The succulent leaves usually have one or more teeth along the leaf margins or near the tip, and are variously arranged; plants usually have short stems. (*Acrodon, Carruanthus, Chasmatophyllum, Faucaria, Hammeria, Odontophorus, Orthopterum, Stomatium, Vanheerdea*)

GROUP 7. TUFTED MESEMBS
The succulent, finger-like leaves are erect and crowded together; they may be cylindrical or three-sided; plants have short stems, often with thickened roots. (*Bergeranthus, Calamophyllum, Cheiridopsis, Cylindrophyllum, Ebracteola, Hereroa, Khadia, Machairophyllum, Marlothistella, Psammophora, Rabiea, Rhombophyllum, Ruschianthus*)

GUIDE TO THE GROUPS

GROUP 8. BEAD-LEAVED MESEMBS
The succulent leaves are arranged so that the newly formed leaf pair differs markedly from the older leaf pair from which it has emerged, thus resulting in a continually alternating series of different leaf pairs. (*Dicrocaulon, Jacobsenia, Meyerophytum, Mitrophyllum, Monilaria*)

GROUP 9. MAT-FORMING MESEMBS
The succulent leaves are variable in size and shape but the plants all creep along the ground, forming dense or sparse mats. (*Antimima, Carpobrotus, Cephalophyllum, Disphyma, Jensenobotrya, Jordaaniella, Malephora, Mossia*)

GROUP 10. DWARF SHRUBBY MESEMBS
The plants are all dwarf shrubs with visible, more or less erect stems; the succulent leaves are variable in size and shape. (*Braunsia, Corpuscularia, Esterhuysenia, Hartmanthus, Sarcozona, Schlechteranthus, Zeuktophyllum*)

GROUP 11. GLITTERING SHRUBBY MESEMBS
The leaves have tiny cells storing water, giving the leaf surfaces a glittering appearance; the plants are all shrubby with visible internodes and creeping or erect stems; fruit capsules are of the *Drosanthemum* or *Delosperma*-type. (*Delosperma, Drosanthemum, Ectotropis, Mestoklema, Trichodiadema*)

GROUP 12. *LAMPRANTHUS*-LIKE SHRUBBY MESEMBS
Shrubby plants with creeping or erect stems with characteristic fruit capsules of the *Lampranthus*-type. (*Amphibolia, Circandra, Enarganthe, Erepsia, Lampranthus, Namaquanthus, Oscularia, Scopelogena, Smicrostigma, Wooleya*)

GROUP 13. *RUSCHIA*-LIKE SHRUBBY MESEMBS
Shrubby plants with creeping or erect stems with characteristic fruit capsules of the *Ruschia*-type. (*Arenifera, Astridia, Eberlanzia, Polymita, Ruschia, Ruschianthemum, Stayneria, Stoeberia*)

GROUP 14. *LEIPOLDTIA*-LIKE SHRUBBY MESEMBS
More or less shrubby plants with creeping or erect stems with characteristic fruit capsules of the *Leipoldtia*-type. (*Hallianthus, Leipoldtia, Octopoma, Ottosonderia, Vanzijlia*)

Mesembryanthemum sp., presently known as *Opophytum aquosum* (left) and *Synaptophyllum juttae* (right)

Aptenia cordifolia

Psilocaulon sp.

Sceletium sp.

Mesembryanthemum guerichianum

Weedy Mesembs

GROUP 1

Distinctive fruit capsule type (see below); plants usually found as weeds in disturbed places such as roadsides or ploughed lands; the stems and leaves are usually glistening or greyish green and soft in texture.

Fruit capsules of the WEEDY MESEMBS are pale in colour with a corky rather than woody texture. They have an axile placenta, i.e. seeds are attached to the central axis of the fruit. The large seeds are easily visible when the capsule is open and may become lodged under the broad wings which are fused to the valves. Fruit of this type usually have four, sometimes five locules (rarely six, as in the case of some *Sceletium* species).

This book groups all the genera with axile placentation (the subfamily Mesembryanthemoideae) in the group WEEDY MESEMBS. In this group, most species have flat or subcylindrical leaves and are somewhat woody. Those species with large flat leaves are herbaceous and are covered with glistening water cells, for example *Mesembryanthemum crystallinum*. The group also includes stem succulents with articulated, green stems and short-lived leaves, such as *Psilocaulon* and *Brownanthus*. In addition, there are numerous species with fleshy, thickened underground parts (geophytes) which include "*Caulipsolon*" and species of *Phyllobolus*.

WEEDY MESEMBS comprise 11 genera and 95 species.

Aptenia (4 species)	"*Caulipsolon*" (1 species)	*Psilocaulon* (13 species)
Aridaria (4 species)	*Mesembryanthemum* (15 species)	*Sceletium* (8 species)
Aspazoma (1 species)	*Phyllobolus* (32 species)	*Synaptophyllum* (1 species)
Brownanthus (10 species)	*Prenia* (6 species)	

Aptenia

Derivation of genus name The name is derived from the Greek word *apten* (wingless), referring to the wingless fruit capsules.

Common names *A. cordifolia* is known as *brakvygie* (brack mesemb).

Description The plants are small shrubs, climbers or grow flat on the ground. They are evergreen or rarely deciduous. The roots sometimes become thick and fleshy. The stems are four-angled or cylindrical, slightly woody, green and fleshy with distinct, closely packed water cells on the surface. The flat or somewhat cylindrical, heart-shaped leaves are usually paired or may become single, alternating with each other towards the flower cluster. The leaf surface has loosely arranged water cells. The stalked, small to medium-sized flowers are white, cream, yellow or pink to purple, occurring singly or in few-flowered clusters. There are four free or slightly fused sepals, of which two are often larger. The petals are united at the base into a short tube. The fruit capsules have four locules. The comparatively large seeds have rough surfaces and are blackish brown.

Distinguishing characters *Aptenia* is characterised by green, succulent stems with closely arranged water cells. This feature is also found in *Brownanthus* and *Aspazoma*. However, in *Aptenia* the seeds are always blackish brown and rather large. The flowers are magenta, rosy pink or sulphur yellow and self-fertile.

Flowering time The plants flower in spring, summer and autumn (August to April in South Africa). Flowers are open during the bright hours of the day.

Distribution and ecology The species occur naturally in the summer rainfall areas of southern Africa. *A. cordifolia* and *A. lancifolia* are frequently found in shady places on rocky outcrops in regions of higher rainfall. In South Africa, the natural distribution of *A. cordifolia* is unclear as it often becomes easily established in areas beyond its natural distribution, as shown by the dotted line on the distribution map.

Cultivation Plants are propagated very easily from stem cuttings. *A. cordifolia* and *A. lancifolia* are very popular garden subjects world-wide and are particularly useful as ground covers on dry slopes. They rapidly form lush green carpets with decorative reddish purple flowers. Plants can also be propagated from seed sown in summer.

Notes *Aptenia cordifolia* has heart-shaped leaves while in *A. lancifolia* the leaves taper gradually towards their bases. Interesting selections of *A. cordifolia* are found in cultivation, varying in leaf and flower colour. As has been suggested, the monotypic genus *Platythyra* belongs in *Aptenia*, but this is not reflected in the species list.

Aptenia N.E.Br. (=*Platythyra* N.E.Br.)

Number of species/subspecies/varieties (4/0/0) Two more species are still to be transferred to *Aptenia*.

Species list and conservation status
A. cordifolia (L.f.) Schwantes
A. lancifolia L.Bolus

Literature
BITTRICH, V. 1986. Untersuchungen zu Merkmalsbestand, Gliederung und Abgrenzung der Unterfamilie Mesembryanthemoideae (Mesembryanthemaceae Fenzl). *Mitteilungen aus dem Institut für Allgemeine Botanik, Hamburg* 21: 5–116.

GERBAULET, M. (In press). *Aptenia* — Mesembryanthemoideae. In Hartmann, H.E.K. (ed.), *IOS Lexicon of Succulent Plants*. Fischer, Jena.

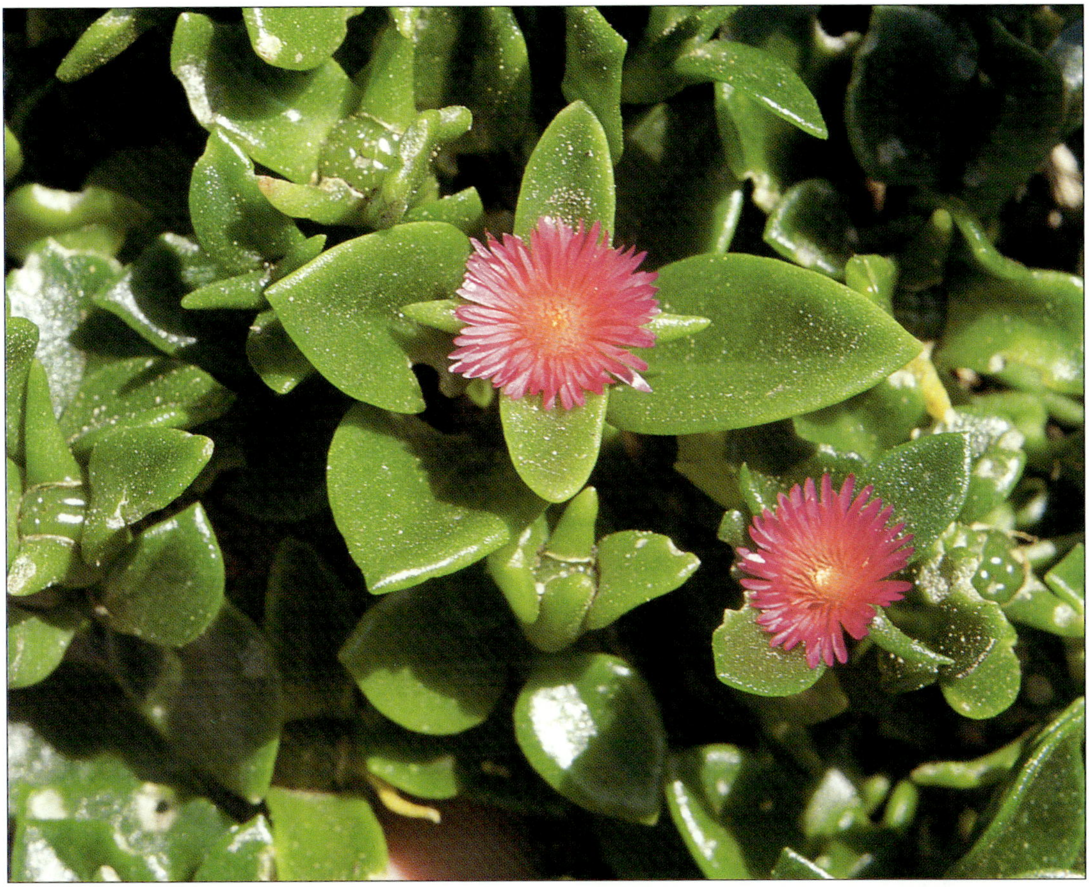
Ben-Erik van Wyk

A. cordifolia

Ernst van Jaarsveld/NBI

A. lancifolia

Ernst van Jaarsveld/NBI

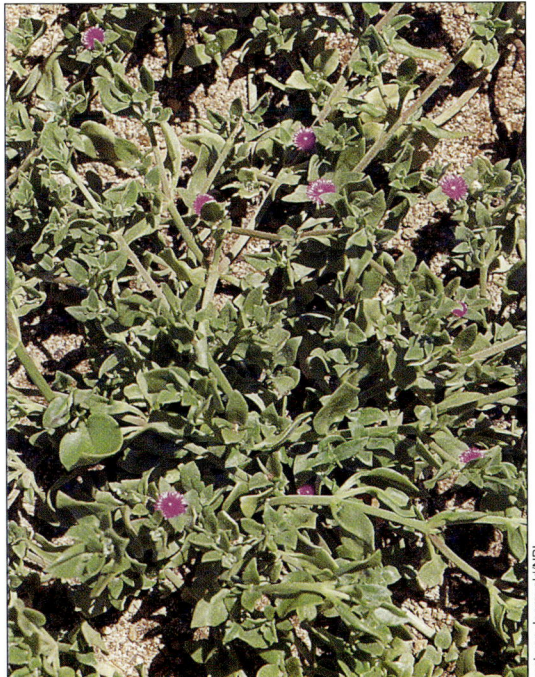

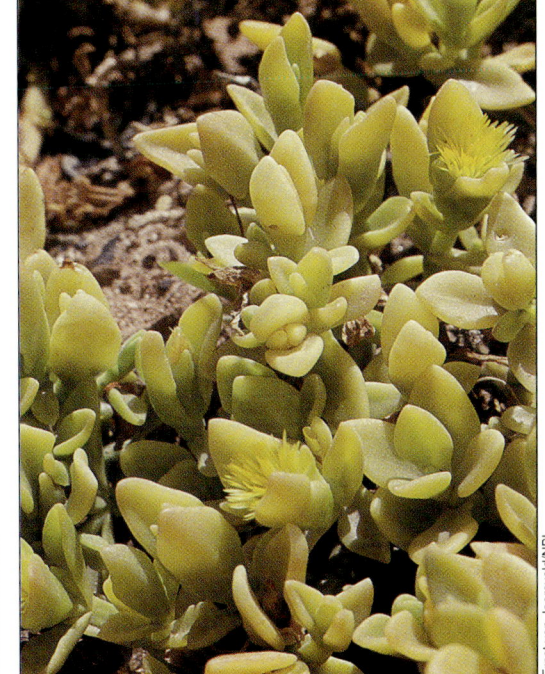

Aptenia sp., presently known as *Platythyra haeckeliana*

ARIDARIA

DERIVATION OF GENUS NAME The name is derived from the Latin word *aridus* (dry), in allusion to the arid natural habitat of the plants.

COMMON NAMES *Brakveldwitvygie* (brack field white mesemb) is a common name for *A. noctiflora*.

DESCRIPTION The plants are erect shrubs of up to 1 m tall or are low growing with woody stems and thickened roots. The smooth, more or less cylindrical leaves are shortly fused at their bases and are deciduous. Flowers are medium to large with petals white or puce on the inside, suffused in different shades of yellow, copper, pink or red on the outside. There are four sepals and nectar glands in the form of four separate grooves. The fruit capsules have four locules, usually opening and closing repeatedly; rarely opening only once and not closing again. The fruits break off easily and the persistent stalks become thorny. The brown or dark brown seeds have rough surfaces.

DISTINGUISHING CHARACTERS *Aridaria* is characterised by leaves with much flattened water cells which are otherwise only found in *Prenia*. It may be distinguished from the latter by its more woody stems and the fact that the upper and lower parts of the fruits are of about equal length.

FLOWERING TIME Flowering occurs mainly from spring to early summer (August to November in southern Africa). The flowers are open during the day, evening or at night, depending on the species.

DISTRIBUTION AND ECOLOGY The plants grow in karroid areas of the Western, Eastern and Northern Cape Provinces, South Africa, and southern and central Namibia. The distribution area is subject to winter or summer rainfall. *A. brevicarpa* and *A. noctiflora* are usually found in sandy places, whereas the remaining two species mostly prefer rocky outcrops.

CULTIVATION Propagation is easy from seed or from cuttings, but the plants are rarely cultivated. Away from their natural habitat, plants should be grown in a greenhouse.

NOTES As a result of recent research, *Aridaria* now includes only *A. noctiflora* and related species. One of the main characters used to differentiate between the species within the genus is the time at which the flowers open. This varies from strictly during the night (*A. noctiflora*) to during the day only (*A. brevicarpa*).

Aridaria N.E.Br. (=*Phyllobolus* subgen. *Aridaria* (N.E.Br.) Bittrich)

NUMBER OF SPECIES/SUBSPECIES/VARIETIES (4/2/0)

SPECIES LIST AND CONSERVATION STATUS
A. brevicarpa L.Bolus
A. noctiflora (L.) Schwantes subsp. defoliata (Haw.) Gerbaulet
A. noctiflora (L.) Schwantes subsp. noctiflora
A. noctiflora (L.) Schwantes subsp. straminea (Haw.) Gerbaulet
A. serotina L.Bolus
A. vespertina L.Bolus

LITERATURE
GERBAULET, M. 1996. Revision of the genus *Aridaria* N.E.Br. (Aizoaceae). *Botanische Jahrbücher* 118: 41–58.

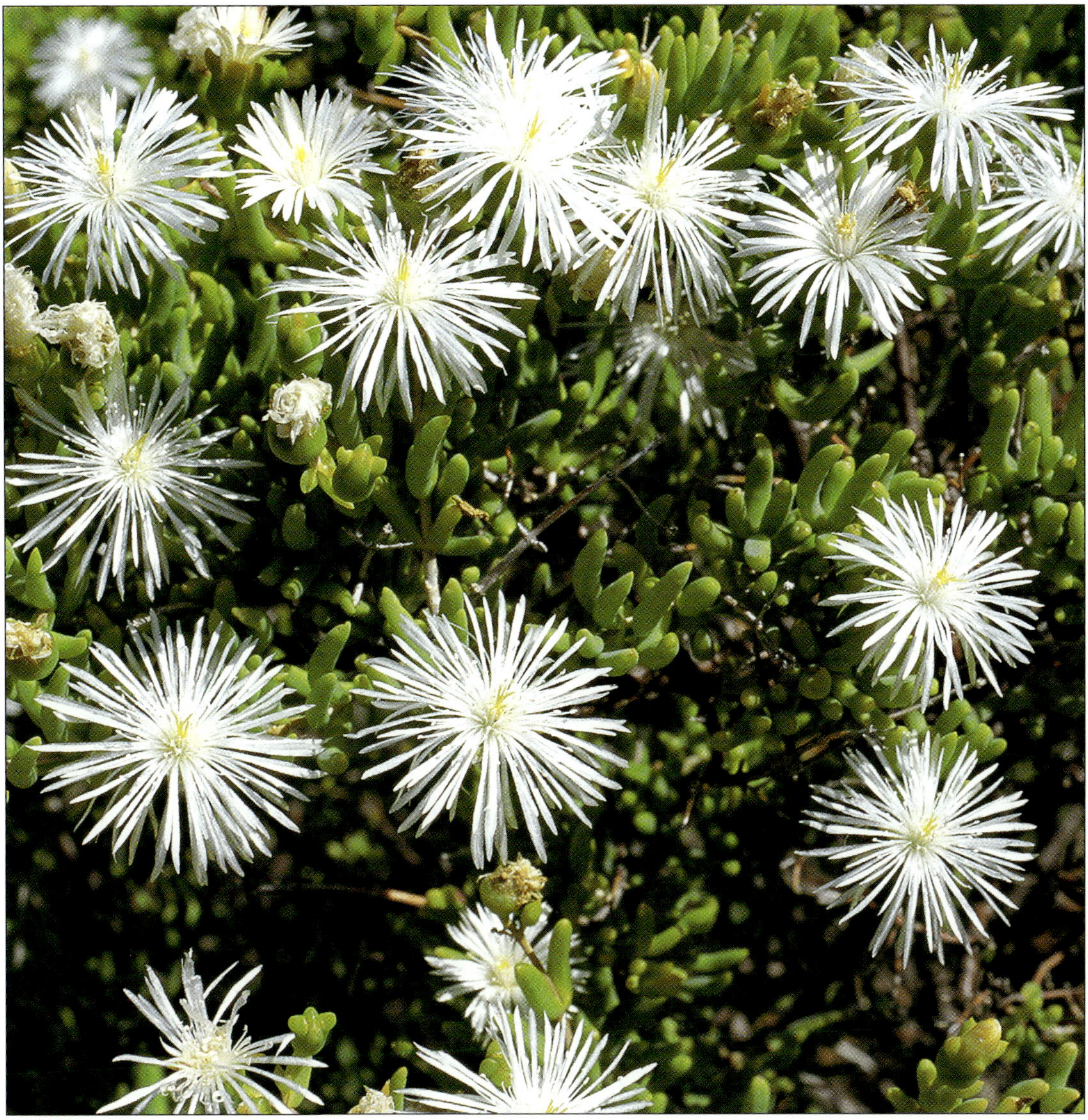

John Manning

A. noctiflora

Ben-Erik van Wyk

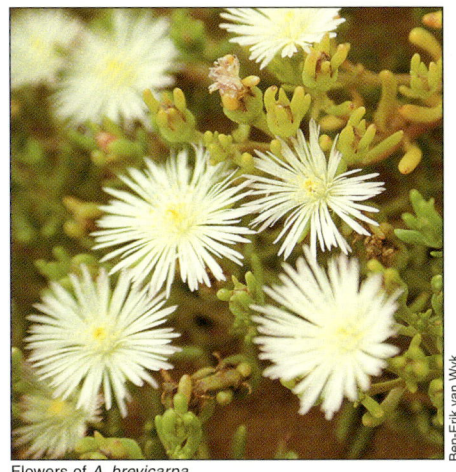

Flowers of *A. brevicarpa*

Ben-Erik van Wyk

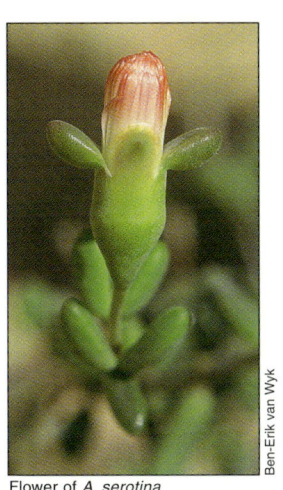

Flower of *A. serotina*

NBI

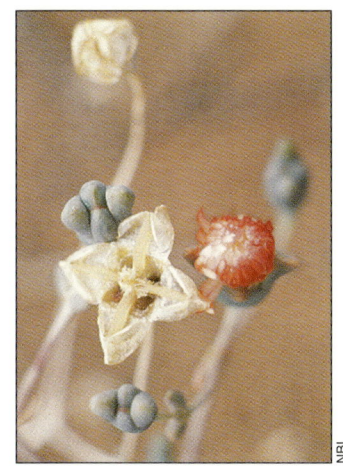

Fruit capsule of *A. brevicarpa*

ASPAZOMA

DERIVATION OF GENUS NAME The name is derived from the Greek word *aspazomai* (to clasp), referring to the way in which the leaf sheath clasps the stem.

COMMON NAMES No vernacular names seem to have been recorded.

DESCRIPTION Plants are bushily branched, up to 150 mm tall, with fibrous roots. Young stems are slightly woody, jointed, green and succulent, with tall and closely packed water cells. Old stems are woody. The deciduous leaves are more or less cylindrical, free towards their bases but they envelop each other, forming a prominent sheath. The large, white to pale yellow flowers occur singly at the tips of the branches. There are four or five sepals and the petals are fused into a short tube towards their bases. Filamentous staminodes are present and the nectar glands are visible as four or five separate grooves. Fruit capsules have four or five locules and each locule mostly contains only two seeds. The large seeds are light brown, with an almost smooth surface.

DISTINGUISHING CHARACTERS The genus is most similar to *Brownanthus* but differs from it mainly in the arrangement of the leaves which clasp the stem and the very large seeds.

FLOWERING TIME Flowering occurs in spring (September to October in South Africa).

DISTRIBUTION AND ECOLOGY The genus is confined to Namaqualand in the Western and Northern Cape Provinces, South Africa, including the Richtersveld, where it occurs on low quartzite hills and gravelly plains of the coastal parts of the dry winter rainfall region.

CULTIVATION Plants may be propagated from seeds or cuttings and care should be taken to water them during the growing season only as they may rot when watered during the resting period. *Aspazoma* species are seldom cultivated.

NOTES *Aspazoma* is closely related to *Brownanthus* and these two genera may eventually be combined.

Aspazoma N.E.Br.

NUMBER OF SPECIES/SUBSPECIES/VARIETIES (1/0/0)

SPECIES LIST AND CONSERVATION STATUS
A. amplectens (L.Bolus) N.E.Br.

LITERATURE
BITTRICH, V. 1986. Untersuchungen zu Merkmalsbestand, Gliederung und Abgrenzung der Unterfamilie Mesembryanthemoideae (Mesembryanthemaceae Fenzl). *Mitteilungen aus dem Institut für Allgemeine Botanik, Hamburg* 21: 5–116.

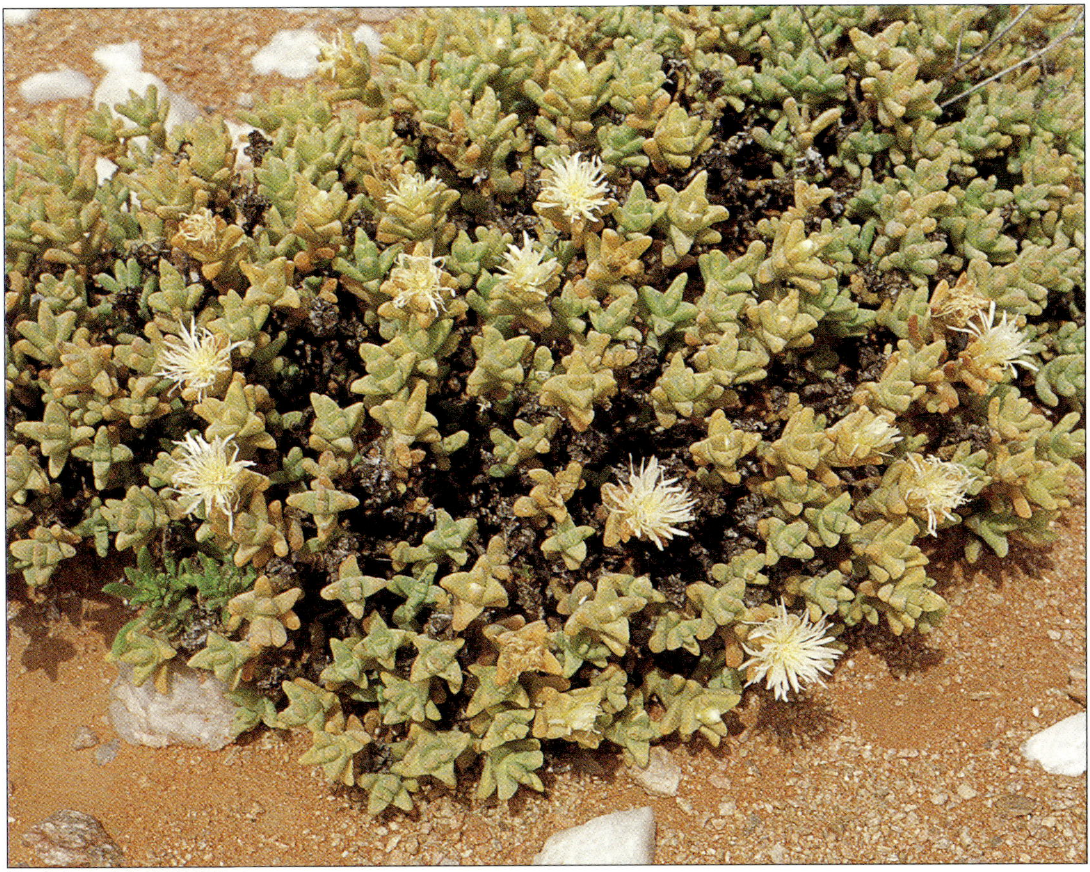

Bruce Bayer/NBI

Growth form of *A. amplectens*

Heidi Hartmann

Flowers of *A. amplectens*

Ernst van Jaarsveld/NBI

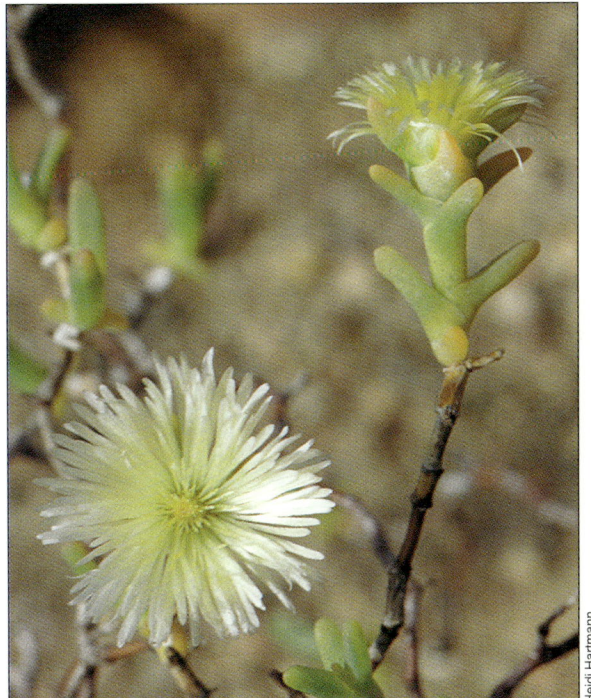

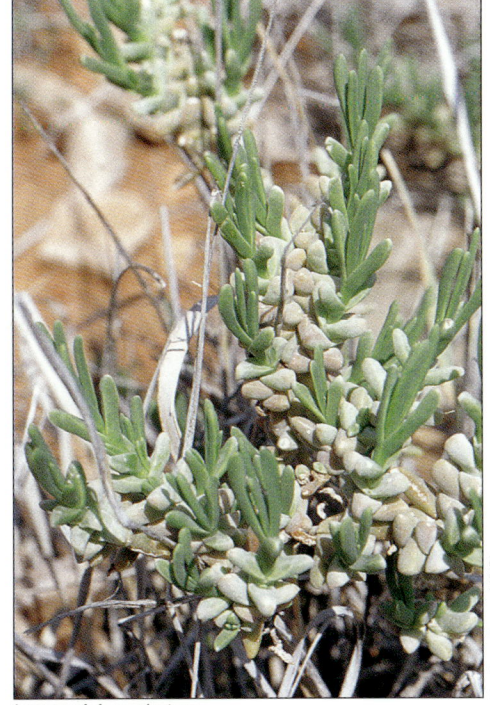

Leaves of *A. amplectens*

BROWNANTHUS

DERIVATION OF THE GENUS NAME The genus was named after the mesemb expert Dr Nicholas Edward Brown (1849 to 1934) who worked at Kew. His name is combined with the latinised form of the Greek word *anthos* (flower).

COMMON NAMES No vernacular names are known.

DESCRIPTION The plants are creeping or erect shrubs of up to 1 m tall, with fibrous roots. Young stems are jointed at first, green and succulent with raised, closely packed water cells, but later become woody. The short-lived leaves are more or less cylindrical and free or shortly fused towards their bases. They sometimes become spiny and the water cells at the leaf bases and edges often become hair-like when they dry out. The small, cream-coloured flowers are solitary or appear in clusters. There are four or five more or less equal sepals which are consistently erect when the flowers open. Fruit capsules have four or five locules, sometimes with a basal seed pocket or they are rarely nut-like. The seeds are smooth, rarely rough, ochre or brown.

DISTINGUISHING CHARACTERS *Brownanthus* is related to *Aptenia* and *Aspazoma* on account of the green succulent stems and the water cells which differ between the stems (where they are very closely packed) and the leaves (where they are loosely arranged). However, in *Aptenia* the old stems never become very woody, something which happens extensively in *Brownanthus*. It differs from *Aspazoma* on account of its smaller seeds and the leaves which are either free or only shortly fused towards the bases.

FLOWERING TIME The flowers appear from spring to early summer (September to December in southern Africa). Flowers are open during daylight hours.

DISTRIBUTION AND ECOLOGY The genus occurs in the winter and summer rainfall regions of Namibia and the Western and Northern Cape Provinces, South Africa, extending into southern Angola. The majority of species occur in and around the Orange River valley. Several species show preferences for certain types of soil, such as *B. corallinus*, which only grows on patches of quartz gravel. *B. ciliatus* appears to be the most tolerant of both rainfall and soil type. It is the most widespread species and is often found as a pioneer in disturbed areas.

CULTIVATION The plants have a short growing period, and should be planted in a warm spot in full sun. Sufficient water should be given during the growing season and the plants should be kept quite dry during the resting period. They do well in sandy soil.

NOTES *Brownanthus* was previously included in *Psilocaulon*, but was later shown to be different from the latter in respect of the fruits, flowers and leaf surfaces.

Brownanthus Schwantes
(=*Pseudobrownanthus* Ihlenf. & Bittrich).

NUMBER OF SPECIES/SUBSPECIES/VARIETIES (10/1/0)

SPECIES LIST AND CONSERVATION STATUS
B. *arenosus* (Schinz) Ihlenf. & Bittrich
B. *ciliatus* (Aiton) Schwantes subsp. *ciliatus*
B. *ciliatus* (Aiton) Schwantes subsp. *schenkii* (Schinz) Ihlenf. & Bittrich
B. *corallinus* (Thunb.) Ihlenf. & Bittrich
B. *kuntzei* (Schinz) Ihlenf. & Bittrich
B. *marlothii* (Pax) Schwantes
B. *namibensis* (Marloth) Bullock [R]
B. *neglectus* Pierce & Gerbaulet
B. *nucifer* (Ihlenf. & Bittrich) Pierce & Gerbaulet
B. *pseudoschlichtianus* Pierce & Gerbaulet
B. *pubescens* (N.E.Br. ex Maas) Bullock

LITERATURE
PIERCE, S.M. & GERBAULET, M. 1997. *Brownanthus* Schwantes (Mesembryanthemoideae, Aizoaceae): two new species and a new combination from the Richtersveld and southwestern Namibia. *Aloe* 34: 42–44.

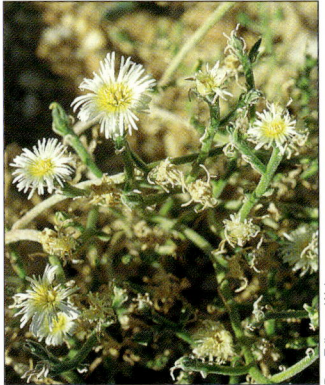

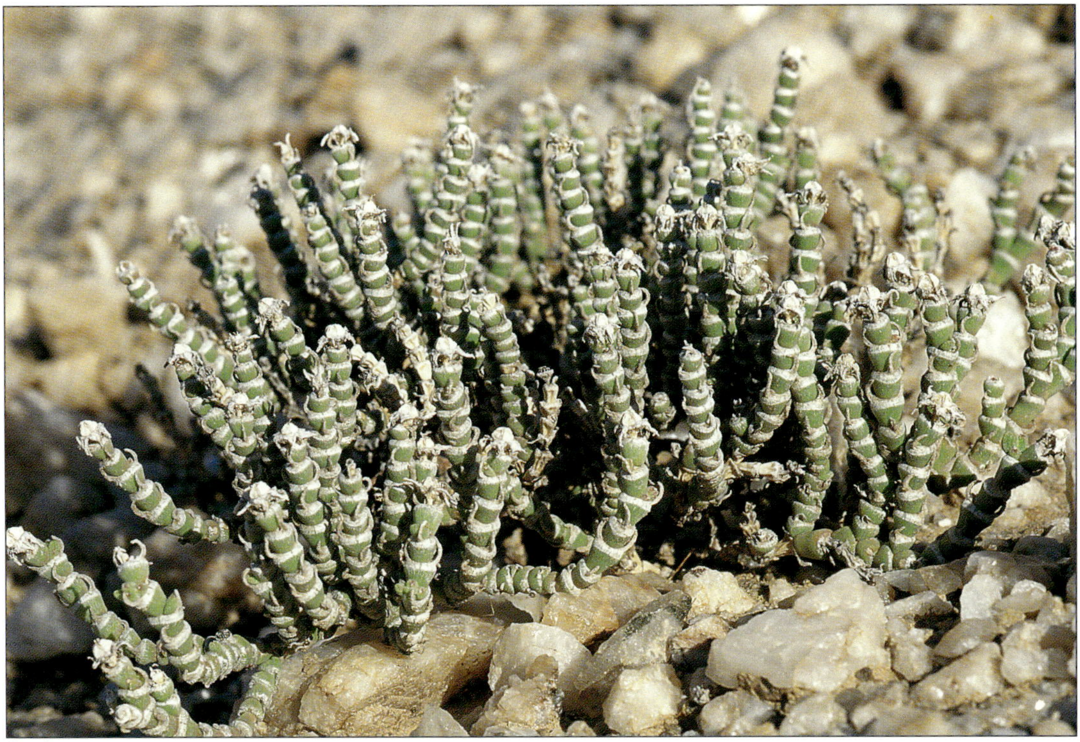
Peter Bruyns

B. pubescens

Ben-Erik van Wyk

Flowers of *B. ciliatus*

Peter Bruyns

B. marlothii

Ernst van Jaarsveld

B. nucifer

Peter Bruyns

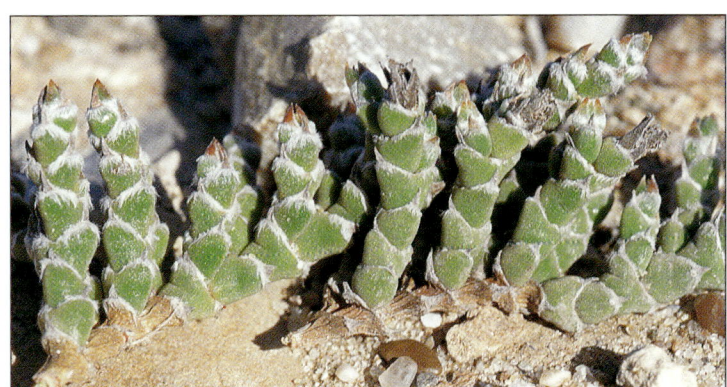

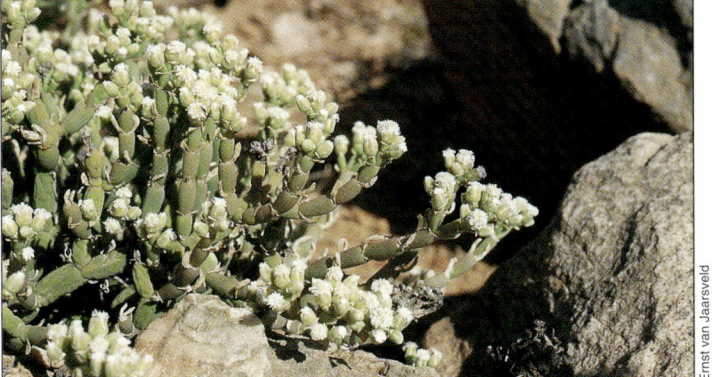

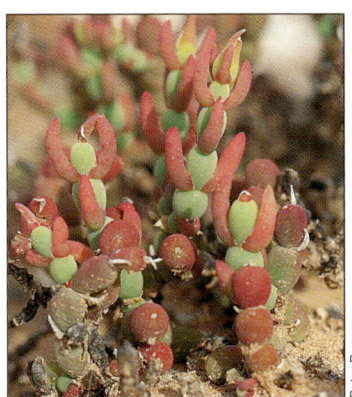

B. corallinus

"CAULIPSOLON"

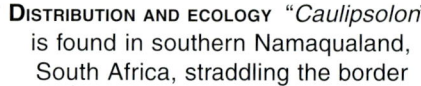

DERIVATION OF GENUS NAME The name is an anagram of *Psilocaulon*, the genus in which it was previously placed.

COMMON NAMES No vernacular names seem to have been recorded.

DESCRIPTION The plant has annual stems which grow flat on the ground from a tuberous rootstock. Stems are herbaceous, jointed, green and succulent with smooth surfaces. The short-lived leaves are semi-cylindrical and smooth in texture. The small to medium-sized flowers are white and occur singly or in few-flowered clusters. The petals decrease in length towards the centre of the flower. There are five sepals and the nectar glands consist of five separate grooves. The fruit capsules have five locules and the valve wings are reflexed and fused in pairs. The fruit base is funnel-shaped and up to 10 mm deep. Seeds are ochre coloured, with rough surfaces.

DISTINGUISHING CHARACTERS The genus is characterised by a tuberous rootstock and creeping annual stems. Stems and leaves have smooth surfaces and fruits have a distinctively funnel-shaped lower part.

FLOWERING TIME The plant flowers in late winter to early spring (July to October in South Africa).

DISTRIBUTION AND ECOLOGY "*Caulipsolon*" is found in southern Namaqualand, South Africa, straddling the border between the Western and Northern Cape Provinces. Its distribution range lies within the winter-rainfall zone, at altitudes between sea level and 1 000 m. It is typically found in disturbed areas, such as along roadsides or in overgrazed areas.

CULTIVATION No preference for a particular soil has been observed. "*Caulipsolon*" should not be watered once the annual stems have dried up. The plants may be watered again two to three months before the onset of the next growing season.

NOTES "*Caulipsolon rapaceum*", the only species in this genus, was previously known as *Psilocaulon rapaceum*. The names "*Caulipsolon*" and "*C. rapaceum*" will be published in *Botanische Jahrbücher*.

"*Caulipsolon*" Klak

NUMBER OF SPECIES/SUBSPECIES/VARIETIES (1/0/0)

SPECIES LIST AND CONSERVATION STATUS
"*C. rapaceum*" (Jacq.) Klak

LITERATURE
KLAK, C. & LINDER, H.P. (In press). Systematics of *Psilocaulon* N.E.Br. (Aizoaceae). *Botanische Jahrbücher*.

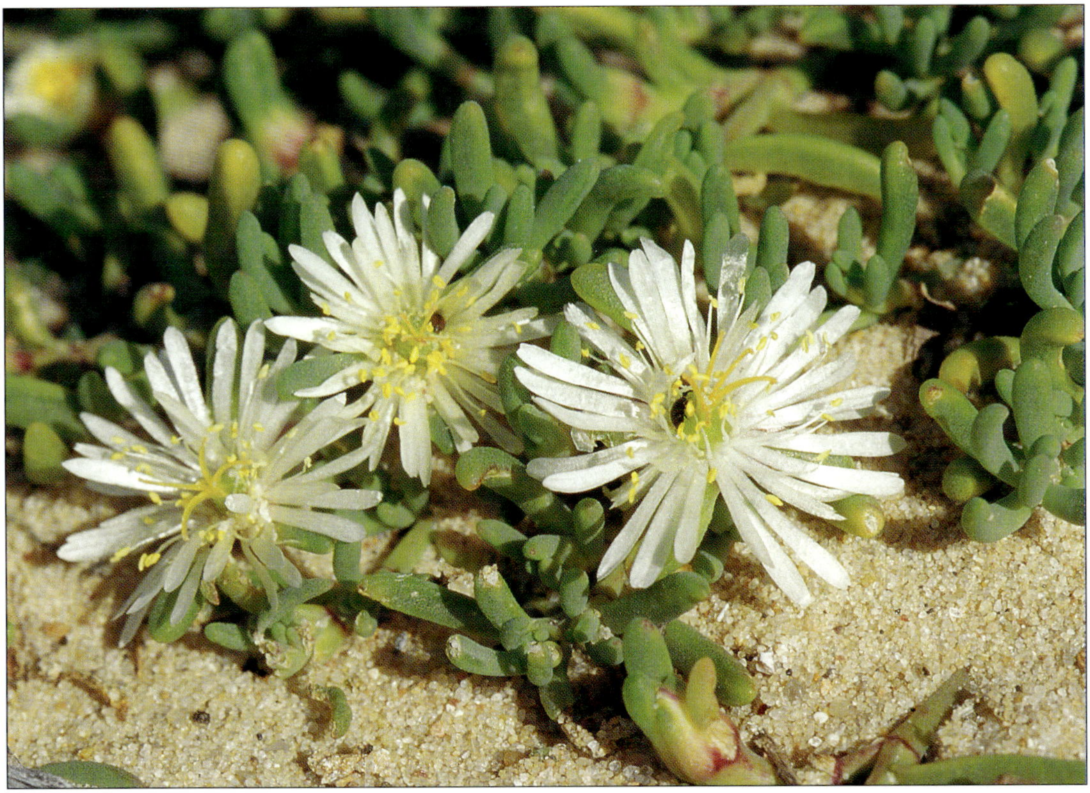

Peter Bruyns

"*C. rapaceum*"

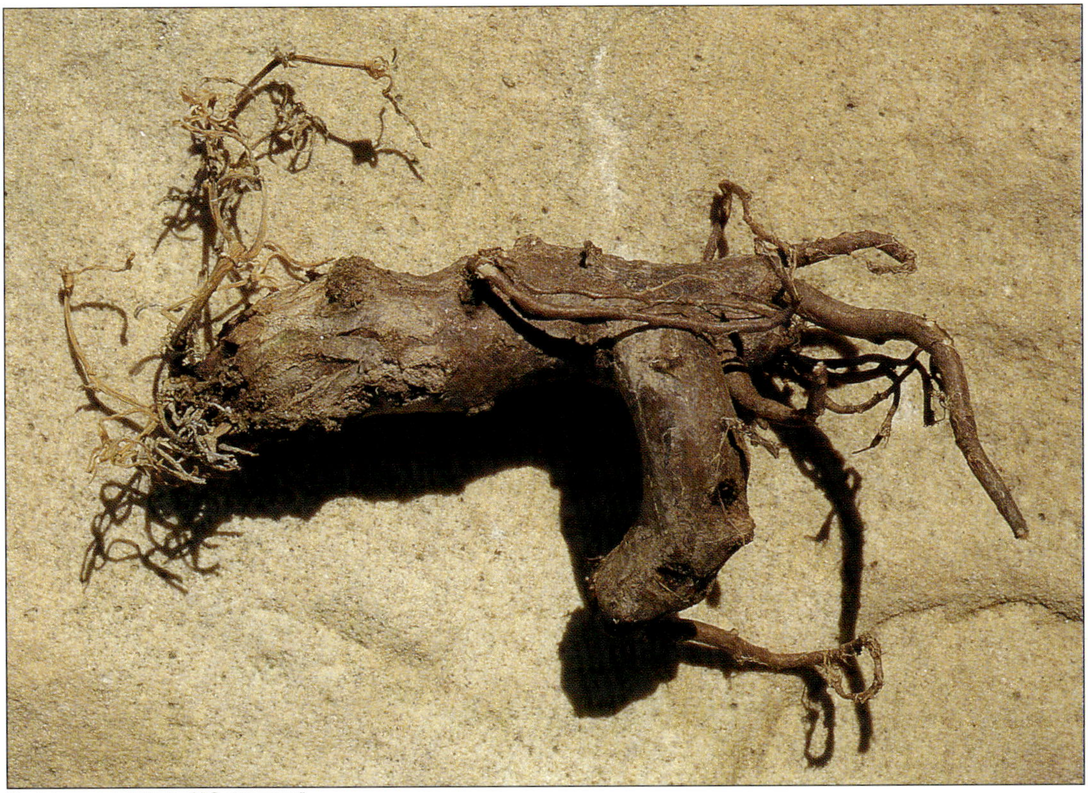

Peter Bruyns

The tuberous rootstock of "*C. rapaceum*"

Mesembryanthemum

Derivation of genus name This old name is derived from the Greek words *mesos* (in the centre), *embryon* (pistil or embryo) and *anthemon* (flower), meaning "a flower with a pistil or embryo in the centre". However, the name *Mesembrianthemum* was originally coined for this genus. The name referred to the opening of the flowers of many mesemb species during midday.

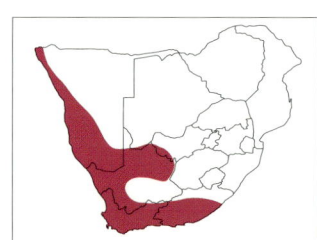

Common names The names ice plant, *brakslaai* (brack salad), *brakvy* (brack mesemb), *olifantslaai* (elephant salad), *soutslaai* (salt salad), *volstruisslaai* (ostrich salad), *slaaibos* (salad bush), *kama*, *nuta* and *rosa-de-jericho* have been recorded.

Description These often large-leaved fleshy herbs are usually annual or, in the case of the larger species, perennial. The creeping or upright stems are herbaceous, cylindrical, angled or winged. The leaves are flat, cylindrical, club-shaped or almost globular, and vary from paired to single. The leaf margins are occasionally wavy. The bases may be shortly fused and the water cells vary from being large and prominent to much flattened and inconspicuous. The pink, yellow or white flowers occur in clusters and vary in size from small to relatively large. There are five sepals, of which two may often be leaf-like in appearance. The sepals and petals may be free or sometimes fused into a short tube. Filamentous staminodes may be present. The nectar gland consists of five shallow or deep grooves. Fruit capsules have five locules, with valve wings which either curl inwards or outwards and which are fused in pairs. Seeds are small, black, brown, ochre or whitish with rough or more or less smooth surfaces.

Distinguishing characters Most species are covered with conspicuous water cells, the stems are herbaceous and there is no thickened rootstock. The often robust, lettuce-like appearance of some species, as well as their abundance at some localities, make them easy to identify.

Flowering time The plants flower mainly in the spring and summer months. Flowers open in the morning and close at night.

Distribution and ecology The genus occurs in southern Angola, Namibia, and western and central South Africa (Western, Eastern, Northern Cape and Northwest Provinces),

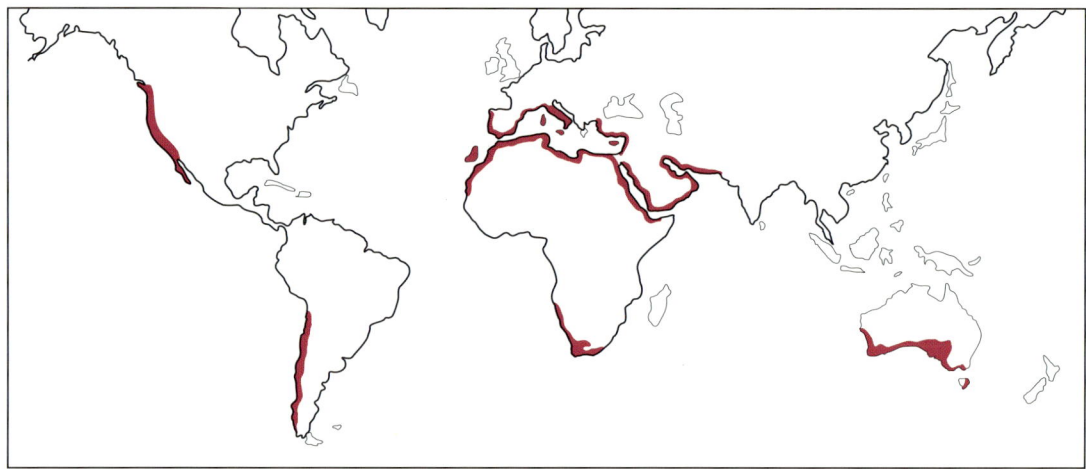

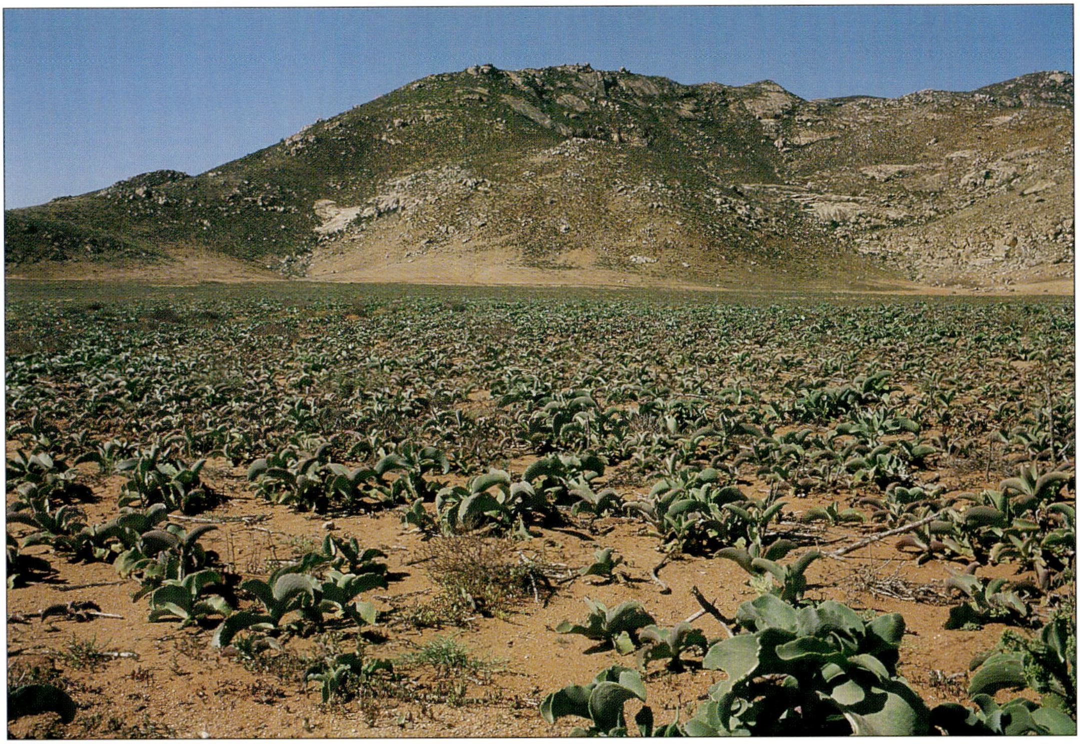

Ernst van Jaarsveld

M. barklyi

Graham Williamson

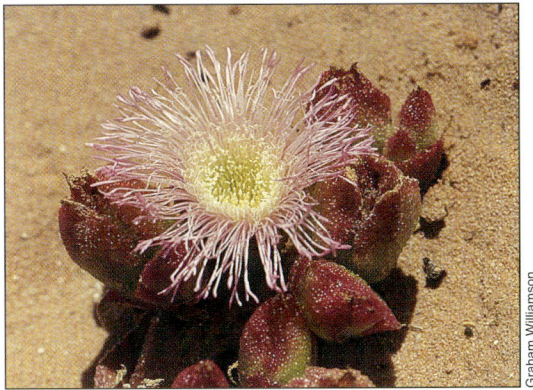

Flower of *M. guerichianum*

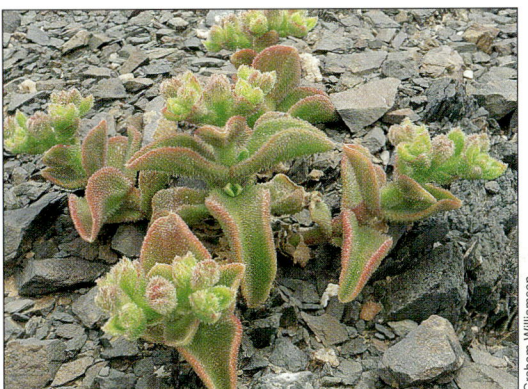

Graham Williamson

M. guerichianum

John Manning

M. alatum

Ernst van Jaarsveld

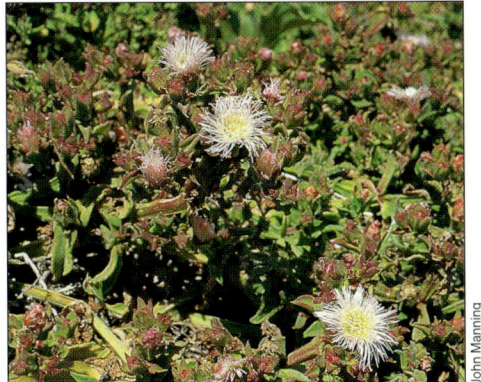

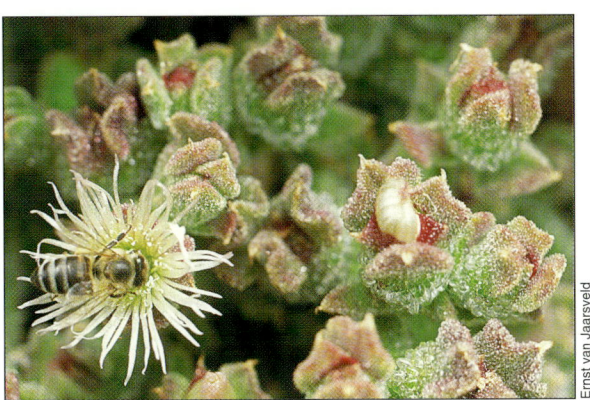

Flower of *M. crystallinum*

with the distribution extending eastwards in a broad band almost as far as the KwaZulu-Natal border. Species of *Mesembryanthemum* also occur in North Africa, the Mediterranean region, California in the United States, along the western coast of South America and in Australia, but were probably introduced to some of these areas. The genus is usually found in disturbed places such as overgrazed areas and along roadsides. Several species, such as *M. cryptanthum*, are tolerant of extreme conditions and often occur in salty habitats.

CULTIVATION These weedy annuals grow easily from seed. With the exception of the ice plant (*M. crystallinum*) they are rarely found in cultivation.

NOTES Until the early 1900s, most mesembs were considered to be members of this genus.

Mesembryanthemum L. (= *Callistigma* Dinter & Schwantes, =*Cryophytum* N.E.Br., =*Derenbergiella* Schwantes, =*Eurystigma* L.Bolus, =*Halenbergia* Dinter, =*Hydrodea* N.E.Br., =*Opophytum* N.E.Br.)

NUMBER OF SPECIES/SUBSPECIES/VARIETIES
(15/0/0)

SPECIES LIST AND CONSERVATION STATUS
M. aitonis Jacq.
M. barklyi N.E.Br.
M. cryptanthum Hook.f. [E on St. Helena]
M. crystallinum L.
M. excavatum (L.Bolus) L.Bolus
M. fastigiatum Thunb.
M. gariusanum Dinter
M. guerichianum Pax
M. hypertrophicum Dinter
M. inachabense Engl.
M. longistylum DC
M. nodiflorum L.
M. pellitum Friedrich
M. stenandrum (L.Bolus) L.Bolus
M. subtruncatum L.Bolus

LITERATURE
GERBAULET, M. (In press). *Mesembryanthemum* — Mesembryanthemoideae. In Hartmann, H.E.K. (ed.), *IOS Lexicon of Succulent Plants*. Fischer, Jena.

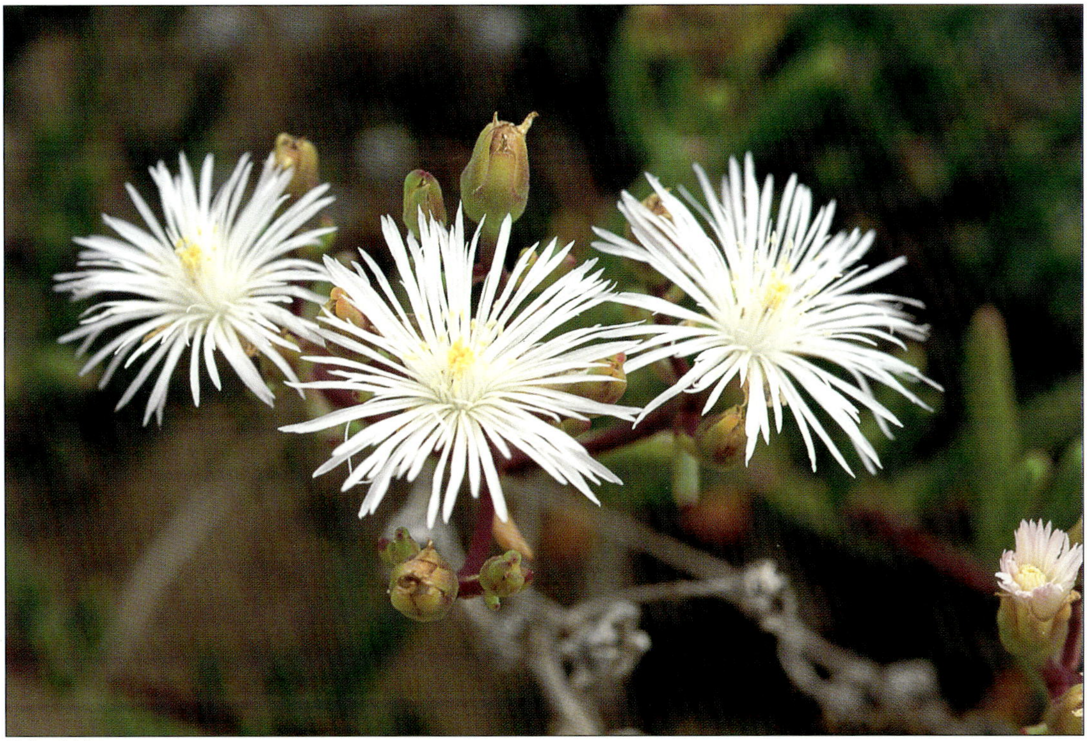
Peter Bruyns

M. longistylum

Ernst van Jaarsveld

M. hypertrophicum

Ernst van Jaarsveld

Mesembryanthemum cryptanthum

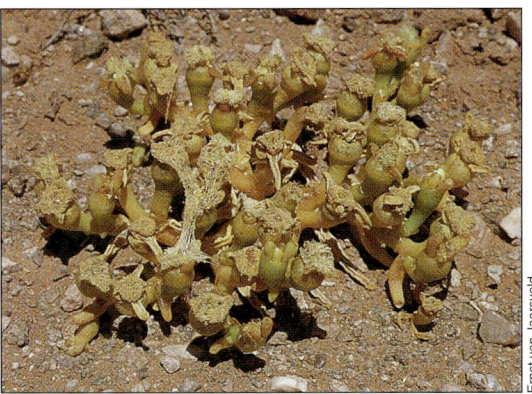

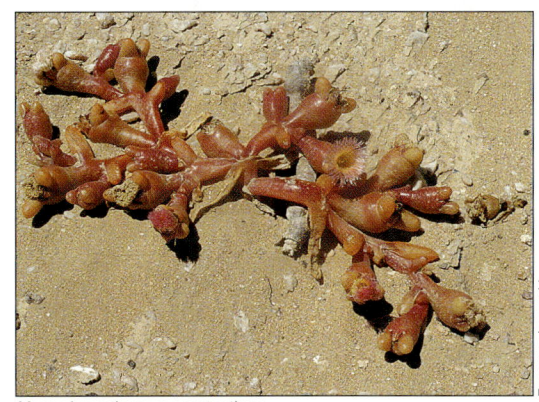

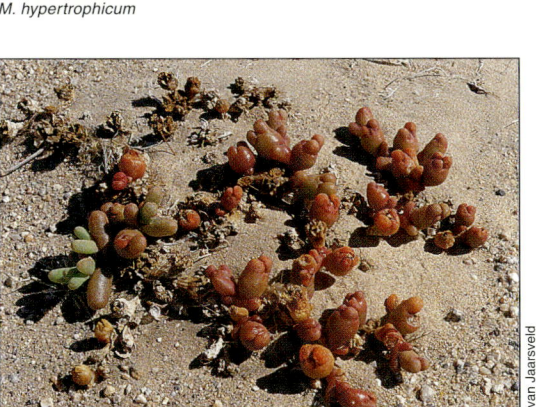

Ernst van Jaarsveld

Mesembryanthemum sp., presently known as *Opophytum aquosum*

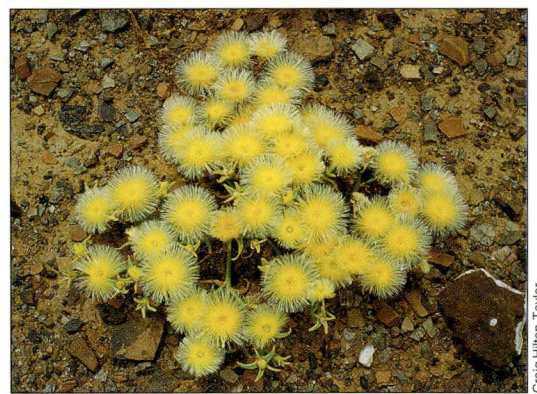

Craig Hilton-Taylor

Mesembryanthemum sp., presently known as *Eurystigma clavatum*

PHYLLOBOLUS

DERIVATION OF GENUS NAME The name is derived from the Greek word *phyllon* (leaf) and from the Latin word *bolus* (to throw or cast), referring to the deciduous leaves of some species.

COMMON NAMES *Vingerkanna* (finger mesemb) and *vingertjie-en-duimpie* (finger and thumb) refer to the well-known *Phyllobolus digitatus*, previously known as *Dactylopsis digitata*. This species is also known in the USA as the hitchhiker plant. *P. rabiei* is known as *oumasepram* (*vygie*) and *brakveldvygie* refers to *P. splendens*.

DESCRIPTION The plants are erect or creeping shrubs, mostly with thickened roots. Stems are corky or woody throughout or only at the base. Water cells are conspicuous on the stem surface and they may be hair-like or rarely flattened. The cylindrical or rarely flattened leaves vary from being paired to single, and alternate with each other. They may be deciduous or may dry out and remain on the stems. The small to medium-sized flowers occur singly or in clusters and vary in colour from green to yellow, orange, red, pink or whitish. There are four or five sepals. Filamentous staminodes are sometimes present and the nectar glands are present as five separate grooves. The fruit capsules have four or five locules. Valve wings are mostly present and folded inwards over the valves or, in rare cases, folded outwards and fused in pairs. Seeds are relatively large, usually black without a crest or brown with a crest, with rough or rarely smooth surfaces.

DISTINGUISHING CHARACTERS Plants often possess a tuberous rootstock. If that is present, the leaves are usually deciduous in summer. Most species of *Phyllobolus* have conspicuous water cells (as is also found in *Mesembryanthemum*) but the latter are annuals or biennials, while *Phyllobolus* species are invariably perennial. *P. digitatus* and several species that were previously included in *Aridaria* lack the conspicuous water cells and their surfaces are more or less smooth. However, the characteristic seeds place them in the genus *Phyllobolus*. Although *P. digitatus* does not have the typical seeds of most species of this genus, it has the characteristic succulence and thick corky stem of *Phyllobolus* species.

FLOWERING TIME The genus flowers mainly from midwinter to mid-summer (June to December in South Africa) but some of the deciduous species flower soon after their resurgence in autumn. In most species, the flowers close at night, although some bear night-opening flowers and others never close.

DISTRIBUTION AND ECOLOGY The genus is widespread in southern Namibia, and in the western and central parts of South Africa (Western, Eastern, Northern Cape and Free State Provinces). Species occur in karroid areas, both in the winter and the summer rainfall regions.

CULTIVATION Plants have distinct growing and resting phases. The growing period starts after the rainy season (winter). During the resting phase (summer), the leaves dry away completely and in some species branches are shed. During this period the plants should be kept dry. Cultivation in sandy loamy soil is recommended. Plants may be propagated from seed or by careful division of older plants.

NOTES The well-known genera *Dactylopsis* and *Sphalmanthus* have recently been included in *Phyllobolus*.

Phyllobolus N.E.Br. (=*Amoebophyllum* N.E.Br., =*Dactylopsis* N.E.Br., =*Sphalmanthus* N.E.Br.)

NUMBER OF SPECIES/SUBSPECIES/VARIETIES (32/2/0)

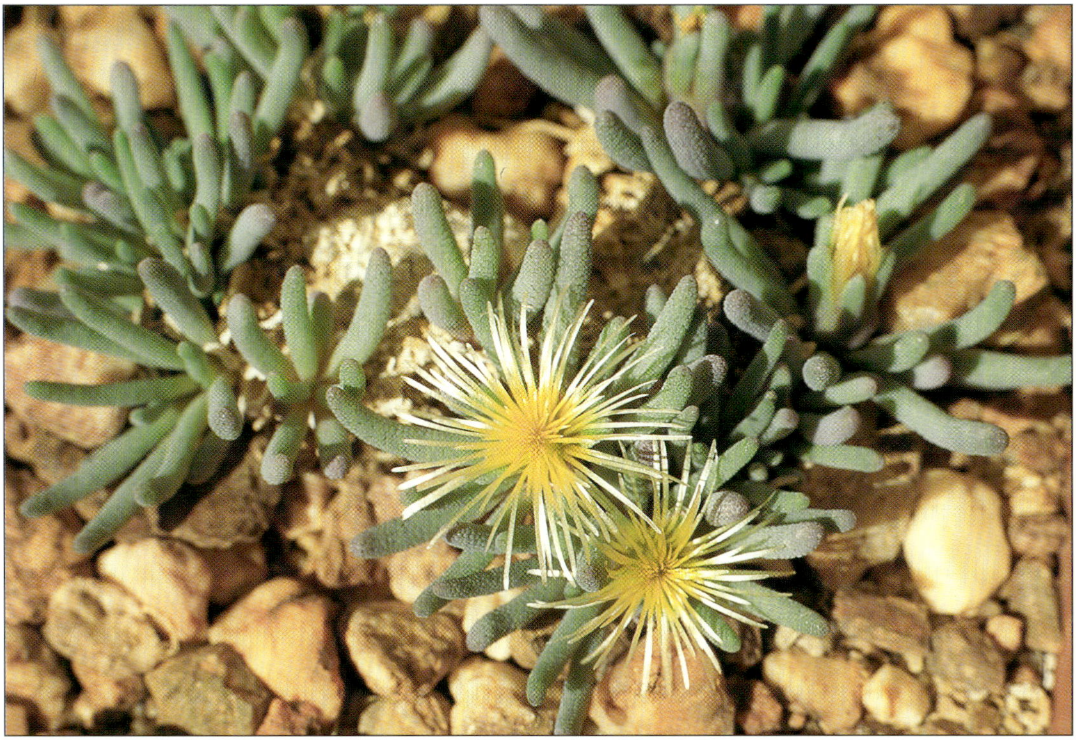

Ben-Erik van Wyk

P. rabiei

Ernst van Jaarsveld/NBI

P. tetragonus

John Manning

P. tenuiflorus

Mike Struck/NBI

Phyllobolus sp.

Ben-Erik van Wyk

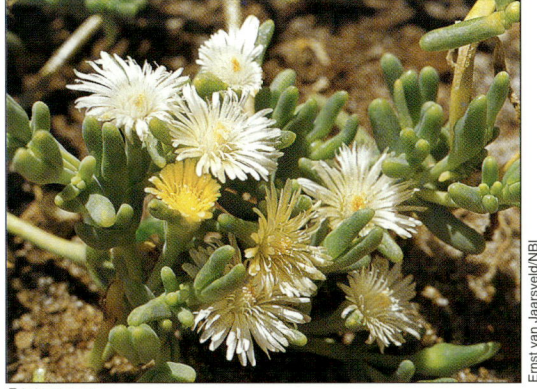

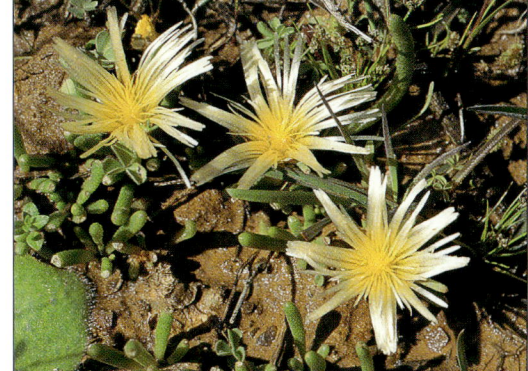

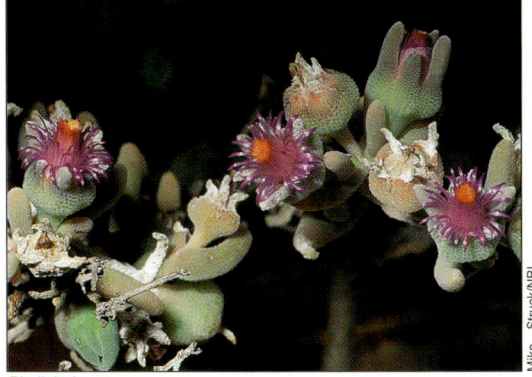

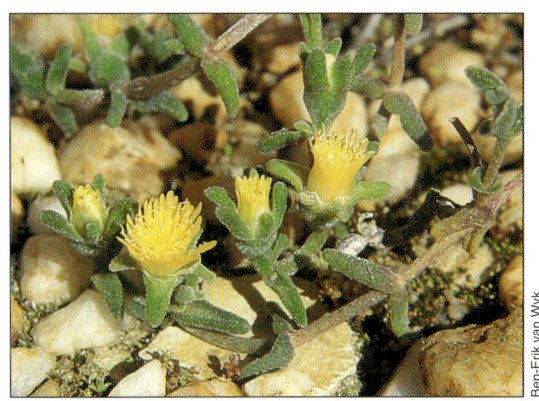

P. abbreviatus

Species list and conservation status

P. abbreviatus (L.Bolus) Gerbaulet
P. amabilis Gerbaulet & Struck
P. canaliculatus (Haw.) Gerbaulet
P. caudatus (L.Bolus) Gerbaulet
P. chrysophthalmus Gerbaulet & Struck
P. congestus (L.Bolus) Gerbaulet
P. deciduus (L.Bolus) Gerbaulet
P. decurvatus (L.Bolus) Gerbaulet
P. delus (L.Bolus) Gerbaulet
P. digitatus (Aiton) Gerbaulet subsp. *digitatus*
P. digitatus (Aiton) Gerbaulet subsp. *littlewoodii* (L.Bolus) Gerbaulet
P. gariepensis Gerbaulet & Struck
P. grossus (Aiton) Gerbaulet
P. herbertii (N.E.Br.) Gerbaulet
P. latipetalus (L.Bolus) Gerbaulet
P. lignescens (L.Bolus) Gerbaulet
P. melanospermus (Dinter & Schwantes) Gerbaulet
P. nitidus (Haw.) Gerbaulet
P. oculatus (N.E.Br.) Gerbaulet
P. prasinus (L.Bolus) Gerbaulet
P. pumilus (L.Bolus) Gerbaulet
P. quartziticus (L.Bolus) Gerbaulet
P. rabiei (L.Bolus) Gerbaulet
P. resurgens (Kensit) Schwantes
P. roseus (L.Bolus) Gerbaulet
P. saturatus (L.Bolus) Gerbaulet
P. sinuosus (L.Bolus) Gerbaulet
P. spinuliferus (Haw.) Gerbaulet
P. splendens (L.) Gerbaulet subsp. *pentagonus* (L.Bolus) Gerbaulet
P. splendens (L.) Gerbaulet subsp. *splendens*
P. suffruticosus (L.Bolus) Gerbaulet
P. tenuiflorus (Jacq.) Gerbaulet
P. trichotomus (Thunb.) Gerbaulet
P. viridiflorus (Aiton) Gerbaulet

Literature

GERBAULET, M. 1995. *Phyllobolus* N.E.Br. emend. Bittrich (Aizoaceae): a reassessment of generic boundaries. *Botanische Jahrbücher* 117: 385–399.

GERBAULET, M. 1997. Revision of the genus *Phyllobolus* N.E.Br. (Aizoaceae). *Botanische Jahrbücher* 119: 145–211.

P. canaliculatus

P. digitatus subsp. *littlewoodii*

P. digitatus subsp. *digitatus*

Phyllobolus sp.

P. roseus

PRENIA

DERIVATION OF GENUS NAME The name is derived from the Greek word *prenes* (prostrate), referring to the creeping habit of the plants.

COMMON NAMES *Prenia vanrensburgii*, which grows close to the shoreline from Hermanus to Bredasdorp, is known as *seepampoen* (sea pumpkin) and *P. sladeniana* is known as *skotteloor* (disc ear).

DESCRIPTION Plants are creeping or erect shrublets with fibrous roots. The branches are slightly woody and smooth, and in time the green parts of the stem surface become dry and remain on the stem as a thick whitish layer. The flat, three-sided or semi-cylindrical leaves are borne in pairs on the plant but those in the flower clusters are single and alternate with each other. They are shortly fused at the bases and are covered with a thick waxy layer which easily rubs off. The medium-sized flowers are borne in clusters and vary in colour from white to yellow or pink. There are four or five sepals and numerous filamentous staminodes. The bases of the sepals and petals are fused into a short tube. Nectar glands occur as four separate grooves. The fruit capsules have four locules and the lower part is funnel-shaped with deep locules. The seeds are black and have rough surfaces.

DISTINGUISHING CHARACTERS *Prenia* usually has soft, smooth stems (*P. sladeniana* has firm stems) and with the exception of *P. tetragona*, all the species are creeping. *Prenia* has fruit capsules with deep locules, in which respect it is similar to *Psilocaulon*.

FLOWERING TIME The plants flower from spring through summer to autumn (September to April in southern Africa). The flowers open by day.

DISTRIBUTION AND ECOLOGY *Prenia* is found in the karroid areas of the Western, Eastern, and Northern Cape Provinces, South Africa, and in southern Namibia. Species are distributed throughout the winter rainfall area, as well as in some of the drier western parts of the summer rainfall region. Most species are found along roadsides or in other disturbed places.

CULTIVATION *Prenia* is a genus of rapid-growing pioneer plants which are useful on embankments in the dry winter rainfall region. Outside of the winter rainfall region plants are best grown in greenhouses. Propagation is from seeds or cuttings, which root quickly and soon develop to the flowering stage. Plants prefer rich, porous soil.

NOTES The species of *Prenia* most likely to be encountered is *P. pallens*, an exceptionally abundant weed along roadsides in the Western Cape Province of South Africa.

Prenia N.E.Br. (=*Phyllobolus* subgen. *Prenia* (N.E.Br.) Bittrich)

NUMBER OF SPECIES/SUBSPECIES/VARIETIES (6/3/0)

SPECIES LIST AND CONSERVATION STATUS
P. englishiae (L.Bolus) Gerbaulet
P. pallens (Aiton) N.E.Br. subsp. lancea (Thunb.) Gerbaulet
P. pallens (Aiton) N.E.Br. subsp. lutea L.Bolus
P. pallens (Aiton) N.E.Br. subsp. namaquensis Gerbaulet
P. pallens (Aiton) N.E.Br. subsp. pallens
P. radicans (L.Bolus) Gerbaulet
P. sladeniana (L.Bolus) L.Bolus
P. tetragona (Thunb.) Gerbaulet
P. vanrensburgii L.Bolus

LITERATURE
GERBAULET, M. 1996. Revision of the genus *Prenia* N.E.Br. (Aizoaceae). *Botanische Jahrbücher* 118: 25–40.

P. sladeniana

Leaves of *P. sladeniana*

P. vanrensburgii

P. tetragona

P. pallens subsp. *lutea*

Psilocaulon

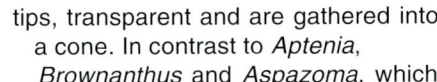

Derivation of genus name The name is derived from the Greek words *psilos* (bare, bald, smooth) and *kaulos* (stalk), referring to the leafless stems.

Common names The most commonly used vernacular name is *asbos* (ash bush), but *loogasbossie* (lye ash bush), *seepbossie* (little soap bush) or *grootlidjies* (large jointlets) are also sometimes used. *Psilocaulon dinteri* is known as *skerpioenvygie* (scorpion mesemb), due to its resemblance to a scorpion.

Description The creeping or erect plants are perennial or annual to biennial. Young stems are jointed, succulent and green, while old ones become woody. Water cells are inconspicuous, hair-like, dome-shaped or much flattened. Leaves are borne in pairs but become single and alternating in the flower clusters. They are cylindrical or semi-cylindrical and shortly fused at their bases, dry and persistent or deciduous, sometimes with a sharp tip. The small flowers are white, pink, puce or rarely pale yellow to greenish and are borne singly or in clusters at the tips of branches. There are four or five sepals which are fused into a short tube. In addition to the normal spreading petals, there are filamentous staminodes which are gathered around the stamens, thus forming a cone. Nectar glands are usually present in the form of four or five narrow grooves, or they may be absent. Fruit capsules have four or five locules, valve wings are curved inwards over the valves and the capsule base is funnel-shaped. Seeds are rough or almost smooth and ochre or brown in colour.

Distinguishing characters *Psilocaulon* may be distinguished from all other genera of weedy mesembs by its peculiar flower structure: the filamentous staminodes are relatively broad, ragged at the tips, transparent and are gathered into a cone. In contrast to *Aptenia*, *Brownanthus* and *Aspazoma*, which also have succulent green stems, the stem surface in *Psilocaulon* is either smooth, or has sparsely scattered minute water cells.

Flowering time Species of *Psilocaulon* flower from early to midsummer (October to January in southern Africa).

Distribution and ecology The genus is widespread in southern, central and western South Africa, western Namibia and southern Angola and occurs in both summer and winter rainfall regions. One species, *P. granulicaule*, is found in Australia. It is thus one of the most widespread mesemb genera. It is mostly found in areas which have been subject to disturbance, such as roadsides and overgrazed areas. Although most species are adapted to a wide variety of soil types and tolerate high salinity, they will not flourish in nutrient-poor acidic soils.

Cultivation Plants require warm sunny positions with adequate water during the growing season (and no water during the resting period). They thrive in rich, porous sandy soil. *P. parviflorum*, however, prefers clayey soils.

Notes In the northwestern parts of the Cape, *Psilocaulon* species are used to construct a *kookskerm* or *asbosskerm* (a traditional shelter around the fireplace) by stacking the plants up to a height of about 2,5 m. The ash of *P. junceum* and *P. coriarium* has been used as lye in soap-making. *P. dinteri* has flowers which smell of cinnamon.

Psilocaulon N.E.Br.

Number of species/subspecies/varieties (13/0/0)

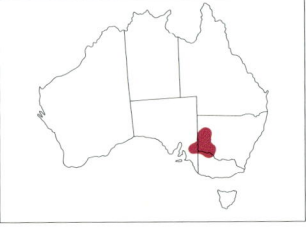

Flat-leaved Mesembs

GROUP 2

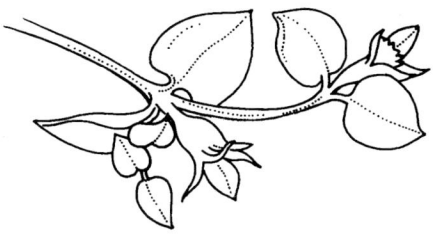

Leafy, mostly annual, not particularly succulent plants, often with short creeping stems, sometimes glistening with surface water cells, flowers mostly yellow or white.

The characteristic flat, stalked leaves are generally considered primitive in the mesembs. Lobed leaves are found in two of the genera: *Aethephyllum* and the closely allied *Cleretum*. The flowers vary in size from large to small and are mostly yellow or white — the multicoloured *Dorotheanthus* is an exception. There are two groups of closely related genera in the FLAT-LEAVED MESEMBS: the *Apatesia*-group (*Apatesia, Carpanthea, Caryotophora, Conicosia, Hymenogyne, Saphesia* and *Skiatophytum*) and the *Dorotheanthus*-group (*Aethephyllum, Cleretum, Dorotheanthus*). The *Apatesia*-group characteristically has fruit capsules which break apart into so-called schizocarps; the fruit is dry and does not split open but break apart at maturity into separate locules (carpels) which contain the seeds.

FLAT-LEAVED MESEMBS comprise 10 genera and 22 species.

Aethephyllum (1 species)
Apatesia (3 species)
Carpanthea (1 species)
Caryotophora (1 species)

Cleretum (3 species)
Conicosia (2 species)
Dorotheanthus (7 species)

Hymenogyne (2 species)
Saphesia (1 species)
Skiatophytum (1 species)

Aethephyllum

Derivation of genus name There are two possible derivations for this name. It may be derived from the Greek word *aither* (ether) which probably refers to the ethereal beauty of the leaves or from *aethes* (irregular or unusual), combined with the word *phyllon* (leaf).

Common names No vernacular names seem to have been recorded.

Description The plant is an annual creeping herb with glistening stems and flat, irregularly lobed leaves which also glisten due to numerous water cells. The small flowers are yellow, and have many narrow and pointed petals in two or three whorls. There are comparatively few stamens. The ovary has five thread-like stigmas. The fruit capsules have five locules with large expanding sheets which taper into loose apical membranes; the covering membranes are reduced to small limbs. Seeds have a rough surface.

Distinguishing characters The plant is an annual herb, easily recognised by the flat-growing habit and small, yellow flowers. The leaves are lobed along the edges (pinnatifid). The fruit capsules are borne on relatively long stalks and are flat on top.

Flowering time *A. pinnatifidum* flowers in mid-winter to early spring (July to October in South Africa). Flowers open in the afternoon, around 16:00.

Distribution and ecology *A. pinnatifidum* is restricted to a narrow north-south band in the southwestern part of the Western Cape Province, South Africa. Although the exact distribution range is insufficiently known, the species has so far been recorded from the Gifberg near Vanrhynsdorp, the Cedarberg, as well as the areas around Tulbagh, Paarl and Stellenbosch. Plants grow in sandy soil and are fireweeds, forming part of the early post-fire vegetation in fynbos.

Cultivation Propagation is easy from seeds although germination may be tricky due to inhibiting factors. Smoke treatment may facilitate germination. Once germinated, plants are rapid growers, flowering and fruiting profusely. They require sandy, acid soil.

Notes This species has fewer stamens than other mesembs; the stamens are often incompletely developed, which may indicate that the plant produces seeds without pollination. The capsule structure is intermediate between those found in *Dorotheanthus* and *Cleretum*.

Aethephyllum N.E.Br.

Number of species/subspecies/varieties (1/0/0)

Species list and conservation status
A. pinnatifidum (L.f.) N.E.Br. [nt]

Literature
IHLENFELDT, H.-D. & STRUCK, M. 1986. Morphologie und Taxonomie der Dorotheanthinae Schwantes (Mesembryanthemaceae). *Beiträge zur Biologie der Pflanzen* 61: 411–453.

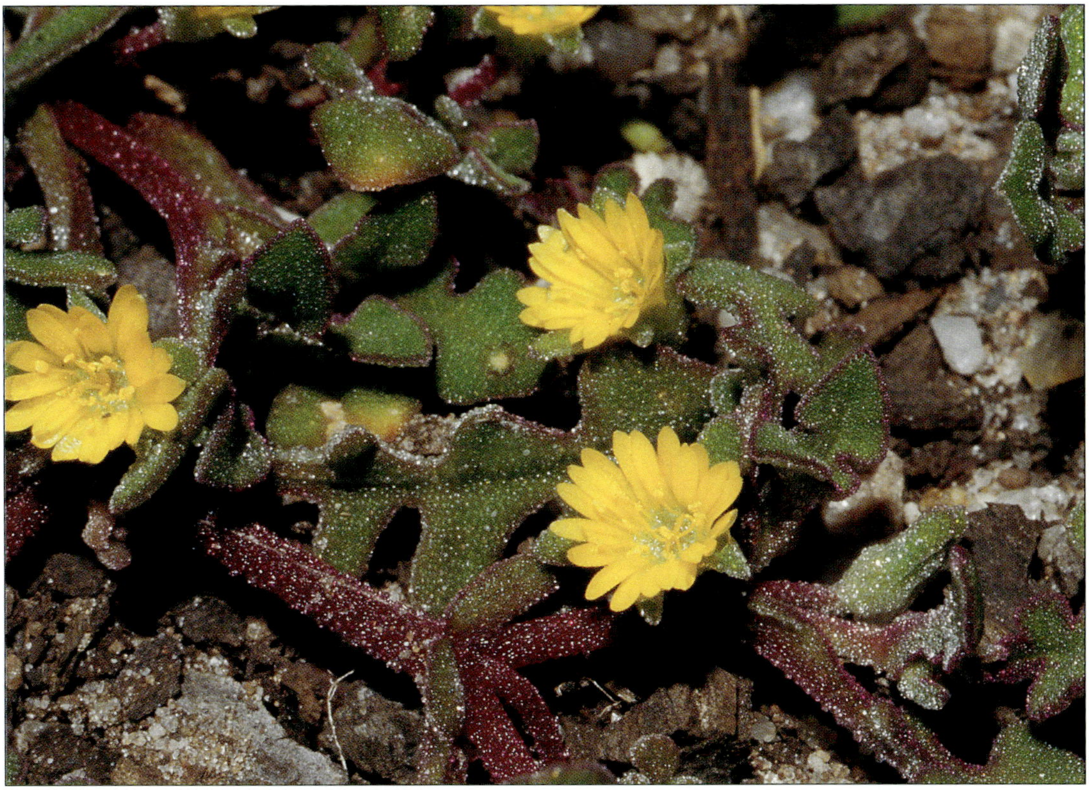

Bob Chinnock

A. pinnatifidum

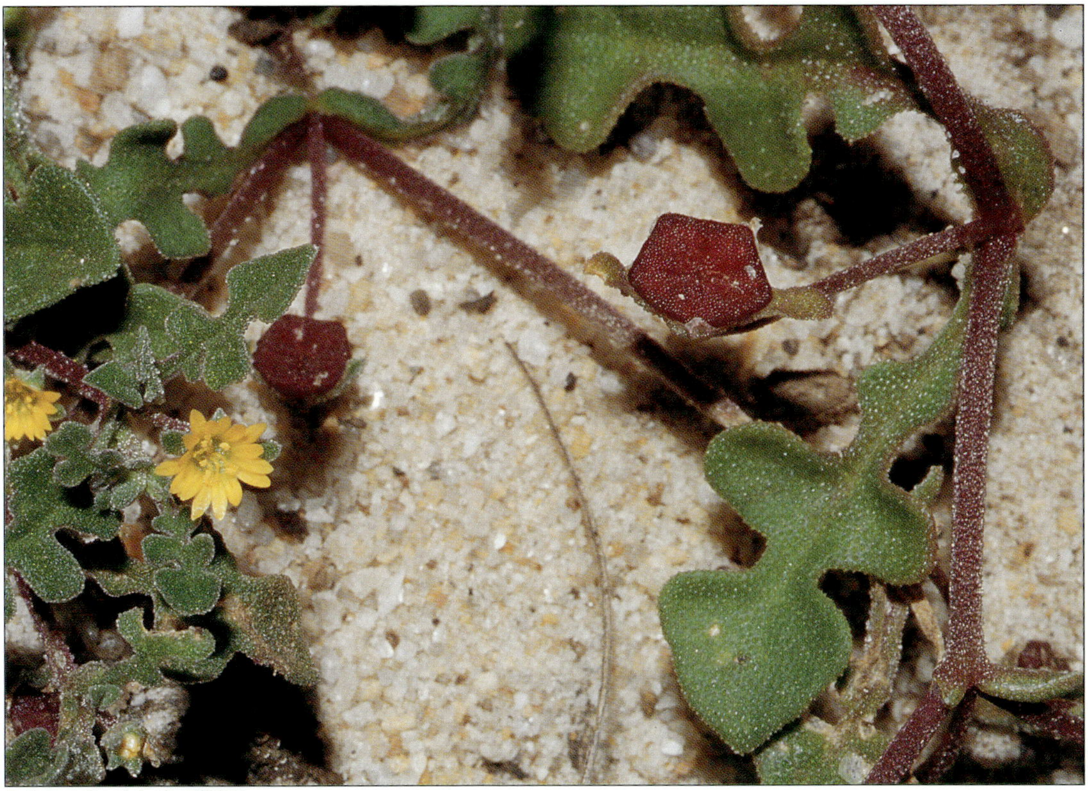

Bob Chinnock

Fruit capsules of *A. pinnatifidum*

APATESIA

DERIVATION OF GENUS NAME The name is derived from the Greek word *apatesis* (deception), due to the resemblance of this plant to *Hymenogyne*.

COMMON NAMES No vernacular names seem to have been recorded.

DESCRIPTION All species are annual herbs with flat-growing branches. The leaves are flat and have a basally sheathing stalk. The solitary flowers are borne on long stalks and have their sepals basally united with the ovary to form a saucer-shaped organ. The numerous yellow petals are narrow and sharply pointed, with minute hairs towards their bases. There are numerous stamens and filamentous staminodes in the flower and the inner stamens are bearded at their bases. The fruit capsules are slightly convex, with expanding keels and very small covering membranes. *A. sabulosa* has valves with a narrow membranous rim, while valve wings are absent in the other two species. Some of the seeds do not develop in the ordinary locule, but beneath the locule in separate chambers or 'seed pockets'. The round seeds are smooth.

DISTINGUISHING CHARACTERS All species are annual herbs with flat, stalked leaves. Their fairly large yellow flowers are borne on long stalks. Only a few seeds develop in seed pockets.

FLOWERING TIME Plants flower in early summer (August to October in South Africa) and the flowers open during the day.

DISTRIBUTION AND ECOLOGY The genus is found in the Western Cape Province of South Africa, in a broad band running parallel to the west coast from Cape Town to Vanrhynsdorp. The species grow mainly in disturbed areas.

CULTIVATION Propagation is from seed, which does not seem to germinate readily. Although members of this genus were grown in Europe during the early 1800s, they are rarely encountered in succulent plant collections.

NOTES The genus seems to be related to *Carpanthea*, *Conicosia* and *Hymenogyne*. These genera are characterised by the progressive development of the seeds in seed pockets, a phenomenon called paraspermy. Associated with this development is the reduction and eventual loss of the expanding tissues of the capsules. It is also worth noting that the fruits of *Apatesia* open more quickly than any other hygroscopic fruit in this family. The three species are almost identical and are differentiated from each other by slight differences in the shape of the fruit.

Apatesia N.E.Br.

NUMBER OF SPECIES/SUBSPECIES/VARIETIES (3/0/0)

SPECIES LIST AND CONSERVATION STATUS
A. helianthoides (Aiton) N.E.Br.
A. pillansii N.E.Br.
A. sabulosa (Thunb.) L.Bolus

LITERATURE
IHLENFELDT, H.-D. & GERBAULET, M. 1990. Untersuchungen zum Merkmals-bestand und zur Taxonomie der Gattungen *Apatesia* N.E.Br., *Carpanthea* N.E.Br., *Conicosia* N.E.Br., *Herrea* Schwantes und *Hymenogyne* Haw. (Mesembryanthemaceae Fenzl). *Botanische Jahrbücher* 111: 457–498.

A. helianthoides

A. sabulosa

CARPANTHEA

DERIVATION OF GENUS NAME The name is derived from the Greek words *carpos* (fruit), and *anthe* (flower), pertaining to their edibility (see under Notes below).

COMMON NAMES The name *vetkousie* (fat little sock) has been recorded.

DESCRIPTION The plant is an annual, succulent herb, with glistening, woolly stems. The leaves are paired and the spoon or lance-shaped blade is shorter than the stalk. The three-nerved leaves have blunt tips and scattered hairs (elongated water cells) on their surfaces and margins. Large flowers are borne singly or in small clusters of up to three. The numerous slender petals are golden yellow with reddish outer surfaces and have minute hairs along the margins. The five sepals are unequal in size: the outer two are leaf-like, while the inner sepals are smaller with membranous margins. There are numerous stamens and filamentous staminodes around the thread-like stigmas. Fruit capsules have 12 to 18 narrow locules, with firm, persistent divisions overarching the seed chambers from the sides. The placenta is raised by a false basal septum, the seed chambers are broader than they are deep and the pair of narrow expanding keels end in awns. The seeds are roundish.

DISTINGUISHING CHARACTERS The plants are annual herbs with woolly stems and flat green leaves, 12 to 18 stigmas and the same number of locules in the capsule. The narrow capsule valves have contiguous expanding keels ending in awns. Flowers are huge, yellow and open during the afternoon.

FLOWERING TIME Plants flower in late spring to early summer. Flowers close up tightly into a twisted tip at night.

DISTRIBUTION AND ECOLOGY *Carpanthea pomeridiana*, the only species in this genus, occurs in the extreme southwestern parts of the Western Cape Province, South Africa. It is found on white sandy soils in strandveld and coastal fynbos vegetation, in full sun or in the light shade of other vegetation. It prefers mainly winter rainfall which ranges between 400 and 500 mm per year. *C. pomeridiana* has also been recorded from southeastern Australia, where it occurs as an exotic.

CULTIVATION Plants are easily propagated from seed sown during autumn. Due to inhibiting factors, the seeds may not germinate readily. Under favourable conditions, however, germination takes place within three weeks and seedlings grow rapidly.

NOTES Unlike many other mesembs, *Carpanthea* also thrives in light shade. The leaves and flowers may be harvested and cooked up as a stew with meat.

Carpanthea N.E.Br.

NUMBER OF SPECIES/SUBSPECIES/VARIETIES (1/0/0)

SPECIES LIST AND CONSERVATION STATUS
C. pomeridiana (L.) N.E.Br.

LITERATURE
IHLENFELDT, H.-D. & GERBAULET, M. 1990. Untersuchungen zum Merkmalsbestand und zur Taxonomie der Gattungen *Apatesia* N.E.Br., *Carpanthea* N.E.Br., *Conicosia* N.E.Br., *Herrea* Schwantes und *Hymenogyne* Haw. (Mesembryanthemaceae Fenzl). *Botanische Jahrbücher* 111: 457–498.

C. pomeridiana

Flowers of *C. pomeridiana*

Open fruit capsules of *C. pomeridiana*

CARYOTOPHORA

DERIVATION OF GENUS NAME The genus name is derived from the Greek words *karyotos* (nut-like) and *phorein* (bearing), alluding to the appearance of the fruits that resemble nuts.

COMMON NAMES No vernacular names seem to have been recorded.

DESCRIPTION Plants are low-growing, smooth, slightly succulent perennials. The very long, stout horizontal roots give rise to numerous well-spaced suckers. Each of these forms a very short stem bearing a tuft of flat, lance to spoon-shaped leaves and creeping annual flowering branches. The attractive white flowers are comparatively large, with numerous, widely spreading petals tapering towards the tips, giving the flowers a rather fluffy appearance. The fruit falls apart into three or four very hard, nut-like parts, each with two chambers containing a single seed.

DISTINGUISHING CHARACTERS The flat, smooth, only slightly succulent, lance to spoon-shaped leaves are characteristic. The genus may also be recognised by the fruit, comprising three or four nut-like parts, the production of as few as six to eight seeds per fruit and its very localised occurrence near Cape Agulhas.

FLOWERING TIME Flowering occurs from spring to early summer (September to November in South Africa) and the flowers open during the day.

DISTRIBUTION AND ECOLOGY *C. skiatophytoides* has a very restricted distribution to the west of Cape Agulhas, the southernmost point of Africa, in the Western Cape Province of South Africa. It is known only from sandy flats with coastal fynbos vegetation in the Brandfontein area, near Bredasdorp.

CULTIVATION *C. skiatophytoides* is not popular amongst mesemb collectors: the plants have a rather untidy appearance and the seeds are scarce and often sterile and almost impossible to germinate unless treated with caustic potash or acids to soften the seed coat. It can, however, be propagated from root cuttings, a rare feature in mesembs.

NOTES *C. skiatophytoides* produces fewer seeds than any other mesemb: a maximum of six to eight per fruit. Plants do not bear many fruits. The species is very similar and closely related to *Skiatophytum tripolium* with which it can be hybridised.

Caryotophora Leistner

NUMBER OF SPECIES/SUBSPECIES/VARIETIES (1/0/0)

SPECIES LIST AND CONSERVATION STATUS
C. skiatophytoides Leistner [R]

LITERATURE
LEISTNER, O.A. 1958. A new monotypic genus of the Mesembryanthemaceae. *Notes on* Mesembryanthemum *and allied genera.* University of Cape Town, Cape Town. Vol. 3: 289–291.

IHLENFELDT, H.-D. & GERBAULET, M. 1990. Untersuchungen zum Merkmalsbestand und zur Taxonomie der Gattungen *Apatesia* N.E.Br., *Carpanthea* N.E.Br., *Conicosia* N.E.Br., *Herrea* Schwantes und *Hymenogyne* Haw. (Mesembryanthemaceae Fenzl). *Botanische Jahrbücher* 111: 457–498.

C. skiatophytoides in its natural habitat

Young fruit capsules of *C. skiatophytoides*

Flower of *C. skiatophytoides*

CLERETUM

DERIVATION OF GENUS NAME It is not clear how the genus name was derived.

COMMON NAMES No vernacular names are known.

DESCRIPTION The plants are flat-growing or erect, annual herbs covered with glistening water cells. The flat, spoon-shaped leaves have smooth or lobed margins. The solitary and shortly-stalked flowers are either small or large. When the flowers are small in size, it means that they are cleistogamous (fertilized in the bud stage, so that the flowers do not develop fully). There are relatively few stamens (often only five), which are at first curved inwards over the ovary and later become erect. Their basal parts have prominent water cells. The stigmas are thread-like or taper to a narrow point. Fruit capsules have five locules with massive, parallel expanding keels and large valve wings. The covering membranes are small or absent. The bluntly triangular, flat seeds have a smooth or slightly rough surface.

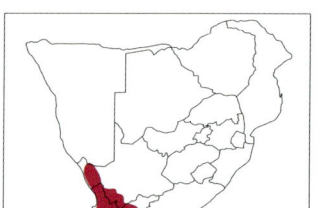

DISTINGUISHING CHARACTERS The plants are ground-hugging annual herbs with flat leaves, small or large yellow to white flowers and firm fruit capsules. The flowers are often cleistogamous. The ridges on the upper surface of the fruit capsule distinguish *Cleretum* from similar genera with flat-topped fruit.

FLOWERING TIME Flowering occurs in late winter to spring (July to October in South Africa) and the flowers open during the day (unless they are cleistogamous).

DISTRIBUTION AND ECOLOGY *Cleretum* species are found in the western parts of South Africa, from the Cape Peninsula in the Western Cape Province, northwards through Namaqualand to the Richtersveld in the Northern Cape Province. They frequently grow in disturbed areas on sandy soil. *C. papulosum* subsp. *papulosum* has been introduced into the southern parts of Australia.

CULTIVATION These rapid-growing annuals are well worth cultivating, mainly for their attractive glistening stems and leaves and not for their rather inconspicuous flowers. Propagation is by seeds, sown during the autumn. Plants prefer sandy soil on well-drained sites, but should be grown in a greenhouse outside the winter rainfall region.

NOTES *Cleretum herrei* is restricted to the Cape Peninsula and has short, curved fruit stalks and black seeds. In contrast, *C. lyratifolium* occurs in the Great Karoo near Matjiesfontein and has longer, straight fruit stalks and brown seeds. These two species have white flowers and the fruit capsules are ribbed on top. The flowers of both subspecies of *C. papulosum* are yellow, but they can be distinguished on the basis of the lengths of the flower stalks: those of the subsp. *schlechteri* are long, whereas those of the subsp. *papulosum* are much shorter. The fruit capsule type of this genus is interpreted as being fairly primitive, with only rudimentary covering membranes and massive expanding keels which, in the closely related *Dorotheanthus*, have evolved into flat surfaces.

Cleretum N.E.Br. (=*Micropterum* Schwantes)

NUMBER OF SPECIES/SUBSPECIES/VARIETIES
(3/1/0)

SPECIES LIST AND CONSERVATION STATUS
C. herrei Schwantes
C. lyratifolium Ihlenf. & Struck [R]
C. papulosum (L.f.) L.Bolus subsp. *papulosum*
C. papulosum (L.f.) L.Bolus subsp. *schlechteri* (Schwantes) Ihlenf. & Struck

LITERATURE
IHLENFELDT, H.-D. & STRUCK, M. 1986. Morphologie und Taxonomie der Dorotheanthinae Schwantes (Mesembryanthemaceae). *Beiträge zur Biologie der Pflanzen* 61: 411–453.

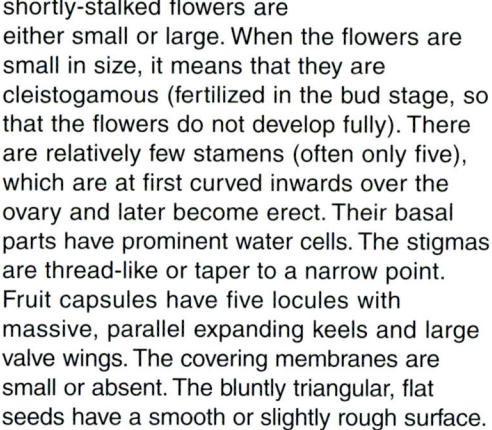

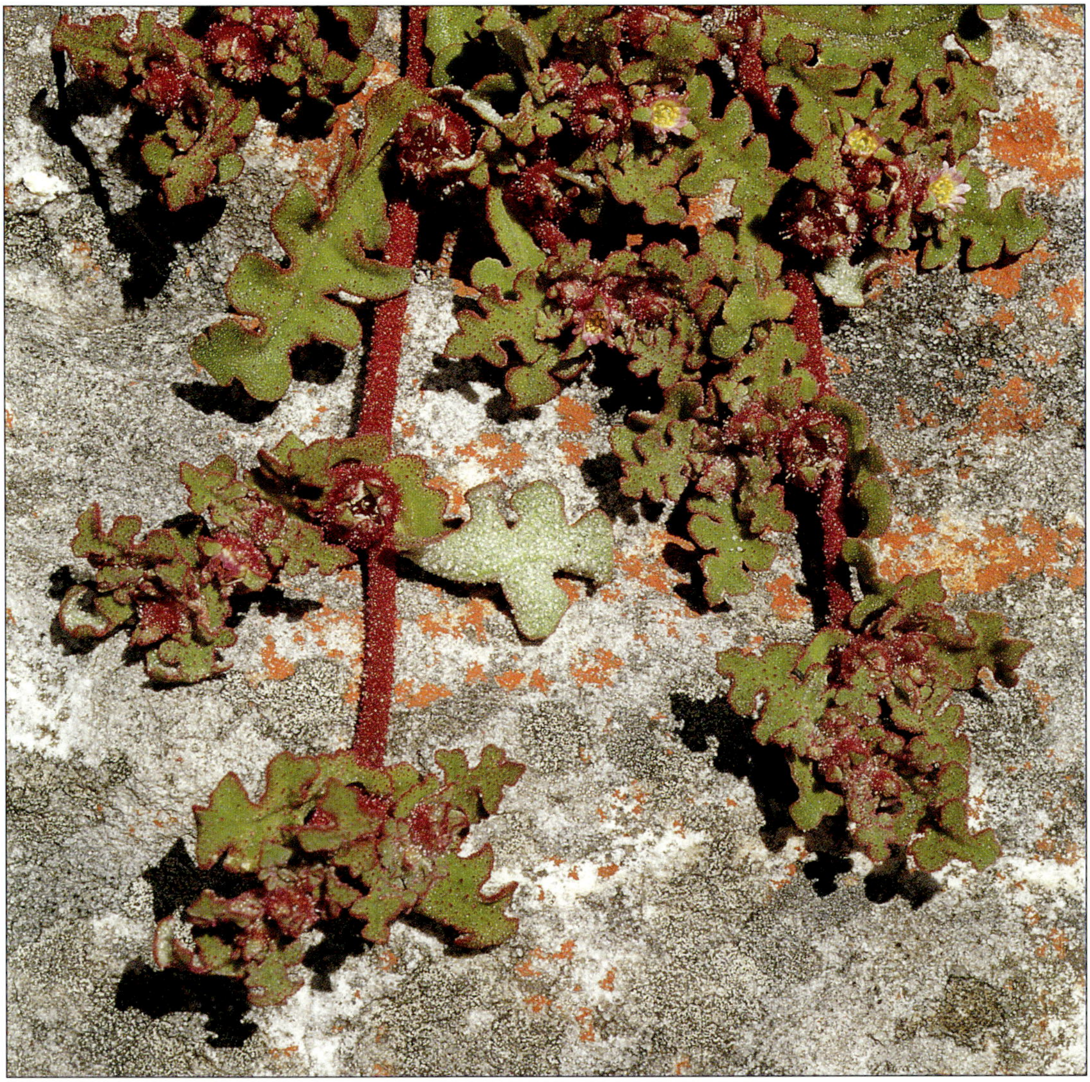

Ernst van Jaarsveld/NBI

C. herrei

Graham Williamson

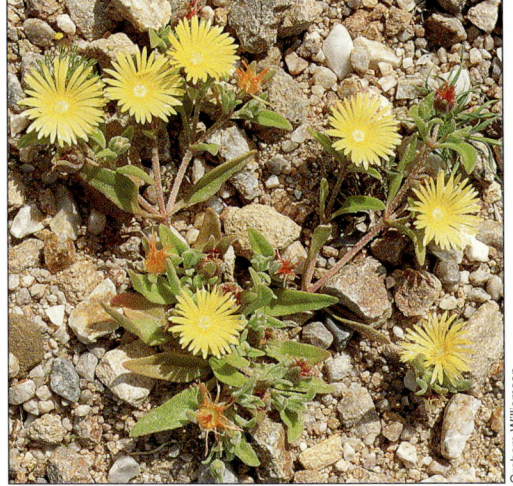

C. papulosum subsp. *schlechteri*

Ernst van Jaarsveld/NBI

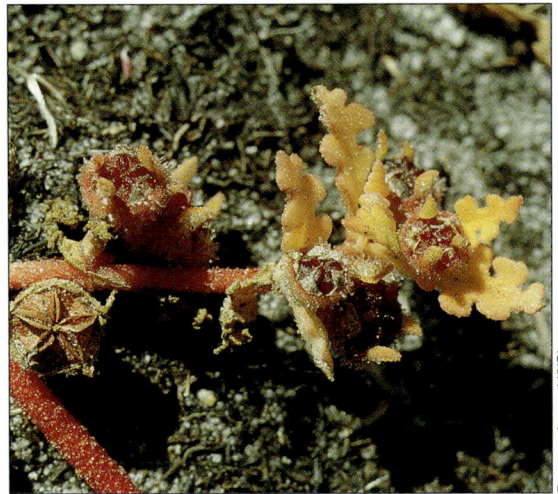

Fruit capsules of *C. herrei*

Conicosia

Derivation of genus name The name is derived from the Greek word *conicos* (cone-shaped), referring to the shape of the fruit.

Common names Several vernacular names such as *gansies* (goslings, used for the flowers), *snotwortel* (slime root), *varkslaai* (pig salad) and *varkswortel* (pig's root) have been recorded.

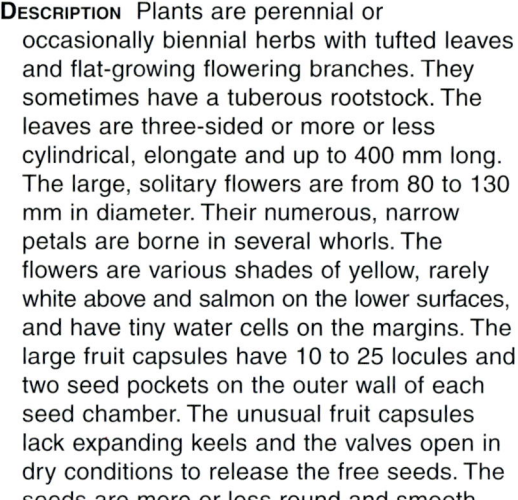

Description Plants are perennial or occasionally biennial herbs with tufted leaves and flat-growing flowering branches. They sometimes have a tuberous rootstock. The leaves are three-sided or more or less cylindrical, elongate and up to 400 mm long. The large, solitary flowers are from 80 to 130 mm in diameter. Their numerous, narrow petals are borne in several whorls. The flowers are various shades of yellow, rarely white above and salmon on the lower surfaces, and have tiny water cells on the margins. The large fruit capsules have 10 to 25 locules and two seed pockets on the outer wall of each seed chamber. The unusual fruit capsules lack expanding keels and the valves open in dry conditions to release the free seeds. The seeds are more or less round and smooth.

Distinguishing characters Plants are perennial or occasionally biennial with elongated leaves and large yellow flowers. The large, many-locular fruit capsules are also distinct.

Flowering time Flowering occurs in spring to early summer (August to December in South Africa). The flowers open in the afternoon and close at sunset.

Distribution and ecology The genus occurs in the western parts of South Africa (Western and Northern Cape Provinces) as well as southern Namibia. *C. pugioniformis* subsp. *muirii* occurs in a distinct belt along the coastal regions as far east as Knysna. Plants often grow in disturbed sites. *C. pugioniformis* has been introduced to Victoria and other parts of southeastern Australia.

Cultivation Plants are rarely cultivated but thrive in strandveld gardens, especially on the Cape Flats. They may be propagated from seeds and require very sandy soil and dry summers.

Notes *Conicosia elongata* can be distinguished from *C. pugioniformis* by its tuberous roots and annual leaves that are rounded in cross-section. *C. pugioniformis*, on the other hand, has a tap root and at least some perennial leaves that are triangular in cross-section. The subspecies of *C. pugioniformis* can be separated from each other mainly on account of differences in plant height and width, consistency of the roots, flower colour, fruit capsule structure and geographical distribution. The genus belongs to a group of genera (*Carpanthea*, *Apatesia*, *Conicosia* and *Hymenogyne*) characterised by fruit capsules with seed pockets which split apart when the capsules ripen and dry out. Associated with this fruit type is the reduction and eventual loss of the expanding tissues which normally open the capsules in mesembs with hygroscopic fruits.

Conicosia N.E.Br. (=*Herrea* Schwantes)

Number of species/subspecies/varieties (2/2/0)

Species list and conservation status
C. elongata (Haw.) N.E.Br.
C. pugioniformis (L.) N.E.Br. subsp. *alborosea* (L.Bolus) Ihlenf. & Gerbaulet
C. pugioniformis (L.) N.E.Br. subsp. *muirii* (N.E.Br.) Ihlenf. & Gerbaulet
C. pugioniformis (L.) N.E.Br. subsp. *pugioniformis*

Literature
IHLENFELDT, H.-D. & GERBAULET, M. 1990. Untersuchungen zum Merkmalsbestand und zur Taxonomie der Gattungen *Apatesia* N.E.Br., *Carpanthea* N.E.Br., *Conicosia* N.E.Br., *Herrea* Schwantes und *Hymenogyne* Haw. (Mesembryanthemaceae Fenzl). *Botanische Jahrbücher* 111: 457–498.

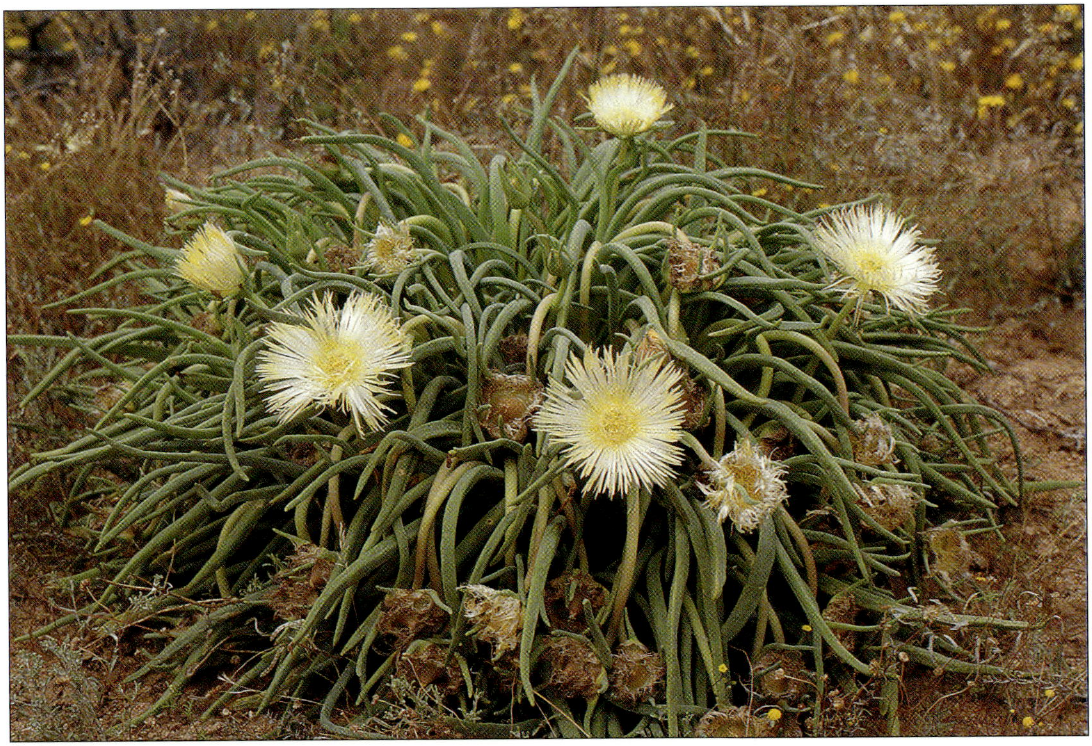

Ben-Erik van Wyk

C. elongata

Bob Chinnock

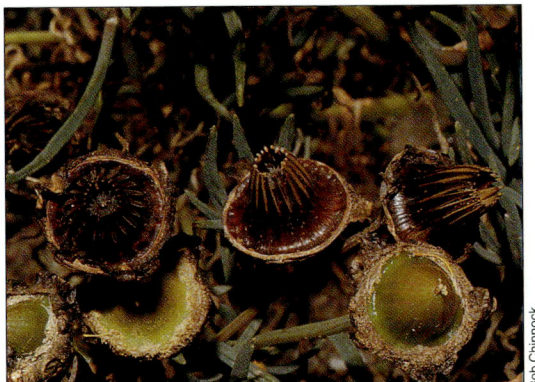

Fruits of *C. pugioniformis*

Ernst van Jaarsveld/NBI

C. pugioniformis

John Manning

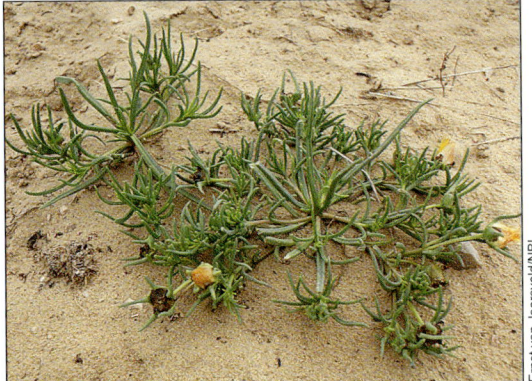

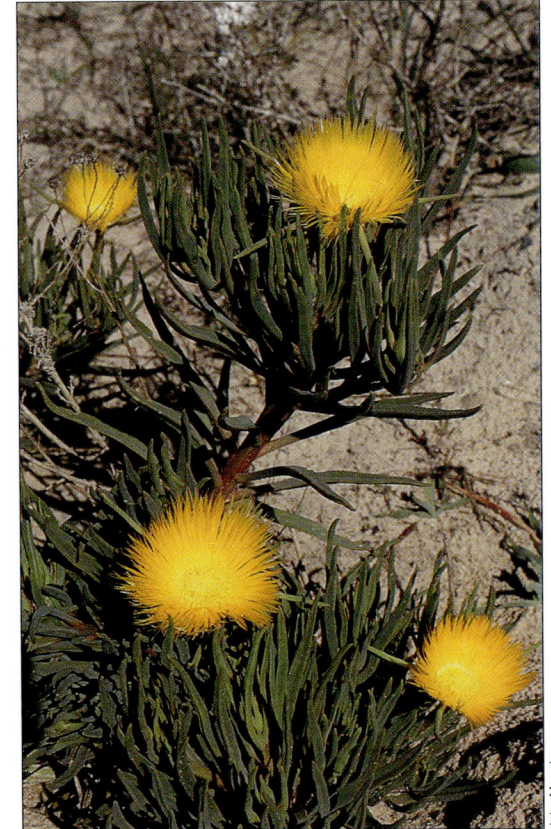

Flowers of *C. pugioniformis*

Dorotheanthus

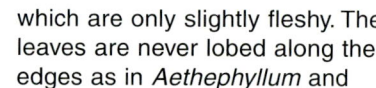

Derivation of genus name Professor Gustav Schwantes named this genus for his mother, Dorothea Schwantes, by combining her name with *anthus*, the Greek word for flower.

Common names In commercial horticulture, *Dorotheanthus bellidiformis* is usually referred to as the Livingstone daisy. In South Africa, it is widely known as the *bokbaaivygie*, Bok Bay vygie or Buckbay-vygie, named after a little bay along the west coast of the Cape called Bokbaai, where it occurs in great abundance in spring. Less commonly used names include *sandvygie* (sand mesemb), *skilpadkos* (tortoise food) and *ysplant* (ice plant).

Description The plants are all dwarf annual herbs of up to 100 mm in height, with prominent, glistening water cells on the leaves and stems. In small plants the leaves form a basal rosette, but later they are dispersed along the stems. The leaves are oblong, more or less flat, and taper sharply towards the narrow, stalk-like bases. Single flowers of up to 40 mm in diameter are borne on stalks of up to 60 mm long. They are large and attractive except in *D. apetalus*, where the petals are inconspicuous, being much reduced in size. The petals are arranged in two whorls and each one is usually lighter towards the base, resulting in a halo effect. Flower colour varies greatly, even within some of the species, ranging from white, yellow and orange to various shades of red, pink and purple. The fruit capsules are five-locular and vary considerably in structure, some with covering membranes and others without. The seeds may be dark brown with coarse, tuberculate surfaces (as in *D. maughanii* and *D. booysenii*) or more usually smooth and whitish in colour.

Distinguishing characters Species of *Dorotheanthus* are easily recognised by the small (annual) growth form and the leaves which are only slightly fleshy. The leaves are never lobed along the edges as in *Aethephyllum* and some species of *Cleretum*. Yellow-flowered plants of *Dorotheanthus* may also be confused with *Cleretum*, but in the latter the flower stalks are curved (s-shaped), while they are straight in *Dorotheanthus*.

Flowering time In nature, plants flower in late winter to spring (July to October in South Africa). Cultivated plants usually flower in spring but the season can be extended in mild climates. Flowers open in the morning and close at night in all the species.

Distribution and ecology All species are restricted to the winter rainfall region of South Africa and are common along the coast and interior of Namaqualand in the Western and Northern Cape Provinces. As short-lived annuals, these plants show interesting adaptations to disturbed sandy places such as inland dunes and river banks. In these environments, the small flat seeds are quickly buried under loose sand, where they survive until the next growing season.

Cultivation *Dorotheanthus bellidiformis* is a well-known and popular garden annual, grown not only in its native South Africa but also in many other parts of the world. Names listed in seed catalogues include *D. gramineus* (sometimes as "*D. tricolor*") and *D. criniflorus* (now considered to be the same species as *D. bellidiformis*). Seeds may be sown in a cold frame or directly in the flower beds in autumn. *D. maughanii*, *D. bellidiformis* subsp. *hestermalensis*, and especially *D. rourkei* are difficult to germinate but germination can be improved if the seeds are sown in a pot in summer, kept dry and hot and then watered liberally in autumn.

Notes The most common and well-known plant in this genus, *D. bellidiformis* subsp. *bellidiformis*, is extremely variable, but there

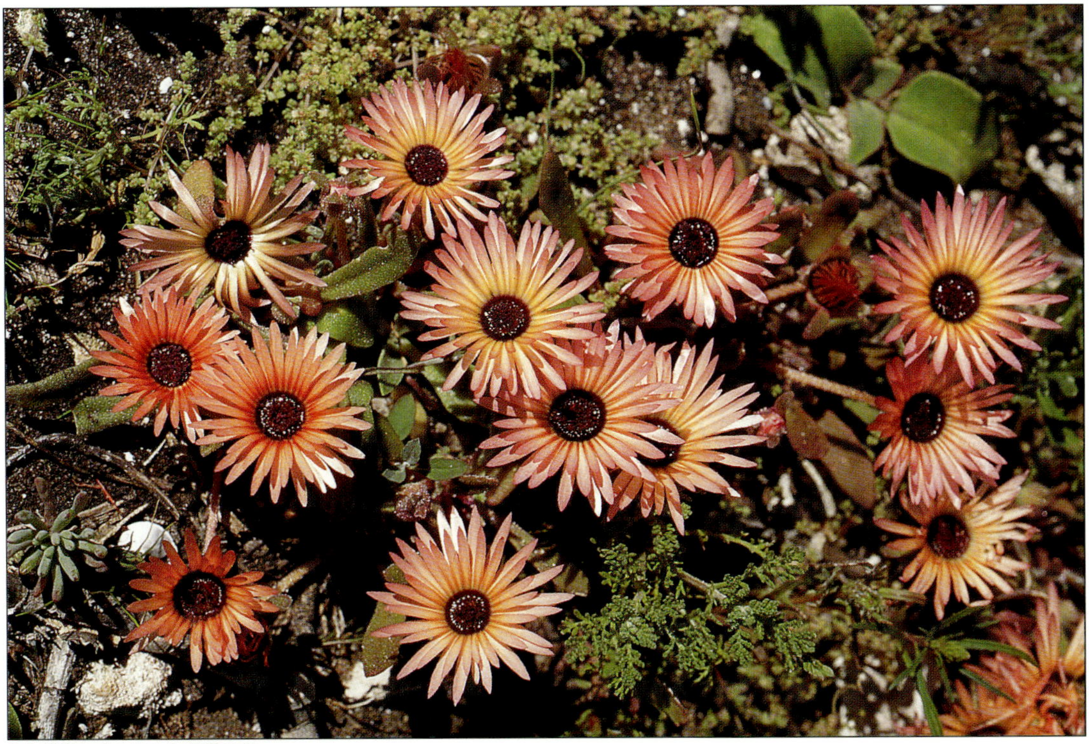

John Manning

D. bellidiformis subsp. *bellidiformis*

Ben-Erik van Wyk

D. bellidiformis subsp. *bellidiformis*

Ernst van Jaarsveld/NBI

D. maughanii

NBI

D. rourkei

NBI

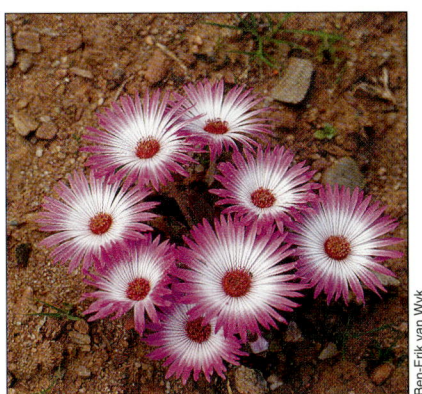

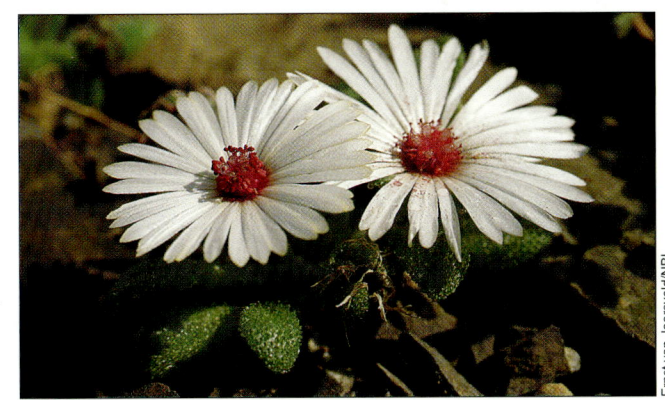

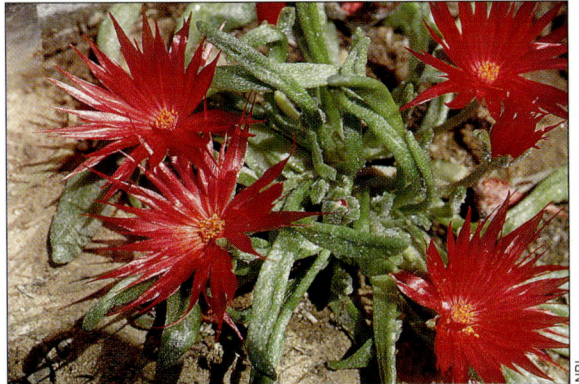

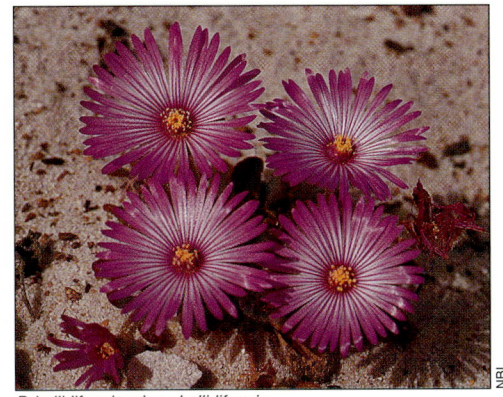

D. bellidiformis subsp. *bellidiformis*

are no distinct differences to separate the various regional colour forms. However, *D. bellidiformis* subsp. *hestermalensis* can easily be recognised by its extremely small size: it only produces a few leaves, and tiny pink flowers. *D. maughanii* has white or yellow flowers with peculiar large fleshy outgrowths in the centre (visible in the photograph). This species was formerly placed in a separate genus, *Pherelobus*. *D. booysenii* is similar to *D. maughanii* but the petals are uniformly coloured rather than transversely striped and they are invariably white or pink. *D. apetalus* is a cleistogamous species (see *Cleretum*). *D. rourkei* has characteristic sharp-pointed petals with a striking range of colours, from salmon pink to brilliant scarlet. *D. gramineus* can be recognised by its more or less upright grassy leaves. *D. ulularis* is a recently described species from the southern Cape coast, similar to *D. bellidiformis* subsp. *bellidiformis* but with textured seeds and somewhat narrower leaves.

Dorotheanthus Schwantes (=*Pherelobus* N.E.Br.; =*Sineoperculum* Van Jaarsv.)

NUMBER OF SPECIES/SUBSPECIES/VARIETIES (7/1/0)

SPECIES LIST AND CONSERVATION STATUS
  D. apetalus (L.f.) N.E.Br. [K]
  D. bellidiformis (Burm.f.) N.E.Br. subsp. *bellidiformis* [nt]
  D. bellidiformis (Burm.f.) N.E.Br. subsp. *hestermalensis* Ihlenf. & Struck
  D. booysenii L.Bolus [nt]
  D. gramineus (Haw.) Schwantes
  D. maughanii (N.E.Br.) Ihlenf. & Struck [nt]
  D. rourkei L.Bolus [nt]
  D. ulularis F.A.Brusse

LITERATURE
IHLENFELDT, H.-D. & STRUCK, M. 1986. Morphologie und Taxonomie der Dorotheanthinae Schwantes (Mesembryanthemaceae). *Beiträge zur Biologie der Pflanzen* 61: 411–453.

BRUSSE, F.A. 1996. A new species of *Dorotheanthus* (Aizoaceae, Ruschioideae) from limestone areas of the south-western Cape, South Africa. *Bradleya* 14: 89–93.

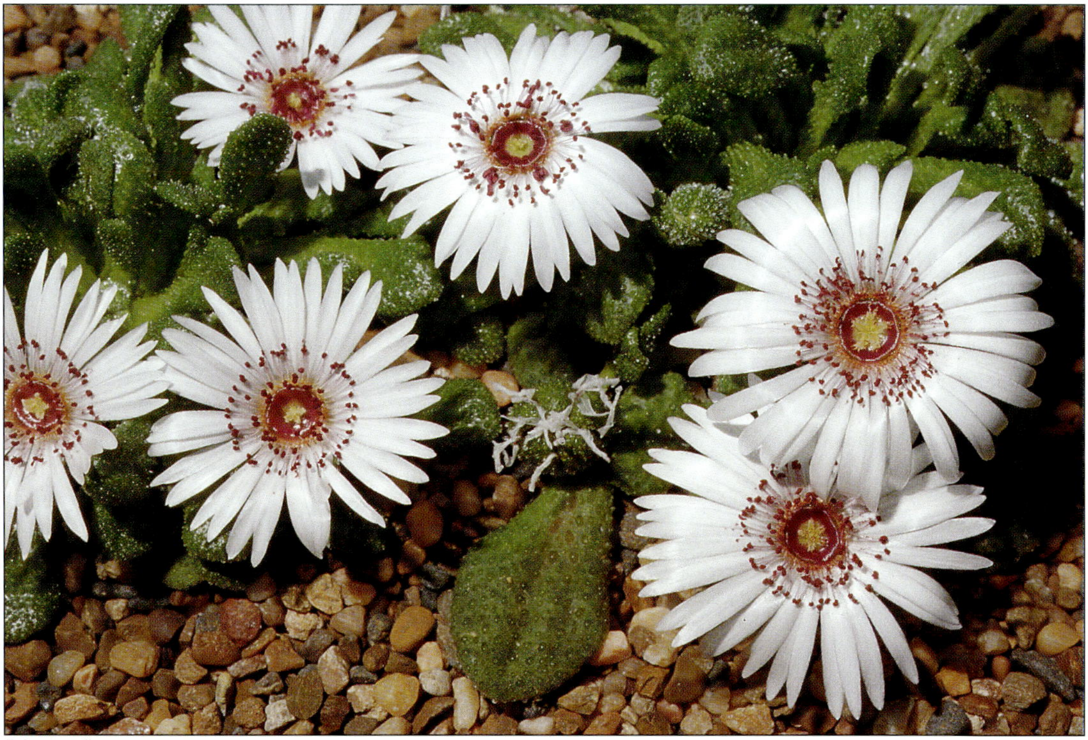

D. booysenii

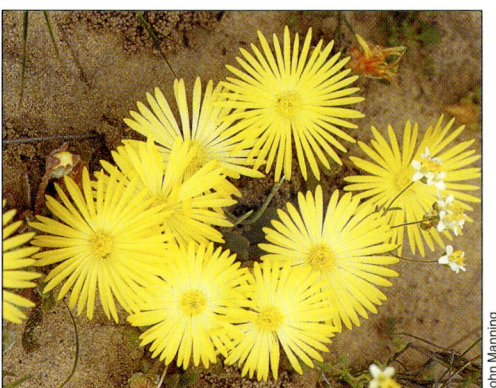

Bob Chinnock

D. bellidiformis subsp. bellidiformis

John Manning

D. gramineus

Bob Chinnock

D. ulularis

Bob Chinnock

D. bellidiformis subsp. bellidiformis

NBI

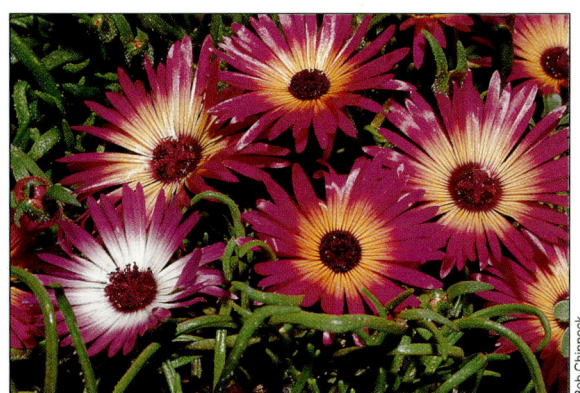

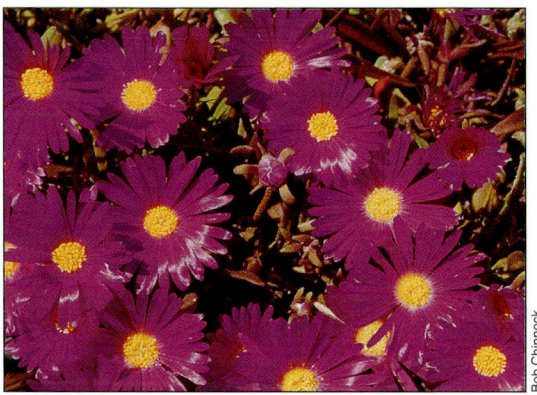

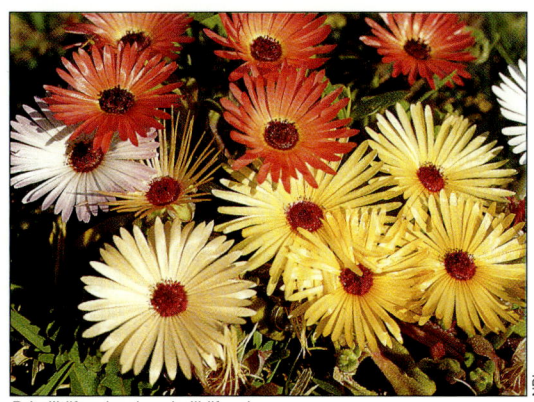

HYMENOGYNE

DERIVATION OF GENUS NAME The name is derived from the Greek words *hymen* (skin) and *gyne* (female), referring to the protective layer enclosing each seed in the fruit capsule.

COMMON NAMES No vernacular names seem to have been recorded.

DESCRIPTION The plant is a hairless annual herb with flat-growing branches and flat, narrowly lance-shaped, stalked leaves. The faintly scented flowers are ivory to yellow, fading to orange. They are solitary, up to 30 mm in diameter, sometimes cleistogamous and are borne on long, erect stalks which lie flat in the fruiting stage. The eight to 12 stigmas are fused to form a funnel-like column. Unlike other mesemb genera, the fruit capsules do not respond to moisture. Opening devices such as expanding keels are not developed. Free seeds are not found as all of the seeds are embedded in seed pockets (a single seed per pocket). At maturity, the fruit capsules break up into circular, disc-like twin segments, and then into winged one-seeded single locules.

DISTINGUISHING CHARACTERS The plants are annual herbs with flat leaves and solitary flowers with a central column of stigmas and characteristic fruit capsules (schizocarps) as described above.

FLOWERING TIME The main flowering season is from spring to early summer (September to November in South Africa). Flowers open in the afternoon and close at night.

DISTRIBUTION AND ECOLOGY The genus occurs in the Western Cape Province of South Africa, from the Cape Peninsula to Clanwilliam and plants are often found in disturbed places.

CULTIVATION Propagation is difficult. Seeds do not germinate readily but the best results can be expected when sown in autumn.

NOTES After pollination the erect flower stalks bend over sideways so that the developing fruits lie on the ground with the upper surface turned upwards. The fruits ripen in this position. The locules of the fruits are dispersed by wind. The flowers of *H. conica* are light yellow and the stigmas are arranged in a cone. The flowers of *H. glabra* are dark yellow and the stigmas are arranged in a funnel. Although the distribution ranges of both species overlap somewhat, *H. glabra* occurs as far south as the Cape Peninsula, whereas *H. conica* has a more northerly distribution, parallel to the west coast, as far as Vanrhynsdorp.

Hymenogyne Haw.

NUMBER OF SPECIES/SUBSPECIES/VARIETIES (2/0/0)

SPECIES LIST AND CONSERVATION STATUS
H. conica L.Bolus
H. glabra (Aiton) Haw.

LITERATURE
IHLENFELDT, H.-D. & GERBAULET, M. 1990. Untersuchungen zum Merkmalsbestand und zur Taxonomie der Gattungen *Apatesia* N.E.Br., *Carpanthea* N.E.Br., *Conicosia* N.E.Br., *Herrea* Schwantes und *Hymenogyne* Haw. (Mesembryanthemaceae Fenzl). *Botanische Jahrbücher* 111: 457–498.

H. glabra

H. glabra

SAPHESIA

DERIVATION OF GENUS NAME The name is derived from the Greek word *saphos* (distinct), since the genus has distinctive fruit capsules.

COMMON NAMES No vernacular names seem to have been recorded.

DESCRIPTION The plants are perennial herbs with more or less flat-growing branches and a tuberous rootstock. The oblong tubers are able to resprout after fire. The flat, stalkless leaves are single and alternate each other along the stems. Medium-sized, solitary flowers are borne on long stalks, and they have narrow, white petals which are shorter than the sepals. Numerous stamens and filamentous staminodes surround the five thread-like stigmas. The fruit capsules are superficially similar to those of carnations (*Dianthus* species): they are cone-shaped and have five locules. The ends of the valves split apart and remain open in this position when they dry out. A few seeds are developed in semi-pockets in the marginal parts of the capsules. The seeds are dimorphic, i.e. there are two kinds of seeds.

DISTINGUISHING CHARACTERS This low-growing, woody shrub has a substantial tuber and solitary, medium-sized, pure white, long-stalked flowers. The fruit capsules are unique in that the locules split apart upon drying and then remain open.

FLOWERING TIME The plant flowers in spring and early summer (September to November in South Africa).

DISTRIBUTION AND ECOLOGY The recorded distribution range of *Saphesia flaccida* indicates that it grows in sandy soil on the plains north of Cape Town, from Malmesbury to the Piketberg Mountains, Western Cape Province, South Africa.

CULTIVATION *Saphesia* is rarely found in cultivation. It is more often grown as a curiosity plant. Plants are best grown in sandy, acidic soil, and require lightly shaded to sunny positions. They should be fertilized only occasionally, preferably with a liquid fertilizer. Propagation is very difficult from seeds, and the smaller of the two seed types is suspected to be sterile. Plants are readily propagated from cuttings and are best grown in coastal fynbos gardens. Seeds should be sown in autumn.

NOTES The species is extremely rare and endangered by urban and agricultural development. It was once thought to be extinct, but has been rediscovered in a protected area (Riverlands Nature Reserve). However, it is endangered throughout its distribution range by infestations of invader plants, particularly alien *Acacia* species. The seeds are dispersed from the permanently open fruit capsules by repeated scattering.

Saphesia N.E.Br.

NUMBER OF SPECIES/SUBSPECIES/VARIETIES (1/0/0)

SPECIES LIST AND CONSERVATION STATUS
S. flaccida (Jacq.) N.E.Br. [E]

LITERATURE
IHLENFELDT, H.-D. & GERBAULET, M. 1990. Untersuchungen zum Merkmalsbestand und zur Taxonomie der Gattungen *Apatesia* N.E.Br., *Carpanthea* N.E.Br., *Conicosia* N.E.Br., *Herrea* Schwantes und *Hymenogyne* Haw. (Mesembryanthemaceae Fenzl). *Botanische Jahrbücher* 111: 457–498.

VAN JAARSVELD, E.J. 1994. *Saphesia flaccida* and its conservation. *Cactus and Succulent Journal of Great Britain* 12: 98–103.

Growth form of *S. flaccida*

Flowers of *S. flaccida*

Leaves of *S. flaccida*

Fruit capsules of *S. flaccida*

SKIATOPHYTUM

DERIVATION OF GENUS NAME The name is derived from the Greek words *skia* (shade) and *phyton* (plant), referring to the shady habitat of the plant.

COMMON NAMES The vernacular name is *platblaarvygie* (flat leaf mesemb).

DESCRIPTION The plants are annuals to short-lived perennials, initially tufted but later developing long branches. The oblong, soft and fleshy leaves have wavy margins when young and are arranged in pairs or single and alternating. The medium-sized, snow-white flowers are borne on long stalks. There are five unequal sepals and numerous petals in several whorls. The five awl-shaped stigmas are longer than the numerous stamens and filamentous staminodes which surround them. The large fruit capsules have five to seven locules and are without expanding keels or marginal wings. Upper distal seed chambers are present and the valves open upon drying and remain open. Seeds are somewhat kidney-shaped, rough in texture and are dispersed in tumble fruits.

DISTINGUISHING CHARACTERS The flat-growing habit and large flat leaves are characteristic. Fruit capsules that remain open and the large kidney-shaped tuberculate seeds are useful characters to distinguish this genus from *Caryotophora*.

FLOWERING TIME Flowering occurs during spring and summer.

DISTRIBUTION AND ECOLOGY The genus is only found in the Western Cape Province of South Africa where it occurs in fynbos and renosterveld in light shady spots, often in depressions below shrubs and trees. It has become very rare near the Cape Peninsula due to urban expansion. After pollination by beetles and bees, the fruit capsules become heavy and topple over.

CULTIVATION The plants are easily cultivated but fresh seeds do not germinate well. They are best stored for a few seasons or treated with smoke before sowing during autumn, with the onset of cooler conditions. Plants grow fast and will last for two years. They are tolerant of full sun and can be grown as curiosity plants or as ground covers on rockeries in semi-shade conditions. Plants require sandy acid soils and dry summers. They are best grown in fynbos gardens.

NOTES *Skiatophytum* is almost indistinguishable from *Caryotophora* when not in fruit. However, the latter has a hard nut-like fruit and is confined to Bredasdorp, whilst *Skiatophytum* has a larger, softer fruit and is more widespread in its distribution.

Skiatophytum L.Bolus

NUMBER OF SPECIES/SUBSPECIES/VARIETIES (1/0/0)

SPECIES LIST AND CONSERVATION STATUS
S. tripolium (L.) L.Bolus

LITERATURE
LEISTNER, O.A. 1958. *Skiatophytum* L.Bol.: 'n morfologies-taksonomiese studie. *Journal of South African Botany* 24: 89–102.

IHLENFELDT, H.-D. & GERBAULET, M. 1990. Untersuchungen zum Merkmalsbestand und zur Taxonomie der Gattungen *Apatesia* N.E.Br., *Carpanthea* N.E.Br., *Conicosia* N.E.Br., *Herrea* Schwantes und *Hymenogyne* Haw. (Mesembryanthemaceae Fenzl). *Botanische Jahrbücher* 111: 457–498.

Leaves and fruit capsules of *S. tripolium*

Growth form of *S. tripolium*

Flower of *S. tripolium*

Pleiospilos nelii

Lithops ruschiorum

Conophytum calculus

Dinteranthus vanzylii

Conophytum lithopsoides

Pleiospilos bolusii

Flowering Stones

GROUP 3

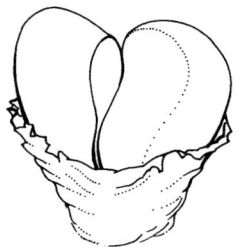

The extremely succulent leaf pairs are fused and rounded (these are called plant bodies); leaves occur in single pairs or in small groups; stems are not visible; the plants are often difficult to detect in the stony places where they grow, sometimes disappearing into the ground or covered by papery leaf remains during the dry season.

FLOWERING STONES are highly compact plants with extremely succulent leaf pairs which are fused and rounded, or slightly elongated. The mimicry plants in this group are often camouflaged and difficult to distinguish from the rocks and pebbles amongst which they grow. The leaves are exceptionally variable, not only in shape but also in colour, and may be various shades of silver, grey, red, brown, yellow or green. The leaf tips are often translucent or variously ornamented. These leaf tips function as windows or lenses for letting in sunlight for photosynthesis to the largely buried leaves. A number of genera disappear underground or are covered by papery leaf remains during the dry season, making them even more difficult to detect. Favourite collectors' genera are included in this artificial group.

FLOWERING STONES comprise 17 genera and 177 species.

Argyroderma (11 species)	*Fenestraria* (1 species)	*Lapidaria* (1 species)	*Oophytum* (3 species)
Conophytum (88 species)	*Frithia* (1 species)	*Lithops* (36 species)	*Pleiospilos* (4 species)
Didymaotus (1 species)	*Gibbaeum* (16 species)	*Muiria* (1 species)	*Tanquana* (3 species)
Dinteranthus (4 species)	*Ihlenfeldtia* (2 species)	*Namibia* (2 species)	*Vlokia* (1 species)
Diplosoma (2 species)			

ARGYRODERMA

DERIVATION OF GENUS NAME The name comes from the Greek words *argyros* (silver) and *derma* (skin), which describe the characteristic silvery grey-green leaves.

COMMON NAMES The vernacular names *bababoudjies* (baby's bottom), *vingervygie* (finger mesemb) and *jakkalsniertjie* (jackal kidney) have been recorded.

DESCRIPTION The plants are compact, dwarf succulents, tufted or single-bodied, rarely sunken into the ground. The silvery to grey-green leaves are more or less rounded or thumb-shaped, opposite in pairs and united at the base, with the upper surface flat or slightly convex and the lower surface rounded. The yellow, purple, red or rarely white flowers are invariably single with a distinctly cup-shaped flower base (hypanthium) into which the stamens bend. The six to eight sepals are fused into a short tube. Stamens occur in a dense ring. The stigmas are united into a small disc-shaped cushion. The fruit capsules have 12 to 25 locules with seed chambers covered by rigid covering membranes. The closing bodies are large; valve wings are narrow; and the expanding keels are stout and awned. The small, brown seeds are more or less round and smooth.

DISTINGUISHING CHARACTERS The plants are easily recognised by their rounded, highly succulent leaf pairs and their distinctive silvery to grey-green colour and smooth texture. The flowers are solitary with distinct cup-shaped bases, and the stigmas are flattened and disc-shaped.

FLOWERING TIME Flowering occurs in autumn and early winter (April to June in South Africa). The flowers open by midday and close in the evening.

DISTRIBUTION AND ECOLOGY *Argyroderma* occurs in the northwestern part of the Western Cape Province, South Africa. It has a restricted distribution in southern Namaqualand in an area known as the Knersvlakte. Plants are locally hyperabundant in succulent karoo, mainly on quartz gravel flats and hillsides, in full sun. They grow in an area with winter rainfall of less than 100 mm per year.

CULTIVATION Plants are popular collectors' items and are best grown in a greenhouse where the moisture regime can be controlled so as to prevent overwatering in the dormant season. Keep them dry in summer and in a slightly shady position to prevent overheating. The soil should preferably be a well-drained clayey-sandy loam. Propagation may be from seed sown in autumn or by the division of large multiheaded plants. Plants grow rapidly and could flower in the second or third season after germination. Water sparingly, even in winter, as leaves may burst from too much water, resulting in unsightly scars.

NOTES *Argyroderma fissum* is the most common and widely distributed species; it occurs throughout the distribution range of the genus. This species varies considerably in the colour and size of the flowers but it can be recognised by its long, finger-like leaves. The other species either have solitary, non-branching bodies or they are variably divided, as can be seen in the illustrations. Features to look for in identifying *Argyroderma* species are the wideness of the gap between the leaves, the presence or absence of a keel and the proportion of the leaf pair which is underground. *A. delaetii* sinks quite deeply; *A. pearsonii* is less deeply sunken and can be recognised by its egg-like shape. Flower colour is also useful to some extent, but where the different species live together they hybridise freely and produce a myriad of flower colours. Beetles are fond of the flowers and sleep in them overnight. *A. delaetii* is one of the most popular species in cultivation, probably because it is so floriferous. *A.*

Bud of *A. testiculare*

Flower of *A. testiculare*

A. framesii subsp. *framesii*

Open fruit capsule of *A. delaetii*

testiculare is notable for its silken multilayered flowers.

Argyroderma N.E.Br.

NUMBER OF SPECIES/SUBSPECIES/VARIETIES (11/1/0)

SPECIES LIST AND CONSERVATION STATUS
- *A. congregatum* L.Bolus
- *A. crateriforme* (L.Bolus) N.E.Br.
- *A. delaetii* Maass
- *A. fissum* (Haw.) L.Bolus
- *A. framesii* L.Bolus subsp. *framesii* [K]
- *A. framesii* L.Bolus subsp. *hallii* (L.Bolus) H.E.K.Hartmann
- *A. patens* L.Bolus
- *A. pearsonii* (N.E.Br.) Schwantes
- *A. ringens* L.Bolus [K]
- *A. subalbum* (N.E.Br.) N.E.Br. [K]

- *A. testiculare* (Aiton) N.E.Br. [K]
- *A. theartii* Van Jaarsv.

LITERATURE

HARTMANN, H.E.K. 1973. New combinations and a key for the genus *Argyroderma* N.E.Br. (Mesembryanthemaceae Fenzl). *National Cactus and Succulent Journal* 28: 48–50.

HARTMANN, H.E.K. 1977. Monographie der Gattung *Argyroderma* N.E.Br. (Mesembryanthemaceae Fenzl). *Mitteilungen aus dem Institut für Allgemeine Botanik, Hamburg* 15: 121–235.

HAMMER, S.A. & LIEDE, S. 1990. Natural and artificial hybrids in Mesembryanthemaceae. *South African Journal of Botany* 56: 356–362.

A. framesii subsp. *hallii*

A. delaetii

A. pearsonii

Purple form of *A. fissum*

A. crateriforme

Orange form of *A. fissum*

Conophytum

Derivation of genus name The name is derived from the Latin word *conus* for cone and the Greek word *phytum* for plant, in reference to the cone-shaped fused leaves of many species.

Common names There are several names for this popular genus such as *knopies* (buttons), *toontjies* (little toes), *waterblasies* (water blisters), *ogies* (small eyes), sphaeroids, conos, cone plants, dumplings, or button plants.

Description The plants are dwarf cushion-forming or single-bodied succulents. The leaves are green to brown, reddish, or whitish blue, usually spotted or lined, often velvety, warted, or windowed, usually united into conical, cylindric, or oblong bodies. The plants are active in autumn and winter, drying into inert papery husks from which new bodies re-emerge in the subsequent autumn; the leaves therefore look distinctly different in summer (deadly white) and winter (lively green or brown). Flowers are solitary or, in rare cases, in delayed groups of two or three. The flowers appear when the bodies begin to expand in autumn or, in a few species, when dormancy approaches in spring or in summer. Many species are highly scented. Bracts are always present but are usually concealed within the leaves. There are four to six succulent sepals, partially united into a tube. The petals are also partially united and in one group of species are connected to the sepal tube. Flower colour varies considerably and may be magenta to blackish or coppery red, yellow, orange, rarely scarlet, pink, white and often bicoloured. Stamens are hidden within the tube or protrude from it. The ovary is usually hidden within the body, which protects the ripening fruit. Nectar glands form a continuous ring around the three to eight feathery stigmas. Fruit capsules are three to eight-locular, without covering membranes or closing bodies. Seeds are minute, some of the smallest amongst the mesembs, pear-shaped, smooth or rough; they are usually attached to the outer walls but sometimes originate in the middle, as in the case of the weedy mesembs.

Distinguishing characters Species of this genus may be distinguished by the dwarf cushions or clusters of conically united leaves and by the petals, which are fused into a basal tube (they are not free to the base as in the case of most mesembs). The plants invariably have bracts, unlike species of *Lithops* with which they can be confused.

Flowering time Most species flower in autumn, but a few flower in other seasons. They vary from day-flowering, twilight-flowering to night-flowering species. In extreme cases, the flowers only open for one hour in the evening (*C. burgeri*) or never close (*C. smorenskaduense*). Some are cleistogamous (self-fertile) and do not bother to open at all (*C. rugosum*).

Distribution and ecology Species of *Conophytum* are widely distributed over the winter rainfall, semi-arid and arid regions of South Africa and southern Namibia. This includes the western parts of the Northern Cape Province, notably Namaqualand, the Richtersveld and Bushmanland, the Western Cape Province, including the Little Karoo, and the southern margin of the Great Karoo. The genus just enters the western part of the Eastern Cape Province. More specifically, the genus occurs northwards from Paarl to Alexander Bay and east to Pofadder. In the south it ranges from Avondrust to Steytlerville. However, it is not found in parts of the Great Karoo which receive rain in summer. The highest concentration of species is found in the Namaqualand region. Plants prefer well-drained rocky sites derived from a variety of substrates. They typically grow in rock crevices, shallow soil or on quartz gravel flats; in pans of detritus or on the steep slopes of rocky hillsides and

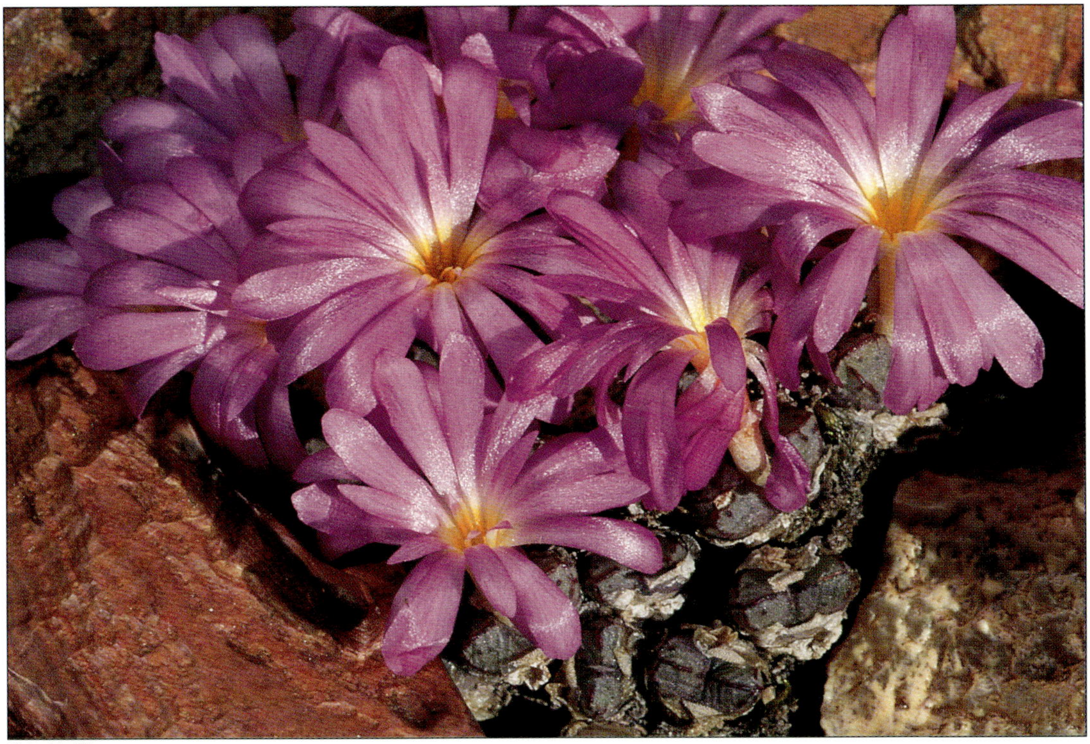
Graham Williamson

C. luckhoffii

Graham Williamson

C. hians

Graham Williamson

C. meyeri

Ernst van Jaarsveld

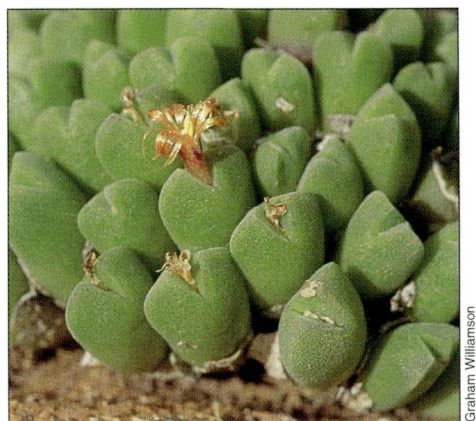

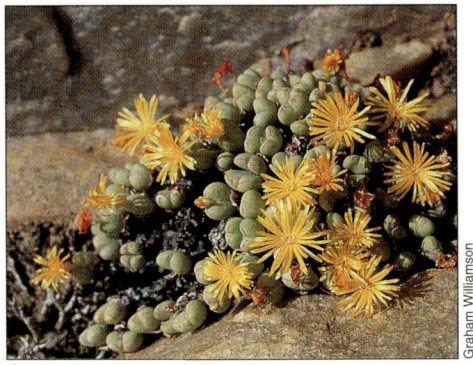

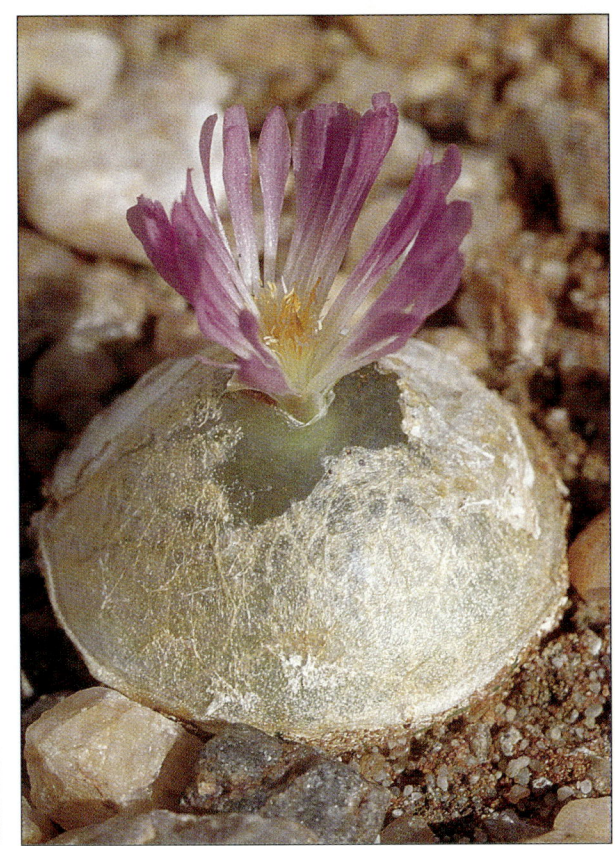

C. burgeri

mountains where they can be anchored by mosses and lichens. The vegetation is mainly succulent karoo or dry fynbos. Plants are winter growers, becoming dormant during the long dry summer months when leaves wither to form protective sheaths around the plant bodies. Most species are cluster or cushion-forming but some are solitary and are sunken with distinct windows on the leaf tips. They thrive in areas where rainfall varies between 50 and 400 mm per year and falls mainly in winter.

CULTIVATION Growing these plants successfully outside their habitat requires a greenhouse, enthusiasm, regular attention, and protection from severe frost. The plants prefer an eastern exposure, with brightness in the morning and some shade in the afternoon. They grow best in a slightly acid soil, preferably a sandy, gravelly clay mixed with ample compost. Plants should be watered from autumn to spring and kept dormant during summer. Ensure that plants are shaded during dormancy as they are easily killed by overheating during that period. Propagation is easy from seed sown during late summer.

NOTES Along with *Lithops*, *Conophytum* species are the most popular and charming of the small mesembs. Some of them have been heavily collected in nature, posing a danger to their continued existence in the wild. The highly sought after fujiyama plant, *C. burgeri*, is highly localised in nature, but it is now quite common in cultivation. Like many species it must be propagated from seed because the plant is single-bodied. *C. obcordellum*, on the other hand, is widely distributed and highly variable.

Conophytum N.E.Br. (=*Berrisfordia* L.Bolus; =*Herreanthus* Schwantes; = *Derenbergia* Schwantes; *Ophthalmophyllum* (Dinter & Schwantes) Schwantes)

NUMBER OF SPECIES/SUBSPECIES/VARIETIES (88/47/11)

SPECIES LIST AND CONSERVATION STATUS
C. achabense S.A.Hammer [R]
C. acutum L.Bolus [V]
C. albiflorum (Rawe) S.A.Hammer
C. angelicae (Dinter & Schwantes) N.E.Br. subsp. *angelicae*
C. angelicae (Dinter & Schwantes) N.E.Br. subsp. *tetragonum* Rawe & S.A.Hammer
C. armianum S.A.Hammer [R]
C. auriflorum Tischer subsp. *auriflorum* [R]
C. auriflorum Tischer subsp. *turbiniforme* (Rawe) S.A.Hammer
C. bachelorum S.A.Hammer subsp. *bachelorum*
C. bachelorum S.A.Hammer subsp. *sponsaliorum* S.A.Hammer
C. bicarinatum L.Bolus [R]
C. bilobum (Marloth) N.E.Br. subsp. *altum* (L.Bolus) S.A.Hammer
C. bilobum (Marloth) N.E.Br. subsp. *bilobum*
C. bilobum (Marloth) N.E.Br. subsp. *gracilistylum* (L.Bolus) S.A.Hammer
C. blandum L.Bolus [R]
C. bolusiae Schwantes subsp. *bolusiae*
C. bolusiae Schwantes subsp. *primavernum* S.A.Hammer
C. breve N.E.Br.
C. burgeri L.Bolus [V]
C. calculus (A.Berger) N.E.Br. subsp. *calculus*
C. calculus (A.Berger) N.E.Br. subsp. *vanzylii* (Lavis) S.A.Hammer
C. caroli Lavis [nt]
C. carpianum L.Bolus [R]
C. chauviniae (Schwantes) S.A.Hammer
C. comptonii N.E.Br.
C. concavum L.Bolus [R]
C. chrisolum S.A.Hammer
C. chrisocruxum S.A.Hammer
C. cylindratum Schwantes
C. depressum Lavis
C. devium G.D.Rowley [nt]
C. ectypum N.E.Br. subsp. *cruciatum* S.A.Hammer
C. ectypum N.E.Br. subsp. *ectypum* var. *brownii* (Tischer) Tischer
C. ectypum N.E.Br. subsp. *ectypum* var. *ectypum*
C. ectypum N.E.Br. subsp. *sulcatum* (L.Bolus) S.A.Hammer

C. schlechteri

C. limpidum

C. smorenskaduense

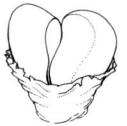

C. ernstii S.A.Hammer subsp. *ernstii* [R]
C. ernstii S.A.Hammer subsp. *cerebellum* S.A.Hammer
C. ficiforme (Haw.) N.E.Br.
C. flavum N.E.Br. subsp. *flavum*
C. flavum N.E.Br. subsp. *novicium* (N.E.Br.) S.A.Hammer
C. fraternum (N.E.Br.) N.E.Br.
C. friedrichiae (Dinter) Schwantes [nt]
C. frutescens Schwantes [R]
C. fulleri L.Bolus
C. globosum (N.E.Br.) N.E.Br.
C. gratum (N.E.Br.) N.E.Br. subsp. *gratum*
C. gratum (N.E.Br.) N.E.Br. subsp. *marlothii* (N.E.Br.) S.A.Hammer
C. halenbergense (Dinter & Schwantes) N.E.Br. [R]
C. hammeri G.Will. & Kennedy
C. herreanthus S.A.Hammer subsp. *herreanthus* [E]
C. herreanthus S.A.Hammer subsp. *rex* S.A.Hammer
C. hians N.E.Br.
C. joubertii Lavis
C. khamiesbergense (L.Bolus) Schwantes [R]
C. klinghardtense Rawe subsp. *baradii* (Rawe) S.A.Hammer [R]
C. klinghardtense Rawe subsp. *klinghardtense*
C. limpidum S.A.Hammer
C. lithopsoides L.Bolus subsp. *arturolfago* S.A.Hammer
C. lithopsoides L.Bolus subsp. *boreale* (L.Bolus) S.A.Hammer
C. lithopsoides L.Bolus subsp. *koubergense* (L.Bolus) S.A.Hammer
C. lithopsoides L.Bolus subsp. *lithopsoides* [K]
C. loeschianum Tischer [R]
C. longum N.E.Br.
C. luckhoffii Lavis [nt]
C. lydiae (Jacobsen) G.D.Rowley [nt]
C. marginatum Lavis var. *karamoepense* (L.Bolus) Rawe
C. marginatum Lavis var. *littlewoodii* (L.Bolus) Rawe
C. marginatum Lavis var. *marginatum*
C. maughanii N.E.Br. subsp. *armeniacum* S.A.Hammer
C. maughanii N.E.Br. subsp. *latum* (Tischer) S.A.Hammer [nt]
C. maughanii N.E.Br. subsp. *maughanii*
C. meyeri N.E.Br.
C. minimum (Haw.) N.E.Br.
C. minusculum (N.E.Br.) N.E.Br. subsp. *leipoldtii* (N.E.Br.) S.A.Hammer
C. minusculum (N.E.Br.) N.E.Br. subsp. *minusculum*
C. minutum (Haw.) N.E.Br. var. *minutum*
C. minutum (Haw.) N.E.Br. var. *nudum* (Tischer) Boom
C. minutum (Haw.) N.E.Br. var. *pearsonii* (N.E.Br.) Boom
C. obcordellum (Haw.) N.E.Br. subsp. *obcordellum* var. *ceresianum* (L.Bolus) S.A.Hammer
C. obcordellum (Haw.) N.E.Br. subsp. *obcordellum* var. *obcordellum*
C. obcordellum (Haw.) N.E.Br. subsp. *rolfii* (de Boer) S.A.Hammer
C. obcordellum (Haw.) N.E.Br. subsp. *stenandrum* (L.Bolus) S.A.Hammer
C. obscurum N.E.Br. subsp. *barbatum* (L.Bolus) S.A.Hammer
C. obscurum N.E.Br. subsp. *obscurum*
C. obscurum N.E.Br. subsp. *vitreopapillum* (Rawe) S.A.Hammer
C. pageae (N.E.Br.) N.E.Br.
C. pellucidum Schwantes subsp. *cupreatum* (Tischer) S.A.Hammer
C. pellucidum Schwantes subsp. *cupreatum* var. *terrestre* (Tischer) S.A.Hammer
C. pellucidum Schwantes subsp. *pellucidum* var. *lilianum* (Littlew.) S.A.Hammer
C. pellucidum Schwantes subsp. *pellucidum* var. *neohallii* S.A.Hammer
C. pellucidum Schwantes subsp. *pellucidum* var. *pellucidum*
C. phoeniceum S.A.Hammer [R]
C. piluliforme (N.E.Br.) N.E.Br. subsp. *edwardii* (Schwantes) S.A.Hammer
C. piluliforme (N.E.Br.) N.E.Br. subsp. *piluliforme*
C. praesectum N.E.Br. [K]
C. pubescens (Tischer) G.D.Rowley
C. pubicalyx Lavis
C. quaesitum (N.E.Br.) N.E.Br. subsp. *densipunctum* (L.Bolus) S.A.Hammer
C. quaesitum (N.E.Br.) N.E.Br. subsp. *quaesitum* var. *quaesitum*
C. quaesitum (N.E.Br.) N.E.Br. subsp. *quaesitum* var. *rostratum* (Tischer) S.A.Hammer

C. rugosum

C. angelicae subsp. *angelicae*

C. ectypum subsp. *ectypum*

C. longum

C. tantillum subsp. *lindenianum*

C. ratum S.A.Hammer
C. reconditum A.R.Mitch. subsp. *reconditum*
C. reconditum A.R.Mitch. subsp. *buysianum* (A.R.Mitch. & S.A.Hammer) S.A.Hammer
C. regale Lavis [R]
C. ricardianum Loesch & Tischer subsp. *ricardianum* [R]
C. ricardianum Loesch & Tischer subsp. *rubriflorum* Tischer [Ex]
C. roodiae N.E.Br.
C. rubrolineatum Rawe
C. rugosum S.A.Hammer subsp. *rugosum* [R]
C. rugosum S.A.Hammer subsp. *sanguineum* S.A.Hammer [R]
C. saxetanum (N.E.Br.) N.E.Br.
C. schlechteri Schwantes [R]
C. semivestitum L.Bolus [Ex]
C. smorenskaduense de Boer subsp. *hermarium* S.A.Hammer [E]
C. smorenskaduense de Boer subsp. *smorenskaduense* [R]
C. stephanii Schwantes subsp. *abductum* S.A.Hammer
C. stephanii Schwantes subsp. *stephanii*
C. stevens-jonesianum L.Bolus
C. subfenestratum Schwantes
C. swanepoelianum Rawe subsp. *proliferans* S.A.Hammer
C. swanepoelianum Rawe subsp. *swanepoelianum* [R]
C. tantillum N.E.Br. subsp. *helenae* (Rawe) S.A.Hammer
C. tantillum N.E.Br. subsp. *inexpectatum* S.A.Hammer
C. tantillum N.E.Br. subsp. *lindenianum* (Lavis & S.A.Hammer) S.A.Hammer
C. tantillum N.E.Br. subsp. *tantillum*
C. taylorianum (Dinter & Schwantes) N.E.Br. subsp. *ernianum* (Loesch & Tischer) de Boer ex S.A.Hammer
C. taylorianum (Dinter & Schwantes) N.E.Br. subsp. *rosynense* S.A.Hammer
C. taylorianum (Dinter & Schwantes) N.E.Br. subsp. *taylorianum* [R]
C. truncatum (Thunb.) N.E.Br. subsp. *truncatum* var. *truncatum*
C. truncatum (Thunb.) N.E.Br. subsp. *truncatum* var. *wiggettiae* (N.E.Br.) Rawe
C. truncatum (Thunb.) N.E.Br. subsp. *viridicatum* (N.E.Br.) S.A.Hammer
C. turrigerum (N.E.Br.) N.E.Br. [nt]
C. uviforme (Haw.) N.E.Br. subsp. *decoratum* (N.E.Br.) S.A.Hammer
C. uviforme (Haw.) N.E.Br. subsp. *rauhii* (Tischer) S.A.Hammer
C. uviforme (Haw.) N.E.Br. subsp. *subincanum* (Tischer) S.A.Hammer [V]
C. uviforme (Haw.) N.E.Br. subsp. *uviforme*
C. vanheerdei Tischer [R]
C. velutinum Schwantes subsp. *polyandrum* (Lavis) S.A.Hammer
C. velutinum Schwantes subsp. *velutinum* [R]
C. verrucosum (Lavis) G.D.Rowley [I]
C. violaciflorum Schick & Tischer
C. wettsteinii (A.Berger) N.E.Br. subsp. *fragile* (Tischer) S.A.Hammer
C. wettsteinii (A.Berger) N.E.Br. subsp. *francoiseae* S.A.Hammer
C. wettsteinii (A.Berger) N.E.Br. subsp. *ruschii* (Schwantes) S.A.Hammer
C. wettsteinii (A.Berger) N.E.Br. subsp. *wettsteinii*

Literature

HAMMER S.A. 1993. *The genus Conophytum: a conograph*. Succulent Plant Publications, Pretoria.

HAMMER, S.A. 1995. The married bean (with apologies to Henry Purcell). *Piante Grasse* Supplement 15: 34–38.

HAMMER, S.A. 1995. Notes on the *Conophytum obscurum* complex. *Piante Grasse* Supplement 15: 23–34.

C. maughanii subsp. *armeniacum*

C. pellucidum var. *neohallii*

C. obcordellum

C. verrucosum

Didymaotus

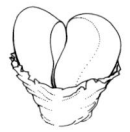

Derivation of genus name The generic name is derived from the Greek words *didymos* (double) and *aotus* (flower), referring to the invariably paired flowers.

Common names *Tweelingvygie* (twin mesemb) or mimicry plant have been used to refer to the single species of this genus.

Description Plants are dwarf, compact, and highly succulent, occurring as single leaf pairs or developing into small clumps with age. One of the two leaves is usually slightly larger than the other. The greyish green to reddish leaves are vaguely dotted, fused at their bases and usually as long as they are broad. The upper leaf surfaces are flat and the lower surfaces are rounded and broadly keeled. A single flower arises from an ear-like structure on one or both sides of a leaf pair, but in extreme cases up to nine flowers have been observed per side. Flowers are white to purplish pink. There are six prominently keeled sepals of which two are larger than the others. Petals are borne in several whorls and are darkest in the centre, or occasionally darker at the tips. Stamens are violet in colour and are bearded at their bases. The nectar glands are fused into a ring and are dark brownish green. The six awl-shaped and somewhat feathery stigmas are shorter than the stamens. The fruit capsules have six locules with covering membranes over the seed chambers. Seeds are rounded and yellowish brown.

Distinguishing characters The plants are unique in producing two flowers simultaneously, one on each side of the leaf pair. The leaves are keeled along the lower sides and are covered with a thick coating of wax.

Flowering time The plant flowers in early summer (October in South Africa). The scented flowers open around noon and are closed by evening.

Distribution and ecology *Didymaotus* grows in the predominantly winter rainfall Tanqua Karoo in the Western Cape Province of South Africa.

Cultivation This extraordinary plant with its strange symmetry is well worth cultivating but it grows with difficulty and requires a bright greenhouse to provide sufficient heat and dryness.

Notes In habitat, the reddish leaves mimic the ironstone pebbles amongst which plants grow.

Didymaotus N.E.Br.

Number of species/subspecies/varieties (1/0/0)

Species list and conservation status
D. lapidiformis (Marloth) N.E.Br. [V]

Literature
GLEN, H.F. 1974. *A revision of the Gibbaeinae (Mesembryanthemaceae)*. Unpublished M.Sc. thesis, University of Cape Town.

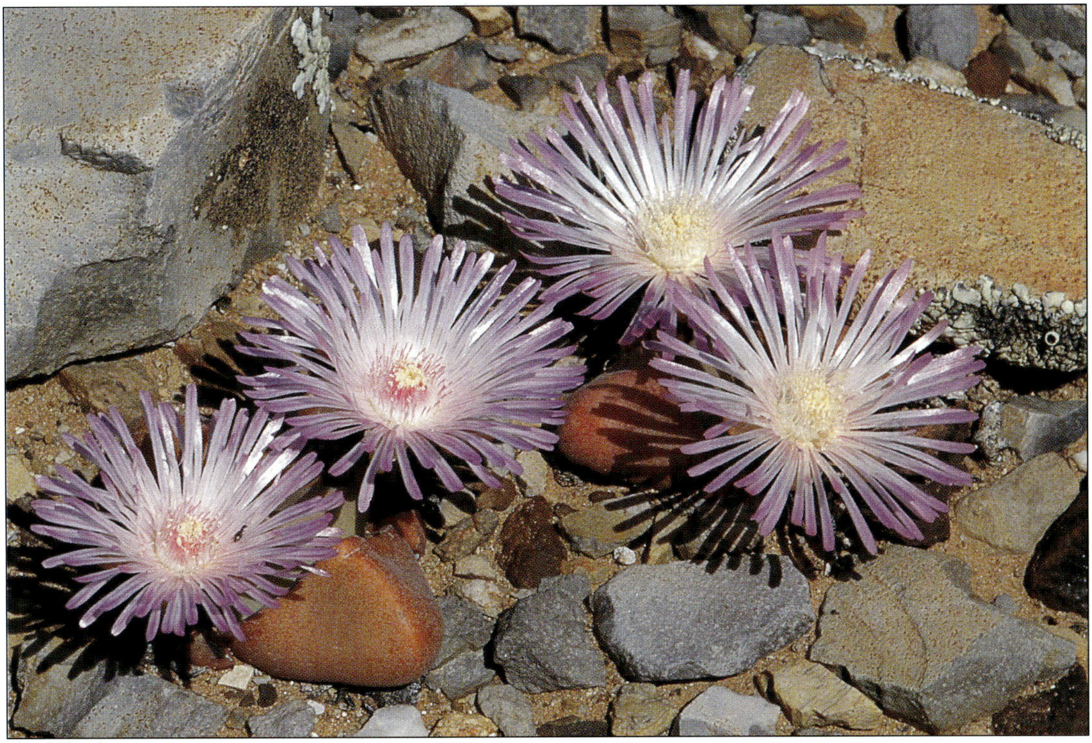

Graham Williamson

D. lapidiformis

Graham Williamson

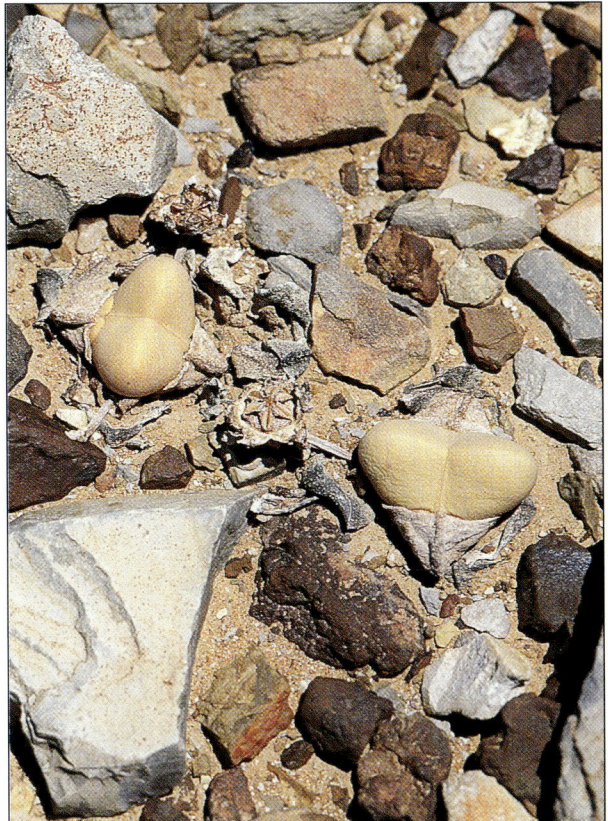

Fruiting plants of *D. lapidiformis*

Frank Stayner/NBI

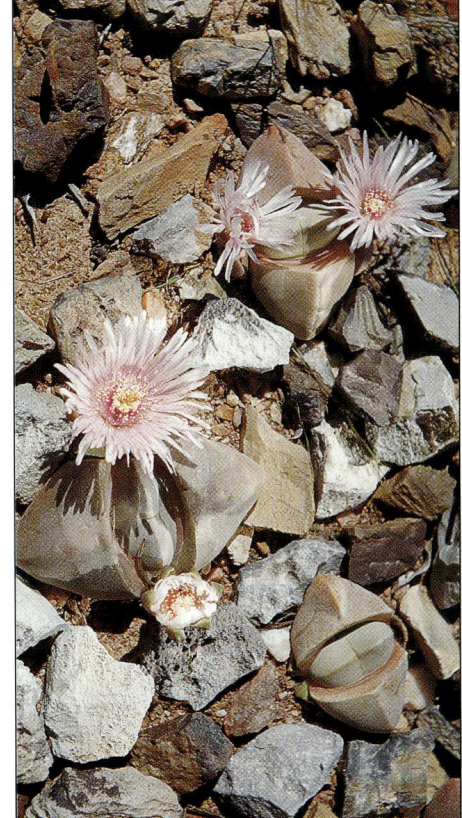

D. lapidiformis

DINTERANTHUS

DERIVATION OF GENUS NAME The genus was named after Moritz Kurt Dinter (1868 to 1945), who contributed extensively to our knowledge of the Namibian flora. The Latinised suffix *-anthus* means flower.

COMMON NAMES *Dinteranthus pole-evansii* is known as the vegetable golf ball.

DESCRIPTION Plants form miniature, fat-leaved, globular bodies reminiscent of other small, sphaeroidal mesembs, particularly *Lithops*. A leaf pair is separated by a shallow or deep fissure from which the flowers arise. The greenish grey leaf surfaces are sometimes rough with variously sized indentations and horny ridges; other species have smooth leaves with scattered dark green or purple spots. Flowers are borne on short, stout stalks and vary in colour from light to dark yellow. As with many other mesembs, the stamens are collected in a cone in the centre of a flower. Fruit capsules have six to 15 locules and the upper surfaces are flattish with slight ridges. The seeds are numerous, slightly coarse in texture and whitish or brown.

DISTINGUISHING CHARACTERS Plants are characterised by variously shaped, greenish grey, fat plant bodies that usually grow above ground level. In contrast to other globular mesembs, the leaves often have a distinct, horny keel along the lower surface, making them appear triangular or boat-shaped. The short, stout flower stalks and robust flower buds resemble clenched fists.

FLOWERING TIME Species of *Dinteranthus* flower from late summer to midwinter (February to July in South Africa). Flowers open in the mid-afternoon and close in the evening.

DISTRIBUTION AND ECOLOGY The genus *Dinteranthus* occurs in the northwestern parts of the Northern Cape Province, South Africa, and the southeastern parts of Namibia. Its distribution area is bounded by Prieska in the southeast, Sperlingspütz in the west and Karasberg in the north. In their natural habitat the plants look very much like the pebbles amongst which they grow, and one can easily see why some species are referred to as mimicry plants. It is worth noting that the distribution areas of the different species hardly overlap.

CULTIVATION Species of *Dinteranthus* will not tolerate overwatering and a distinct summer resting period should be observed. Plants look their best if grown under conditions of high light intensity. Seeds germinate with difficulty and need high temperatures.

NOTES The strange growth form of these plants is very popular among collectors and some populations have suffered from excessive collecting. In nature, the seeds germinate under semi-translucent quartz pebbles and grow with amazing rapidity until they eventually emerge.

Dinteranthus Schwantes

NUMBER OF SPECIES/SUBSPECIES/VARIETIES (4/2/0)

SPECIES LIST AND CONSERVATION STATUS
D. microspermus (Dinter & Derenb.) Schwantes subsp. *microspermus* [R]
D. microspermus (Dinter & Derenb.) Schwantes subsp. *puberulus* (N.E.Br.) N.Sauer
D. pole-evansii (N.E.Br.) Schwantes [I]
D. vanzylii (L.Bolus) Schwantes [V]
D. wilmotianus L.Bolus subsp. *impunctatus* N.Sauer [R]
D. wilmotianus L.Bolus subsp. *wilmotianus*

LITERATURE
SAUER, N. 1975. *Dinteranthus. Aloe* 13: 9–24.

SAUER, N. 1978. *Dinteranthus pole-evansii, D. microspermus* subsp. *microspermus, D. microspermus* subsp. *puberulus, D. wilmotianus* subsp. *wilmotianus* and *D. wilmotianus* subsp. *impunctatus. Flowering plants of Africa* 45: t. 1778A–1780B.

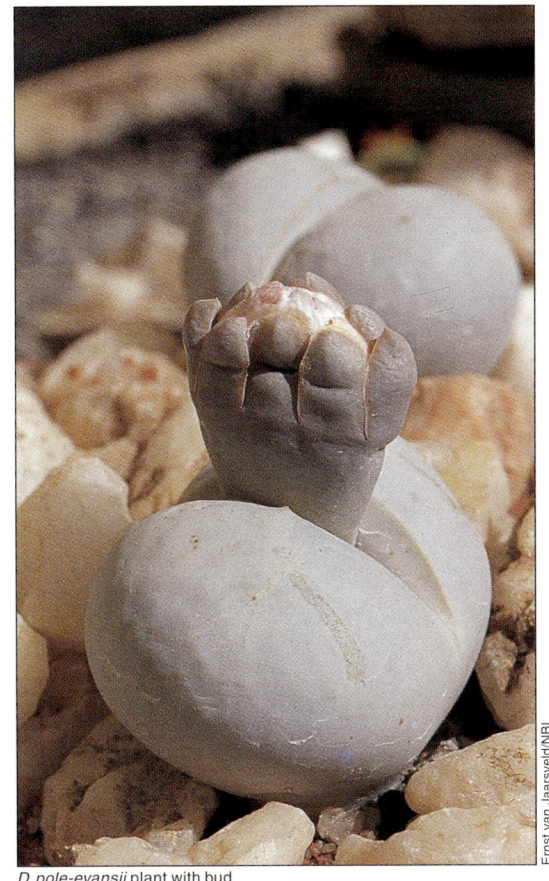

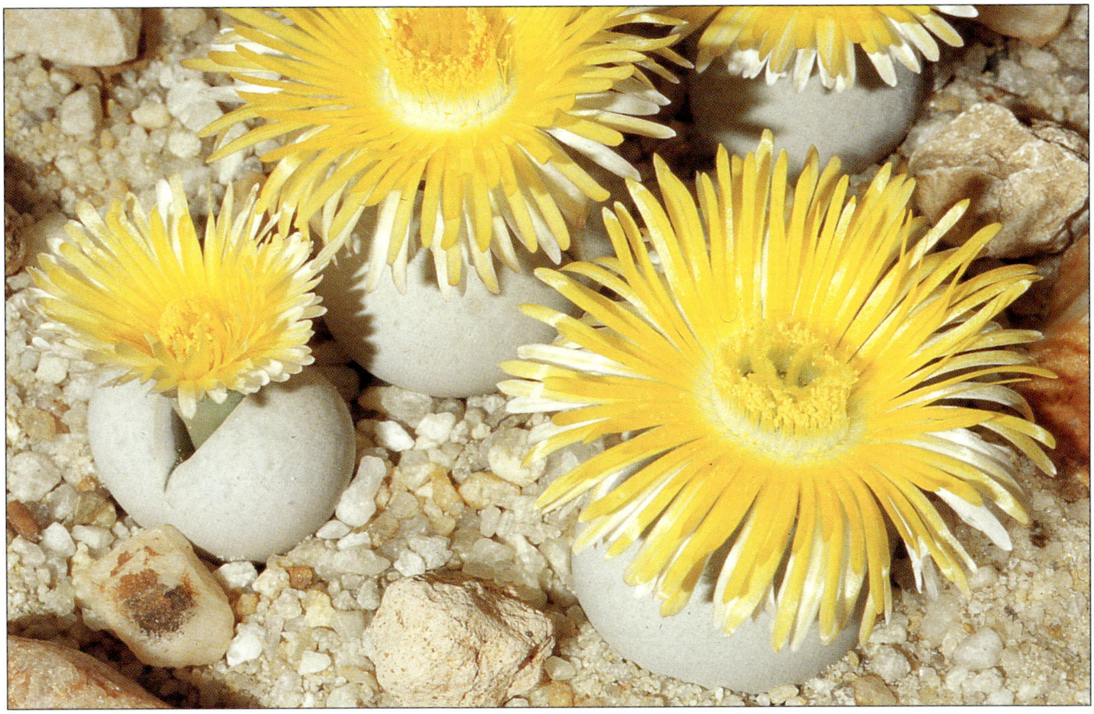
D. pole-evansii
Bob Chinnock

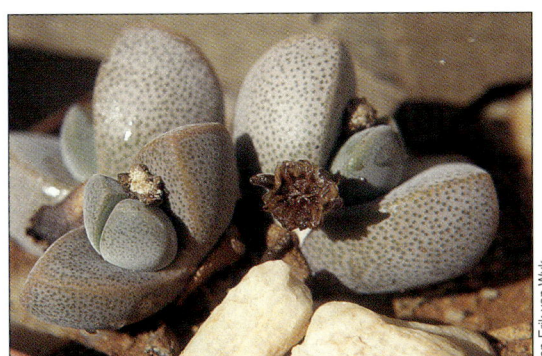

D. microspermus subsp. *puberulus*
Ben-Erik van Wyk

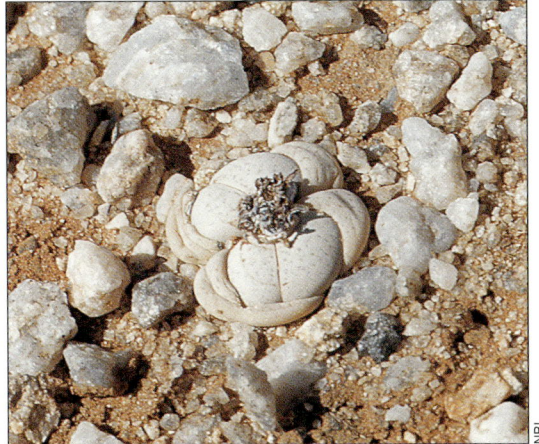

D. vanzylii in its habitat
NBI

D. pole-evansii plant with bud
Ernst van Jaarsveld/NBI

Diplosoma

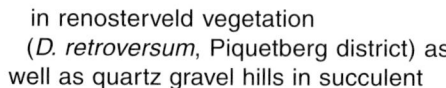

Derivation of genus name The name is derived from the Greek words *diplos* (double) and *soma* (body), referring to the two-lobed leaves.

Common names The vernacular name is *eendvoetvygie* (duck foot mesemb).

Description The dwarf, highly succulent plants are solitary or form small clumps with age. They have a distinctive dry, dormant stage. Each stem annually produces two opposite leaves enclosed by brown to black dry sheaths derived from the leaf bases of the previous year. The soft and tender leaves, with their reddish colour and distinctive dots, are upright or spread horizontally and are unequally united at their bases. The relatively small purple and white flowers are without stalks. They occur singly or sometimes in twos or threes, and have five to seven unequal roundish sepals. The numerous narrow petals are arranged in two or three whorls around the equally numerous filamentous staminodes and stamens (stamens are purple-black in *D. luckhoffii*). The dark green nectar glands are fused into a ring around the five to seven thread-like stigmas. Fruit capsules have five to seven locules and they have no closing devices. Seeds are small and pear-shaped.

Distinguishing characters *Diplosoma* is easily recognised by the soft, pulpy leaf texture, blistery surfaces and peculiar leaf shape.

Flowering time Species of *Diplosoma* flower from winter to spring (June to September in South Africa). Flowers open after noon and close in the evening.

Distribution and ecology The plants are confined to a few localities in the northwestern parts of the Western Cape Province of South Africa. They occur in quartz and limestone gravel on hills and flats in renosterveld vegetation (*D. retroversum*, Piquetberg district) as well as quartz gravel hills in succulent karoo (*D. luckhoffii*, Knersvlakte). They thrive in areas where rainfall occurs from autumn to spring and ranges between 75 and 200 mm per year.

Cultivation This genus is easy to cultivate but specialised greenhouse conditions are needed. They can flower within six months from seed but are short-lived, lasting only five or six years. Plants should be grown in well-drained mineral-rich soil, watered during the cool winter growing season and kept completely dry during the summer months. Provide ample shade during summer when plants become dormant.

Notes There are only two species in the genus. They both grow on quartz gravel flats but in different areas. *Diplosoma retroversum*, the southern species, is a localised and threatened species in need of conservation measures. *D. luckhoffii* is more common and widely distributed. It is easily distinguished from *D. retroversum* by the smaller, rounder leaf pairs which do not lie flat on the ground.

Diplosoma Schwantes (=*Maughaniella* L.Bolus; *Maughania* N.E.Br.)

Number of species/subspecies/varieties (2/0/0)

Species list and conservation status
  D. luckhoffii (L.Bolus) Schwantes ex. Ihlenf. [R]
  D. retroversum (Kensit) Schwantes [E]

Literature
  IHLENFELDT, H.-D. 1988. Morphologie und Taxonomie der Gattungen *Diplosoma* Schwantes und *Maughaniella* L. Bolus (Mesembryanthemaceae). Monographie der Mitrophyllinae Schwantes IV). *Beiträge zur Biologie der Pflanzen* 63: 375–401.

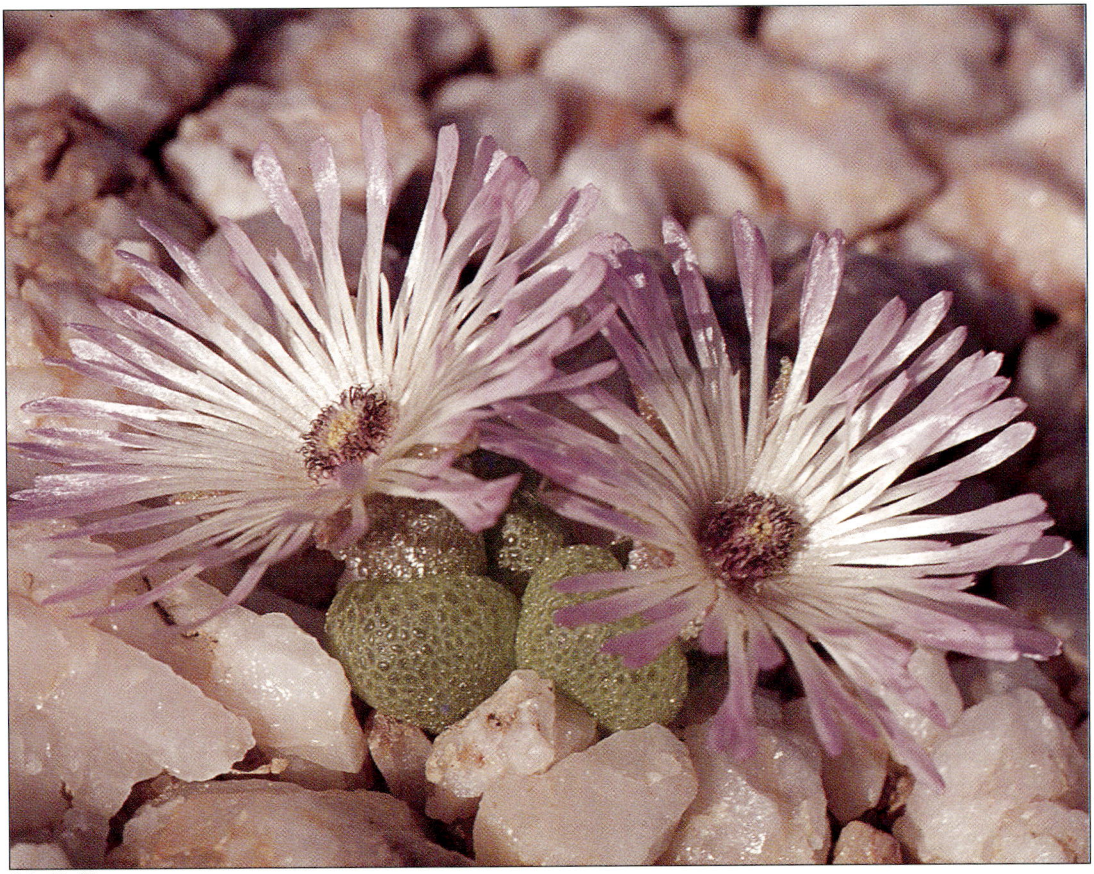

Ernst van Jaarsveld

D. luckhoffii

Bruce Bayer/NBI

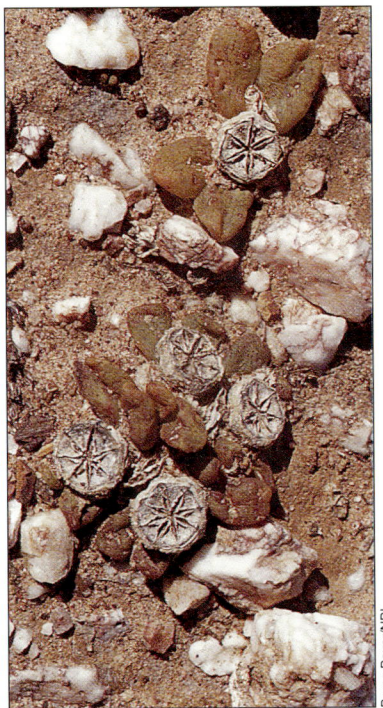

Fruiting plants of *D. retroversum* commencing their annual retreat

Peter Bruyns

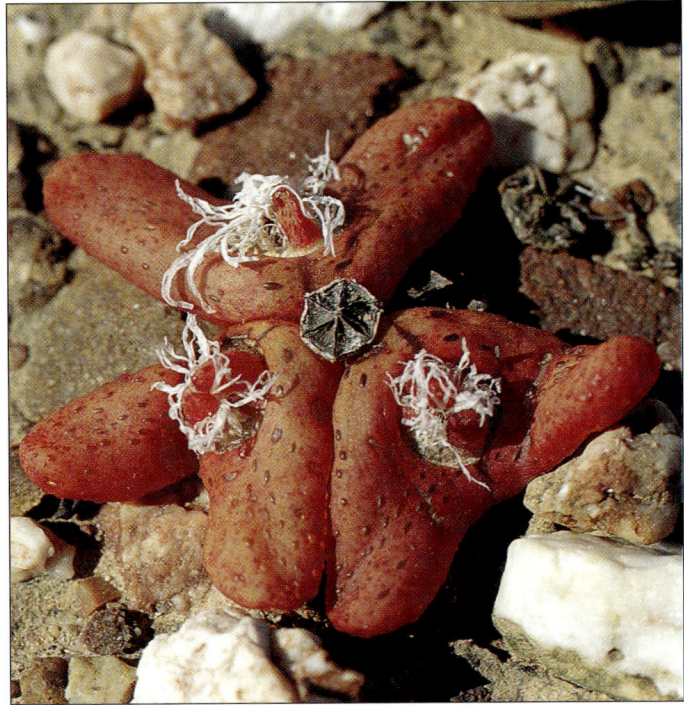

D. retroversum

FENESTRARIA

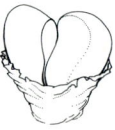

DERIVATION OF GENUS NAME The name is derived from the Latin word *fenestra* (window) or *fenestraria* (a number of windows) alluding to the characteristic windowed leaves. These plants are submerged in the sand of their natural habitat, having only the windows of the leaf tips exposed to allow sunlight to reach the chlorophyll in the lower parts of the leaves.

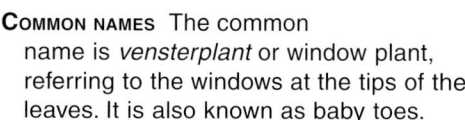

COMMON NAMES The common name is *vensterplant* or window plant, referring to the windows at the tips of the leaves. It is also known as baby toes.

DESCRIPTION The dwarf plants may form mats or single clumps. Roots and short stems are fleshy and the club-shaped leaves have a smooth waxy surface. The leaves are highly succulent, distinctly flat-topped and windowed. The medium to large flowers are white or yellow, and appear singly or in groups of up to three. They are borne on long flower stalks. The five blunt, rounded sepals have membranous margins. Petals are free, and occur in one to several whorls. There are numerous stamens. Stigmas spread from the base and are feathery. The fruit capsules are eight to 16-locular and become detached with age, acting as tumble fruits. There are small closing bodies and small valve wings. Expanding keels are parallel and have teeth. Covering membranes are well developed. The small whitish seeds are somewhat cone-shaped, with brown tips.

DISTINGUISHING CHARACTERS In *Fenestraria*, a fleshy underground stem is absent but roots are thickened. The leaves are smooth along their sides and the leaf tips are wholly to half windowed, with clear windows. The flowers are white to yellow or salmon, never magenta. Fruits are very shallow, pedicels are slender and seeds are whitish.

FLOWERING TIME Flowering occurs from mid-winter to spring (July to September in South Africa) after good rains. The flowers open at midday and remain open into the late afternoon.

DISTRIBUTION AND ECOLOGY *Fenestraria* is found from the environs of Lüderitz, southwards along the coast of Namibia as well as in the Richtersveld, South Africa, in an area receiving winter rainfall of 100 mm per year or less. The plants grow in full sun in wind-blown sand and are found no further than 40 km inland from the sea. The two subspecies have distinct distributions, the subspecies *aurantiaca* preferring the littoral sands of the southernmost parts of the distribution, while the subspecies *rhopalophylla* seems to prefer the calciferous sands further to the north.

CULTIVATION Propagation is best from seed although the rooting of cuttings can also be successful. Very sandy soils should be used as a growing medium and watering should be withheld in summer and sparse in winter.

NOTES The two subspecies can be distinguished by the size and colour of their flowers. The subspecies *aurantiaca* has large flowers (about 70 mm in diameter) that are golden yellow in colour. The subspecies *rhopalophylla* has smaller, white flowers (18 to 30 mm in diameter).

Fenestraria N.E.Br.

NUMBER OF SPECIES/SUBSPECIES/VARIETIES (1/1/0)

SPECIES LIST AND CONSERVATION STATUS
F. *rhopalophylla* (Schltr. & Diels) N.E.Br. subsp. *aurantiaca* (N.E.Br.) H.E.K.Hartmann
F. *rhopalophylla* (Schltr. & Diels) N.E.Br. subsp. *rhopalophylla* [R]

LITERATURE
HARTMANN, H.E.K. 1982. Monographien der Subtribus Leipoldtiinae. III. Monographie der Gattung *Fenestraria*. *Botanische Jahrbücher* 103: 145–183.

F. rhopalophylla subsp. *rhopalophylla*

F. rhopalophylla subsp. *aurantiaca*

Leaf tips of *F. rhopalophylla*

FRITHIA

DERIVATION OF GENUS NAME This genus was named after Mr Frank Frith of South Africa, who was an enthusiastic collector of succulent plants.

COMMON NAMES *Frithia* is commonly known as fairy elephant's feet as only the leaf tips are visible above the sand, resembling tiny elephant's feet.

DESCRIPTION These tiny stemless plants grow well sunken into the soil. The roots are fleshy and the tiny leaves are single and alternating in their arrangement. They have a distinct, flat, windowed top and tiny water cells on their textured surfaces. The small, solitary, stalkless or shortly stalked flowers vary from purple to white. The five unequal sepals are united into a short, cup-shaped tube and resemble the leaves. Numerous petals are arranged in several whorls and are fused at their bases. Filamentous staminodes surround three whorls of stamens which in turn enclose five short, stout, sharply pointed stigmas. Fruit capsules have five or six locules and are very fragile, breaking up when ripe or moistened or rolling as tumble fruits to disperse the seeds. The valves rise to a vertical position only. The expanding keels are parallel with short diverging tips. Seeds are small and slightly rough-textured.

DISTINGUISHING CHARACTERS In *Frithia* the leaves are alternate and the rootstock thickened. The leaves are not smooth along their sides but have water cells arranged in distinct rows. Flowers are tricoloured (magenta, white and orange-yellow) in the west, or pure white or pale pink in the east. Fruit capsules are fragile and short-lived.

FLOWERING TIME Flowering occurs from spring to summer (October to February in South Africa). The flowers open during mid-morning and are open until mid-afternoon.

DISTRIBUTION AND ECOLOGY *Frithia* is a summer rainfall species from the northeastern part of South Africa. It has been collected on the Magaliesberg range to the west of Gauteng in the North-West Province and is also known from the vicinity of Bronkhorstspruit and Witbank straddling the boundary between Gauteng and Mpumalanga. A link has yet to be found between these separated distribution areas. Plants usually grow in patches of coarse quartzitic soils, surrounded by typical grassland vegetation. The area receives rainfall of between 700 and 800 mm per year.

CULTIVATION Propagation is easy from seed if sown in sandy, gravelly, acidic soil. It is a summer grower.

NOTES *Frithia pulchra* has contractile leaves which shrink into the soil when conditions become unfavourable during the dry winter season. A small form, known as var. *minor*, differs mainly in its smaller size, more tuberous roots, white to pale pink (rather than rich magenta) flower colour, undulating window margins and its more easterly distribution. *Frithia* and *Fenestraria* are superficially similar but not closely related.

Frithia N.E.Br

NUMBER OF SPECIES/SUBSPECIES/VARIETIES (1/0/0)

SPECIES LIST AND CONSERVATION STATUS
F. *pulchra* N.E.Br. [R]

LITERATURE
HAMMER, S.A. 1989. Big and little Frithias, and a fiery *Fenestraria*. *Cactus and Succulent Journal (US)* 61: 199–201.

ZIMMERMANN, N.F.A. 1996. *Frithia pulchra* N.E.Br. Eine Reise zu zwei Populationen im Transvaal mit Besprechung der Sukkulenten Begleitvegetation. *Kakteen und andere Sukkulenten* 47: 7.

F. pulchra in its winter retreat

F. pulchra

White form of F. pulchra

Small form of F. pulchra together with the normal form

Gibbaeum

Derivation of genus name The name is derived from the Latin word *gibba* (hump) referring to the two leaves of each leaf pair which differ from each other in shape and size.

Common names *Gibbaeum* is known by several vernacular names including *volstruistone* (ostrich toes), mimicry plants, *duimpie-snuif* (thumb snuff), *papegaaibek* (parrot beak), *vinger-en-duim* (finger and thumb), *visbekvygie* (fishmouth vygie) and *volstruiswater* (ostrich water), in reference to the water contained in the leaves. The leaves are eaten by ostriches in times of drought.

Description The plants are dwarf succulents forming small, dense clumps or tufts of up to 200 mm in height. Stems, if present, are short and upright or grow horizontally. Roots may be fleshy, but a stout, short, woody rootstock is prevalent among most species. The branches each end in an unequal or subequal pair of leaves united into rounded units called bodies. These bodies may be divided into lobes, often with an oblique fissure, resembling a shark's mouth. The leaves may be light to dark green, bluish green, reddish, metallic grey, whitish green or grey-white in colour. The surfaces are smooth, minutely hairy or may have long dense hairs. The small to relatively large, solitary flowers appear in various shades of pink to dark purple or white and are borne on stalks. Six or sometimes up to nine unequal sepals surround the flowers. Petals occur in one to three whorls which surround the numerous stamens. Filamentous staminodes may or may not be present. In the centre of the flower are six to nine feathery stigmas. Fruit capsules have six to nine locules with triangular valves and broad valve wings. Covering membranes are present and closing bodies absent. Many seeds are found in each seed chamber. They are egg-shaped and smooth.

Distinguishing characters In species of *Gibbaeum*, the leaves are often velvety, as a result of minute, sometimes forked hairs (visible with a magnifying glass). Petals are purple or white, never yellow. The often radical inequality of the two leaves is highly distinctive, unlike the near perfect symmetry of *Argyroderma*.

Flowering time Flowering occurs from late winter to early summer (July to December in South Africa). The flowers open in the afternoon and close in the evening.

Distribution and ecology The genus is mostly found growing in the Little Karoo, Western Cape Province and adjacent areas, South Africa. It only extends peripherally into the Northern Cape Province. This is a winter and summer rainfall area with between 100 and 700 mm per year. The plants prefer sunny, open rocky areas and often grow amongst white quartzite pebbles, which make them difficult to detect as their colour blends in with the surrounding rocks.

Cultivation Species of *Gibbaeum* are propagated by means of seeds or cuttings which grow quite easily. Water sparingly at all times otherwise the leaves may burst.

Notes The species are distinguished by the shape, colour and texture of the leaves; they can be egg-smooth and rounded as in *G. heathii*, to velvety and oblong as in *G. velutinum* and *G. pachypodium*. With age, the latter forms an exposed woody trunk. *G. album* is one of the most striking of the quartz mimicry plants in its angularity, firm texture and ash-white colour. The leaves of *Gibbaeum* species are generally silvery in colour; those that are green can turn brownish or deep red, as in *G. dispar* and *G. gibbosum*. Some resemble *Conophytum* in that they produce a papery leaf sheath in the dormant season. Some species flower with difficulty. *G. heathii*, for example, does not

G. album

G. dispar

G. gibbosum

G. esterhuyseniae

G. petrense

G. heathii

flower at low light intensities. Different plants from the same population may differ markedly in the number and size of flowers produced under the same conditions (see illustration of *G. petrense*). *G. heathii* and *G. album* display the full range of colour variation (white through pink to purple) found in the genus as a whole. Uniquely, the flowers of *G. pachypodium* start out as white and darken with time to a reddish pink.

Gibbaeum (Haw.) N.E.Br. (=*Imitaria* N.E.Br.)

NUMBER OF SPECIES/SUBSPECIES/VARIETIES (16/1/1)

SPECIES LIST AND CONSERVATION STATUS
 G. album N.E.Br.
 G. angulipes (L.Bolus) N.E.Br. [K]
 G. cryptopodium (Kensit) L.Bolus
 G. dispar N.E.Br. [K]
 G. esterhuyseniae L.Bolus [E]
 G. geminum N.E.Br.
 G. gibbosum (Haw.) N.E.Br.
 G. haagei Schwantes var. *haagei*
 G. haagei Schwantes var. *parviflorum* L.Bolus
 G. heathii (N.E.Br.) L.Bolus
 G. johnstonii Van Jaarsv. & S.A.Hammer
 G. nebrownii Tischer [K]
 G. pachypodium (Kensit) L.Bolus [K]
 G. petrense (N.E.Br.) Tischer
 G. pilosulum (N.E.Br.) N.E.Br.
 G. pubescens (Haw.) N.E.Br. subsp. *pubescens*
 G. pubescens (Haw.) N.E.Br. subsp. *shandii* (N.E.Br.) Glen
 G. velutinum (L.Bolus) Schwantes

LITERATURE

NEL, C.G. 1953. *The Gibbaeum Handbook. A genus of highly succulent plants, native to South Africa*. Eds. P.G. Jordaan & E.W. Shurly. Blandford Press, London.

GLEN, H.F. 1974. *A revision of the Gibbaeinae (Mesembryanthemaceae)*. Unpublished M.Sc. thesis, University of Cape Town.

VAN JAARSVELD, E.J. & HAMMER, S.A. 1996. *Gibbaeum johnstonii*, a new species from the Little Karoo. *Bradleya* 14: 14–16.

G. pubescens subsp. *pubescens*

G. johnstonii

G. velutinum

G. pilosulum

G. haagei var. *parviflorum*

G. pachypodium

IHLENFELDTIA

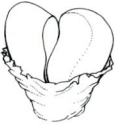

DERIVATION OF THE GENUS NAME The genus is named after Prof. Dr. H.-D. Ihlenfeldt, long-time student of mesembs and professor of Botany at the University of Hamburg, Germany.

COMMON NAMES No vernacular names seem to have been recorded for this genus.

DESCRIPTION The compact plants have keeled, three-sided leaves, sometimes with tiny often reddish teeth along the margins and keels. The solitary flowers are usually yellow and may sometimes be tinged with purple or red, particularly on the outside of the petals. However, the petals may also be creamy or pure white in different combinations with purple, and all of these colours may turn red with age. The stamens form an erect ring in the centre of the flower. They are mostly paler in colour than the petals. The fruit capsules have 10 to 15 locules with the seed chambers covered by thin, flexible covering membranes which collapse into the empty seed chambers once the seeds have been dispersed. The valve wings taper from their broad bases towards the tips, and the closing bodies are small. The seeds are smooth.

DISTINGUISHING CHARACTERS Superficially, the plants look similar to *Cheiridopsis* species but the capsules resemble those of *Titanopsis*. The closing bodies are much smaller and are often absent. The fruit stalks are robust; petals vary from deep egg-yellow (similar to *Vanheerdea*) to rose or pale orange-violet, colours rarely seen in *Cheiridopsis*.

FLOWERING TIME Flowering occurs in winter and the flowers are open from midday until dusk.

DISTRIBUTION AND ECOLOGY Both species occur in the Northern Cape Province, South Africa, and prefer quartzitic soils in Bushmanland and the adjacent areas of Namaqualand. These areas receive about 125 mm rainfall per year.

CULTIVATION The plants grow easily from seed and flower readily in cultivation. When grown in a greenhouse, they should be planted in sandy, well-drained soil. The main growing season is autumn, when watering should be abundant to ensure bud formation for the next flowering season. Outside of their habitat, plants are best grown in a greenhouse.

NOTES *Ihlenfeldtia excavata* may be distinguished from *I. vanzylii* by the invariably sharper leaf margins and leaf tips, the frequent presence of small reddish teeth and the more western distribution. The flowers of *I. vanzylii* are always a rich yellow colour, while those of *I. excavata* have unusual tints, even extending into grey (sometimes with lilac pollen).

Ihlenfeldtia H.E.K.Hartmann

NUMBER OF SPECIES/SUBSPECIES/VARIETIES (2/0/0)

SPECIES LIST AND CONSERVATION STATUS
I. excavata (L.Bolus) H.E.K.Hartmann
I. vanzylii (L.Bolus) H.E.K.Hartmann

LITERATURE
HARTMANN, H.E.K. 1992. *Ihlenfeldtia*, a new genus in Mesembryanthema (Aizoaceae). *Botanische Jahrbücher* 114: 29–50.

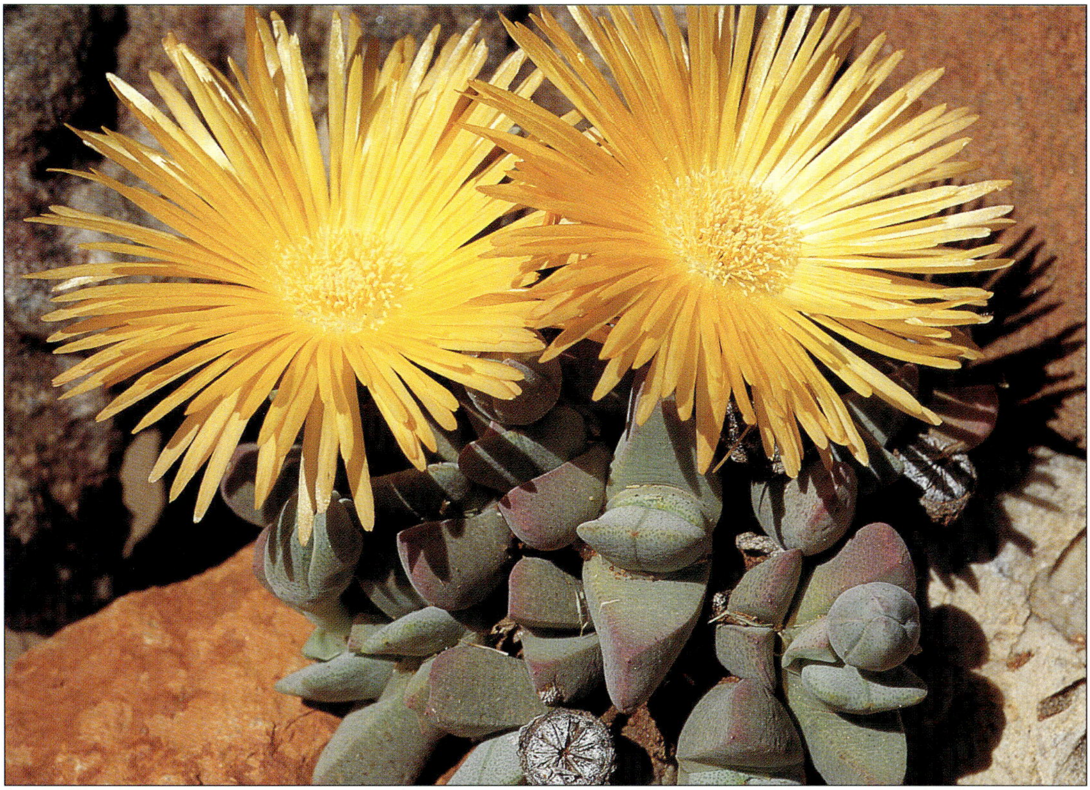

I. excavata

Graham Williamson

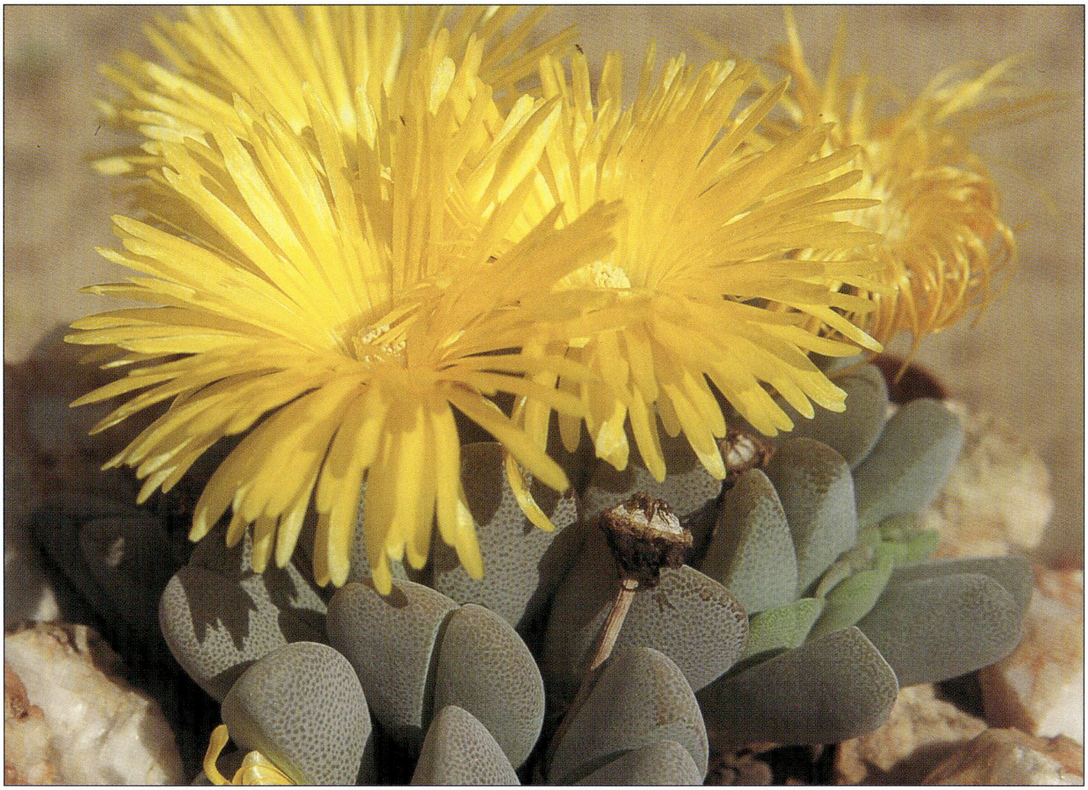

I. vanzylii

Ben-Erik van Wyk

LAPIDARIA

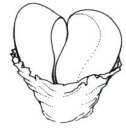

DERIVATION OF GENUS NAME The name is derived from the Latin word *lapis* (stone): *lapidaria* means group of stones.

COMMON NAMES The name *klipvygie* (rock mesemb) has been used for this genus.

DESCRIPTION The plants are compact, stemless and highly succulent with one to four leaf pairs per main branch. Leaves are three-sided, pale pinkish brown or whitish, with a minutely rough texture. The golden yellow flowers are solitary (or up to three in succession) and borne on short, winged stalks. There are numerous stamens, and filamentous staminodes are also present. The nectar glands are fused into a ring but are not clearly visible. There are six to eight awl-shaped, curved stigmas. The fruit capsules have six to eight locules and are flat on top with distinct ridges. The expanding keels have broad, membranous margins. Covering membranes are present and closing bodies are absent. The small, light brown seeds are pear-shaped.

DISTINGUISHING CHARACTERS Plants bear at least two pairs of keeled leaves. The keels are usually dark pink or orange while the rest of the leaf surfaces are paler. Flower stalks are winged and the unripe fruit capsules are metallic grey on top. The leaves appear somewhat shrunken even under optimal conditions, giving the plant a characteristically collapsed appearance. This is probably due to the prominence of their off-centred keels.

FLOWERING TIME Plants flower in winter (March to September in South Africa). The flowers open during the day.

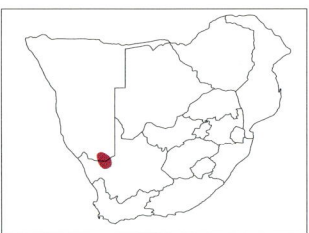

DISTRIBUTION AND ECOLOGY The single species of this genus occurs in the Warmbad area of southern Namibia and in adjacent parts of the Northern Cape Province, South Africa. It occurs in quartzite fields and fissures of gneiss.

CULTIVATION The plant is extremely easy to cultivate in a greenhouse, provided that care is taken with watering. The main growing season is autumn when watering should be generous. The soil should be sandy and mineral rich.

NOTES Previously *Lapidaria* was included in *Dinteranthus,* but it differs mainly in the larger number of active leaf pairs per branch and the larger seeds. Young seedlings are identical to those of *Lithops* in their conical shape, while those of *Dintheranthus* look like tiny green raspberries.

Lapidaria (Dinter & Schwantes) N.E.Br. (=*Dinteranthus* Schwantes subgen. *Lapidaria* Dinter & Schwantes)

NUMBER OF SPECIES/SUBSPECIES/VARIETIES (1/0/0)

SPECIES LIST AND CONSERVATION STATUS
  L. margaretae (Schwantes) Dinter & Schwantes

LITERATURE
  BROWN, N.E. 1928. *Mesembryanthemum* and some new genera separated from it. *Gardener's Chronicle* 84: 472, 492.

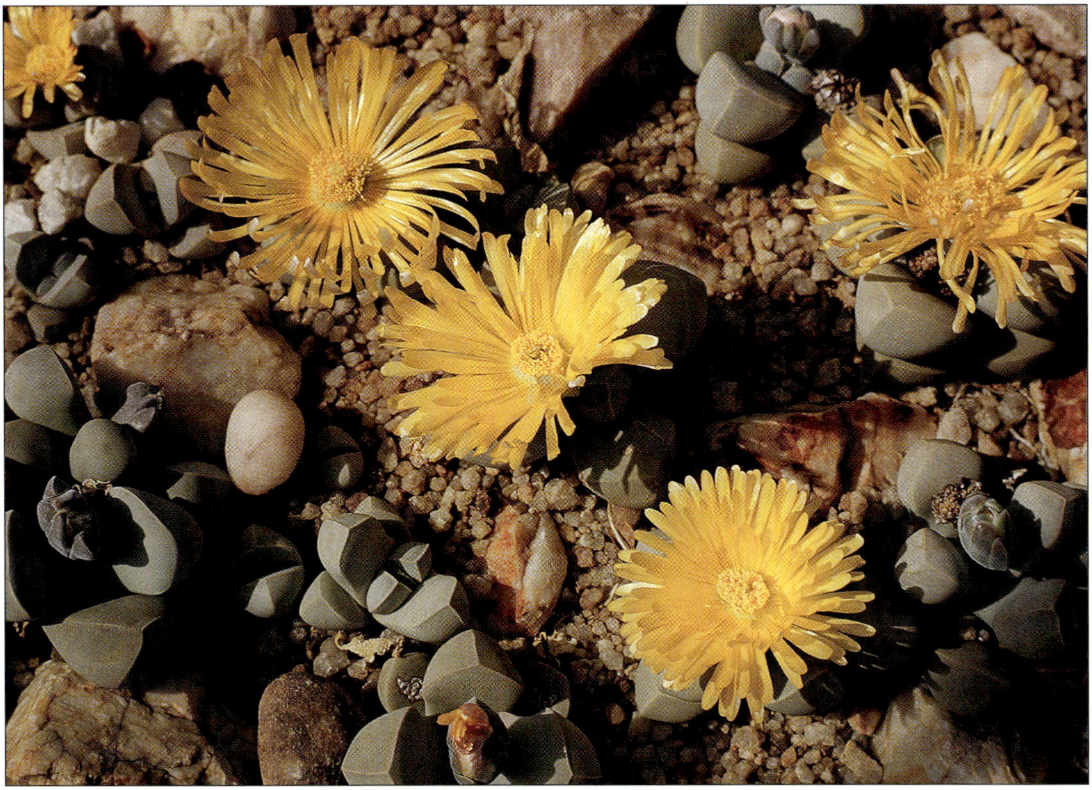

Flowering plants of *L. margaretae*

Schleben/NBI

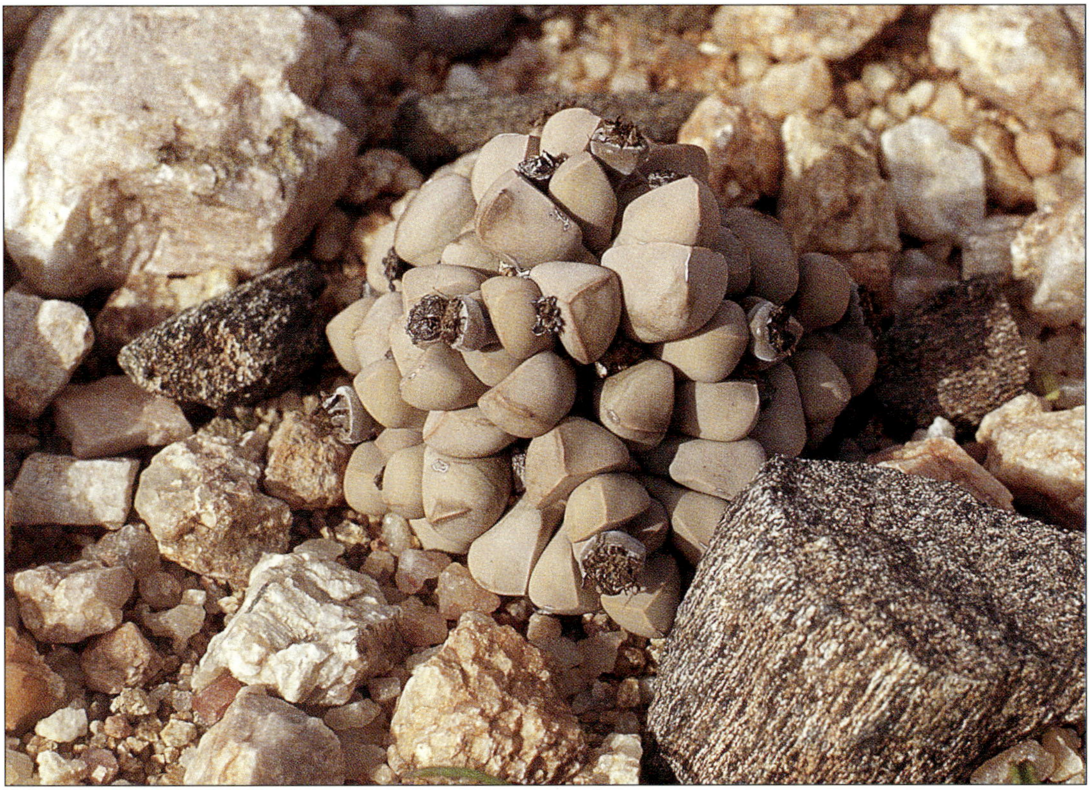

Fruiting plant of *L. margaretae*

Ernst van Jaarsveld

LITHOPS

DERIVATION OF GENUS NAME The name is derived from the Greek words *lithos* (stone) and *opsis* (like), alluding to the pebble-like appearance of the plants.

COMMON NAMES The vernacular names *beeskloutjie* (cattle hoof), *perdeklou* (horse's hoof), stone faces and stone plant are used for *Lithops* species.

DESCRIPTION The dwarf, practically stemless plants are partly sunken into the soil, where they occur as solitary or branched leaf pairs (plant bodies). The leaves are fused, forming cones with flat tops which are variously marked or windowed. Old leaves turn into more or less hard sheaths which enclose the new bodies. The surfaces of the plants are more or less smooth, or textured with ridges and warts, or adorned with water cells. The small to medium-sized, white, yellow or orange flowers arise on short stalks from the centre of the fissure, and are solitary or in rare cases occur in twos or threes. There are four to seven sepals and numerous petals arranged in one to four whorls. Numerous stamens surround the four to nine thread-like stigmas. Fruit capsules have four to nine locules and the valves have a single, stout, central, expanding keel which is divided at the tip. Covering membranes are reduced to a limb or are absent and the valve wings are broad. The seeds are highly variable in size, shape, colour and texture.

DISTINGUISHING CHARACTERS *Lithops* are easily recognised by the paired leaves which are fused into an oblong tapering body with a flat or rounded top. In habitat only the flat or rounded leaf tips are exposed above the rocky surface, which they often resemble. Flowers appear between the leaves. The fruit capsules have four to nine locules and the covering membranes are strongly reduced or absent. Bracts are not present, whereas in *Conophytum* bracts are present but hidden.

FLOWERING TIME The plants flower in autumn to early winter (April to July in southern Africa). The flowers, which are often strongly scented, open in the mid-afternoon and close at sunset.

DISTRIBUTION AND ECOLOGY The genus is widespread in southern Africa (Namibia, Botswana and South Africa, but excluding KwaZulu-Natal) and occurs mainly in the Nama Karoo, succulent karoo and in rare cases in semi-arid savanna regions. Plants are mostly confined to gravelly flats and hills and grow in soils derived from granite, quartz, shale, schist and limestone. The plants are difficult to detect in habitat when they are not in flower.

CULTIVATION *Lithops* species are often difficult to cultivate but once their special requirements are understood the art of growing them is easily mastered. Outside of their natural habitat plants are best grown in a greenhouse or on a sunny window sill. The soil should be gravel-rich and a mixture of sandy-clay-loam with ample compost added. Full-grown plants should be watered during the summer and autumn, whereas seedlings and young plants need more regular watering, from spring until early winter. Seeds should preferably be sown during the warmer months, and the initial use of a fungicide may be advisable. Seeds germinate readily and plants may flower in the first year after sowing. When grown in light shade the bodies become drawn out and when taken too rapidly from a low to high intensity sunlight, the plants are easily burnt and killed. *Lithops* can be grown out of doors in arid gardens.

NOTES *Lithops* is one of the most popular succulent plant groups. All known species and many attractive cultivars have been introduced into the succulent plant trade, mainly through the efforts of Professor Desmond Cole. Plants are now widely

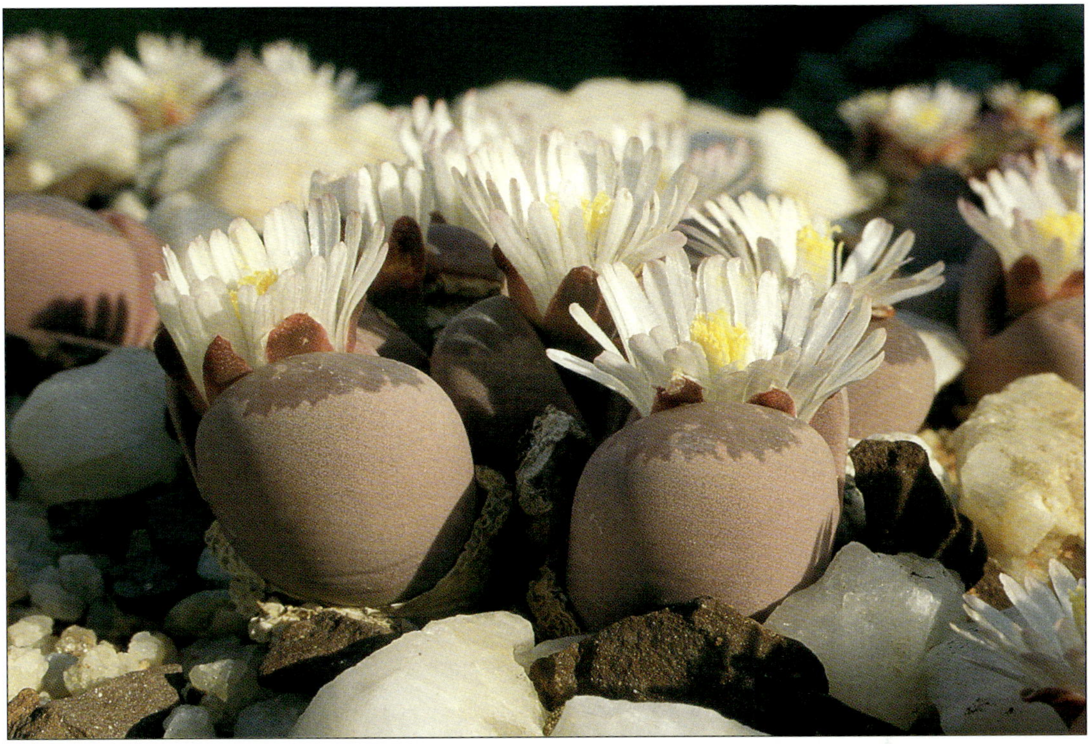

Jan van der Pol/NBI

L. optica

John Trager

L. divergens

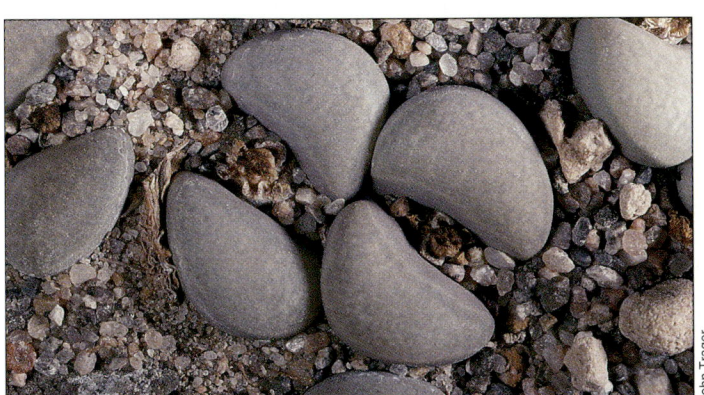

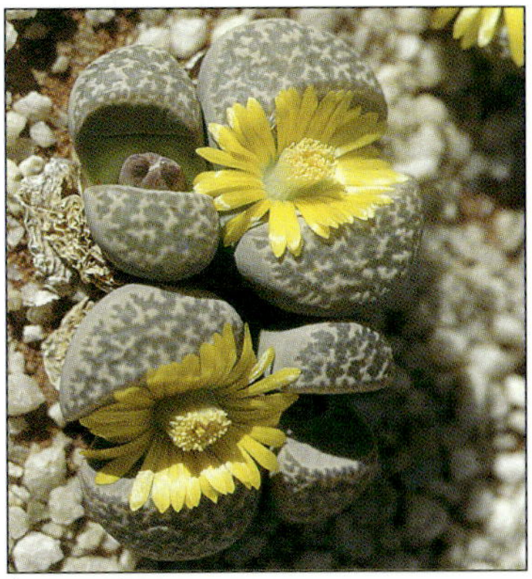

NBI

L. naureeniae

NBI

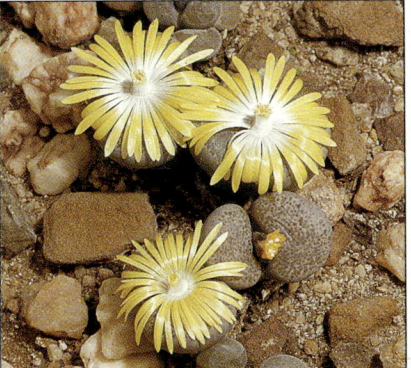

L. localis, sometimes also known as *L. terricolor*

NBI

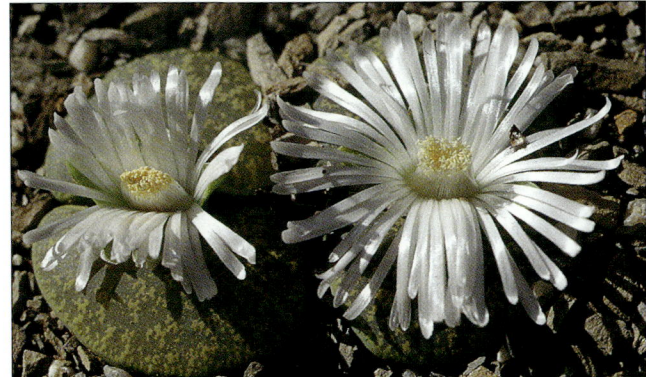

L. lesliei var. *lesliei*

cultivated throughout the world and private growers all contribute to *ex situ* conservation. *L. steineckeana* is a greenhouse hybrid which arose in Germany. There are numerous other horticultural freaks and forms of *Lithops* not detailed in the list below.

Lithops N.E.Br.

NUMBER OF SPECIES/SUBSPECIES/VARIETIES
(36/15/36)

SPECIES LIST AND CONSERVATION STATUS
  L. aucampiae L.Bolus subsp. *aucampiae* var. *aucampiae*
  L. aucampiae L.Bolus subsp. *aucampiae* var. *koelemanii* (de Boer) D.T.Cole [K]
  L. aucampiae L.Bolus subsp. *euniceae* (de Boer) D.T.Cole var. *euniceae* [R]
  L. aucampiae L.Bolus subsp. *euniceae* (de Boer) D.T.Cole var. *fluminalis* D.T.Cole [R]
  L. bromfieldii L.Bolus var. *bromfieldii*
  L. bromfieldii L.Bolus var. *glaudinae* (de Boer) D.T.Cole [I]
  L. bromfieldii L.Bolus var. *insularis* (L.Bolus) B.Fearn
  L. bromfieldii L.Bolus var. *mennellii* (L.Bolus) B.Fearn
  L. coleorum S.A.Hammer & R.Uijs [V]
  L. comptonii L.Bolus var. *comptonii* [E]
  L. comptonii L.Bolus var. *weberi* (Nel) D.T.Cole [R]
  L. dinteri Schwantes subsp. *dinteri* var. *brevis* (L.Bolus) B.Fearn [K]
  L. dinteri Schwantes subsp. *dinteri* var. *dinteri* [K]
  L. dinteri Schwantes subsp. *frederici* (D.T.Cole) D.T.Cole [K]
  L. dinteri Schwantes subsp. *multipunctata* (de Boer) D.T.Cole [K]
  L. divergens L.Bolus var. *amethystina* de Boer [I]
  L. divergens L.Bolus var. *divergens* [V]
  L. dorotheae Nel [V]
  L. francisci (Dinter & Schwantes) N.E.Br. [R]
  L. fulviceps (N.E.Br.) N.E.Br. var. *fulviceps*
  L. fulviceps (N.E.Br.) N.E.Br. var. *lactinea* D.T.Cole [R]
  L. gesineae de Boer var. *annae* (de Boer) D.T.Cole [I]
  L. gesineae de Boer var. *gesineae* [E]
  L. geyeri Nel [R]
  L. gracilidelineata Dinter subsp. *brandbergensis* (de Boer) D.T.Cole [R]
  L. gracilidelineata Dinter subsp. *gracilidelineata* var. *gracilidelineata*
  L. gracilidelineata Dinter subsp. *gracilidelineata* var. *waldroniae* de Boer [K]
  L. hallii de Boer var. *hallii*
  L. hallii de Boer var. *ochracea* (de Boer) D.T.Cole
  L. helmutii L.Bolus [R]
  L. herrei L.Bolus
  L. hookeri (A.Berger) Schwantes var. *dabneri* (L.Bolus) D.T.Cole [K]
  L. hookeri (A.Berger) Schwantes var. *elephina* (D.T.Cole) D.T.Cole
  L. hookeri (A.Berger) Schwantes var. *hookeri*
  L. hookeri (A.Berger) Schwantes var. *lutea* (de Boer) D.T.Cole
  L. hookeri (A.Berger) Schwantes var. *marginata* (Nel) D.T.Cole [K]
  L. hookeri (A.Berger) Schwantes var. *subfenestrata* (de Boer) D.T.Cole [K]
  L. hookeri (A.Berger) Schwantes var. *susannae* (D.T.Cole) D.T.Cole [R]
  L. julii (Dinter & Schwantes) N.E.Br. subsp. *fulleri* (N.E.Br.) B.Fearn var. *brunnea* de Boer
  L. julii (Dinter & Schwantes) N.E.Br. subsp. *fulleri* (N.E.Br.) B.Fearn var. *fulleri*
  L. julii (Dinter & Schwantes) N.E.Br. subsp. *fulleri* (N.E.Br.) B.Fearn var. *rouxii* (de Boer) D.T.Cole
  L. julii (Dinter & Schwantes) N.E.Br. subsp. *julii*
  L. karasmontana (Dinter & Schwantes) N.E.Br. subsp. *bella* (N.E.Br.) D.T.Cole
  L. karasmontana (Dinter & Schwantes) N.E.Br. subsp. *eberlanzii* (Dinter & Schwantes) D.T.Cole
  L. karasmontana (Dinter & Schwantes) N.E.Br. subsp. *karasmontana* var. *aiaisensis* (de Boer) D.T.Cole [K]
  L. karasmontana (Dinter & Schwantes) N.E.Br. subsp. *karasmontana* var. *karasmontana*
  L. karasmontana (Dinter & Schwantes) N.E.Br. subsp. *karasmontana* var. *lericheana* (Dinter & Schwantes) D.T.Cole
  L. karasmontana (Dinter & Schwantes) N.E.Br. subsp. *karasmontana* var. *tischeri*

L. julii

L. julii 'hotlips'

L. otzeniana

L. karasmontana var. *tischeri*

L. schwantesii var. *marthae*

L. lesliei subsp. *burchellii*

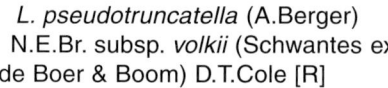

D.T.Cole [K]
L. lesliei (N.E.Br.) N.E.Br. subsp. burchellii D.T.Cole [R]
L. lesliei (N.E.Br.) N.E.Br. subsp. lesliei var. hornii de Boer [K]
L. lesliei (N.E.Br.) N.E.Br. subsp. lesliei var. lesliei
L. lesliei (N.E.Br.) N.E.Br. subsp. lesliei var. mariae D.T.Cole [K]
L. lesliei (N.E.Br.) N.E.Br. subsp. lesliei var. minor de Boer
L. lesliei (N.E.Br.) N.E.Br. subsp. lesliei var. rubrobrunnea de Boer
L. lesliei (N.E.Br.) N.E.Br. subsp. lesliei var. venteri (Nel) de Boer & Boom
L. localis (N.E.Br.) N.E.Br.
L. marmorata (N.E.Br.) N.E.Br. var. elisae (de Boer) D.T.Cole
L. marmorata (N.E.Br.) N.E.Br. var. marmorata
L. meyeri L.Bolus [R]
L. naureeniae D.T.Cole [R]
L. olivacea L.Bolus var. nebrownii D.T.Cole [R]
L. olivacea L.Bolus var. olivacea
L. optica (Marloth) N.E.Br. [nt]
L. otzeniana Nel [R]
L. pseudotruncatella (A.Berger) N.E.Br. subsp. archerae (de Boer) D.T.Cole
L. pseudotruncatella (A.Berger) N.E.Br. subsp. dendritica (Nel) D.T.Cole
L. pseudotruncatella (A.Berger) N.E.Br. subsp. groendrayensis (H.Jacobsen) D.T.Cole
L. pseudotruncatella (A.Berger) N.E.Br. subsp. pseudotruncatella var. elisabethiae (Dinter) de Boer & Boom [R]
L. pseudotruncatella (A.Berger) N.E.Br. subsp. pseudotruncatella var. pseudotruncatella
L. pseudotruncatella (A.Berger) N.E.Br. subsp. pseudotruncatella var. riehmerae D.T.Cole [R]

L. pseudotruncatella (A.Berger) N.E.Br. subsp. volkii (Schwantes ex de Boer & Boom) D.T.Cole [R]
L. ruschiorum (Dinter & Schwantes) N.E.Br. var. lineata (Nel) D.T.Cole
L. ruschiorum (Dinter & Schwantes) N.E.Br. var. ruschiorum
L. salicola L.Bolus [V]
L. schwantesii Dinter subsp. gebseri (de Boer) D.T.Cole
L. schwantesii Dinter subsp. schwantesii var. marthae (Loesch & Tischer) D.T.Cole
L. schwantesii Dinter subsp. schwantesii var. rugosa (Dinter) de Boer & Boom [R]
L. schwantesii Dinter subsp. schwantesii var. schwantesii
L. schwantesii Dinter subsp. schwantesii var. urikosensis (Dinter) de Boer & Boom
L. steineckeana Tischer
L. vallis-mariae (Dinter & Schwantes) N.E.Br.
L. verruculosa Nel var. glabra de Boer
L. verruculosa Nel var. verruculosa
L. villetii L.Bolus subsp. deboeri (Schwantes) D.T.Cole
L. villetii L.Bolus subsp. kennedyi (de Boer) D.T.Cole
L. villetii L.Bolus subsp. villetii
L. viridis H.A.Lückh. [R]
L. werneri Schwantes & H.Jacobsen [R]

LITERATURE

COLE, D.T. 1988. Lithops. *Flowering stones*. Acorn Books, Randburg.

HAMMER, S.A. & UIJS, R. 1994. *Lithops coleorum* S.A.Hammer & R Uijs sp. nov., a new species of *Lithops* N.E.Br. from the Northern Transvaal. *Aloe* 31: 36–38.

CLARK, J.Y. 1996. A key to *Lithops* (Aizoaceae). *Bradleya* 14: 1–9.

L. schwantesii

L. ruschiorum

L. marmorata

L. herrei

L. karasmontana subsp. *eberlanzii*

L. meyeri

MUIRIA

DERIVATION OF GENUS NAME The genus was named after the prolific plant collector Dr John Muir (1874 to 1947).

COMMON NAMES The vernacular names *muiskopvygie* (mouse head mesemb) and schmoo plant (USA) are used.

DESCRIPTION These peculiar stemless little plants are single-bodied or sparsely clustered. The leaf pairs are all curved inwards and are seasonally covered by white, withered, felt-like leaf sheaths. The leaves form oblong or spherical soft bodies with a small fissure near the tip and a velvety surface of long hairs. The small whitish to purple flowers are solitary and rupture the tips of the plant bodies when they emerge. There are six oblong sepals with membranous tips and numerous petals and stamens. The six nectar glands are almost completely fused into a ring and surround the six short, stout, curved stigmas. Fruit capsules have six locules and no covering membranes or closing devices. The smooth brown seeds are egg-shaped and slightly flattened.

DISTINGUISHING CHARACTERS *Muiria* has a very distinctive appearance and is unlike any other mesemb. Its leaves are so completely fused that their junction is not discernable. The fine velvety hairs are amongst the longest of any mesemb.

FLOWERING TIME Plants flower during early summer (November to December in South Africa). The flowers are open during the day.

DISTRIBUTION AND ECOLOGY *Muiria* is confined to a small region on the northern side of the Langeberg range in the Barrydale district of the Little Karoo, Western Cape Province, South Africa. Plants occur on quartz gravel hills in succulent karoo vegetation and are locally abundant but within a very small, exceedingly saline area. It thrives in a habitat with winter and summer rainfall which ranges from 200 to 300 mm per year.

CULTIVATION Outside of its habitat the plants require greenhouse conditions. Water sparingly but regularly during the growing season and allow to dry completely when the plants turn yellow. This happens immediately after flowering. Do not withhold water for long periods (even during the brief summer dormancy) as the plants may lose their roots and deteriorate. Grow in a sunny spot in clay-loam, mixed with gravel. Propagation is by seed, which may be sown fresh. Seeds will also germinate well after several years of storage. Plants are difficult to cultivate in regions of low winter light conditions such as found in Europe.

NOTES *Muiria hortenseae* and *Gibbaeum album* co-occur and frequently form natural hybrids. These can be recognised by the incomplete fusion of the leaves. The hybrid was described as X*Muirio-gibbaeum*.

Muiria N.E.Br.

NUMBER OF SPECIES/SUBSPECIES/VARIETIES (1/0/0)

SPECIES LIST AND CONSERVATION STATUS
M. hortenseae N.E.Br. [E]

LITERATURE
GLEN, H.F. 1974. *A revision of the Gibbaeinae (Mesembryanthemaceae).* Unpublished M.Sc. thesis, University of Cape Town.

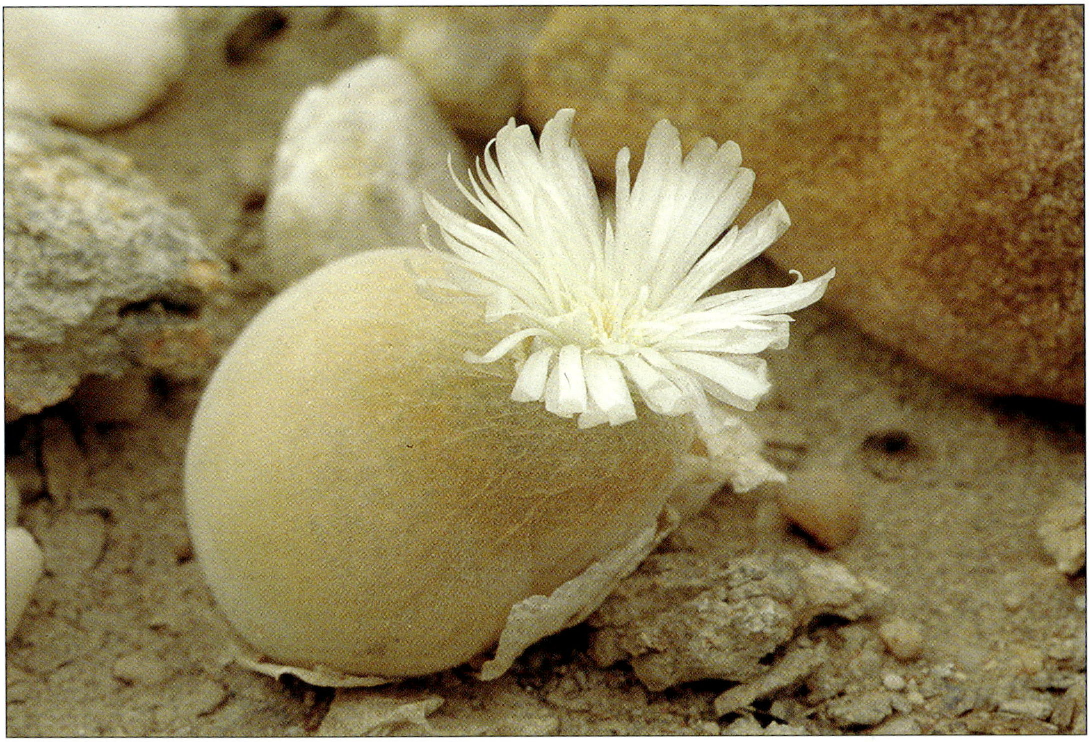

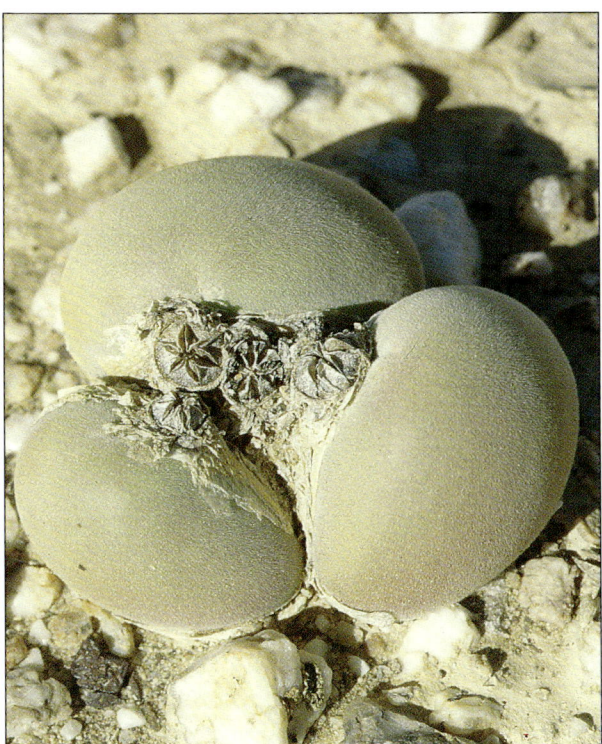

NBI

M. hortenseae

NBI

Fruiting plant of *M. hortenseae*

Deon Viljoen

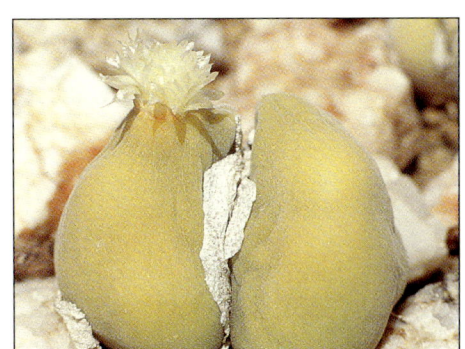

Flowering plant of *M. hortenseae*

Ernst van Jaarsveld

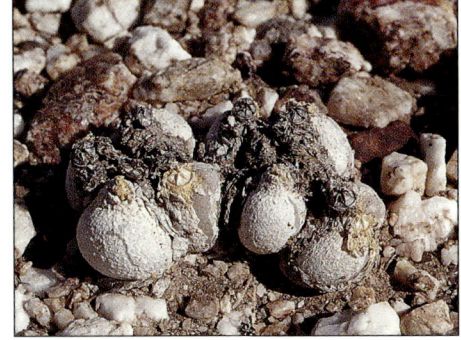

Dormant plant of *M. hortenseae*

Namibia

Derivation of genus name The genus is named after the Namib Desert which is situated along the coast of Namibia.

Common names No vernacular names seem to have been recorded.

Description The plants form dense, compact clumps or cushions of up to 100 mm in height and 200 mm in diameter. There are four to six opposite leaves per branch. The leaves are fused at their bases, thick and swollen, slightly three-sided, and velvety with a greyish flaky, waxy surface. The pink or white flowers are solitary at the tip of each branch and often not perfectly round but somewhat four-sided in appearance. The five sepals are unequal in size and have minute water cells on their surfaces. The numerous stamens are grouped together in the form of a cone in the centre of the flower. Nectar glands are fused into a ring around the numerous (up to 28) awl-shaped stigmas. Fruit capsules have between nine and 28 locules. Covering membranes are strongly reduced or absent; closing devices are absent.

Distinguishing characters The leaves are large, soft and puffy, much more so than in *Juttadinteria* and *Dracophilus* and never toothy. *Namibia* is distinguished from similar genera by the leaf surfaces that are densely covered with minute rounded tubercles and by the square appearance of the flowers. It is easily distinguished from *Juttadinteria* by the number of sepals (*Namibia* has five, whereas *Juttadinteria* has four). It differs from *Dracophilus* and *Juttadinteria* in the extremely thick and more compact, whitish grey leaves.

Flowering time Flowering occurs from mid-winter to late spring (June to October). One species (*N. ponderosa*) opens in the morning and closes in the evening; the other (*N. cinerea*) opens in the morning and remains open thereafter.

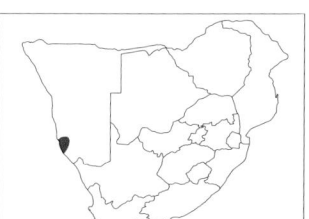

Distribution and ecology *Namibia* occurs in the surroundings of Lüderitz in southwestern Namibia and also a few kilometres east of Prince of Wales Bay, about 60 km south of Lüderitz.

Cultivation Plants should be cultivated in sandy loam and propagation is from seed, which grow readily. *Namibia* species require high intensities of light and rarely thrive in cultivation. Keep dry during the summer.

Notes The main difference between the species lies in the non-closing of the flowers of *Namibia cinerea*, which has an excessive number of locules (25 to 28), whereas *N. ponderosa* has about 10 locules. Otherwise the two species are difficult to distinguish when not in flower or fruit. Unfortunately there is uncertainty about the correctness of the name *N. ponderosa*.

Namibia (Schwantes) Schwantes
 (=*Juttadinteria* subg. *Namibia* Schwantes)

Number of species/subspecies/varieties (2/0/0)

Species list and conservation status
 N. cinerea (Marloth) Dinter & Schwantes
 N. ponderosa (Dinter) Dinter & Schwantes

Literature
WALGATE, M. 1939. A revision of *Juttadinteria* Schwant., *Dracophilus* Dint. et Schwant., and *Namibia* Dint. et Schwant. Notes on Mesembryanthemum *and allied Genera*. University of Cape Town, Cape Town. Vol. 3: 171–188.

SCHWANTES, G. 1927. Zur Systematik der Mesembrianthemen. *Zeitschrift für Sukkulentenkunde* 3: 106.

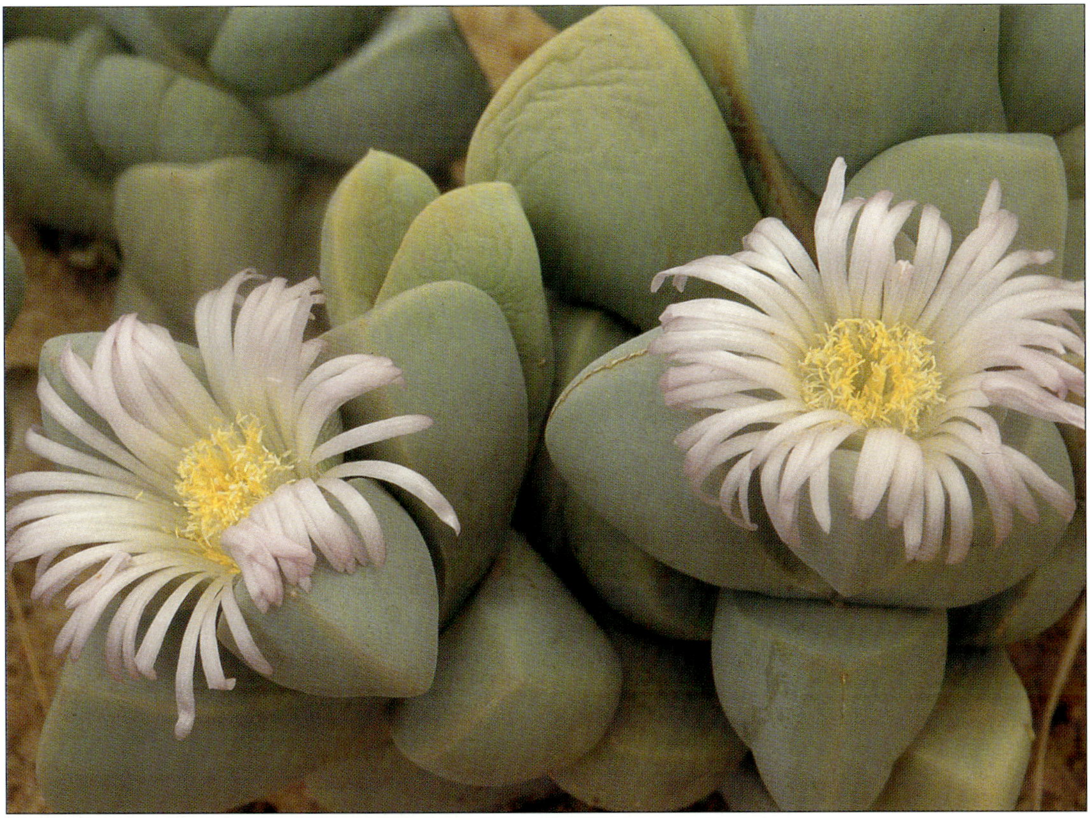

Graham Williamson

N. cinerea

Graham Williamson

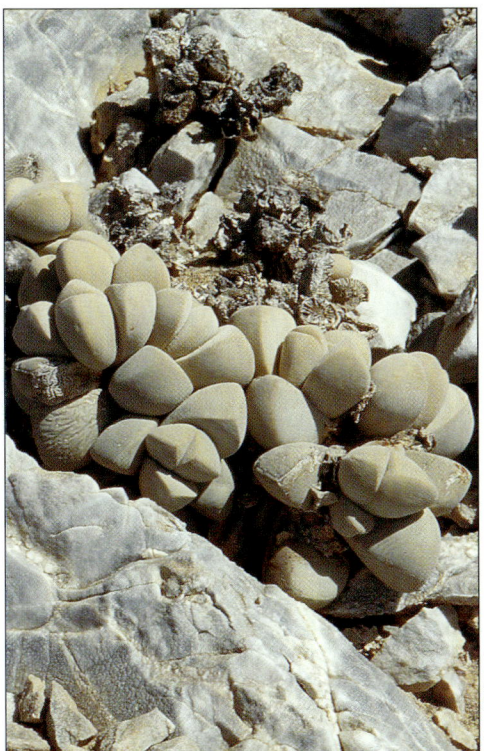

Fruiting plant of *N. ponderosa* in habitat

Ernst van Jaarsveld

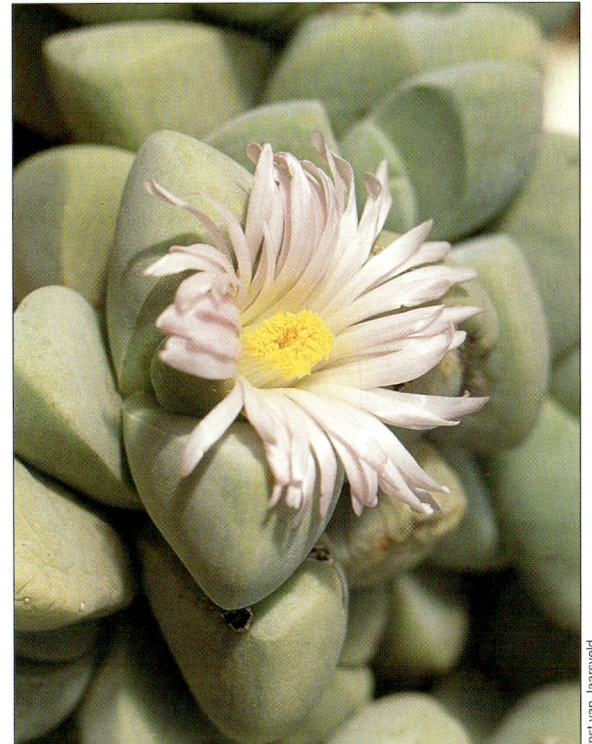

N. ponderosa

Oophytum

Derivation of the genus name The name comes from the Greek words *oon* (egg) and *phyton* (plant), in reference to the egg-shaped plant bodies.

Common names No vernacular names are known.

Description The compact plants develop two leaf pairs per season on each of up to about 20 branches per individual. The outer leaf pair turns into papery sheaths which partly or fully envelop the young leaf pair which emerges. The leaves are soft, with visible water cells on their surfaces. They turn bright brownish red towards the end of the growing season, otherwise they are dark green to brownish green or yellowish green. Flowers emerge (at different times of year in different species) after active growth has commenced. Six sepals surround the petals which, like the staminodes, can be white, pink or pale violet. The fruit capsules develop directly on top of the drying plant body. They have broad valve wings. Closing bodies are absent. Expanding keels are almost radial, and covering membranes are either absent or present as narrow rims over the locules.

Distinguishing characters Plants are stemless clusters of cigar-shaped, fused, papillate leaves. The formation of a dry, white sleeve enclosing a fresh leaf pair during the summer, with ripe fruits on top of the branches of the compact plants, are also characteristic.

Flowering time Flowering occurs in winter (July to September in South Africa). The flowers open from midday but close at about 17:00.

Distribution and ecology The distribution area is restricted to the quartz fields of the Knersvlakte north of Vanrhynsdorp in the Western Cape Province of South Africa, where about 125 mm of rainfall are expected per year, mainly in winter. Plants grow in rich stands, adding to the spring flower sensation of the region despite the small size of individuals. *O. nanum* and *O. oviforme* have a widely overlapping distribution in the middle of the Knersvlakte, while *O. nordenstamii* is restricted to a few hills near Vredendal.

Cultivation *Oophytum* species grow easily from seed and from cuttings. Like all plants with differently shaped leaves (two different sorts of leaf pairs are produced per season), *Oophytum* plants need a strict summer resting period. Cultivation is only possible if plants are given strong winter light and some warmth. This allows them to accept enough water to ensure flowering and to preserve the correct sequence of dimorphic leaves. Given too little light, the plants lose their natural growth cycle and produce too many leaves. The plants are prone to rot in horticulture, and also during wet winters in nature.

Notes The distinction of the species is mainly based on differences in the timing and degree of emergence of the new leaves from the old leaves and leaf sheaths. The well-known *O. oviforme* has relatively large, cigar-shaped plant bodies, while the tiny *O. nanum* has shorter, fatter ones. *O. nordenstamii* has globular rather than cigar-shaped plant bodies and the leaves are very dark glossy green. Its bracts are short and enclosed within the plant body, and its petals are larger and far broader than in the other species. The flowers appear earlier and they are pure white, not magenta.

Oophytum N.E.Br.

Number of species/subspecies/varieties (3/0/0)

Species list and conservation status
O. nanum (Schlechter) L.Bolus [nt]
O. nordenstamii L.Bolus
O. oviforme (N.E.Br.) N.E.Br.

Literature
IHLENFELDT, H.-D. 1978. Morphologie und Taxonomie der Gattung *Oophytum* N.E.Br. (Mesembryanthemaceae). *Botanische Jahrbücher* 99: 303–328.

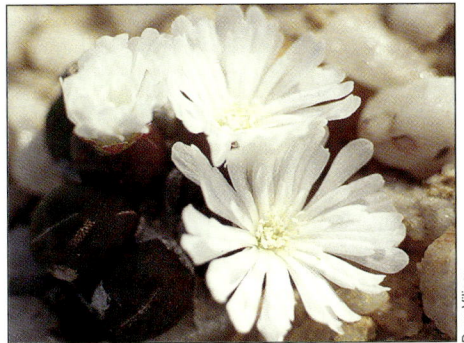

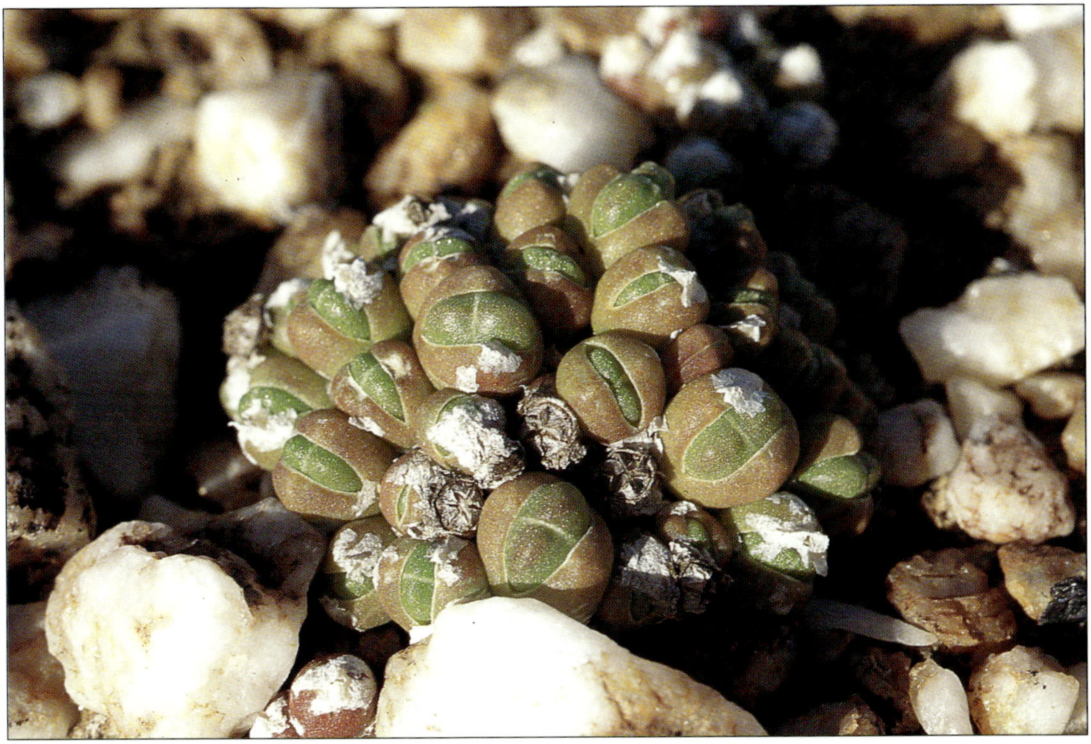
O. *nanum*

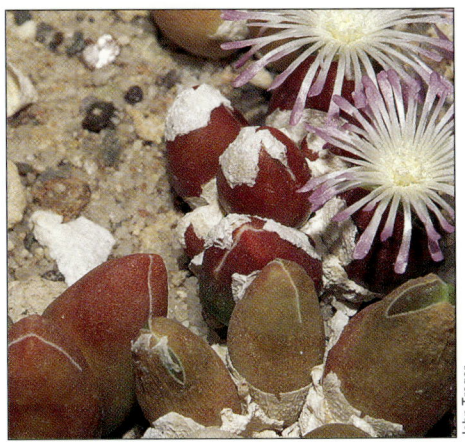

Ernst van Jaarsveld

O. nordenstamii

Deon Viljoen

O. oviforme

John Trager

Flowering plant of *O. nanum*

John Trager

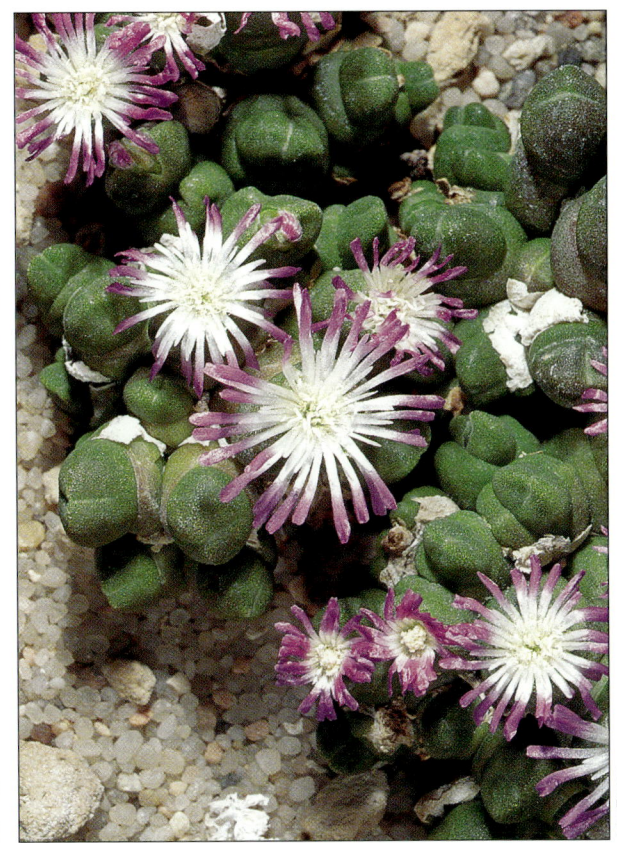

PLEIOSPILOS

DERIVATION OF GENUS NAME The name is derived from the Greek words *pleios* (full) and *spilos* (dots) and refers to the many small dots on the plants.

COMMON NAMES Vernacular names for *Pleiospilos* include *kwaggavy* (quagga mesemb), *lewerplant* (liver plant, alluding to the reddish purple colour of the leaves of some species), *lewervygie*, *klipplant* (stone plant) and split rock.

DESCRIPTION Plants are compact to clustered, sometimes unbranched, with one to four pairs of leaves per branch. Leaves are fused at their bases, broad, thick and firmly fleshy, rounded, often roughly lumpy, grey-green or brownish, conspicuously dotted and covered with a thick wax layer which may be flaky. The large flowers are yellow to coppery orange, rarely white or pinkish and they are borne on short stalks with bracts. They may be solitary or emerge in small groups. The numerous petals are surrounded by five to six sepals, with up to eight sepals in rare cases, which are sometimes united into a short tube above the ovary. Dark green, ring-shaped nectar glands surround the nine to 15 thread-like stigmas. Fruit capsules are nine to 15-locular with broad apically tapering valve wings, very firm covering membranes and large white to brown closing bodies. Seeds are ovoid and dark brown.

DISTINGUISHING CHARACTERS The plants are easily recognised by their invariably well-dotted, often purplish brown and excessively swollen, rounded, rock-like leaves.

FLOWERING TIME Plants flower in winter (*P. nelii*) or in autumn (all other species). The extremely fragrant flowers open in the mid-afternoon and stay open a little past dusk.

DISTRIBUTION AND ECOLOGY The genus has its main distribution in the Little Karoo, straddling the border between the Western and Eastern Cape Provinces of South Africa. It also extends further north into the arid south-central parts of the Great Karoo in the Northern Cape Province. An outlying species known either as *P. compactus* subsp. *canus* or *P. borealis* has been recorded from somewhat further northeast, near Bloemfontein in the Free State. However, it has not been found again at this locality in recent years. Plants occur in stony semi-arid, shale or sandstone flats in typical karroid vegetation where rainfall is mainly in summer, with occasional winter rain. The total annual rainfall is between 150 and 300 mm. The plants are difficult to detect against the stony background as the leaves are often the same colour as the stones amongst which they grow (see illustration of *P. bolusii*).

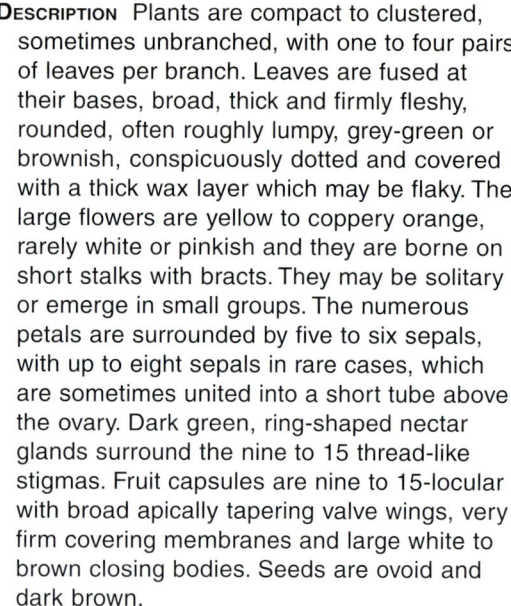

CULTIVATION For successful cultivation, a greenhouse or sunny windowsill is required or the plants may be grown in rockeries in arid regions. Plants are highly susceptible to red spider mites which scar the leaf surface. With the exception of *P. nelii*, which will not flower if kept too dry, this genus should be kept dry during winter. Soils should be sandy and mineral rich. Plants are easily propagated by seed sown during the summer months.

NOTES *Pleiospilos* is a very popular group amongst succulent collectors. The most commonly cultivated species are *P. bolusii* and *P. nelii*, the famous split rocks, *P. simulans*, the true liver plant, and the various hybrids of these. Cultivars of *P. bolusii* have been created since the 1930s, some having pure white flowers and others, such as the bright green cultivar 'envy', having remarkable leaf colours. The various subspecies of *P. compactus*, including the tiny subsp. *minor*, are encountered in collections.

Pleiospilos N.E.Br.

NUMBER OF SPECIES/SUBSPECIES/VARIETIES (4/4/0)

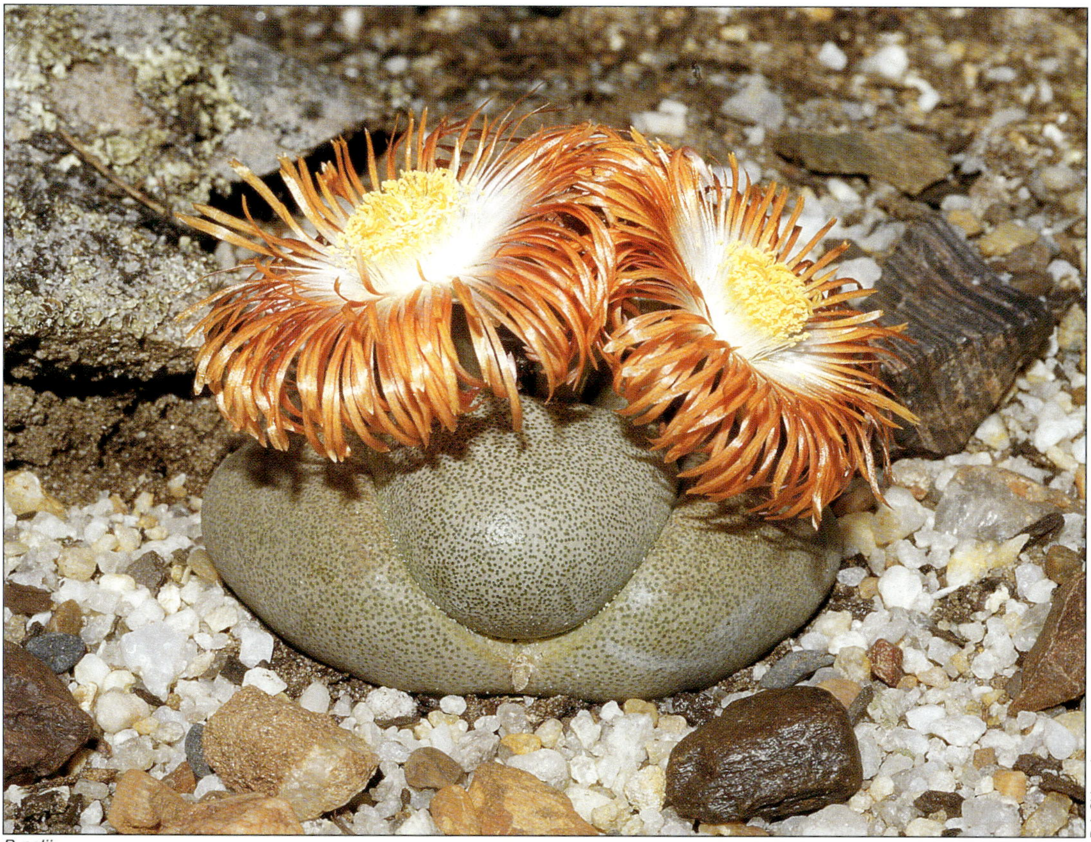

Bob Chinnock

P. nelii

NBI

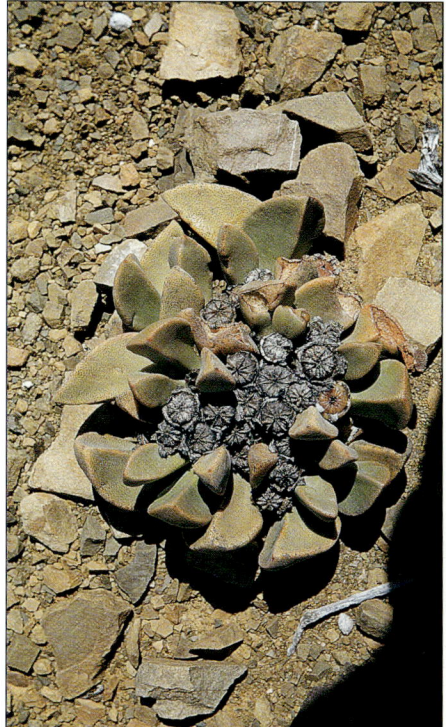

Fruiting plant of *P. compactus* subsp. *sororius*

NBI

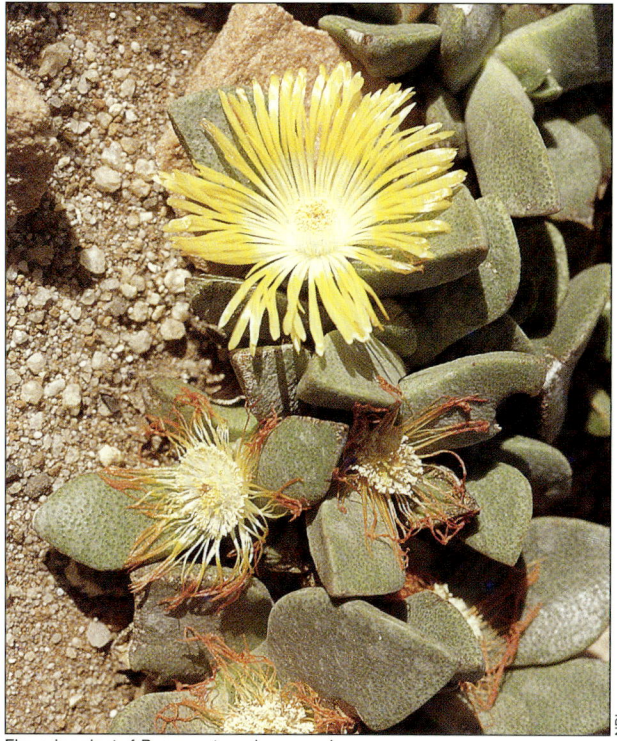

Flowering plant of *P. compactus* subsp. *sororius*

Pleiospilos

Species list and conservation status
P. bolusii (Hook.f.) N.E.Br.
P. compactus (Aiton) Schwantes subsp. canus (Haw.) H.E.K.Hartmann & Liede [nt]
P. compactus (Aiton) Schwantes subsp. compactus
P. compactus (Aiton) Schwantes subsp. fergusoniae (L.Bolus) H.E.K.Hartmann & Liede
P. compactus (Aiton) Schwantes subsp. minor (L.Bolus) H.E.K.Hartmann & Liede [R]
P. compactus (Aiton) Schwantes subsp. sororius (N.E.Br.) H.E.K.Hartmann & Liede
P. nelii Schwantes [R/V]
P. simulans (Marloth) N.E.Br. [E]

Literature
HARTMANN, H.E.K. & LIEDE, S. 1986. Die Gattung *Pleiospilos* s.lat. (Mesembryanthemaceae). *Botanische Jahrbücher* 106: 433–485.

P. bolusii 'envy'

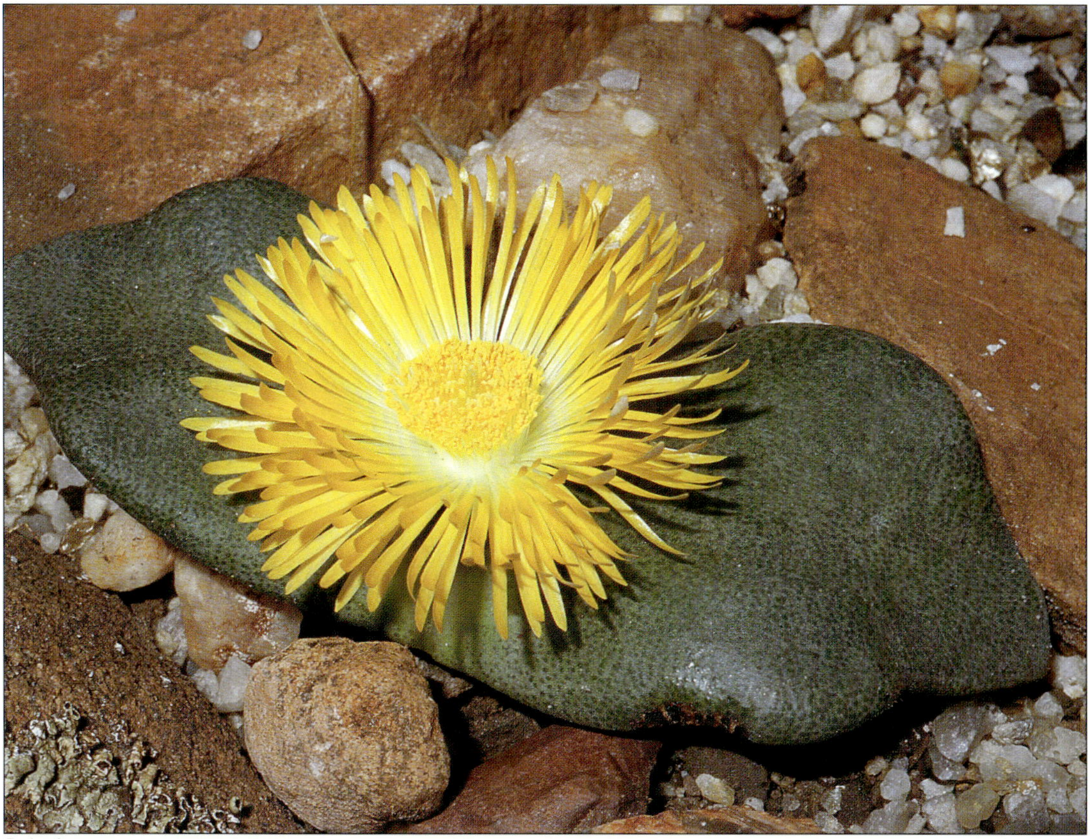

P. simulans

Bob Chinnock

P. bolusii in its habitat

Ernst van Jaarsveld

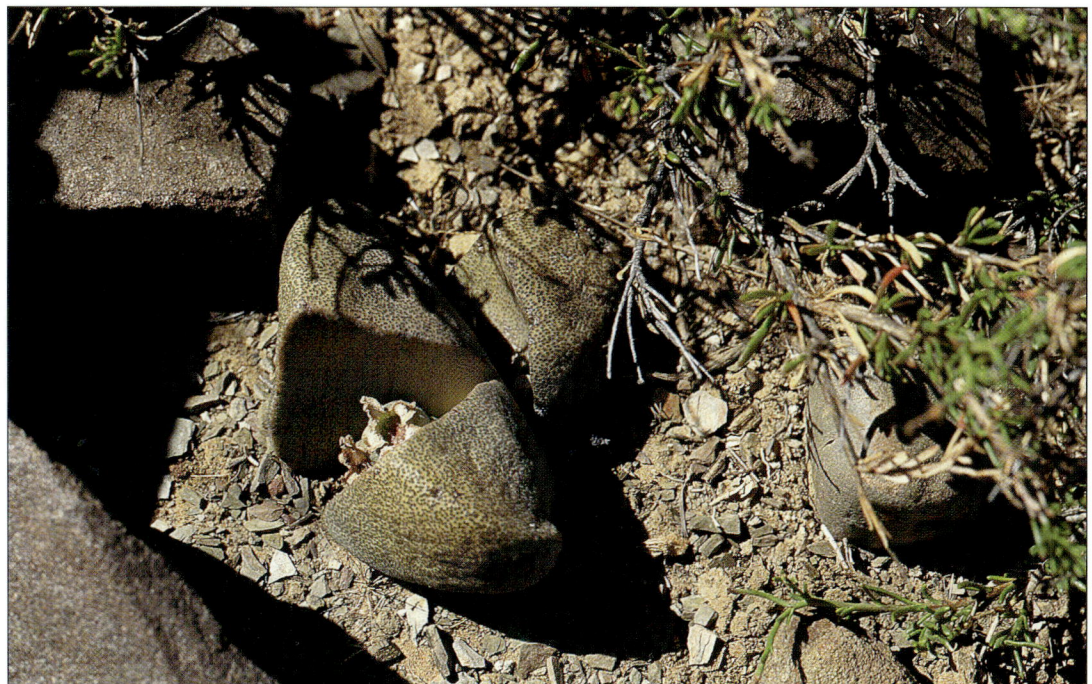

Tanquana

Derivation of genus name The genus is named after the Tanqua Karoo, an arid area in the Western and Northern Cape Provinces, South Africa.

Common names Plants are known by the vernacular names *klipvygie* (rock mesemb) and *tankwa-beesklou* (tanqua cattle hoof).

Description These highly branched to unbranched plants have smooth, polished, bright green to purplish leaves. The invariably yellow, rather small flowers are very shortly stalked so that they may often appear nestled between the leaves. There are four or five sepals and numerous slender petals surrounding the 10 thread-like stigmas. The 10-locular fruit capsules have covering membranes and tiny closing bodies.

Distinguishing characters This genus is easily distinguished from the superficially similar *Pleiospilos* by its smooth, unkeeled leaves. The capsules are quite different from those of *Pleiospilos*. They are much smaller, convex on top and have tiny closing bodies.

Flowering time Plants flower in autumn (around April to May in South Africa). The highly scented flowers open in the afternoon and close again at night.

Distribution and ecology *Tanquana* occurs in the central part of the Western Cape Province, South Africa, where it grows in the arid Tanqua Karoo and southwestern Great Karoo, just entering the Little Karoo south of Laingsburg. It favours stony ground where plants resemble their surroundings and are difficult to detect when not in flower. It thrives in areas with a summer and winter rainfall which ranges between 100 and 150 mm per year.

Cultivation Cultivation may not be easy for beginners. Outside its natural habitat the plants need to be kept from rain (greenhouse or window sill) and grown in well-drained, mineral-rich soil. Water sparingly, especially during winter when the new leaves absorb moisture from the ageing ones, and also in summer when the plants become dormant. Seeds germinate easily and should be sown during the warmer months.

Notes Of the three species, *Tanquana hilmarii* is by far the smallest and has a strong tendency to remain single-bodied. The other two form clusters, very large in the case of *T. prismatica*, and sparse in the case of *T. archeri*. Only *T. prismatica* is widespread. The others are highly localised. This genus was previously included in *Pleiospilos*.

Tanquana H.E.K.Hartmann & Liede

Number of species/subspecies/varieties/formas (3/0/0)

Species list and conservation status
T. archeri (L.Bolus) H.E.K.Hartmann & Liede [R]
T. hilmarii (L.Bolus) H.E.K.Hartmann & Liede [R]
T. prismatica (Marloth) H.E.K.Hartmann & Liede [nt]

Literature
HARTMANN, H.E.K. & LIEDE, S. 1986. Die Gattung *Pleiospilos* s.lat. (Mesembryanthemaceae). *Botanische Jahrbücher* 106: 433–485.

HARTMANN, H. & LIEDE, S. 1986. Bemerkungen zu Verbreitung und Ökologie von *Pleiospilos* Schwantes *s. str.* und *Tanquana* H.E.K.Hartmann & Liede (Mesembryanthemaceae). *Mitteilungen aus dem Institut für Allgemeine Botanik, Hamburg* 21: 117–125.

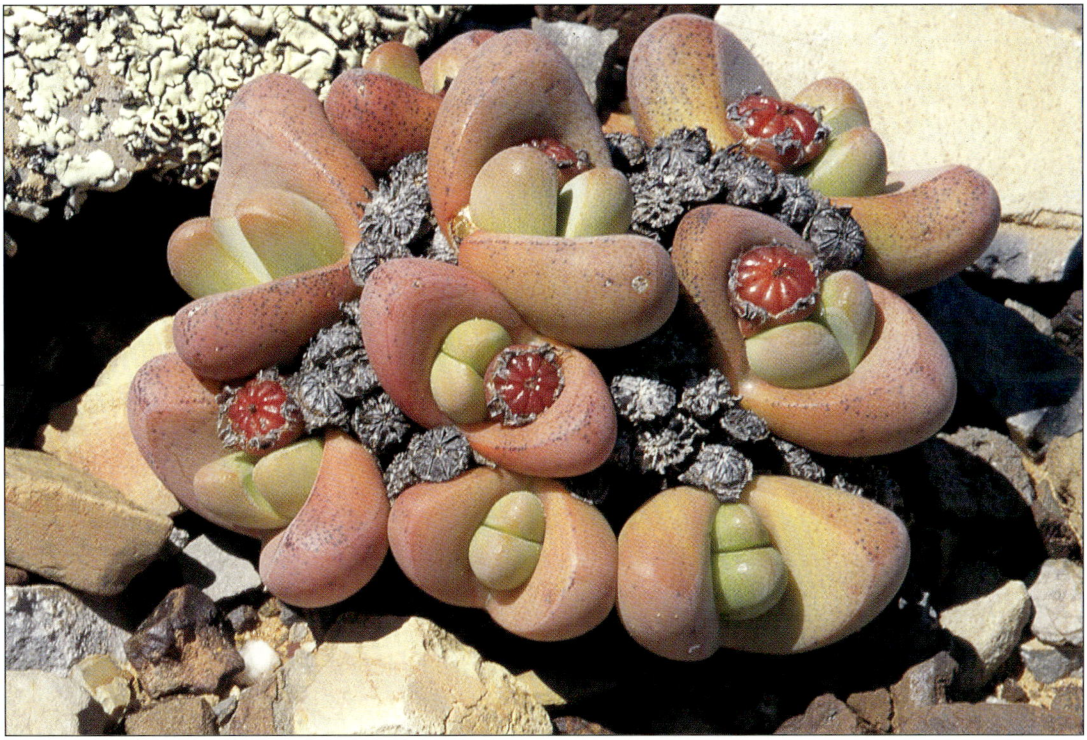

Fruiting plant of *T. prismatica*

Graham Williamson

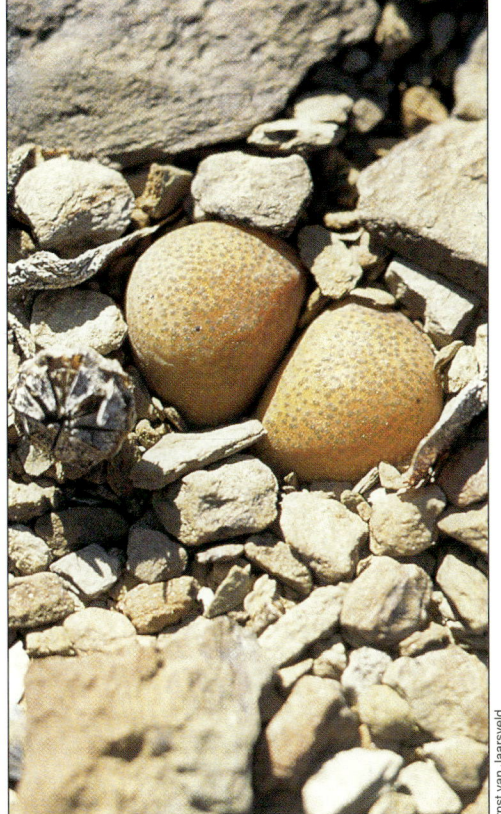

T. archeri in habitat

Ernst van Jaarsveld

T. hilmarii

Bob Chinnock

Flowering plant of *T. prismatica*

NBI

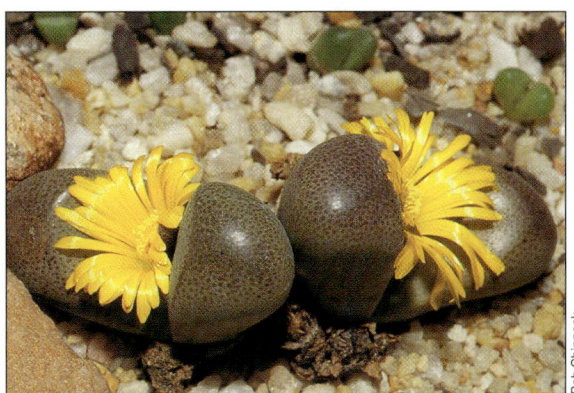

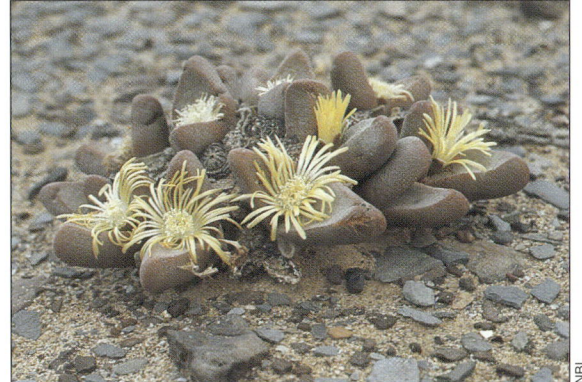

Vlokia

Derivation of genus name This recently described genus is named after Mr Jan Vlok, a South African botanist specialising in fynbos, who was the first to collect the plant.

Common names No vernacular names seem to have been recorded.

Description *Vlokia* is a dwarf succulent creeper with long trailing stems and thin lateral roots. The leaves are fused and clasp the stems, with one or two new pairs forming per year. They are fat, boat-shaped and have a central blister. The leaves are dark green and spotted along the reddish margins. Old leaves are long-persistent, crumpled, black in colour (hence the specific name, *ater*) and shrunken, so that parts of the stems are exposed (this is not visible in the young plants seen in the illustrations). Flowers are pink, solitary and scentless, borne above two bracts on short flower stalks. There are six sepals surrounding the sparse petals, with numerous filamentous staminodes concealing the short stamens. The ovary is convex on top with nectar glands forming a ring around the five to seven stigmas. Fruit capsules are five to seven-locular, with covering membranes and without closing bodies. Seeds are pear-shaped and very hard.

Distinguishing characters *Vlokia* is easily recognised by the persistent trailing stems which rarely branch and support only a few active leaves. Other useful features are the old crumpled leaves and capsules, both of which are black.

Flowering time Flowers appear from late winter to early spring (August to September in habitat). The short-lived flowers open at midday and are closed by dusk.

Distribution and ecology The genus is known from only a few populations on a mountain near Montagu in the western Little Karoo of the Western Cape Province, South Africa. It grows in very shallow pans of finely eroded quartzite near fynbos.

Cultivation Plants are not difficult to grow if given a steady supply of moisture and bright light. The very thin roots will die if kept dry for extended periods and the stems do not readily form new roots. One-year-old seedlings can flower.

Notes *Vlokia* was only recently discovered. It is a peculiar genus of only one species and its affinities with other genera are not clear.

Vlokia S.A. Hammer

Number of species/subspecies/varieties (1/0/0)

Species list and conservation status
V. ater S.A. Hammer [R]

Literature
HAMMER, S.A. 1994. *Vlokia*, a new genus in Aizoaceae. *Cactus and Succulent Journal (US)* 66: 255–258.

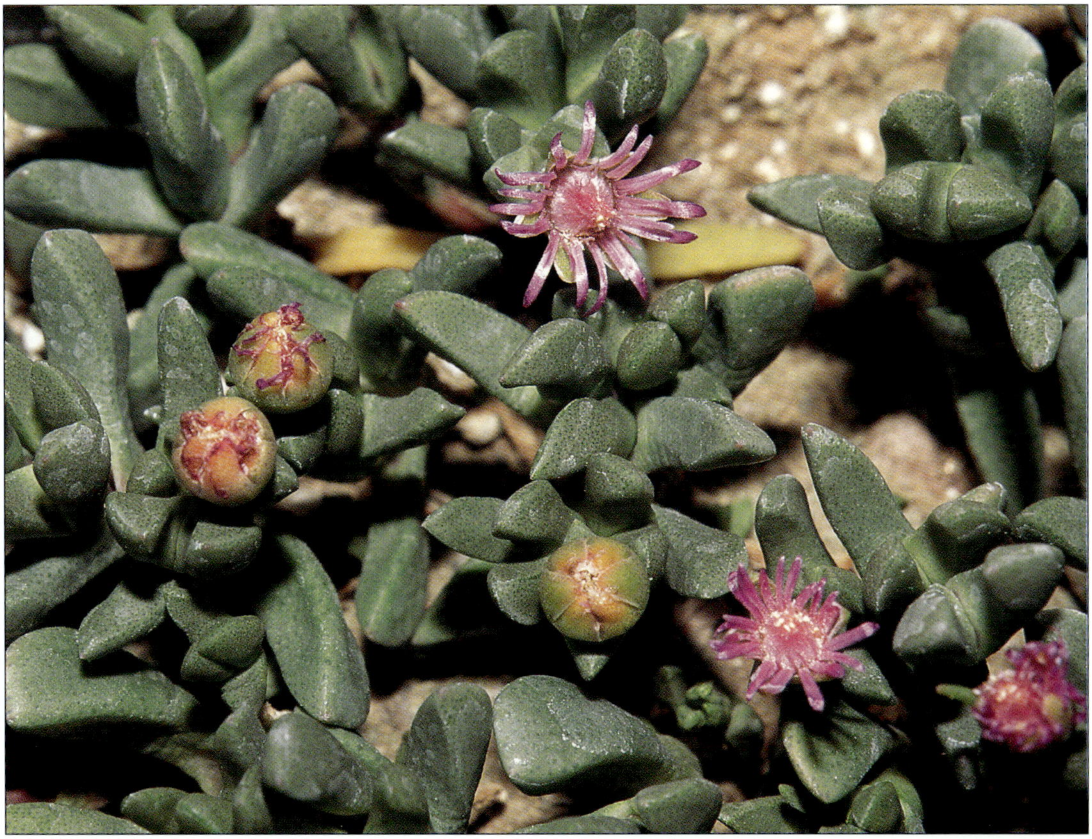
Chris Barnhill
Flowering plant of *V. ater*

Heidi Hartmann
Fruiting plant of *V. ater*

John Trager
Leaves of *V. ater*

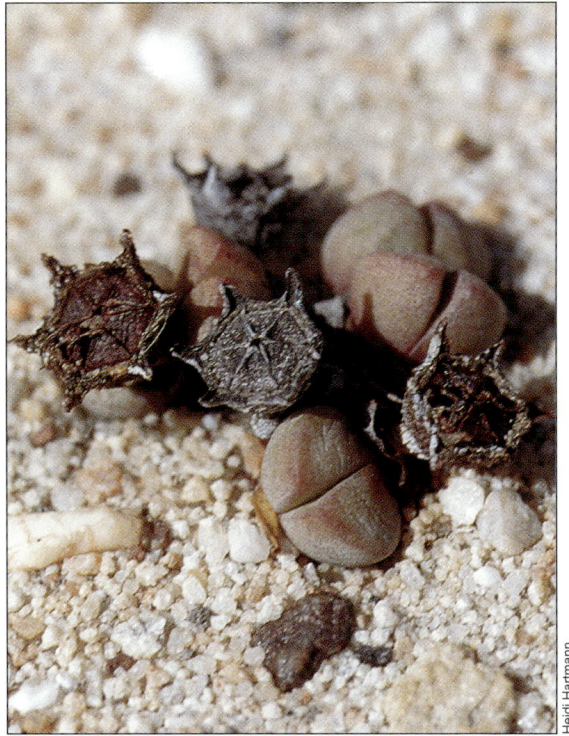

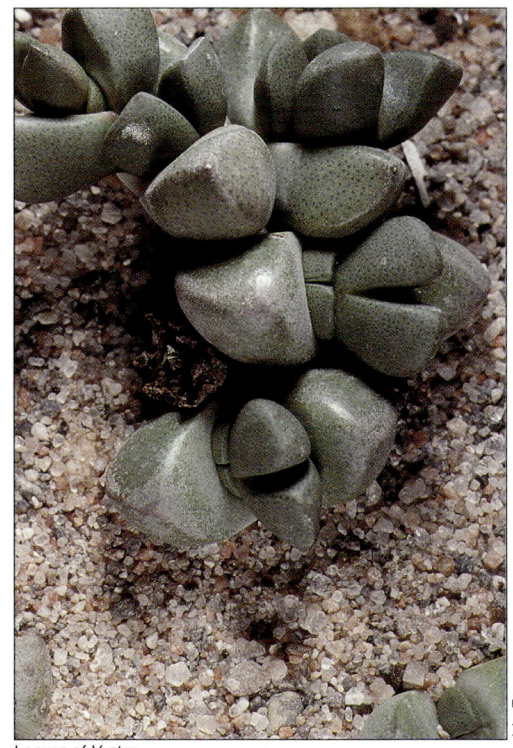

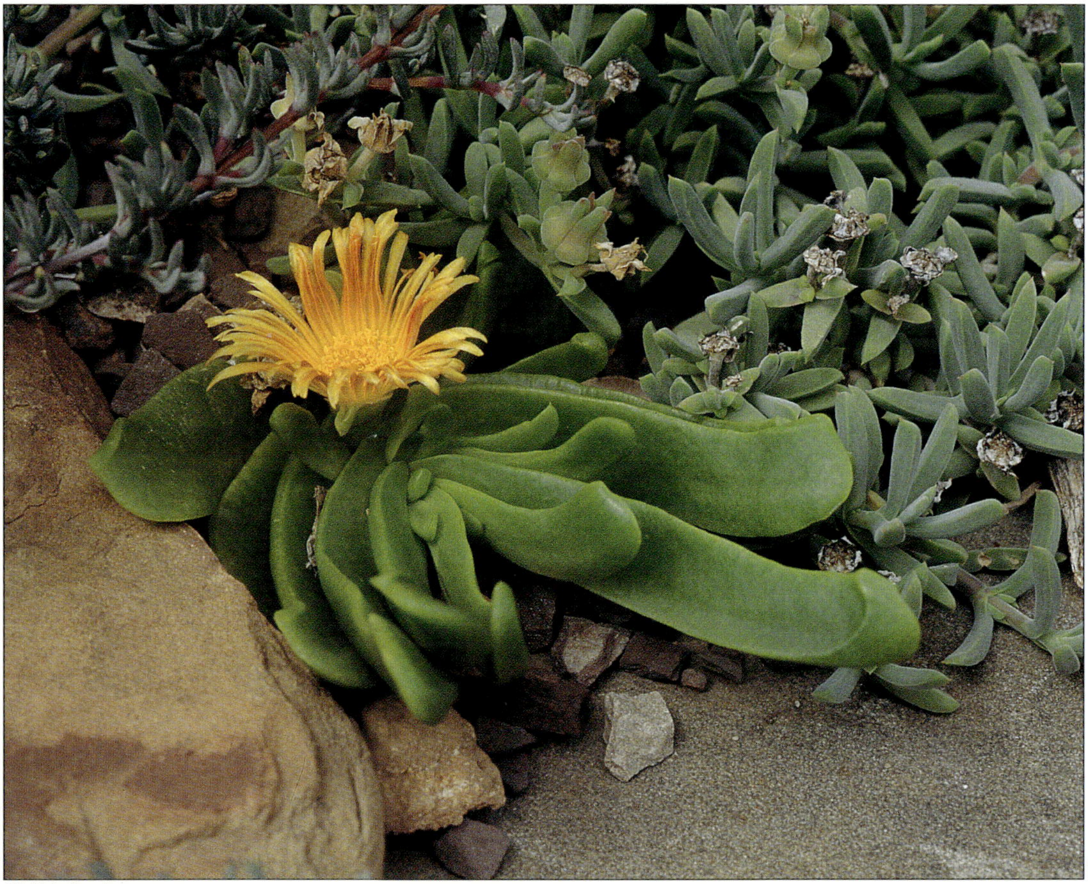

Glottiphyllum longum

NBI

Bijlia dilatata

Ernst van Jaarsveld

Glottiphyllum nelii

John Trager

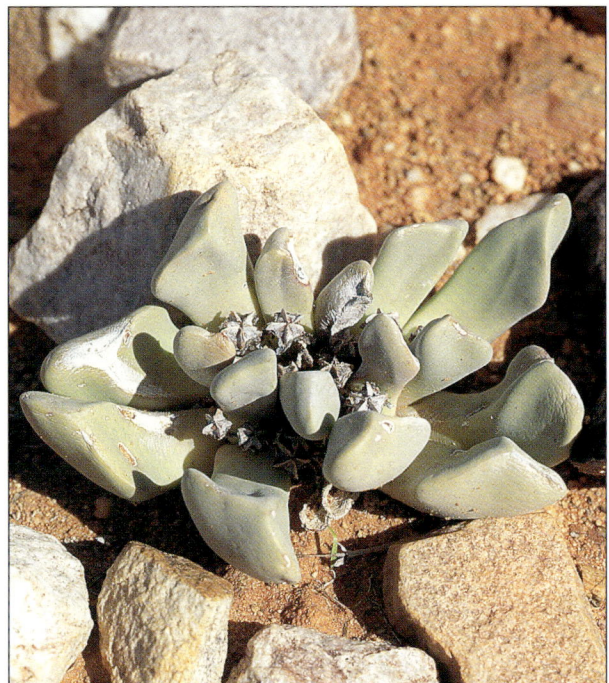

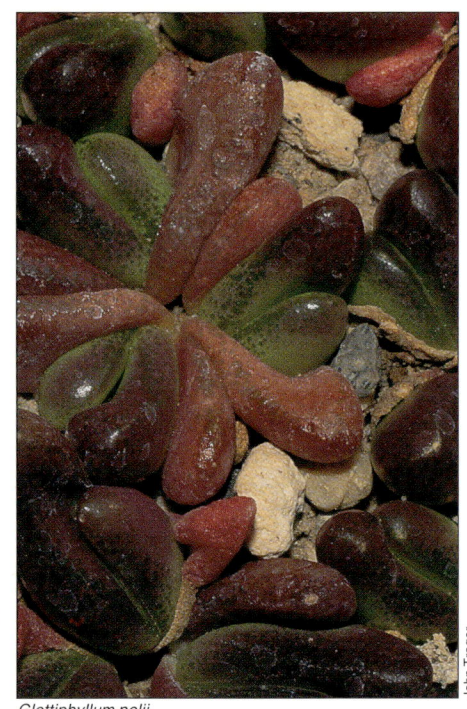

Tongue-leaved Mesembs

GROUP 4

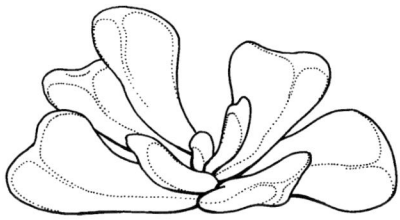

The highly succulent leaves are oblong and slightly flattened; stems usually not clearly visible; distinguished from the previous group simply by the more oblong leaves and being mostly clump-forming rather than single bodied.

The TONGUE-LEAVED MESEMBS is an artificial group of unrelated plants. They are all highly succulent and clump-forming. The flowers are yellow, white, pink or purple.

TONGUE-LEAVED MESEMBS comprise 9 genera and 48 species.

Antegibbaeum (1 species)	*Dracophilus* (4 species)	*Juttadinteria* (10 species)
Bijlia (2 species)	*Drosanthemopsis* (1 species)	*Nelia* (4 species)
Cerochlamys (1 species)	*Glottiphyllum* (17 species)	*Schwantesia* (8 species)

ANTEGIBBAEUM

DERIVATION OF GENUS NAME The name is derived from the Latin word *ante* (before) and the genus name *Gibbaeum*, implying that the plant is a primitive *Gibbaeum*, which is in fact not the case.

COMMON NAMES Common names are *volstruistone* (ostrich toes), or mimicry plants.

DESCRIPTION Plants are dwarf shrublets with highly succulent leaves borne on loose, upright or flat-growing stems of about 100 mm in length. The leaves are flat on the upper surfaces and keeled on the lower surfaces and are grey-green, turning reddish in the winter months. Leaf surfaces are smooth and the tips are blunt. The purple or deep pink flowers are large (about 60 mm in diameter) and are borne on short stalks with two pairs of bracts below each flower. Six unequal, keeled sepals with membranous margins are present, of which two are longer than the others. There are no filamentous staminodes in the flowers. In addition, it has an uninterrupted ring of glands which separates it from *Gibbaeum*. The fruit capsules have six or seven locules and have a conical shape. They display broad valve wings upon opening. Covering membranes are present and closing bodies are absent. The seeds are brown, with rough surfaces.

DISTINGUISHING CHARACTERS The flowers of *Antegibbaeum* are huge, with crinkled petals. The presence of fused nectar glands and the two pairs of bracts below the flowers distinguish this genus from *Gibbaeum*. The latter has filamentous staminodes and separate nectar glands.

FLOWERING TIME Flowers appear in late winter or early spring (August to September in South Africa) and are open from midday until evening.

DISTRIBUTION AND ECOLOGY *Antegibbaeum* is restricted to the semi-arid Little Karoo in the Western Cape Province of South Africa. Winter rainfall predominates in this region and is less than 300 mm per year. This genus grows in soils derived from shales and quartzites. These plants prefer growing in full sunlight.

CULTIVATION The plant is relatively easy to grow from cuttings or seed and readily produces its showy flowers.

NOTES *Antegibbaeum* has previously been included in *Gibbaeum*. However, despite the superficial similarities, the two genera are believed not to be closely related.

Antegibbaeum Schwantes ex C.Weber

NUMBER OF SPECIES/SUBSPECIES/VARIETIES (1/0/0)

SPECIES LIST AND CONSERVATION STATUS
 A. *fissoides* (Haw.) Schwantes ex C.Weber

LITERATURE
 WEBER, C. 1968. Notes on the nomenclature of some Aizoaceae. *Baileya* 16: 9–12.

 GLEN, H.F. 1974. *A revision of the Gibbaeinae (Mesembryanthemaceae).* Unpublished M.Sc. thesis, University of Cape Town.

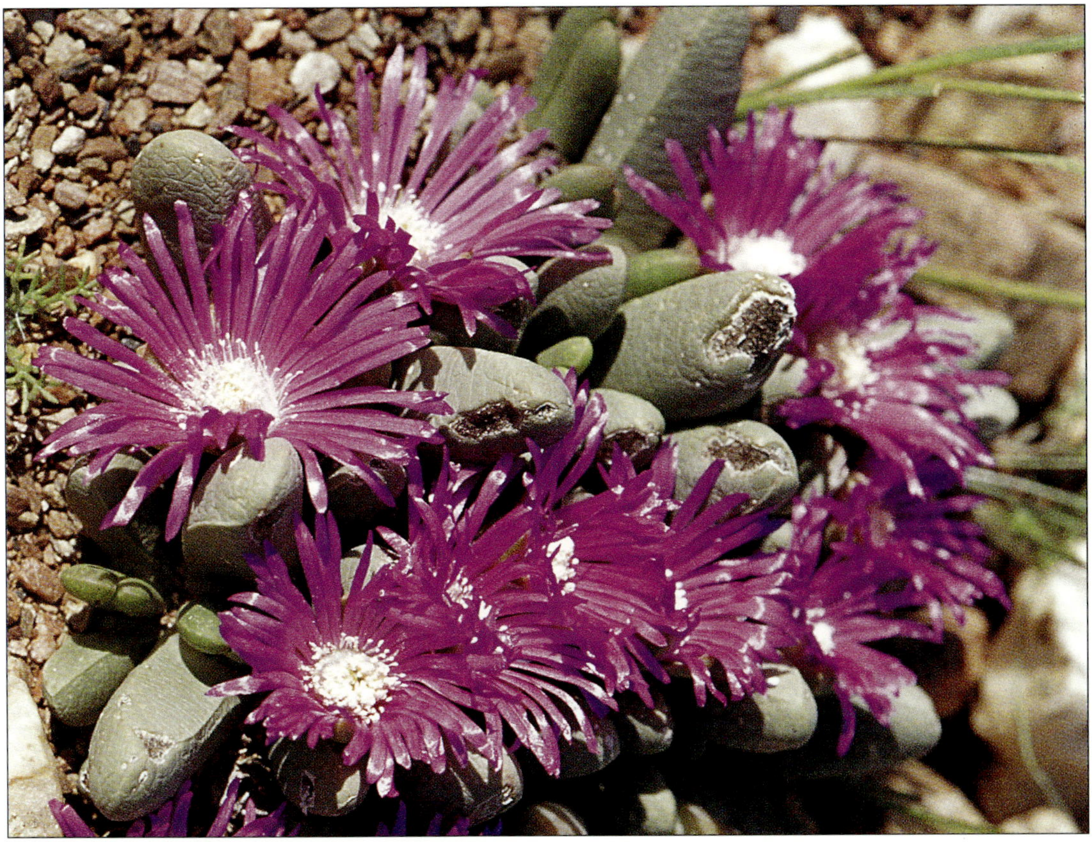
Bruce Bayer/NBI

Flowering plant of *A. fissoides*

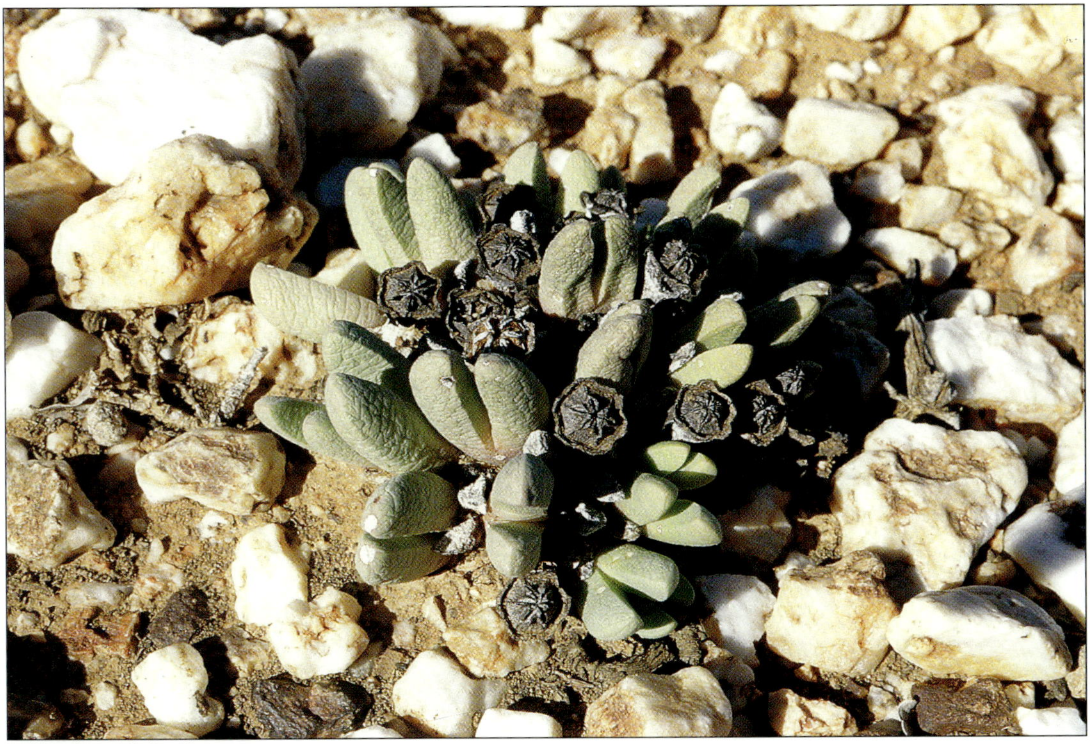
NBI

Fruiting plant of *A. fissoides*

BIJLIA

DERIVATION OF GENUS NAME The genus was named after Mrs Deborah Susanna van der Bijl, who corresponded with N.E. Brown, sending him specimens of undescribed species from her succulent collection near Prince Albert.

COMMON NAMES Prince Albert vygie.

DESCRIPTION Plants grow in small clumps which may have up to three branches. At the ends of the branches the small, rather open rosettes of diagonally arranged leaf pairs are borne close to the ground. The yellowish white leaves are shaped somewhat like the bow of a man-of-war with an off-centre keel or they are flattened and paddle-like. The terminal, rounded leaf portions are especially well developed and look like the flattened nose of a whale. Although smooth, the leaves are often indented with numerous small depressions. The bright yellow flowers are shortly stalked and are borne very close to the leaves. The fruit capsules are fairly large and tend to remain attached to the centre of the rosettes. They have five locules and covering membranes are present.

DISTINGUISHING CHARACTERS The boat-like, highly asymmetric leaves, one of each pair bearing the intimate impress of the other, are characteristic of *B. dilatata*. Both species have an unusual yellowish white leaf colour and are completely dotless. The bright yellow flowers and large, persistent fruits further characterise the genus.

FLOWERING TIME The plants flower in autumn (around April and May in South Africa) and have a relatively short flowering season. The flowers open around noon and close at night.

DISTRIBUTION AND ECOLOGY The genus *Bijlia* has a rather restricted distribution in the Western Cape Province of South Africa where it occurs near the small town of Prince Albert. This is one of the arid areas of the Little Karoo which receives precipitation mainly in early summer and autumn.

CULTIVATION The plants may be grown from seed and are often seen in succulent plant collections. They like bright light otherwise they turn lanky and lose their distinctive shape.

NOTES Although this genus is often considered to comprise a single species, *Bijlia dilatata*, a second closely related species, *B. tugwelliae*, was recently transferred to *Bijlia*. This species has thinner, more upright leaves.

Bijlia N.E.Br.

NUMBER OF SPECIES/SUBSPECIES/VARIETIES (2/0/0)

SPECIES LIST AND CONSERVATION STATUS
B. dilatata H.E.K. Hartmann
B. tugwelliae (L.Bolus) S.A. Hammer

LITERATURE
HARTMANN, H.E.K. 1992. A new name for *Bijlia cana* (Mesembryanthema). *Cactus and Succulent Journal (US)* 64: 173–180.

HAMMER, S.A. 1995. Mastering the art of growing mesembs. *Cactus and Succulent Journal (US)* 67: 195–247 (see p. 219).

B. dilatata

B. dilatata

B. tugwelliae

B. tugwelliae

Cerochlamys

Derivation of genus name The name is derived from the Greek words *ceros* (wax) and *chlamys* (mantle), in reference to the waxy leaf surface.

Common names The vernacular name is *pronkvingertjies* (showy fingers).

Description The plants slowly form small tufts. The leaf pairs are shortly united at their bases and occur in up to three pairs on each branch. The leaves are three-angled, club-shaped, have rounded keels and are twisted sideways at the tips, almost at right angles to the base. They sometimes have irregular wavy margins. A waxy layer covers the dotless purplish green to yellowish leaves which are firm in texture. The usually purple but sometimes white or rarely yellow flowers are borne singly or in small clusters of up to three. The flower stalks are very short so that the flowers hardly protrude above the small clumps of leaves. The five sepals are almost equal in length and surround the narrow petals which are arranged in up to three flat whorls. Numerous filamentous staminodes and stamens are carried to the inside of the petals in this sequence. Each flower has five nectar glands as well as five short, sharply pointed stigmas. The fruit capsules have five locules, each with broad expanding keels, but no valve wings. The valves open to a vertical position only. Seeds are somewhat egg-shaped and minutely rough.

Distinguishing characters *Cerochlamys* is characterised by its showy flowers which, unlike those of most other mesembs, appear in midwinter. They can also be distinguished from other similar-looking genera by their irregularly shaped, purplish green leaves which have an often partially peeling waxy covering.

Flowering time Flowering occurs during the autumn and winter months (May to August in South Africa). The white-flowered form opens earlier in the season. Flowers open in the late morning, stay open all day and are usually very long lasting, probably because the flowering period is in the mild winter months.

Distribution and ecology Plants grow in reasonably large populations in fully exposed positions on rocky outcrops of sandstone and conglomerate in the Little Karoo, Western Cape Province, South Africa. Rainfall is both in winter and summer and ranges from 200 to 300 mm per year.

Cultivation *Cerochlamys pachyphylla* is easily propagated from seed or by rooting cuttings obtained from dividing clumps. Plants should be sparingly watered during summer. Outside their natural habitat they should be protected from excessive rainfall. They do well in small containers on a sunny window sill.

Notes The flower colour is usually purple but some populations from around Calitzdorp have white or rarely yellow flowers and are further distinguished by their wavy leaf margins. The species can interbreed with *Bijlia*.

Cerochlamys N.E.Br.

Number of species/subspecies/varieties (1/0/0)

Species list and conservation status
C. pachyphylla (L.Bolus) L.Bolus

Literature
BROWN, N.E. 1928. *Cerochlamys*: a new genus of Mesembryanthemeae. *Journal of Botany* 66: 171.

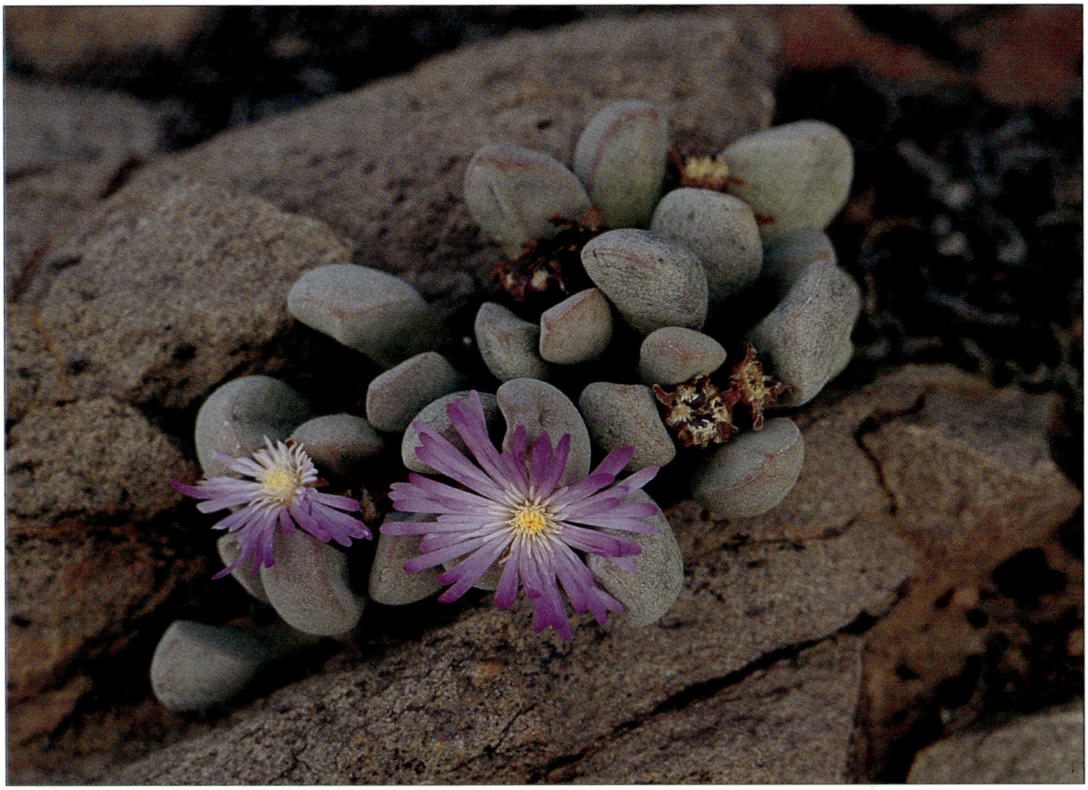
Ernst van Jaarsveld

C. pachyphylla in its habitat

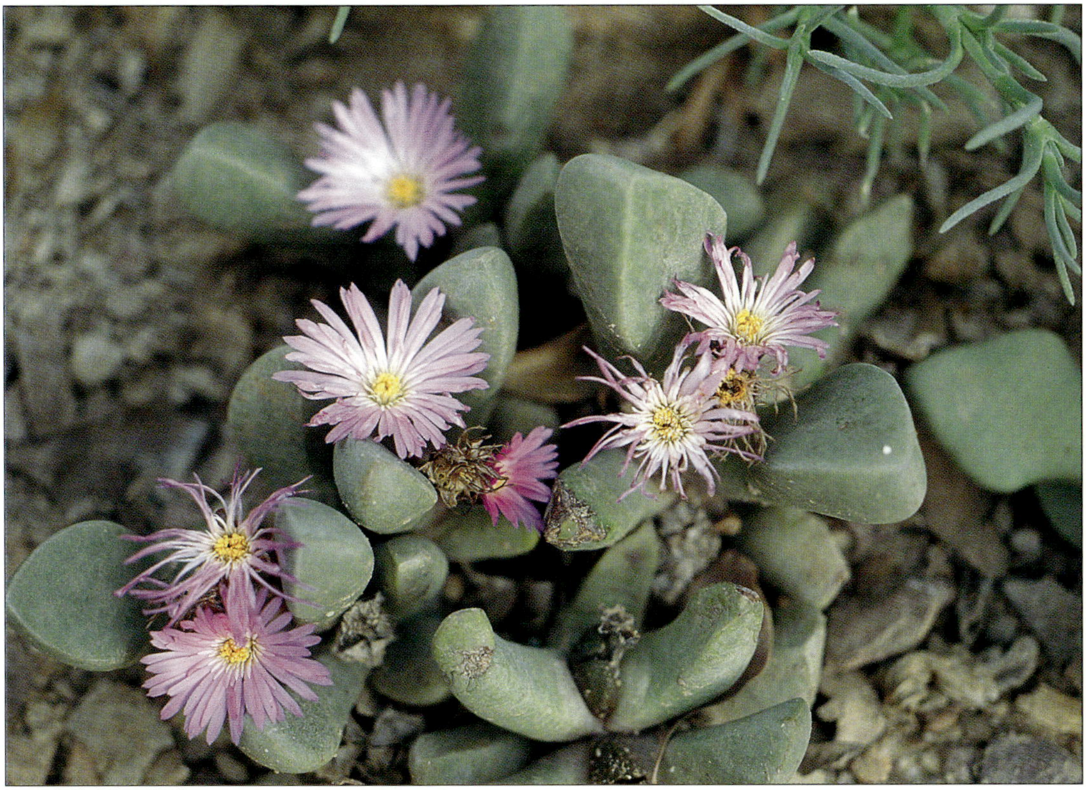
Ernst van Jaarsveld

C. pachyphylla in cultivation

DRACOPHILUS

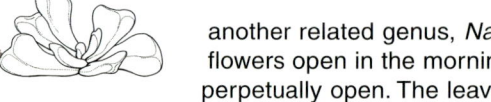

DERIVATION OF GENUS NAME Due to the presence of one species on the Drachenfels (Dragon rock), the genus was named after this locality. The genus name is derived from the Greek words *draco* (dragon) and *philos* (friend).

COMMON NAMES No vernacular names seem to have been recorded.

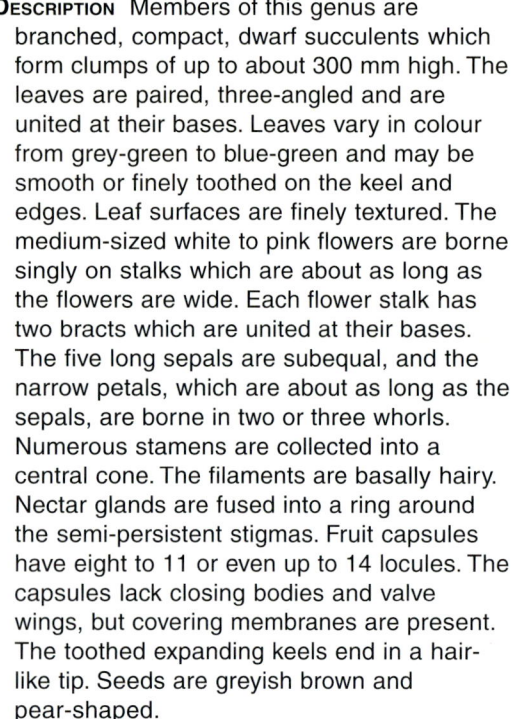

DESCRIPTION Members of this genus are branched, compact, dwarf succulents which form clumps of up to about 300 mm high. The leaves are paired, three-angled and are united at their bases. Leaves vary in colour from grey-green to blue-green and may be smooth or finely toothed on the keel and edges. Leaf surfaces are finely textured. The medium-sized white to pink flowers are borne singly on stalks which are about as long as the flowers are wide. Each flower stalk has two bracts which are united at their bases. The five long sepals are subequal, and the narrow petals, which are about as long as the sepals, are borne in two or three whorls. Numerous stamens are collected into a central cone. The filaments are basally hairy. Nectar glands are fused into a ring around the semi-persistent stigmas. Fruit capsules have eight to 11 or even up to 14 locules. The capsules lack closing bodies and valve wings, but covering membranes are present. The toothed expanding keels end in a hair-like tip. Seeds are greyish brown and pear-shaped.

DISTINGUISHING CHARACTERS This genus is closely related to *Juttadinteria* but can be distinguished from it by its five sepals (*Juttadinteria* has four); the bracts on the flower stalks (*Juttadinteria* has no bracts); and the presence of covering membranes on the fruit capsules (which are poorly developed in *Juttadinteria*). In *Dracophilus* growth is usually (but not always) stemless, with flowers opening in the evening (in *Juttadinteria* flowers open in the morning). In another related genus, *Namibia*, flowers open in the morning or are perpetually open. The leaves of all three genera smell like rhubarb and all three are difficult to propagate from cuttings.

FLOWERING TIME The species flower practically all year round. The honey-scented flowers open in the late afternoon and stay open well into the night.

DISTRIBUTION AND ECOLOGY This genus is found in very arid areas from Lüderitz in southwestern Namibia, southwards to the Richtersveld in South Africa. In these regions rainfall is less than 100 mm per year and falls mainly in the winter months. *Dracophilus* species grow in rock crevices or on plains in full sun.

CULTIVATION The plants need ample sunlight and warmth to flourish. Seeds germinate readily.

NOTES *Dracophilus* plants are not very popular but are nevertheless often seen in collections. When well grown, they can be very beautiful. In theory, the species can be distinguished by the prominence or absence of teeth on the leaf margins and other subtle differences in habit and leaf shape; in practise, they seem to merge.

Dracophilus (Schwantes) Dinter & Schwantes

NUMBER OF SPECIES/SUBSPECIES/VARIETIES (4/0/0)

SPECIES LIST AND CONSERVATION STATUS
D. *dealbatus* (N.E.Br.) Walgate
D. *delaetianus* (Dinter) Dinter & Schwantes
D. *montis-draconis* (Dinter) Dinter & Schwantes
D. *proximus* (L.Bolus) Walgate

LITERATURE
WALGATE, M. 1939. A revision of *Juttadinteria* Schwant., *Dracophilus* Dint. et Schwant., and *Namibia* Dint. et Schwant. *Notes on* Mesembryanthemum *and allied genera*. University of Cape Town, Cape Town. Vol. 3: 171–188.

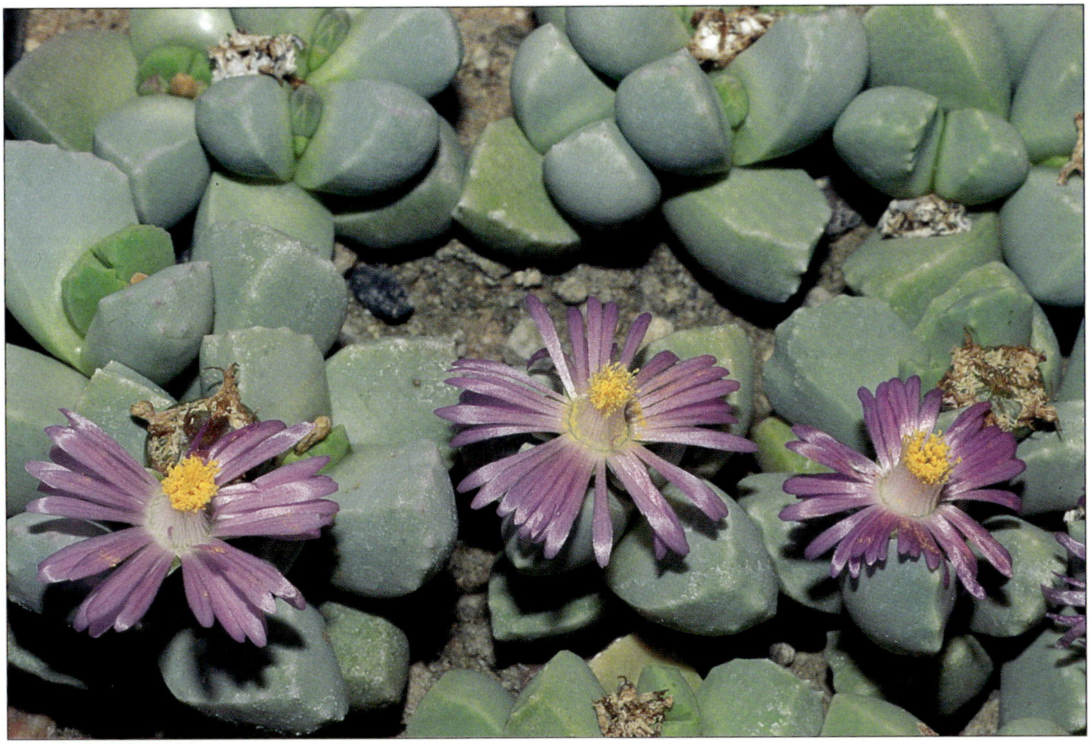
Chris Barnhill

D. montis-draconis

Bruce Bayer/NBI

D. proximus

Schleben/NBI

D. delaetianus

John Trager

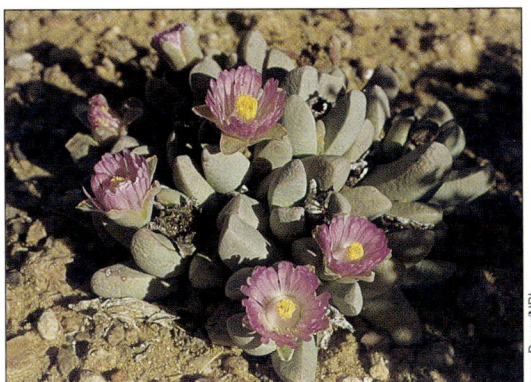

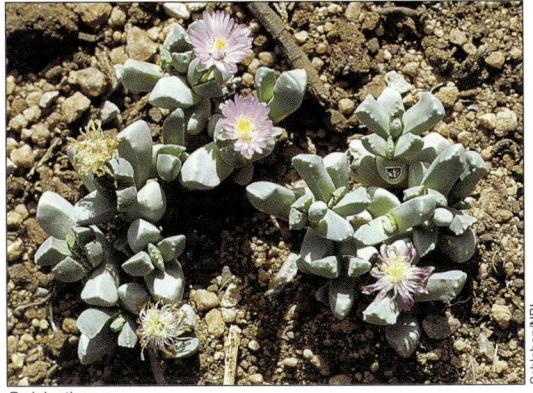

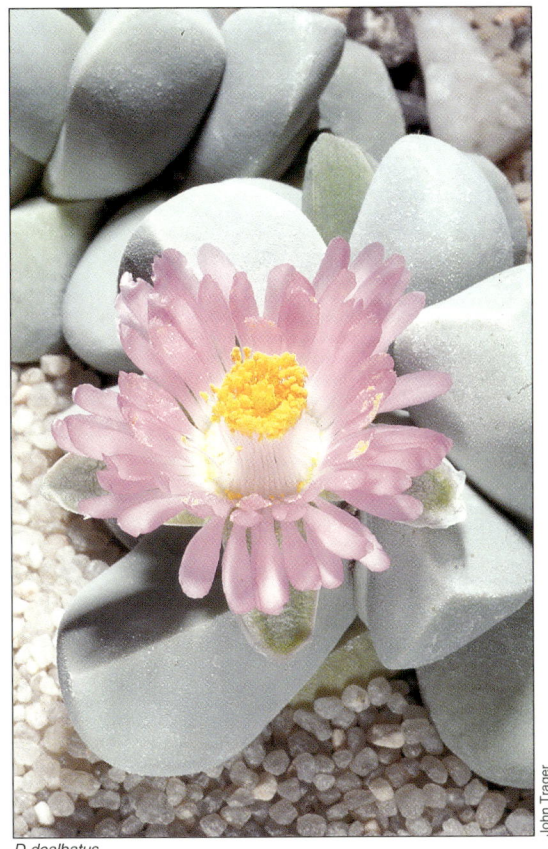

D. dealbatus

Drosanthemopsis

Derivation of genus name The genus name means "with the appearance of a *Drosanthemum*". This name was presumably chosen because the plant was originally placed in *Drosanthemum*; the actual resemblance is slight.

Common names No vernacular names seem to have been recorded.

Description The plants are succulent shrublets of up to 150 mm in height. The greyish green, more or less rounded leaves are fused at their bases and have minute branched hair-like structures on their surfaces, giving them a velvety texture. The upper sides of the leaves are flat, while the lower sides have roundish keels running to the leaf tips. Cream-coloured flowers are borne singly on stalks which have bracts. Six to eight unequal sepals surround several whorls of long narrow petals. The stamens are borne erectly in a cone-shaped structure. The nectar glands are fused into a ring around the eight to ten stigmas in the centre of the flower, which coincides with the eight to ten fruit locules. Valve wings are present. The expanding keels are irregularly toothed, ending in awns. Covering membranes are present and closing bodies are absent. The large seeds are brown in colour.

Distinguishing characters The plant is an irregularly humped succulent with strangely textured leaves which turn orange in the dry season. Flowers are cream-coloured, never purple. The strongly brackish taste of the leaves may help to identify this plant.

Flowering time Flowering occurs during spring (August to September in South Africa). Flowers open around midday and close in the evening.

Distribution and ecology The species is plentiful in stony brack areas near Riethuis in Namaqualand in the Northern Cape Province of South Africa. It grows fully exposed.

Cultivation Although *Drosanthemopsis* is rarely cultivated, it does well under greenhouse conditions provided it is given mineral-rich sandy soil. Sow seed during autumn. Growth may be relatively fast in a rockery or large container.

Notes One of the older names for this species, *Anisocalyx salarius*, refers to the sepals which are unequal in size and number (*Anisocalyx*) and also to the salty habitat and salty taste of the leaves (*salarius* means belonging to salt).

Drosanthemopsis Rauschert (=*Anisocalyx* L. Bolus)

Number of species/subspecies/varieties (1/0/0)

Species list and conservation status
D. vaginata (L.Bolus) Rauschert

Literature
BOLUS, H.M.L. 1958. *Notes on Mesembryanthemum and allied genera.* University of Cape Town, Cape Town. Vol. 3: 385–386.

RAUSCHERT, S. 1982. Nomina nova generica et combinationes novae Spermatophytorum et Pteridophytorum. *Taxon* 31: 555.

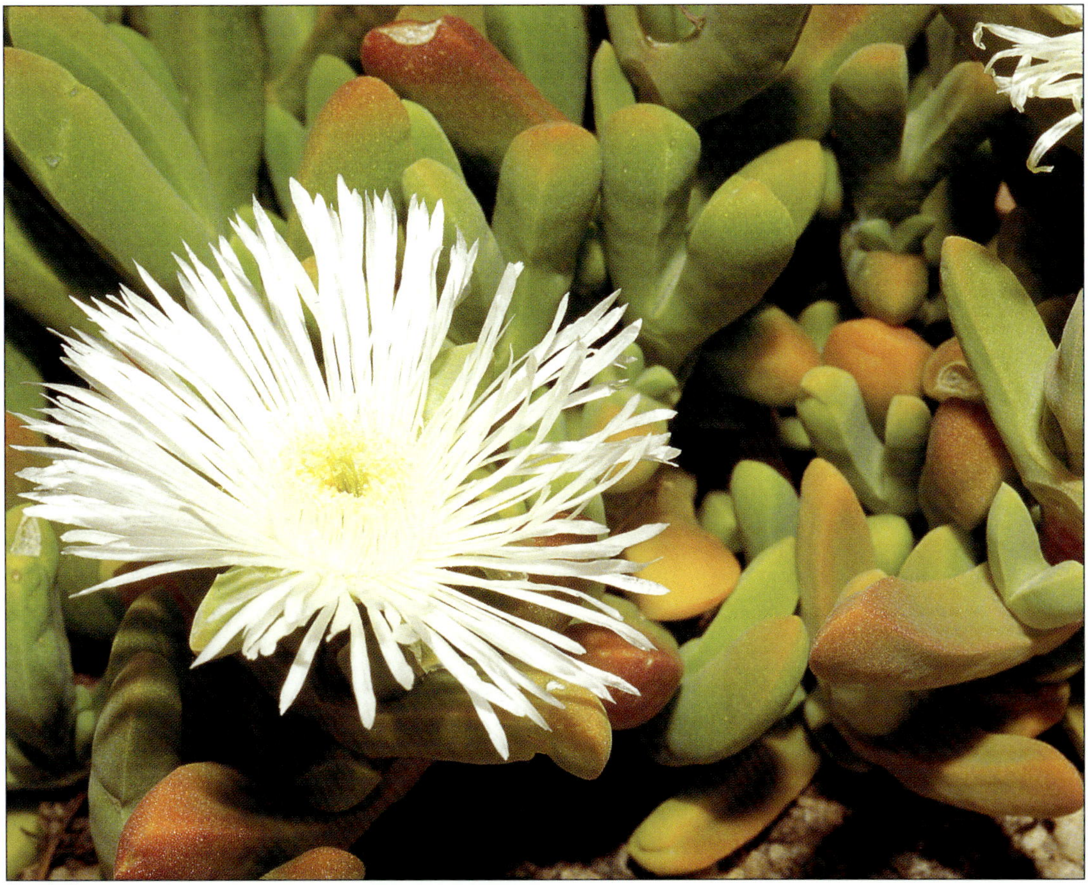
Bob Chinnock

D. vaginata

John Trager

D. vaginata

John Trager

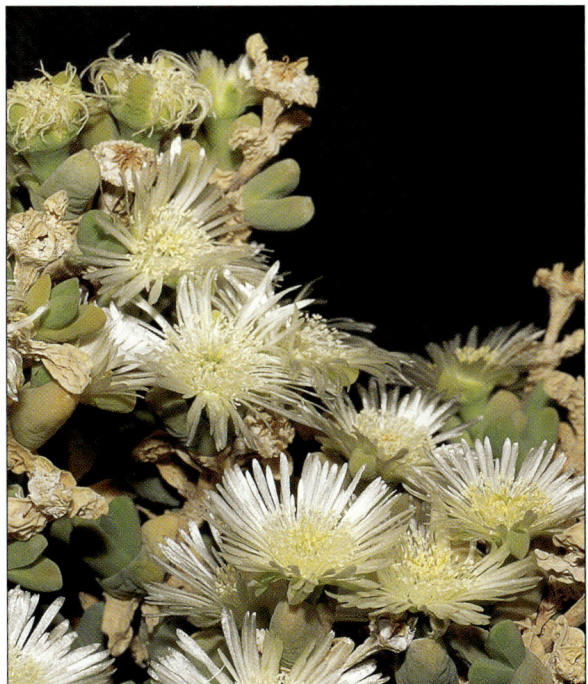

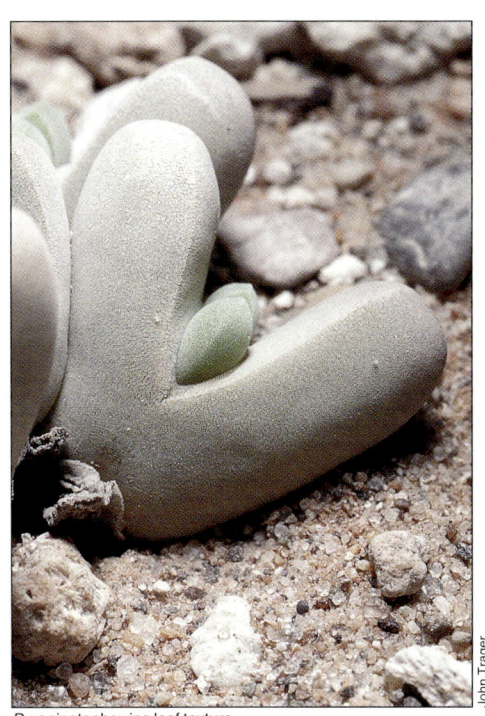

D. vaginata showing leaf texture

GLOTTIPHYLLUM

DERIVATION OF GENUS NAME The name of the genus refers to the tongue-shaped leaves, and is taken from the Greek words *glottis* (mouth of the wind-pipe, perhaps referring to the uvula) and *phyllon* (leaf).

COMMON NAMES The following common names have been recorded: *skilpadkos* (tortoise food) for *G. depressum* (=*G. latum* var. *cultratum*) and *G. longum*; *volstruiskos* (ostrich food) for *G. depressum* (=*G. muirii*); and *tongblaarvygie* (tongue-leaved mesemb) for *G. linguiforme*.

DESCRIPTION Plants are low-growing and, in time, form small clumps of rather compact rosettes. In some species branching is more extensive, plants becoming multi-headed on weak, sprawling stems. The tongue or strap-shaped leaves are borne either in a single row opposite one another with the new ones emerging from the centre, or in small erect rosettes. Leaves are mostly smooth and soft to the touch and those of many of the long-leaved species rest on the ground, which give them a rather untidy appearance. In other species the leaves are short and are borne erectly. One of the leaves in a leaf pair is often slightly larger than the other. The leaves are often bright green in colour but may also be various shades of red or purple. Some are whitish from a heavy surface layer of wax. The leaf margins of many species are minutely hairy. In some species they are windowed and in others they may have opaque white lines (see illustration of *G. peersii*). The large flowers of all species are bright yellow (very rarely white) and are borne on stalks of varying lengths. The fruits are large, pale and soft, with clearly demarcated valves and large closing bodies, easily visible in the open capsules. Seeds are relatively large and variable in texture.

DISTINGUISHING CHARACTERS The large, strap-shaped leaves that are generally soft and easily damaged, as well as the large yellow flowers with their four sepals and the soft fruits, assist in the recognition of *Glottiphyllum* species. The fruit capsules of some species become detached entirely and these tumble fruits ensure the wider dispersal of seeds.

FLOWERING TIME Species of *Glottiphyllum* usually flower in spring (September in South Africa) or autumn (March), but depending on watering in cultivation, plants can be stimulated into flowering at any time of year. Flowers open in the morning and stay open all day.

DISTRIBUTION AND ECOLOGY *Glottiphyllum* species are found in the southern parts of South Africa. They occur over much of the southern Cape interior, particularly the Little Karoo where their distribution straddles the border between the Western and Eastern Cape Provinces. Outliers are found in the Tanqua Karoo and Great Karoo in the west and north, and further east in the Eastern Cape Province. Plants tend to be smaller where they grow in exposed, often rocky places, as opposed to the more robust plants which grow in the shade of surrounding vegetation. Their natural habitats are regions where rainfall ranges from as little as 150 mm per year in the west to regions that receive as much as 500 mm in the east. Species included in *Glottiphyllum* cross the border between areas with winter and summer rainfall peaks, with most species occurring in regions with spring and autumn maximums.

CULTIVATION Species of *Glottiphyllum* are mostly very easy to cultivate. Owing to their soft leaves that can be easily damaged mechanically or by insects, the plants are often unsightly and do not make the best of horticultural subjects. Like many other mesembs, they tend to die back from the centre of a clump. This results in the rather unkempt look of plants growing in open beds. However, much can be tolerated of these plants because of their tendency to produce

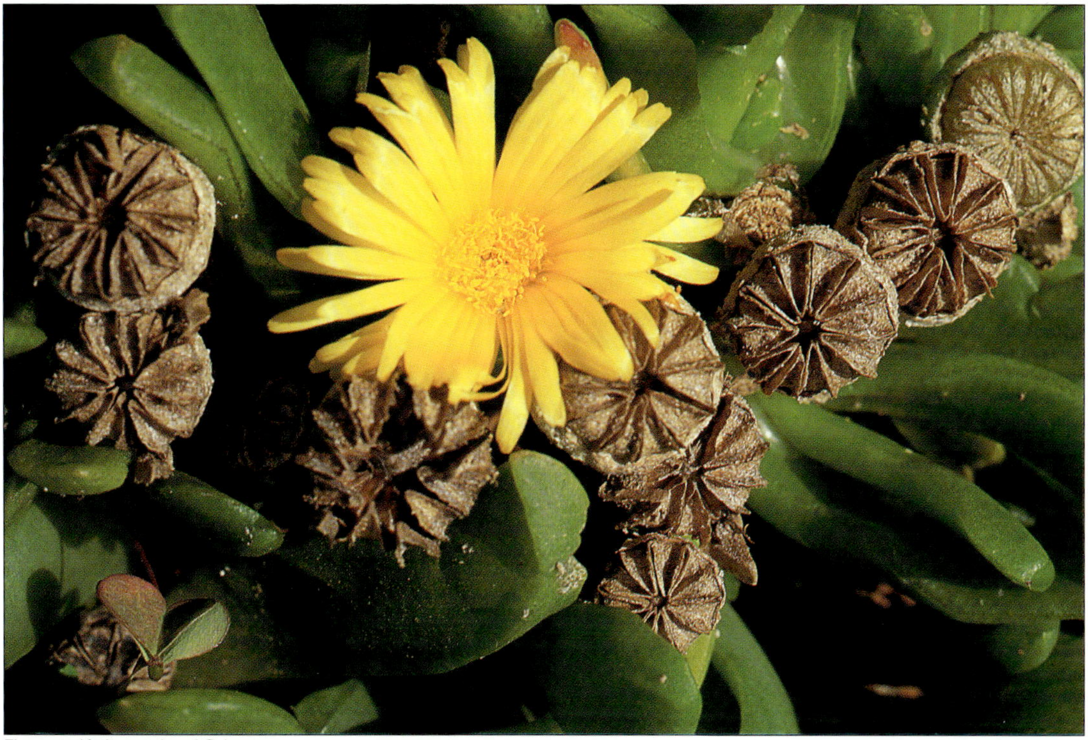
Ernst van Jaarsveld

Flower and fruit capsules of *G. longum*

Ben-Erik van Wyk

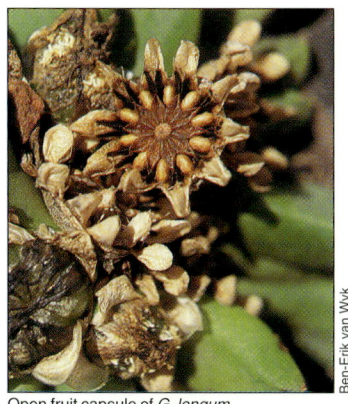

Open fruit capsule of *G. longum*

John Trager

G. linguiforme

Ernst van Jaarsveld/NBI

Glottiphyllum sp.

John Trager

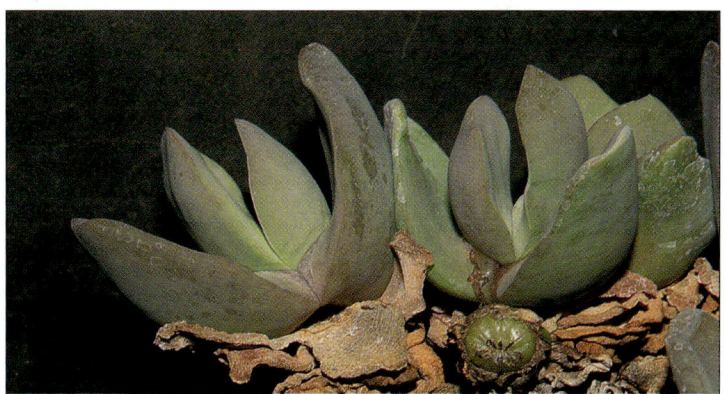

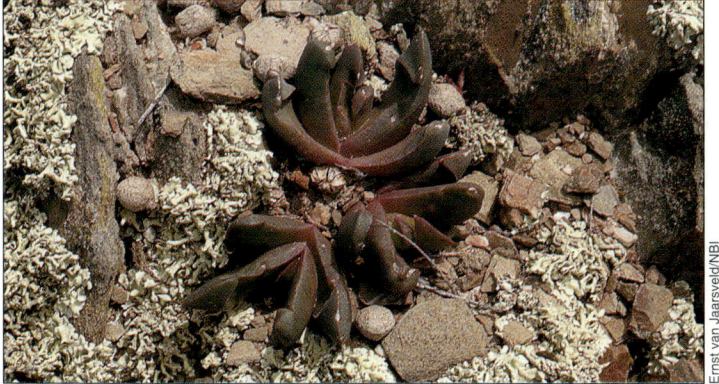

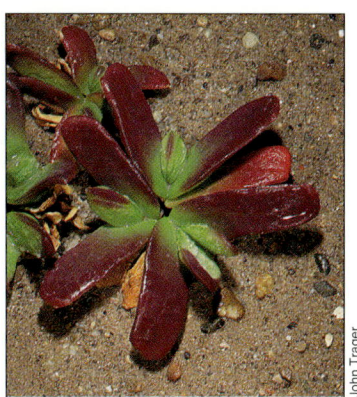

G. longum

a profusion of large yellow flowers. Seeds can be sown at any time of the year, with the quickest results obtained from summer sowing.

NOTES *Glottiphyllum* species respond to variation in light intensity by changing their leaf shape and colour enormously. Cultivated plants bear little resemblance to counterparts of the same species growing under harsher environmental conditions. As a result, no less than 50 species have been described, many of which were later found to be extreme variants of the same species. Some of these variants are illustrated here. It has been suggested that *G. ochraceum*, with its peculiar leaf shape and unusual orange flowers, is not correctly placed in *Glottiphyllum*; it may be an intergeneric hybrid. A white-flowered form of *G. grandiflorum* was recently seen near Alexandria in the Eastern Cape and other species also occasionally show white mutants.

Glottiphyllum N.E.Br.

NUMBER OF SPECIES/SUBSPECIES/VARIETIES (17/0/0)

SPECIES LIST AND CONSERVATION STATUS
G. carnosum N.E.Br.
G. cruciatum (Haw.) N.E.Br.
G. depressum (Haw.) N.E.Br. [nt]
G. difforme (L.) N.E.Br.
G. fergusoniae L.Bolus
G. grandiflorum (Haw.) N.E.Br.
G. linguiforme (L.) N.E.Br.
G. longum (Haw.) N.E.Br.
G. neilii N.E.Br.
G. nelii Schwantes
G. ochraceum (A.Berger) N.E.Br.
G. oligocarpum L.Bolus
G. peersii L.Bolus
G. regium N.E.Br.
G. salmii (Haw.) N.E.Br.
G. suave N.E.Br.
G. surrectum (Haw.) L.Bolus

LITERATURE
HARTMANN, H.E.K. & GÖLLING, H. 1993. A monograph of the genus *Glottiphyllum* (Mesembryanthema, Aizoaceae). *Bradleya* 11: 1–49.

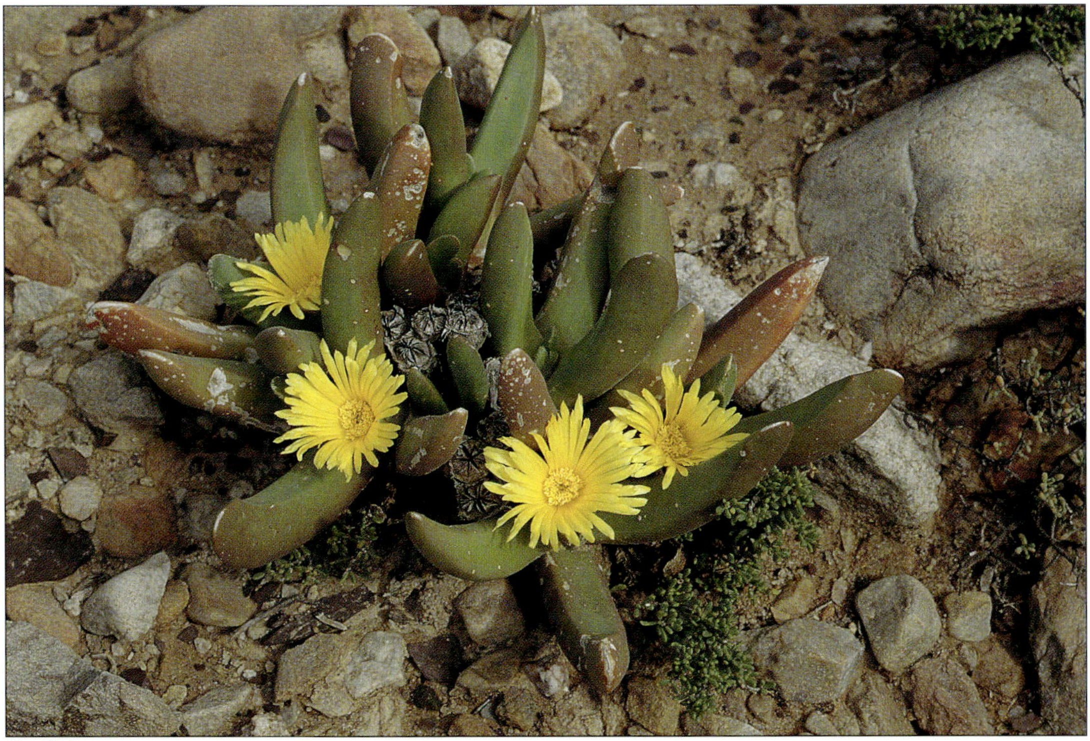

G. regium

NBI

G. peersii

John Trager

G. nelii

John Trager

G. ochraceum

Bruce Bayer/NBI

G. neilii

John Trager

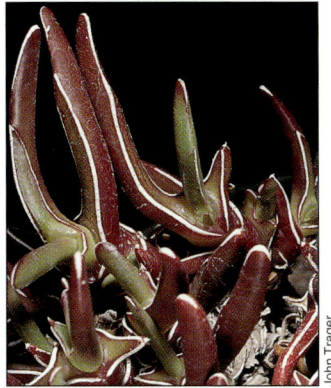

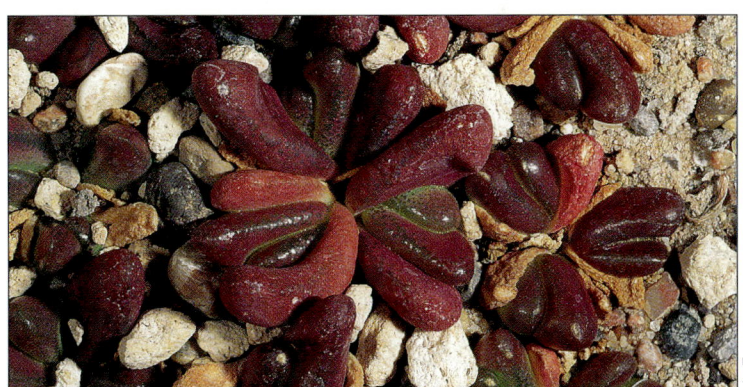

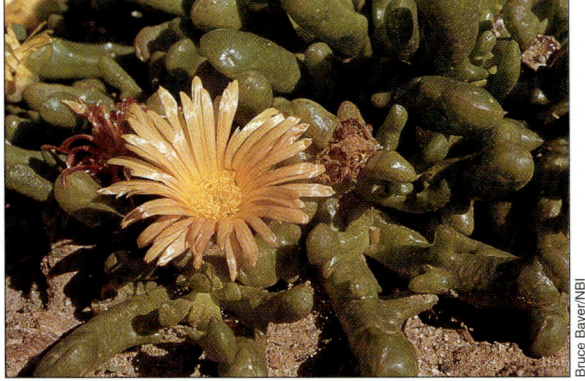

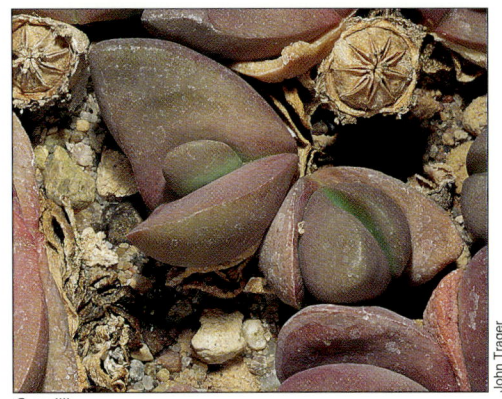

JUTTADINTERIA

DERIVATION OF GENUS NAME The genus was named in honour of Mrs Jutta Dinter, wife of Prof. M.K. Dinter.

COMMON NAMES No vernacular names are known.

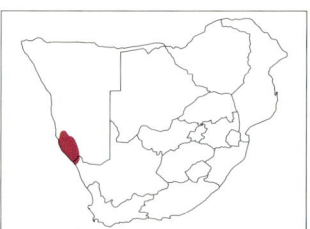

DESCRIPTION The plants are small succulents forming irregular clumps, or they are more upright and shrubby. Leaves are pale grey-green or blue-green, mostly short and fleshy, fused, almost cylindrical, broadly boat-shaped or triangular, sometimes with reddish or brownish teeth on the margins and keels. Leaf surfaces are minutely textured and are somewhat waxy. Flowers are mostly white, solitary, terminal, shortly stalked, about 50 mm in diameter and always scented. There are four sepals with conspicuous non-succulent edges and one to three whorls of petals. The numerous stamens are grouped into a cone. A ring of nectar glands surrounds the usually eight but up to 11 awl-shaped stigmas. The fruit capsules are mostly eight-locular (up to 11), with valve-wings and expanding keels diverging only near their tips. Covering membranes are rarely well developed and mostly reduced to a limb. There are no closing bodies. Seeds are spherical or pear-shaped, and wrinkled or warty.

DISTINGUISHING CHARACTERS Plants of the genus *Juttadinteria* smell like rhubarb. They also have the following combination of features: erect or somewhat spreading branches; leaves covered with minute flattish or rounded white tubercles. Flowers are white, diurnal, have no bracts, four sepals, one to three whorls of reflexed petals and usually eight stigmas. Fruit capsules have rudimentary covering membranes.

FLOWERING TIME Plants flower in winter (July to August in southern Africa). The strongly and peculiarly perfumed flowers open in the early afternoon and close at night.

DISTRIBUTION AND ECOLOGY In Namibia the genus ranges from Lüderitz in the northwest to Aus in the northeast; it is common in the Sperrgebiet. In South Africa it is confined to the northern Richtersveld. The plants are common and grow on rocky or sandy plains.

CULTIVATION Growth occurs mainly in winter. Plants grow well in sandy loam in a bright and warm greenhouse. Do not allow them to freeze. Propagation is by seed, which germinates readily.

NOTES There are three main groups of species. The first, comprising *Juttadinteria ausensis*, *J. elizae*, *J. kovisimontana*, *J. suavissima* and *J. simpsonii*, has fierce teeth along the leaf margins; they are difficult to distinguish from one another. The second group, comprising *J. attenuata*, *J. deserticola*, *J. insolita* and *J. tetrasepala*, has smooth rounded leaves and an often flat and spreading growth form. The third group, comprising only *J. albata*, has a strictly upright habit and large paddle-shaped leaves which are usually toothless but which in old age can develop teeth.

Juttadinteria Schwantes

NUMBER OF SPECIES/SUBSPECIES/VARIETIES
(10/0/0)

SPECIES LIST AND CONSERVATION STATUS
J. albata L.Bolus [nt]
J. attenuata Walgate
J. ausensis (L.Bolus) Schwantes
J. deserticola (Marloth) Schwantes
J. elizae (Dinter & A.Berger) L.Bolus
J. insolita (L.Bolus) L.Bolus
J. kovisimontana (Dinter) Schwantes
J. suavissima (Dinter) Schwantes
J. simpsonii (Dinter) Schwantes
J. tetrasepala L.Bolus [nt]

LITERATURE
WALGATE, M. 1939. A revision of *Juttadinteria* Schwant., *Dracophilus* Dint. et Schwant., and *Namibia* Dint. et Schwant. *Notes on* Mesembryanthemum *and allied genera*. University of Cape Town, Cape Town. Vol. 3: 171–188.

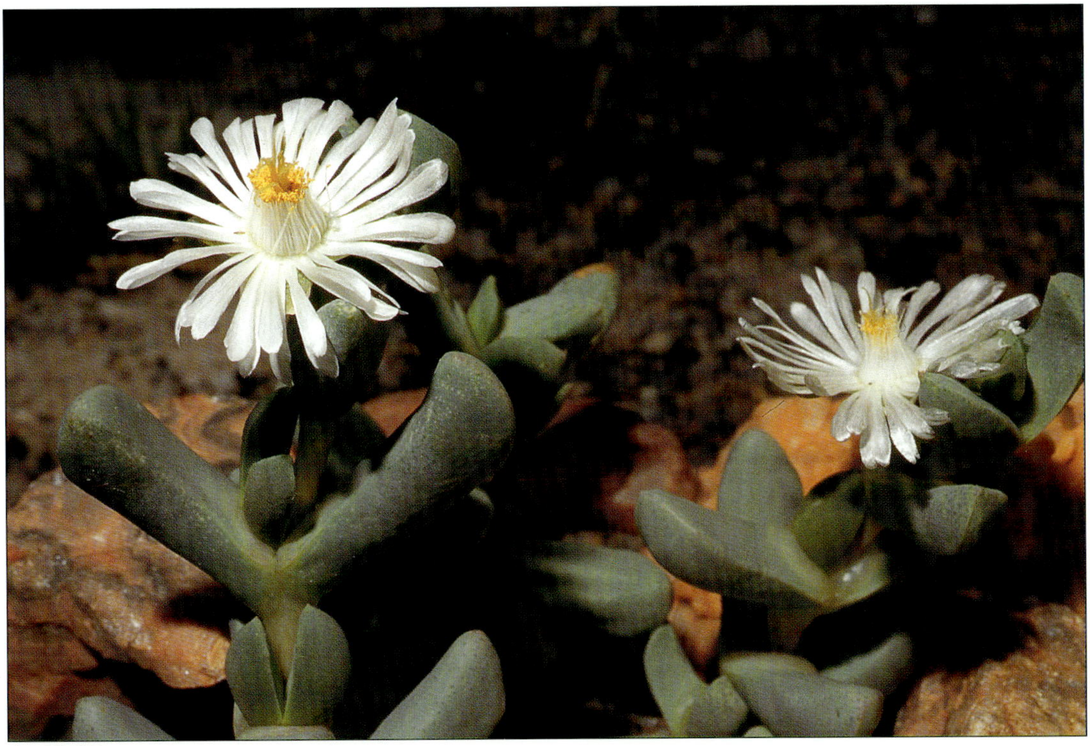

Bob Chinnock

J. suavissima

Graham Williamson

J. deserticola

Ben-Erik van Wyk

J. ausensis

Bob Chinnock

J. simpsonii

Ben-Erik van Wyk

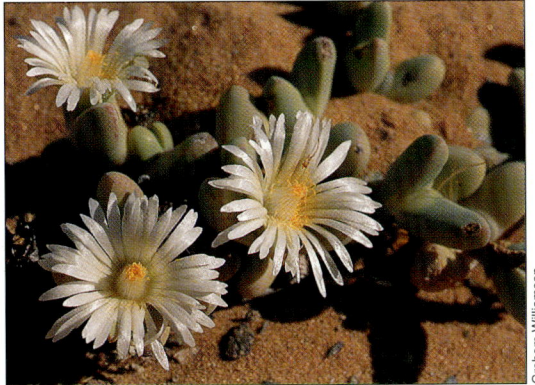

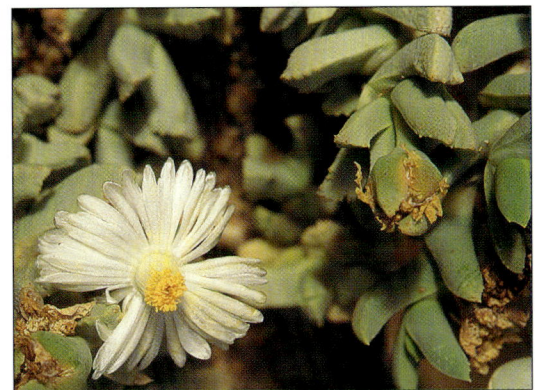

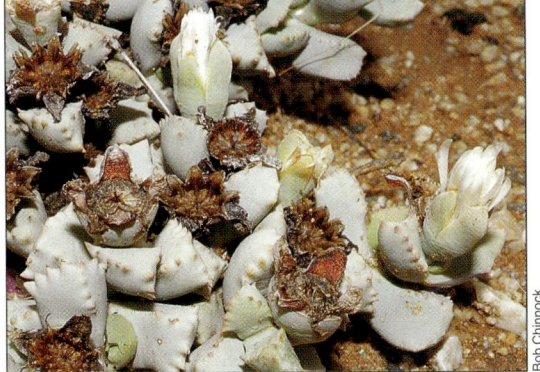

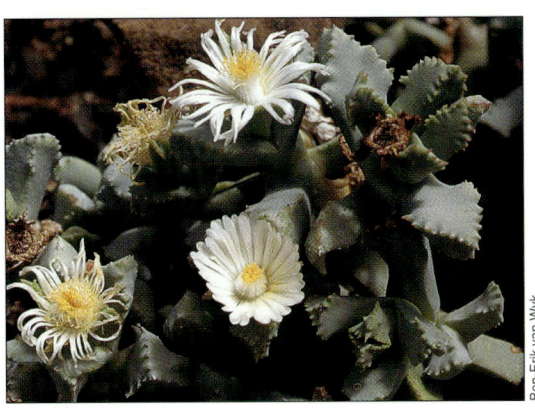

J. kovisimontana

Nelia

Derivation of genus name Plants of this genus were discovered by Rev. G. Meyer and named after the late Prof. Dr Gert Cornelius Nel of Stellenbosch University, South Africa, who published books on *Lithops* and *Gibbaeum*.

Common names The name soap-leaved mesemb has been used for this genus.

Description Plants are dwarf, tufted succulents with greyish to light green leaves. The fat leaves are joined at the base, three-angled, elongated with rounded tips and a smooth surface. Flowers are ivory white or rarely pale pink and occur singly or in clusters of three on a short stalk with two bracts. They are about 20 mm in diameter and remain open continuously for about three weeks. The five sepals are unequal and slightly fused above the ovary. Petals are sparse, often scarcely longer than the sepals, linear to spoon-shaped and notched at the tips, with short hairs at their bases, passing into white filamentous staminodes surrounding the five short stigmas. The fruit capsule is five-locular, with covering membranes present as rather broad rectangular rims of the dividing walls (septa). Expanding keels are stout, yellow, contiguous, with broad valve wings and no closing bodies. Seeds are about 1 mm in length, pear to heart-shaped, light yellow and minutely textured.

Distinguishing characters The plants are compact, clump-forming, with thick, smooth, pale bluish green, elongated or short, very firm leaves. The petals are white or pale yellow and remarkably stiff. Very short stigmas are hidden at the base of a short tube formed by the calyx above the ovary, which is conspicuously nectar-laden. Visible nectar is rare in mesembs.

Flowering time Plants flower sporadically over a long period during the cool winter months (May to August in habitat in South Africa). Flowers open for the first time in the morning and never close again.

Distribution and ecology The genus is found in the Richtersveld in the northwestern part of the Northern Cape Province, South Africa, and extends southwards into Namaqualand. It grows on quartz fields near the coast and very often the plants are covered with a black lichen.

Cultivation *Nelia* species are most active in winter and are best grown under greenhouse conditions. They should be watered only sparingly during summer. Seed should be sown in summer or autumn and the seedlings transplanted into a sandy, mineral-rich soil because humus-rich soils tend to make the plants bloat and rupture.

Notes The number of species in this genus is possibly only two, namely *Nelia schlechteri*, with its extremely dwarf leaves and serrate margins, and *N. pillansii*, with its long, straight-margined leaves. *N. robusta*, with its very large, broad leaves appears to be merely a robust form of *N. pillansii*. *N. meyeri* probably also belongs to *N. pillansii*. The var. *longipetala*, which differed from the others by its pale pink petals, has not been found again.

Nelia Schwantes (=*Sterropetalum* N.E.Br.)

Number of species/subspecies/varieties (4/0/1)

Species list and conservation status
 N. meyeri Schwantes var. *longipetala* L.Bolus
 N. meyeri Schwantes var. *meyeri*
 N. pillansii (N.E.Br.) Schwantes [I]
 N. robusta Schwantes
 N. schlechteri Schwantes [R]

Literature
 SCHWANTES, G. 1928. *Nelia* Schwantes gen. nov. *Möllers Deutsche Gärtner-Zeitung* 43: 92.

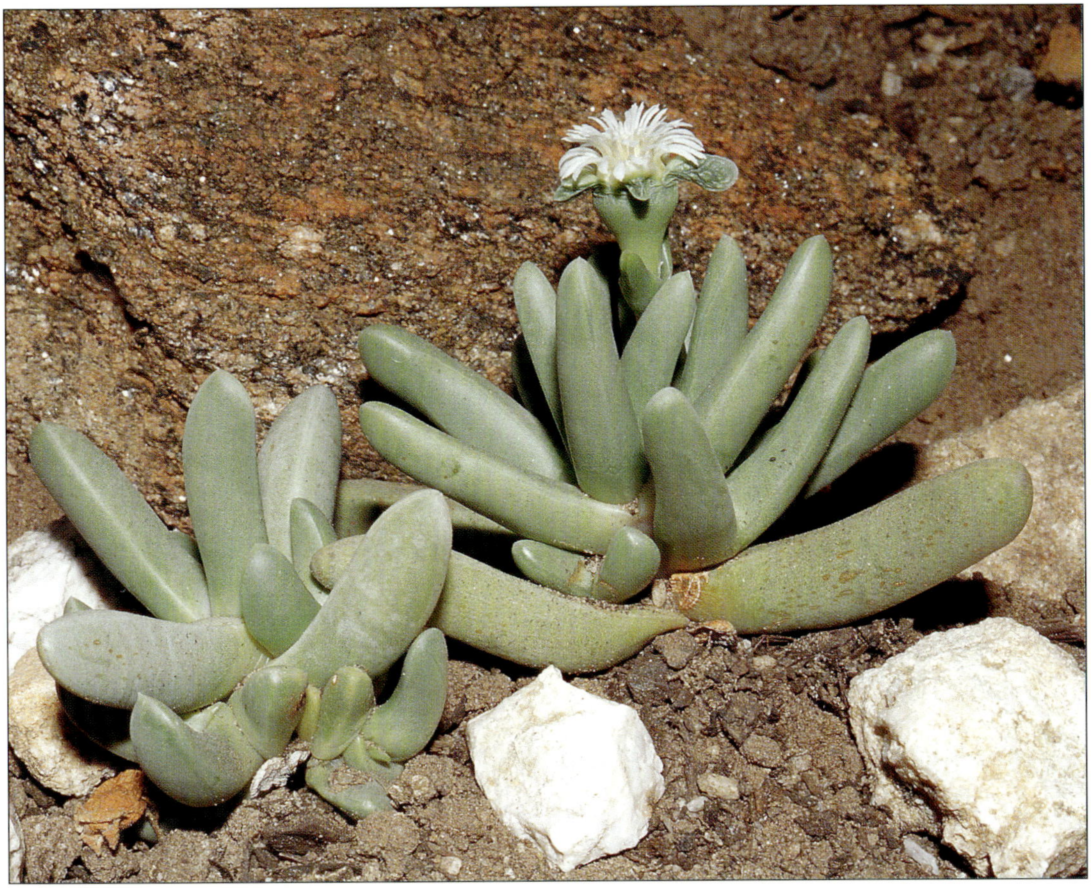
Bob Chinnock

N. pillansii

John Trager

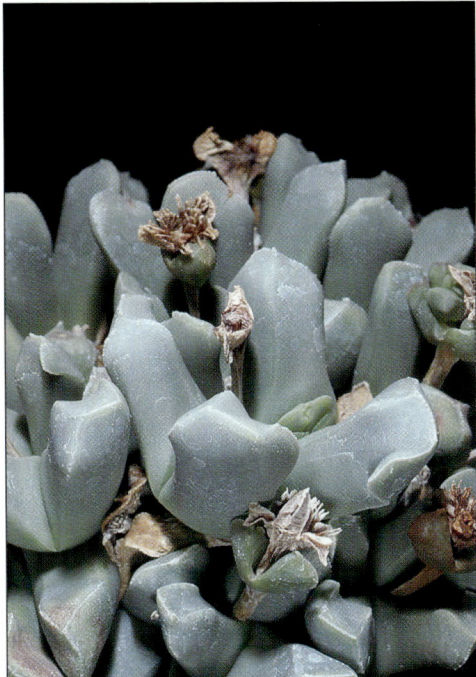

Leaves of *N. schlechteri*

Graham Williamson

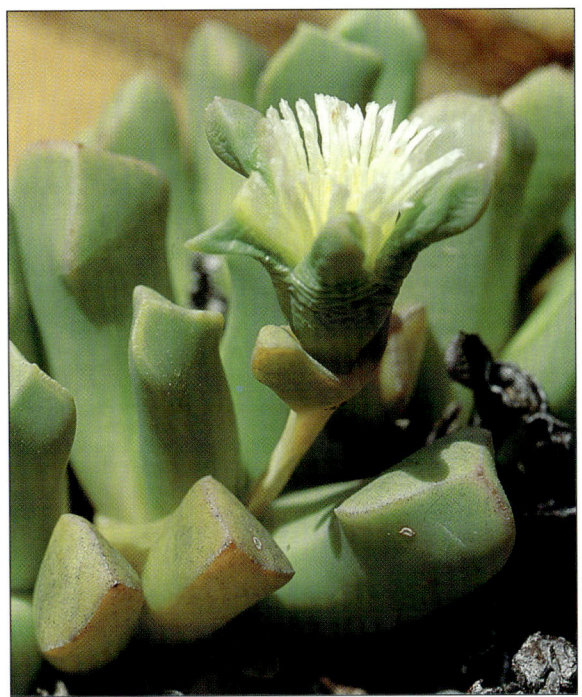

Flower of *N. schlechteri*

Schwantesia

Derivation of genus name The genus was named after Dr Martin Heinrich Gustav (Georg) Schwantes (1881 to 1960), professor of Prehistory at Kiel and student of mesembs.

Common names No vernacular names seem to have been recorded.

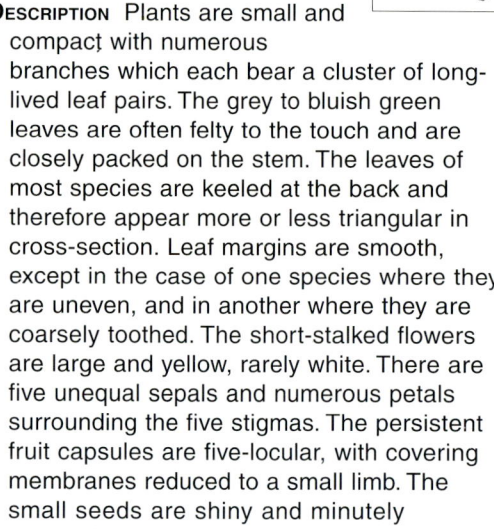

Description Plants are small and compact with numerous branches which each bear a cluster of long-lived leaf pairs. The grey to bluish green leaves are often felty to the touch and are closely packed on the stem. The leaves of most species are keeled at the back and therefore appear more or less triangular in cross-section. Leaf margins are smooth, except in the case of one species where they are uneven, and in another where they are coarsely toothed. The short-stalked flowers are large and yellow, rarely white. There are five unequal sepals and numerous petals surrounding the five stigmas. The persistent fruit capsules are five-locular, with covering membranes reduced to a small limb. The small seeds are shiny and minutely textured.

Distinguishing characters The leaves are asymmetric, more or less erect, sometimes toothed and whitish. As in *Lapidaria*, to which this genus is closely related, the leaves enclose the stem internodes, obscuring the stem from view.

Flowering time Species tend to flower in autumn (March to May in Namibia and South Africa). Flowers open around mid-afternoon, and stay open until nightfall.

Distribution and ecology *Schwantesia* occurs in the Northern Cape Province of South Africa, from Bushmanland in the east, through the Richtersveld, to southern Namibia. The species are peculiar in that they cross the border between winter and summer rainfall. Plants generally grow in cracks in rocks, on stony outcrops or shady, south-facing cliffs.

Cultivation *Schwantesia* species occur in extremely arid areas and care should be taken not to over-water plants in cultivation. Sow seeds during summer or autumn. Plants are best grown in a greenhouse in sandy gravel.

Notes *Schwantesia herrei* is widespread in habitat and in cultivation, and can generally be recognised by its smooth, broad leaves. Another commonly cultivated species, *S. pillansii*, has narrow, fuzzy leaves and is better known (at least in horticulture) as *S. triebneri*. *S. borcherdsii*, the easternmost species, is found at Upington and can be recognised by its wavy-margined and beautifully marbled leaves. Only one species, *S. ruedebuschii*, has leaf tips with small, brown teeth and the genus was therefore not grouped with the TOOTH-LEAVED MESEMBS.

Schwantesia Dinter

Number of species/subspecies/varieties (8/0/1)

Species list and conservation status
S. *acutipetala* L.Bolus [R]
S. *borcherdsii* L.Bolus [V]
S. *constanceae* N.Zimm. [R]
S. *herrei* L.Bolus var. *herrei*
S. *herrei* L.Bolus var. *minor* L.Bolus
S. *pillansii* L.Bolus [nt]
S. *ruedebuschii* Dinter [nt]
S. *speciosa* L.Bolus [nt]
S. *succumbens* (Dinter) Dinter [R]

Literature
ZIMMERMANN, N.F.A. 1996. The genus *Schwantesia* Dinter. *Cactus and Succulent Journal (US)* 68: 257–266.

ZIMMERMANN, N.F.A. 1996. *Schwantesia constanceae* N.Zimmermann spec. nov. — eine neue Art aus dem südlichen Namibia. *Kakteen und andere Sukkulenten* 47: 81–86.

S. borcherdsii

S. herrei var. herrei

S. pillansii

S. ruedebuschii

Aloinopsis setifera

Aloinopsis malherbei

Neohenricia sibbettii

Titanopsis primosii

Rough-leaved Mesembs

GROUP 5

The succulent leaves usually have rough surfaces and are often arranged in rosettes; stems usually not clearly visible; flowers have various shades of yellow, occasionally pink, often with a darker line along the middle of the petals; stamens are usually grouped together into a cone-like structure in the centre of the flower.

Perhaps the best example of ROUGH-LEAVED MESEMBS is *Titanopsis calcarea*, where the leaves are highly sculptured with verrucose white structures, making the plants almost undetectable amongst the white calcareous rocks of its natural habitat. In contrast, some species have leaves which are only slightly roughened by minute bumps as in *Rhinephyllum*. The leaves are often arranged in rosettes and may be spoon-shaped or pointed at the tips. The flowers in this group are relatively uniform in colour and usually open in the afternoon, evening or at night. The ROUGH-LEAVED MESEMBS is a group of more or less related dwarf mesembs, several of which occur in the summer rainfall areas of South Africa.

ROUGH-LEAVED MESEMBS comprise 6 genera and 45 species.

Aloinopsis (14 species)	*Neohenricia* (2 species)
Deilanthe (2 species)	*Rhinephyllum* (12 species)
Nananthus (10 species)	*Titanopsis* (5 species)

ALOINOPSIS

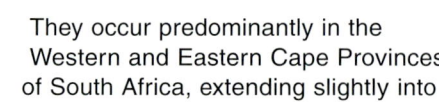

DERIVATION OF GENUS NAME The genus name means "with the appearance of an *Aloe*", an allusion to the similarity of some species to miniature species of *Aloe*.

COMMON NAMES The name *streepvygie* (striped mesemb) has been recorded for *Aloinopsis rubrolineata*, while *A. rosulata* is known under the general name: mimicry plant.

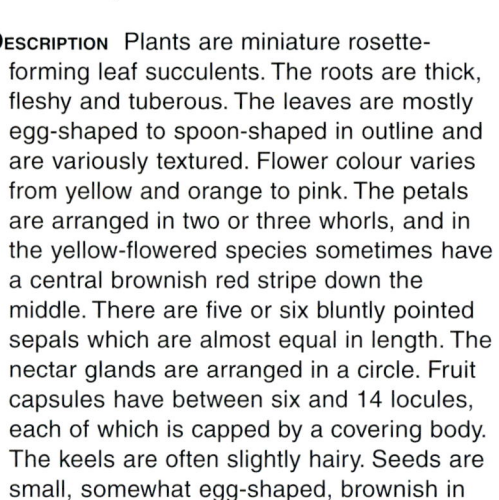

They occur predominantly in the Western and Eastern Cape Provinces of South Africa, extending slightly into the southern Northern Cape Province. A single outlier, *Aloinopsis orpenii*, occurs in the northeastern corner of the Northern Cape Province. Rainfall is sparse throughout their distribution range. The plants grow in rocky places and they are often difficult to see.

DESCRIPTION Plants are miniature rosette-forming leaf succulents. The roots are thick, fleshy and tuberous. The leaves are mostly egg-shaped to spoon-shaped in outline and are variously textured. Flower colour varies from yellow and orange to pink. The petals are arranged in two or three whorls, and in the yellow-flowered species sometimes have a central brownish red stripe down the middle. There are five or six bluntly pointed sepals which are almost equal in length. The nectar glands are arranged in a circle. Fruit capsules have between six and 14 locules, each of which is capped by a covering body. The keels are often slightly hairy. Seeds are small, somewhat egg-shaped, brownish in colour and have slightly rough surfaces.

DISTINGUISHING CHARACTERS The usually rough surfaces of the more or less spoon-shaped leaves are characteristic of the species of *Aloinopsis*. Roots are usually tuberous and flowers are often striped with red; striped petals are not found in *Titanopsis*, with which this genus can easily be confused. *Nananthus* also has striped petals, but its seeds are flat, not egg-shaped.

FLOWERING TIME Species of *Aloinopsis* mostly flower from late winter to early spring (July to September in South Africa). The honey-scented flowers expand in the late afternoon and close at sunset.

DISTRIBUTION AND ECOLOGY Species of *Aloinopsis* are typical Great and Little Karoo mesembs.

CULTIVATION Plants thrive under greenhouse conditions. They should be given a bright, sunny spot to ensure the retention of leaf colour and shape. Water sparingly during summer and keep them dry in winter. Plants are sensitive to red spider and mealy bug attacks.

NOTES Of all the mesembs, species of *Aloinopsis* are some of the most desirable to cultivate. Although there are other mesembs with more colourful flowers, the strange and beautiful leaf shapes and rough surfaces of *Aloinopsis* species more than compensate for what they might lack florally. One of the most widely cultivated species is *A. rubrolineata*, with its delicately striped petals and evenly warted, dagger-shaped leaves. The similar *A. rosulata* has spoon-shaped leaves. Uniquely, *A. spathulata* has rosy pink flowers. *A. orpenii*, better known as *Prepodesma orpenii*, has felty, upright leaves. It grows around Kuruman. *A. schooneesii* has tiny toe-shaped leaves which are dwarfed by its enormous tuber. *A. villetii* is known by its fine, closely spaced, not very prominent warts. *A. luckhoffii* is very similar, but often has more colourful brownish leaves.

Aloinopsis Schwantes

NUMBER OF SPECIES/SUBSPECIES/VARIETIES (14/0/0)

SPECIES LIST AND CONSERVATION STATUS
A. *acuta* L.Bolus [R]

A. rubrolineata

A. spathulata

A. malherbei

A. luckhoffii

ALOINOPSIS

A. *hilmarii* (L.Bolus) L.Bolus
A. *jamesii* L.Bolus [R]
A. *lodewykii* L. Bolus
A. *loganii* (L.Bolus) L.Bolus [R]
A. *luckhoffii* (L.Bolus) L.Bolus
A. *malherbei* (L.Bolus) L.Bolus
A. *orpenii* (N.E.Br.) L.Bolus
A. *rosulata* (Kensit) Schwantes
A. *rubrolineata* (N.E.Br.) Schwantes

A. *schooneesii* L.Bolus
A. *setifera* (L.Bolus) L.Bolus [R]
A. *spathulata* (Thunb.) L.Bolus
A. *villetii* (L.Bolus) L.Bolus [R]

LITERATURE
BOLUS, H.M.L. 1958. *Aloinopsis* Schwantes. *Notes on* Mesembryanthemum *and allied genera*. University of Cape Town, Cape Town. Vol. 3: 372–381.

A. *loganii*

A. *orpenii*

A. villetii

A. schooneesii

A. lodewykii

A. rosulata

DEILANTHE

DERIVATION OF GENUS NAME The name is derived from the Greek words *deile* (evening) and *anthos* (flower), in reference to the night-opening flowers of the first known species.

COMMON NAMES No vernacular names seem to have been recorded.

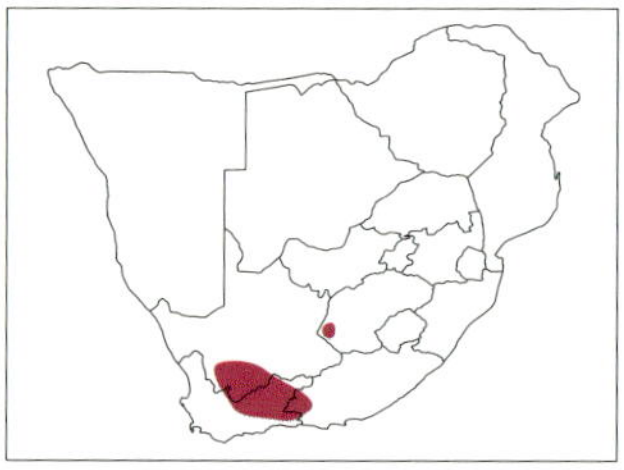

DESCRIPTION Plants are succulent with tuberous rootstocks. The leaves are flat on the ground or erect, sparse, long-triangular, thick, dark green to reddish brown, velvety and inconspicuously spotted. Flowers occur solitary or in delayed sets of three. They are pale to deep yellow or a strange greyish salmon colour. The thick flower stalk is erect and bears leaf-like bracts. Five equal sepals surround the numerous linear to narrowly spoon-shaped petals, which can be obscurely striped with white. There are numerous erect, hairy stamens around the convex and glassy ovary. Nectar glands are large and united into a deep ring around the eight to 12 stigmas. The robust, persistent fruit capsules have eight to 12 locules and small closing bodies. Seeds are pear-shaped and smooth.

DISTINGUISHING CHARACTERS Plants have large fleshy rootstocks and velvety (never warted), often dust-covered leaves. The flowers are showy, opening in the late afternoon or evening.

FLOWERING TIME Plants flower from winter to early spring (June to September in South Africa). The long-lasting, highly scented flowers smell like freesias or honey and they open in the late afternoon (freesia-smell) or evening (honey-smell).

DISTRIBUTION AND ECOLOGY Species of *Deilanthe* are widespread in the Little and Great Karoos in the central Western Cape Province, extending slightly into the Northern and Eastern Cape Provinces, South Africa. They grow from Anysberg in the southwest to Luckhoff in the northeast. A single species, *D. peersii*, occurs in the Free State. Plants are often buried up to the leaf tips in the quartz scree, shale rubble, or silt in which they grow.

CULTIVATION *Deilanthe* species prefer a very well-drained, gravelly soil; the tuberous taproot can easily rot if kept too damp.

NOTES One of the species illustrated here as *Aloinopsis hilmarii* is soon to be moved to the genus *Deilanthe*. This species and *D. peersii* grow together near Laingsburg but they are separated by flowering time (*A. hilmarii* in the day; *D. peersii* at night). The flowers of *D. thudichumii* have a delightful freesia-like perfume.

Deilanthe N.E.Br.

NUMBER OF SPECIES/SUBSPECIES/VARIETIES
(2/0/0)

SPECIES LIST AND CONSERVATION SATUS
D. peersii (L.Bolus) N.E.Br.
D. thudichumii (L.Bolus) S.A.Hammer

(*Aloinopsis hilmarii* will soon be transferred to *Deilanthe*).

LITERATURE
BROWN, N.E. 1931. *Deilanthe*. *Gardener's Chronicle* 89: 137.

HAMMER, S.A. 1995. New nothogenera, and a new combination in Mesembryanthema. *Cactus and Succulent Journal (US)* 67: 172–173.

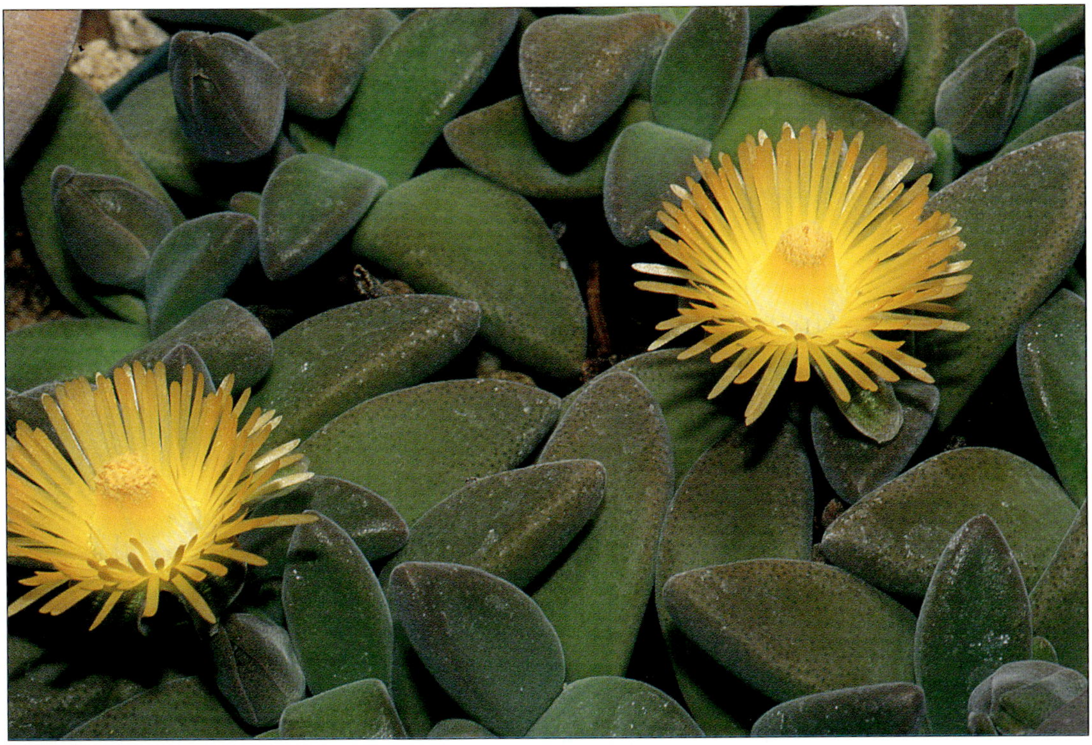

Chris Barnhill

D. peersii

Ernst van Jaarsveld

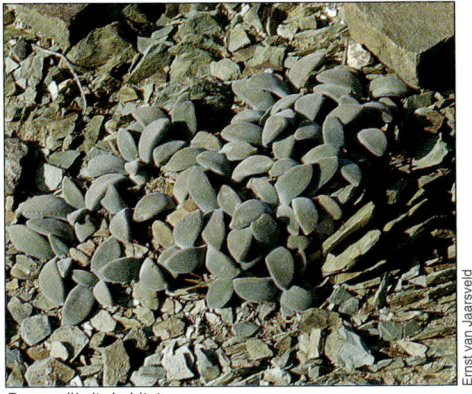

D. peersii in its habitat

Ernst van Jaarsveld

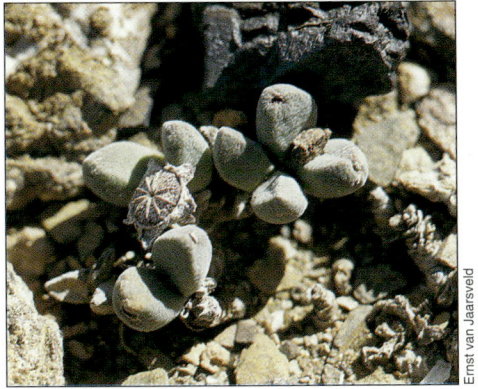

Deilanthe sp., presently known as *Aloinopsis hilmarii*

John Trager

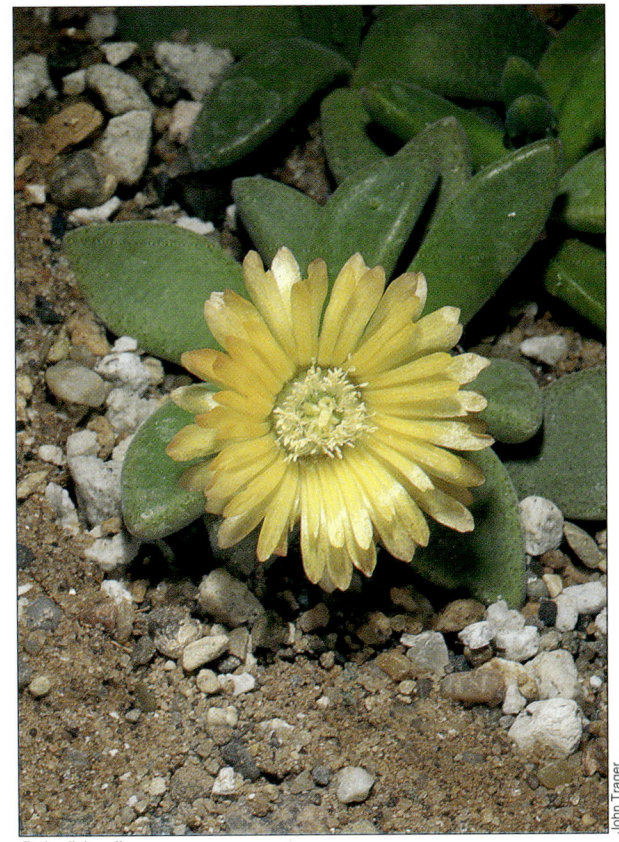

D. thudichumii

NANANTHUS

DERIVATION OF GENUS NAME The genus name literally means "small flowers", derived from a combination of the Greek word *nanos* (small) and the Latinised form of the Greek word *anthos* (flower). The name is not very appropriate since the flowers of species of *Nananthus* are not particularly small.

COMMON NAMES Three species of *Nananthus* have multiple common names: *N. aloides* is known as *brakvygie* or *vlaktevygie* (brack mesemb or flats mesemb); *N. vittatus* as *brakveldvygie* or *stryvygie* (quarrel mesemb); and *N. wilmaniae* as *brakvygie* or *moervygie* (yeast mesemb).

DESCRIPTION Plants form small, aloe-like, multi-headed tufts. The roots are thick and tuberous. The boat-shaped, keeled leaves taper to both ends and are ground-hugging. White dots often occur on the leaf surfaces. The margins of the smooth and shiny or dull green leaves are virtually always without teeth. The shortly stalked flowers are yellow, sometimes with a central reddish stripe down the middle of the petals. White flowers are rarely seen. The five free sepals are often fringed with a soft papery margin. Fruit capsules have between six and ten locules, each supporting a thin, tapering, dried stigma which is slightly roughened towards its tip. The valves are strongly curved backwards when the fruit open, and the expanding keels open into an upright position. The more or less oval-shaped seeds are smooth and dark brown.

DISTINGUISHING CHARACTERS The sharp, but softly pointed leaves, generally with lighter green spots, arranged in aloe-like rosettes, are typical of the species. Tuberous roots and the yellowish flowers, usually with a central, darker, reddish brown or even a yellow stripe, are also distinctive for species of *Nananthus*. The seeds are flat like miniature pancakes, unlike the fatter seeds of *Aloinopsis*.

FLOWERING TIME Flowering is almost throughout the year, but in habitat plants will not flower in the cold, dry winter months. The honey-scented flowers typically open just after midday and they close at dusk.

DISTRIBUTION AND ECOLOGY Species of *Nananthus* are widespread in the climatically severe southern African interior. In South Africa, the distribution area stretches from near Graaff-Reinet in the Great Karoo (Eastern Cape Province), northward through the Northern Cape Province and the central Free State to around Potchefstroom in the North-West Province, extending into the south of Gauteng. To the west the distribution stretches to near Upington and Bushmanland in the Northern Cape Province. The occurrence of an apparently isolated population in Namibia, due east of Windhoek, near the border with Botswana, highlights a lack of collections from the Northern Cape Province and Botswana. Some species of *Nananthus* can tolerate widely differing rainfall regimes, from extreme aridity to seasonally flooded river banks or pans.

CULTIVATION Outside of its habitat, *Nananthus* is best grown in a greenhouse but also thrives in rockeries because it is so hardy. Plants can be grown in a variety of soils, but they will retain their natural colour and form in clayey soil. Seed should be sown during spring or summer.

NOTES The species of *Nananthus* are difficult to identify and a detailed taxonomic treatment of the group is urgently required. Some species have rather shiny, dark green leaves with strikingly pale marginal zones or stubby greyish leaves. Others have two raised ridges along the length of the upper leaf surface.

Nananthus N.E.Br.

NUMBER OF SPECIES/SUBSPECIES/VARIETIES (10/0/4)

N. broomii

N. transvaalensis

N. aloides

N. aloides in its habitat

NANANTHUS

Species list and conservation status

- *N. aloides* (Haw.) Schwantes var. *aloides*
- *N. aloides* (Haw.) Schwantes var. *latus* L.Bolus
- *N. aloides* (Haw.) Schwantes var. *striatus* (L.Bolus) L.Bolus
- *N. broomii* (L.Bolus) L.Bolus
- *N. cibdelus* (N.E.Br.) Schwantes
- *N. gerstneri* (L.Bolus) L.Bolus
- *N. margaritiferus* L.Bolus
- *N. pallens* (L.Bolus) L.Bolus
- *N. pole-evansii* N.E.Br. [K]
- *N. transvaalensis* (Rolfe) L.Bolus var. *griquensis* L.Bolus
- *N. transvaalensis* (Rolfe) L.Bolus var. *latus* L.Bolus
- *N. transvaalensis* (Rolfe) L.Bolus var. *transvaalensis*
- *N. vittatus* (N.E.Br.) Schwantes [K]
- *N. wilmaniae* (L.Bolus) L.Bolus

Literature

BOLUS, H.M.L. 1958. *Nananthus* N.E.Br. *Notes on* Mesembryanthemum *and allied genera*. University of Cape Town, Cape Town. Vol. 3: 382–385.

Nananthus sp. (Karoo Gardens, Worcester)

Nananthus sp. (Tabaksberg)

Nananthus sp. (Virginia mine)

NEOHENRICIA

DERIVATION OF GENUS NAME The name is derived from the Latin word *neo* (new) and Henricia, and was chosen as a replacement for the genus name *Henricia*, which could not be used because it had already been applied to another group of plants. *Neohenricia* honours Dr Marguerite Henrici (1892 to 1971), a plant physiologist who worked at Fauresmith in the Free State Province of South Africa.

COMMON NAMES The name coral plant has been used for *N. sibbettii*.

DESCRIPTION Plants are dwarf, compact, creeping succulents, less than 20 mm in height. The minute leaves occur in four to six leaf pairs per branch, which often root at the nodes. The leaves are reddish green, orange or greyish green, erect, club-shaped, warted like coral or with tiny raised spikes. The yellowish, pinkish or greenish flowers are solitary on curved, wiry stalks. There are five sepals and relatively few filamentous staminodes. The stamens are erect and hairless. The nectar glands are united into a ring around the four to six stigmas. The very fragile, flattish fruit capsules have four to six locules and no closing bodies. The seeds are pear-shaped, slightly coarse-textured and are attached to the central axis of the capsule as in the WEEDY MESEMBS.

DISTINGUISHING CHARACTERS Plants are cryptic and minute, with soft tender and strongly textured leaves. The flower stalks are very long and slender, so that the nocturnal, highly scented flowers are held above the plant.

FLOWERING TIME Plants flower over a long period during summer (South Africa) and the flowers open at twilight, closing before dawn.

DISTRIBUTION AND ECOLOGY The species of *Neohenricia*, a Great Karoo genus, occur in a belt stretching from Victoria West in the Northern Cape Province to Fauresmith in the Free State. The recent discovery of a single, isolated population of a new species, *N. spiculata*, near Sterkstroom, could reflect poor botanical exploration in the past in this part of the Eastern Cape. *N. sibbettii* grows and hides in grit pans, whereas *N. spiculata* grows in fissures of dolerite domes.

CULTIVATION Both species grow well in shallow trays and both are extremely active in summer. They are prone to red spider mite, so it is wise to mist the plants frequently.

NOTES The flowers are delightfully scented, like pineapple (*N. sibbettii*) or licorice (*N. spiculata*). They are most redolent on warm evenings.

Neohenricia L.Bolus

NUMBER OF SPECIES/SUBSPECIES (2/0/0)

SPECIES LIST AND CONSERVATION STATUS
N. sibbettii (L.Bolus) L.Bolus [V]
N. spiculata S.A.Hammer [R]

LITERATURE
BOLUS, H.M.L. 1931. *Mesembryanthemum* (?) *sibbettii*. Notes on Mesembrianthemum *and allied genera*. University of Cape Town, Cape Town. Vol. 2: 251–252.

BOLUS H.M.L. 1936. *Henricia* gen. nov. Notes on Mesembryanthemum *and allied genera*. University of Cape Town, Cape Town. Vol.3: 39–40.

BOLUS H.M.L. 1938. *Neohenricia*. Journal of South African Botany 4: 51.

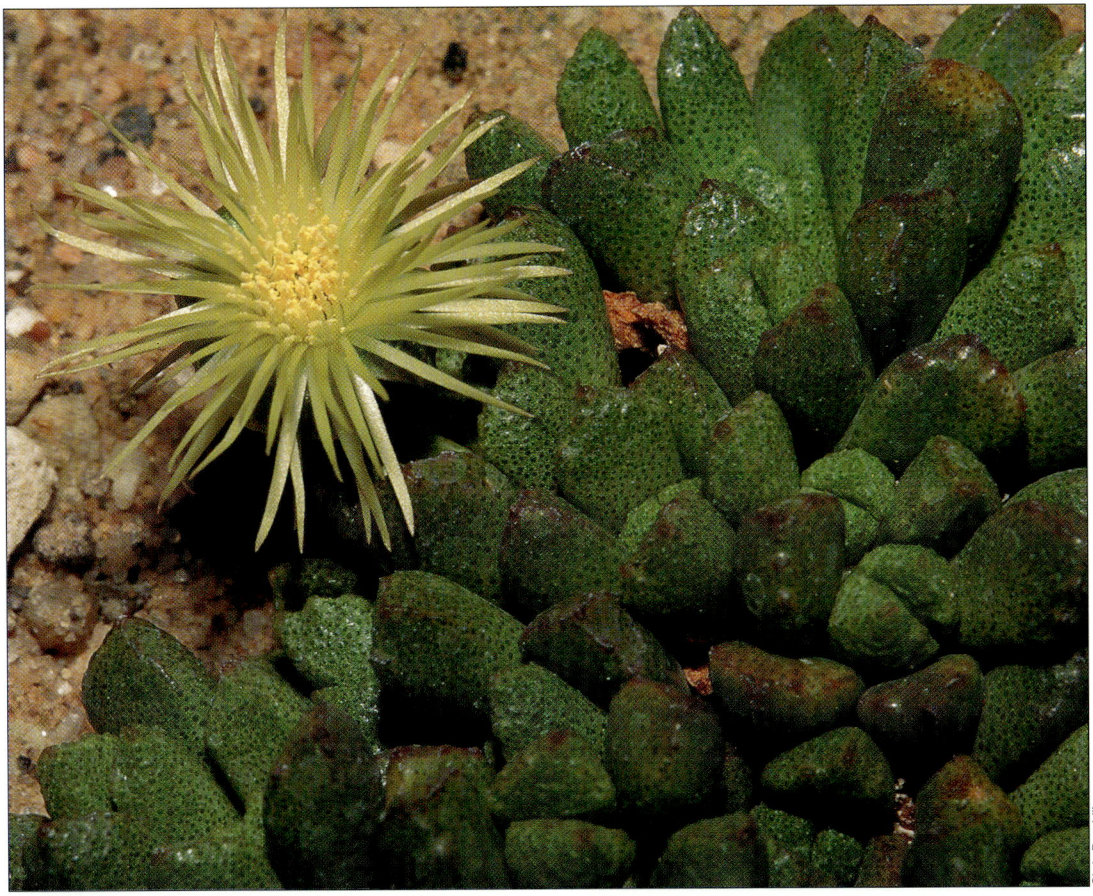
Chris Barnhill

N. spiculata

Ernst van Jaarsveld

Leaves of *N. sibbettii*

Bob Chinnock

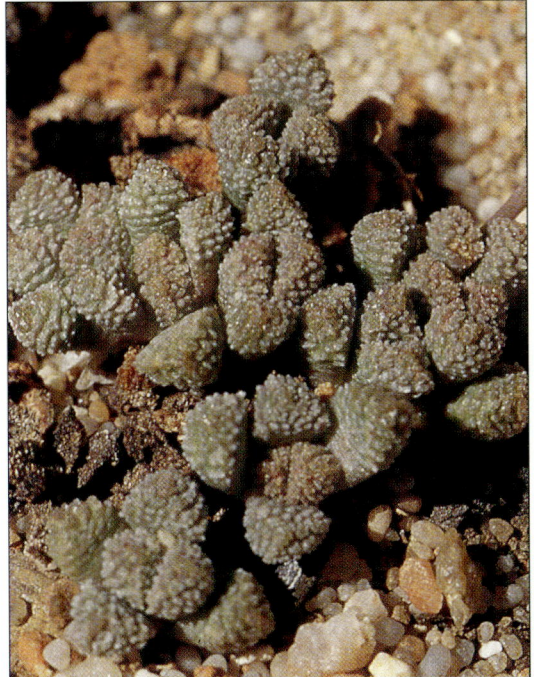

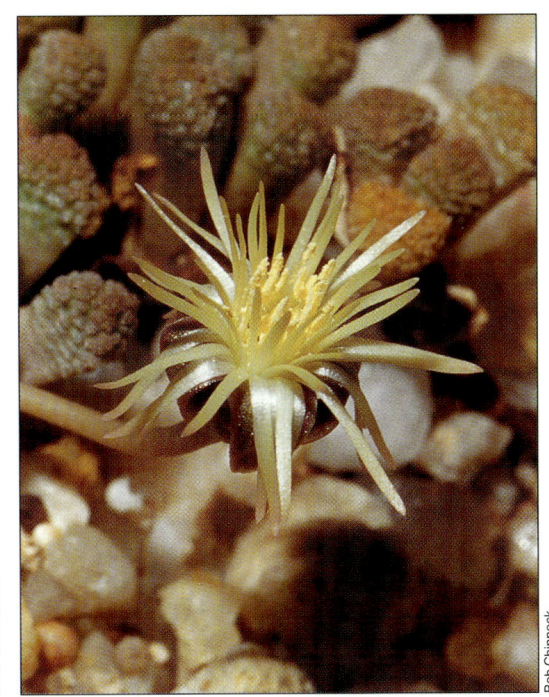

Flower of *N. sibbettii*

RHINEPHYLLUM

DERIVATION OF GENUS NAME The name was derived from the Greek words *rhine* (file) and *phyllon* (leaf), an allusion to the rough leaf surfaces found in this genus.

COMMON NAMES As is the case for many other mesembs, the general term mimicry plant is sometimes used.

DESCRIPTION Plants are dwarf succulents of up to 100 mm in height, and some have tuberous rootstocks. They are often much branched by means of thin, cylindrical, elongated, woody branches. The pointed or slightly club-shaped leaves are borne more or less upright with the older ones tending to lie on the ground. Upper leaf surfaces are flattened while the lower ones are rounded or slightly keeled. The leaf surfaces lack hairs, but are rough to the touch on account of the small raised, hard, white protuberances that cover them. The leaf margins are smooth, or rarely toothed. Single flowers are borne on bractless stalks. They are medium sized and white to yellow in colour. The five thick sepals may be free or form a short tube. Petals are free and are arranged in one to five whorls. Some species have the stamens bearded at their bases. Five thin, elongated stigmas are present. The fruit capsules have five locules and diverging expanding keels with marginal wings. Covering membranes are often absent or much reduced. The gaping seed exits are slightly raised, the valves are triangular and bent backwards when expanded. Closing bodies are absent. The seeds are globular and more or less smooth.

DISTINGUISHING CHARACTERS In habit the plants loosely resemble species of *Stomatium*, *Titanopsis* or *Psammophora*. The structure of the fruits and flowers, however, is quite different. *Chasmatophyllum* is also allied to *Rhinephyllum*, but differences include the absence of covering membranes in the fruit capsules of *Chasmatophyllum* and the absence of teeth on the leaf margins of most species of *Rhinephyllum* (except *R. rouxii*, *R. broomii*, *R. inaequale* and *R. comptonii*), which are found on most *Chasmatophyllum* species.

FLOWERING TIME The scented flowers appear during spring and midsummer (October to December in South Africa). Depending on the species, flowers open at midday, in the late afternoon or in the evening.

DISTRIBUTION AND ECOLOGY *Rhinephyllum* occurs mainly in the Little and Great Karoos of the Northern, Western and Eastern Cape Provinces, South Africa, where rain falls in winter and summer. However, rainfall is always less than 500 mm per year. Plants can be found growing on rocky slopes or on flats in full sun.

CULTIVATION Propagation is easy from seeds or cuttings. Plants thrive in a variety of soils, but care must be taken not to over-water during the winter months. They do well in a greenhouse or on rockeries in suitable climatic regions. Although species of *Rhinephyllum* are not as often cultivated as might be expected, these rough-leaved plants are real gems amongst the mesembs. They certainly deserve a place in any collection of succulent plants.

NOTES *Rhinephyllum* species can be distinguished by their flowering habits and by their leaves and roots. The thick-rooted *R. parvifolium* has butter-yellow flowers which open in the late afternoon and require cross-pollination; the thin-rooted, short-lived *R. broomii* is superficially similar, but its flowers fertilize themselves. Some other species open their flowers at night. The thick-rooted *R. muirii* has waxy-looking white flowers, while the thin-rooted *R. pillansii* produces yellow powder-puffs. *R. macradenium* and *R. frithii* used to be grouped together in the old genus *Peersia*, but one of them (*R. macradenium*) opens its wide-petalled flowers at night, while the other

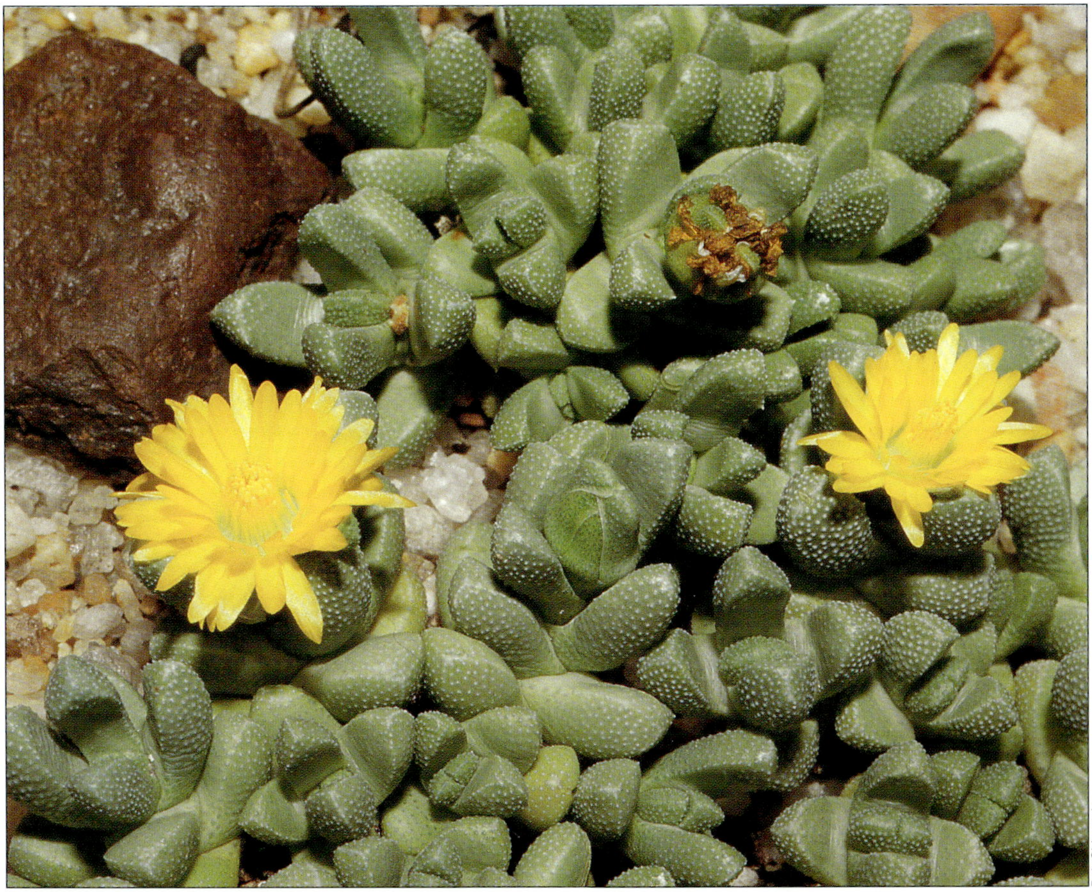

R. parvifolium

Bob Chinnock

R. muirii

Ben-Erik van Wyk

R. macradenium

Ben-Erik van Wyk

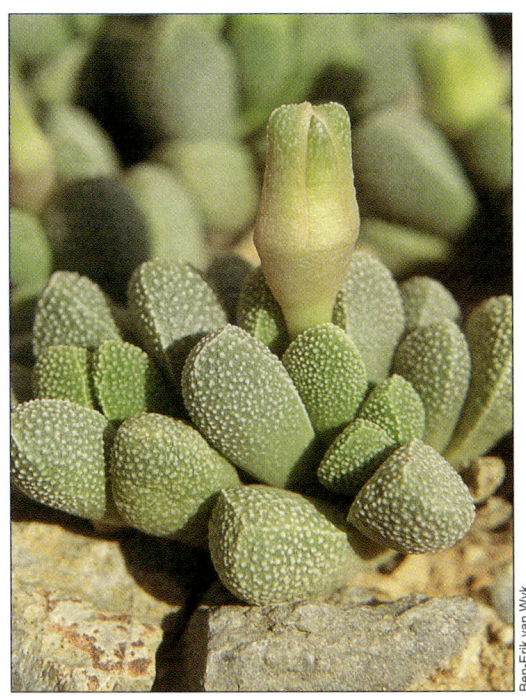

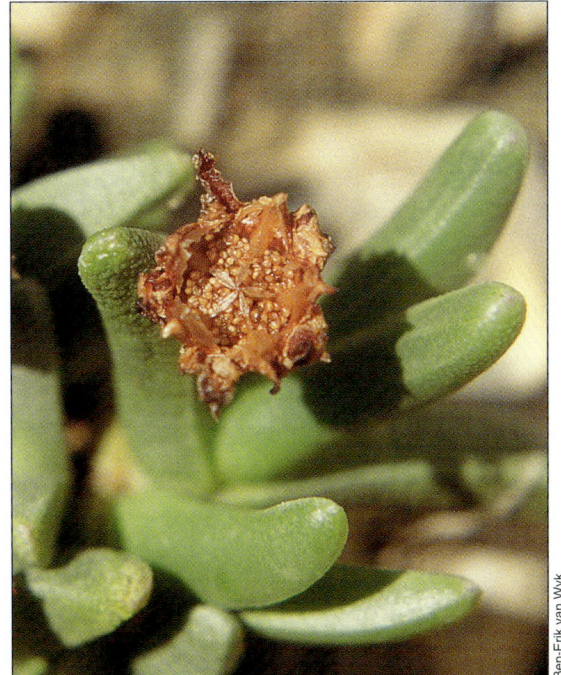

flowers in the afternoon. Neither has warty leaves, so that they do not fit very well in *Rhinephyllum*.

Rhinephyllum N.E.Br. (=*Peersia* L.Bolus)

NUMBER OF SPECIES/SUBSPECIES/VARIETIES (12/0/1)

SPECIES LIST AND CONSERVATION STATUS
  R. broomii L.Bolus
  R. comptonii L.Bolus
  R. frithii (L.Bolus) L.Bolus
  R. graniforme (Haw.) L.Bolus
  R. inaequale L.Bolus var. *inaequale* [I]
  R. inaequale L.Bolus var. *latipetalum* L.Bolus [I]
  R. luteum (L.Bolus) L.Bolus

  R. macradenium (L.Bolus) L.Bolus
  R. muirii N.E.Br.
  R. obliquum L.Bolus
  R. parvifolium L.Bolus
  R. pillansii N.EBr
  R. rouxii L.Bolus

LITERATURE
BROWN, N.E. 1927. *Mesembryanthemum and some new genera separated from it. Gardener's Chronicle* 82: 92.

BOLUS, L. 1936. *Notes on* Mesembryanthemum *and allied genera.* University of Cape Town, Cape Town. Vol. 3: 40–43.

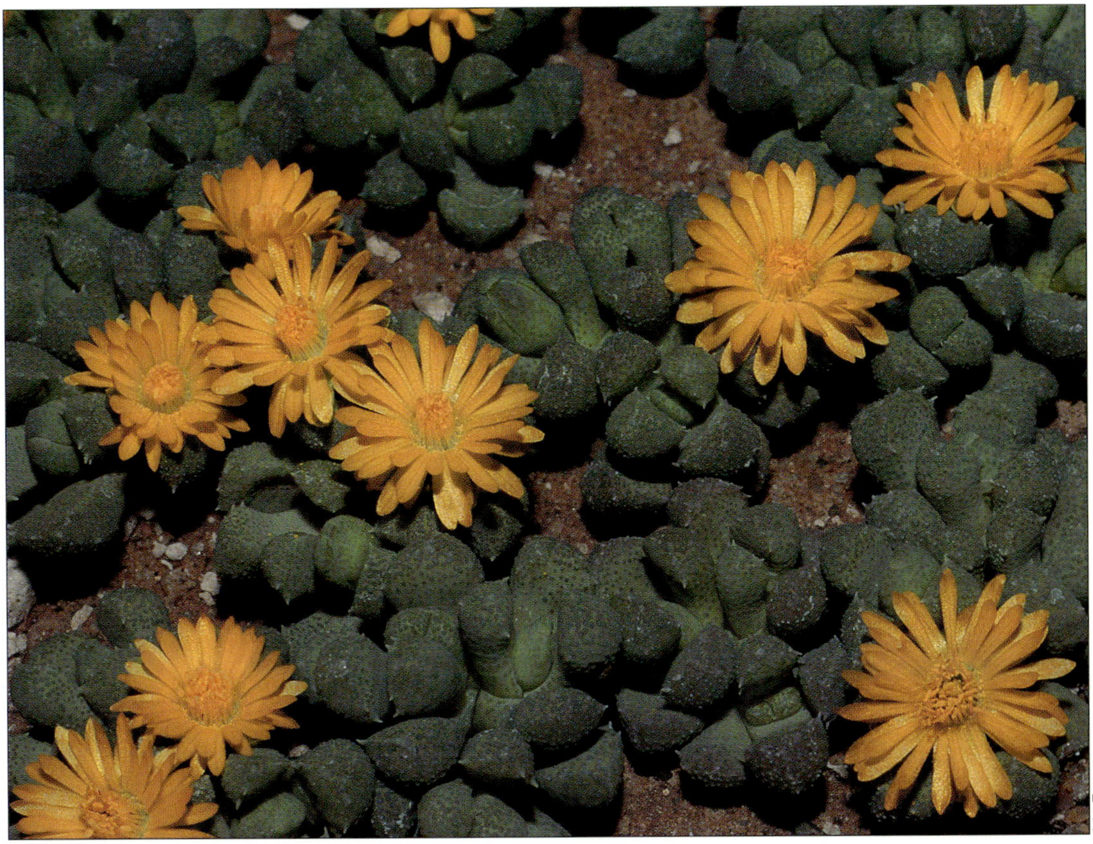

R. broomii

John Trager

R. frithii

Ben-Erik van Wyk

R. pillansii

John Trager

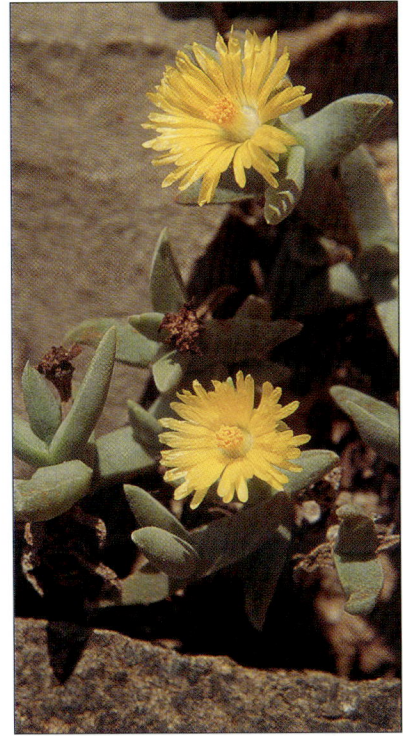

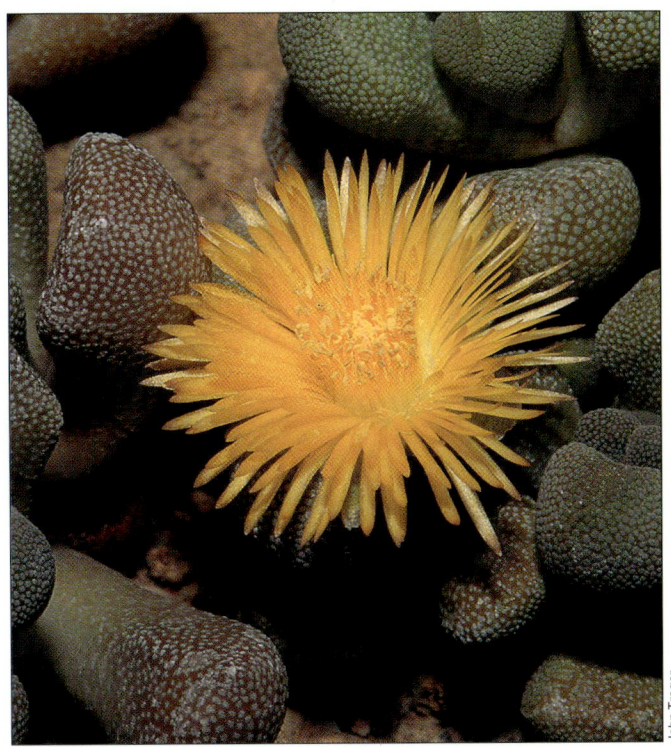

Titanopsis

Derivation of genus name The genus name was derived from a combination of the Greek words *titanos* (chalk) and *opsis* (appearance), alluding to the raised calcium-filled warts on the leaf surfaces.

Common names *Titanopsis calcarea* is known as kalk(veld)vygie (chalk field mesemb), *skilpadvoetjies* (little tortoise feet) or sheep's tongue.

Description Plants form small clumps of closely clustered rosettes. The rootstock is somewhat thickened, but only occasionally tuberous in old plants. The leaves are spoon to club-shaped and have conspicuous warts which can be various shades of white, green, copper and pink or reddish. These structures also occur on the sepals. The flowers are shortly stalked or stalkless. The petals are narrow and are arranged in one or two whorls. Flower colour is mostly yellow or amber, but pink or white-flowered forms of some species are also known. Fruit capsules have five to ten (usually six) locules. Each locule is covered by a papery covering membrane. The expanding keels are short and rather stiff. The seeds are more or less globular in shape, minutely textured and characteristically pale.

Distinguishing characters The warty, spoon-shaped leaves and the warty sepals are useful characters to separate these species from other similar-looking mesembs.

Flowering time There are two basic flowering periods. The western species flower in winter, the easterners flower in late spring and summer. The honey-scented flowers open in the late afternoon (even later in the day in the western species) and close at dusk.

Distribution and ecology Species of *Titanopsis* are distributed in three main areas of southern Africa. In southern Namibia they grow as far north as Lüderitz. To the east, the range straddles the Orange River in a broad north-south belt extending into Bushmanland. In central South Africa, the distribution extends from the Northern Cape Province into the Free State. This is an arid part of southern Africa where either winter (Lüderitz) or summer rainfall (Kimberley and surroundings) can predominate.

Cultivation *Titanopsis* species thrive in cultivation. Plants should be kept dry in winter. It is intriguing to hide the plants amongst calcareous pebbles, but contrary to common belief, alkaline soil is not essential. Seed is best sown during the summer.

Notes *Titanopsis calcarea* is by far the best known and most beloved species of all. This species has a random mixture of small and large pearly warts. In the other species the warts are arranged in various patterns; the reddish warts of *T. hugo-schlechteri* have the disconcerting appearance of human warts; *T. schwantesii* has an elegantly graded arrangement of pearly warts. *T. fulleri*, which may simply be an eastern form of *T. calcarea*, concentrates its large warts towards the tips of the unusually upright, narrow leaves. "*T. primosii*", though widely seen in cultivation, has never been properly named. In any case, it is probably a southern form of *T. schwantesii*, localised to western Bushmanland, whereas *T. luederitzii* seems to be a western, almost coastal, finely warted form of that species. *T. hugo-schlechteri* has a lovely whitish green form, described as var. *alboviridis*, and in some aberrant southern forms, *T. hugo-schlechteri* seems to intergrade with *T. schwantesii* (see illustration).

Titanopsis Schwantes

Number of species/subspecies/varieties (5/0/1)

Species list and conservation status
 T. calcarea (Marloth) Schwantes

T. calcarea

T. hugo-schlechteri var. *hugo-schlechteri*

T. calcarea

T. hugo-schlechteri var. *alboviridis*

T. fulleri

T. calcarea

TITANOPSIS

T. *fulleri* Tischer
T. *hugo-schlechteri* (Tischer) Dinter & Schwantes var. *alboviridis* Dinter
T. *hugo-schlechteri* (Tischer) Dinter & Schwantes var. *hugo-schlechteri*

T. *luederitzii* Tischer
T. *schwantesii* (Dinter) Schwantes

LITERATURE
SCHWANTES, G. 1926. Zur Systematik der Mesembrianthemen. *Zeitschrift für Sukkulentenkunde* 2: 178.

T. *calcarea*

T. *hugo-schlechteri* var. *alboviridis*

Fruit capsules of T. calcarea

T. schwantesii

T. fulleri

T. hugo-schlechteri var. hugo-schlechteri

Faucaria felina

Faucaria longidens

Odontophorus angustifolius subsp. *protoparcoides*

Carruanthus peersii

Tooth-leaved Mesembs

GROUP 6

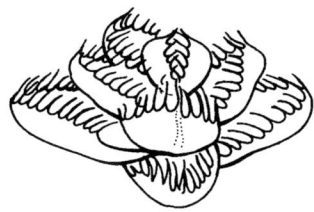

The succulent leaves usually have one or more teeth along the leaf margins or near the tip, and are variously arranged; plants usually have short stems.

The TOOTH-LEAVED MESEMBS is a group characterised by the presence of conspicuous teeth along the leaf margins, at the leaf tips or sometimes also along the keels of the leaves. The possession of toothed leaves does not necessarily imply close relationship, although *Orthopterum* and *Faucaria* are closely allied genera. The purpose of gathering together such a diverse group of genera is merely to allow easy identification.

TOOTH-LEAVED MESEMBS comprise 9 genera and 95 species.

Acrodon (4 species)	*Faucaria* (33 species)	*Orthopterum* (2 species)
Carruanthus (2 species)	*Hammeria* (2 species)	*Stomatium* (39 species)
Chasmatophyllum (6 species)	*Odontophorus* (5 species)	*Vanheerdea* (2 species)

Acrodon

Derivation of genus name The name was derived from the Greek words *akros* (point) and *odus* (tooth), referring to the toothed leaves of the genus.

Common names No vernacular names seem to have been recorded.

Description Plants are branched from the base and form low-growing or rarely creeping, compact clumps. The recurved leaves are triangular in cross-section, and the leaf keels and margins are often toothed. Leaves are dark green or greyish blue-green, smooth and dotless. Flowers are medium sized and solitary, with bracts on their stalks. Colour ranges from white to fleshy pink or even orange pink with a darker central line and reddish margins to the petals. There are five or six nearly equal sepals which bear tiny teeth. The narrow, notched petals are arranged in two whorls around the more or less erect group of stamens with distinctive reddish anthers. Nectar glands are dark green and shortly toothed, forming a ring around the five short and feathery stigmas. The fruit capsules are somewhat woody with five deep locules. Valves open incompletely into a more or less erect position. Valve wings are lacking, expanding keels are pointed, covering membranes are stiff and complete, closing bodies are large. The dark brown seeds are pear-shaped, with a rough surface.

Distinguishing characters This distinctive genus is characterised by its sharply three-angled leaves with small teeth on the keel near the tip and by its solitary, medium-sized flowers with a central longitudinal stripe of a darker shade than the rest of the petals.

Flowering time Flowering is in early spring (August in South Africa). The flowers open at midday and close by evening.

Distribution and ecology The species of *Acrodon* are restricted to a broad band along the southwestern and southern coast of the Western and Eastern Cape Provinces, South Africa. The genus also extends into the wetter parts of the Little Karoo.

Cultivation Species of *Acrodon* are not at all difficult to grow, especially from seed which should be sown in summer. Plants require acid sandy soil and should be watered sparingly during winter and summer.

Notes Plants of this small genus were among the first highly succulent mesembs to become known in Europe. *A. bellidiflorus* was well known in England and the Netherlands early in the eighteenth century. It appears that this species was introduced into cultivation as early as the end of the seventeenth century, as the first descriptions of it date from 1700.

Acrodon N.E.Br.

Number of species/subspecies/varieties (4/0/0)

Species list and conservation status
A. bellidiflorus (L.) N.E.Br.
A. parvifolius R.du Plessis
A. quarcicola H.E.K.Hartmann
A. subulatus (Miller) N.E.Br.

Literature
GLEN, H.F. 1986. Numerical taxonomic studies in the subtribe Ruschiinae (Mesembryanthemaceae) — *Astridia*, *Acrodon* and *Ebracteola*. *Bothalia* 16: 203–226.

HARTMANN, H.E.K. 1996. Miscellaneous taxonomic notes on Aizoaceae. *Bradleya* 14: 29–56.

A. bellidiflorus

Carruanthus

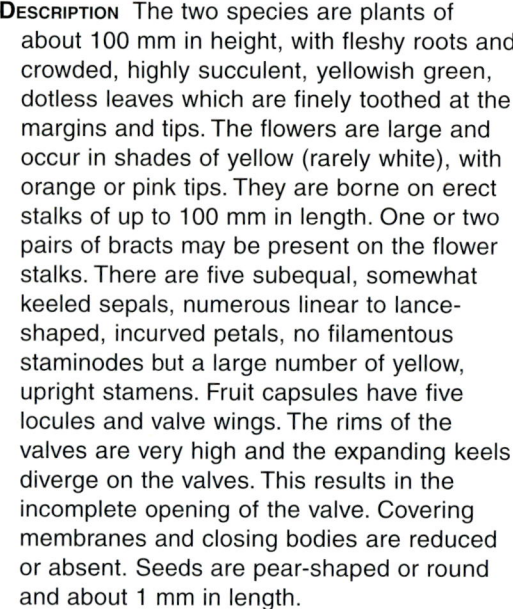

Derivation of genus name The genus name is derived from *carru* (karoo) and the Greek word *anthos* (flower), as it occurs in the Karoo.

Common names The common name *tierbekvygie* (tiger mouth mesemb) is shared with another mesemb genus, *Faucaria*. This name alludes to the tooth-like outgrowths on the leaf margins.

Description The two species are plants of about 100 mm in height, with fleshy roots and crowded, highly succulent, yellowish green, dotless leaves which are finely toothed at the margins and tips. The flowers are large and occur in shades of yellow (rarely white), with orange or pink tips. They are borne on erect stalks of up to 100 mm in length. One or two pairs of bracts may be present on the flower stalks. There are five subequal, somewhat keeled sepals, numerous linear to lance-shaped, incurved petals, no filamentous staminodes but a large number of yellow, upright stamens. Fruit capsules have five locules and valve wings. The rims of the valves are very high and the expanding keels diverge on the valves. This results in the incomplete opening of the valve. Covering membranes and closing bodies are reduced or absent. Seeds are pear-shaped or round and about 1 mm in length.

Distinguishing characters *Carruanthus* has strongly toothed, bright yellowish green leaves. These resemble those of some *Faucaria* species, but the *Faucaria* fruit is stalkless, not long-stalked as in *Carruanthus*.

Flowering time *Carruanthus ringens* flowers in late winter or early spring, while *C. peersii* flowers from spring to early summer (September to November in South Africa). In both species, the flowers open in the late afternoon and stay open a little after dark.

Distribution and ecology *Carruanthus* is restricted to areas around Willowmore, straddling the border between the Western and Eastern Cape Provinces, South Africa. Plants generally grow in rocky soil in full sunlight (*C. peersii*) or on partly shaded steep cliffs (*C. ringens*). The habitat receives relatively high rainfall (about 300 mm per year) which falls in winter and summer.

Cultivation Propagation of this genus can be from cuttings, which root easily, or from seeds that germinate readily. Outside of their habitat *Carruanthus* species are best grown in a greenhouse. Water sparingly in winter and summer.

Notes The two species are distinguished by fruit characters (absence of covering membranes in *C. peersii*) and by subtle differences in the leaves. *C. ringens* usually has much longer flower stalks, at least in cultivation.

Carruanthus (Schwantes) Schwantes
(=*Tischleria* Schwantes)

Number of species/subspecies/varieties (2/0/0)

Species list and conservation status
C. peersii L.Bolus
C. ringens (L.) Boom

Literature
SCHWANTES, G. 1927. Zur Systematik der Mesembrianthemen. *Zeitschrift für Sukkulentenkunde* 3: 106.

BOLUS, L. 1936. *Notes on Mesembryanthemum and allied genera*. University of Cape Town, Cape Town. Vol. 3: 4.

WEBER, C. 1968. Notes on the nomenclature of some Aizoaceae. *Baileya* 16: 12.

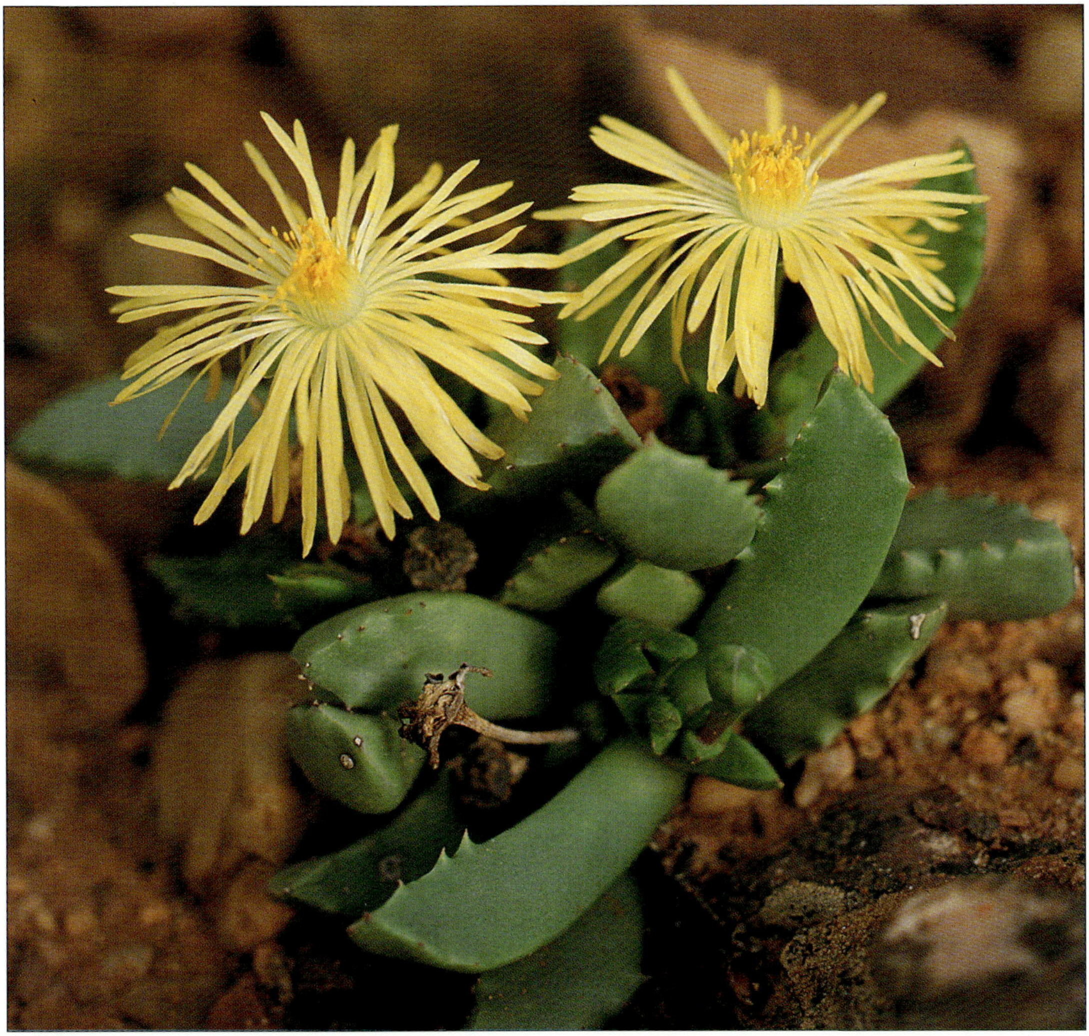

C. ringens

Ernst van Jaarsveld

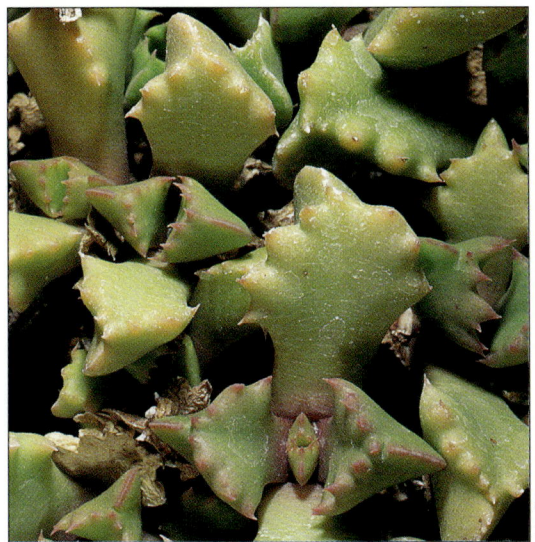

Leaves of *C. peersii*

John Trager

Flowering plant of *C. peersii*

Bob Chinnock

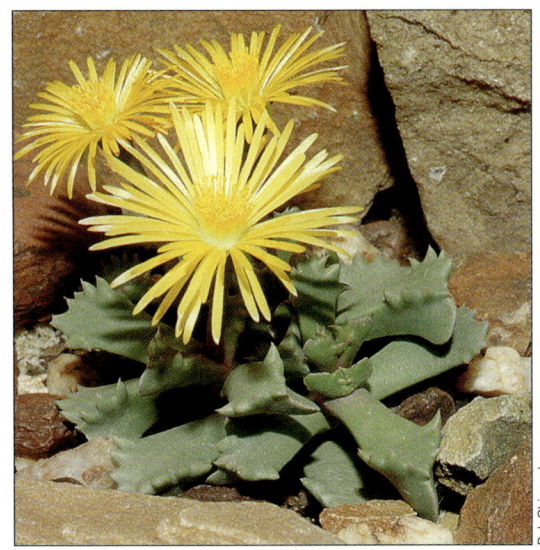

CHASMATOPHYLLUM

DERIVATION OF GENUS NAME The genus name was derived from the Greek words *chasma* (gaping mouth) and *phyllon* (leaf), referring to the mouth-like appearance of the clustered, toothed leaves.

COMMON NAMES The names *geelbergvygie* (yellow mountain mesemb) and *geelswaelstertvygie* (yellow swallowtail mesemb) have been recorded for the most common species of *Chasmatophyllum*, *C. musculinum*.

DESCRIPTION Plants are small, compact creepers, spreading by means of short, thick underground rhizomes that rapidly take root. The above-ground stems are short and tend to be somewhat woody, rather than soft and herbaceous as in many mesembs. The leaves are borne erectly and are rough to the touch, especially in the dry season. A fairly prominent off-white protuberant tooth occurring terminally on the back of the leaves gives them a laterally flattened appearance. The stalked flowers are yellow with a slight reddish tinge and are borne more or less at the same level as the leaves. The fruit capsules are usually five-locular with valve wings and partial covering membranes; closing bodies are usually absent.

DISTINGUISHING CHARACTERS The short, somewhat woody, creeping stems and the leaves that carry soft whitish teeth terminally on the back distinguish the genus from other mesembs.

FLOWERING TIME Species of *Chasmatophyllum* flower in summer (from December to February in South Africa). The flowers open from late afternoon to early evening.

DISTRIBUTION AND ECOLOGY *Chasmatophyllum* has a wide, but sparse distribution in parts of the southern African interior, stretching from the Karasberg Mountains in Namibia in the west, scattered in the Great and Little Karoos, the Eastern Cape Province, North-West Province and Gauteng to Mpumalanga in South Africa. The growth of these plants are stunted where they grow in rocky places, but they form large ground covers in deeper soils. One of the peculiarities of the genus is that it has representatives that grow in areas with rainfall ranging from as little as 150 mm per year in the west to as much as 750 mm in the east.

CULTIVATION *Chasmatophyllum* species are fairly easy to cultivate from stem cuttings or whole plants and will grow happily in full sun or semi-shade in the southern African summer rainfall region. However, many of the species are prone to attack by scale insects that will leave plants disfigured with small, circular spots on the leaves, even after successful treatment of the pest. Red spider mite can also be troublesome.

NOTES The genus is in need of detailed taxonomic treatment, which would more than likely reduce the number of species currently recognised.

Chasmatophyllum Dinter & Schwantes

NUMBER OF SPECIES/SUBSPECIES/VARIETIES (6/0/1)

SPECIES LIST AND CONSERVATION STATUS
  C. braunsii Schwant. var. *braunsii* [K]
  C. braunsii Schwant. var. *majus* L.Bolus [K]
  C. maninum L.Bolus [K]
  C. musculinum (Haw.) Dinter & Schwant.
  C. nelii Schwant.
  C. verdoorniae (N.E.Br.) L.Bolus
  C. willowmorense L.Bolus [K]

LITERATURE
  SCHWANTES, G. 1927. Zur Systematik der Mesembrianthemen. *Zeitschrift für Sukkulentenkunde* 3: 14, 17.

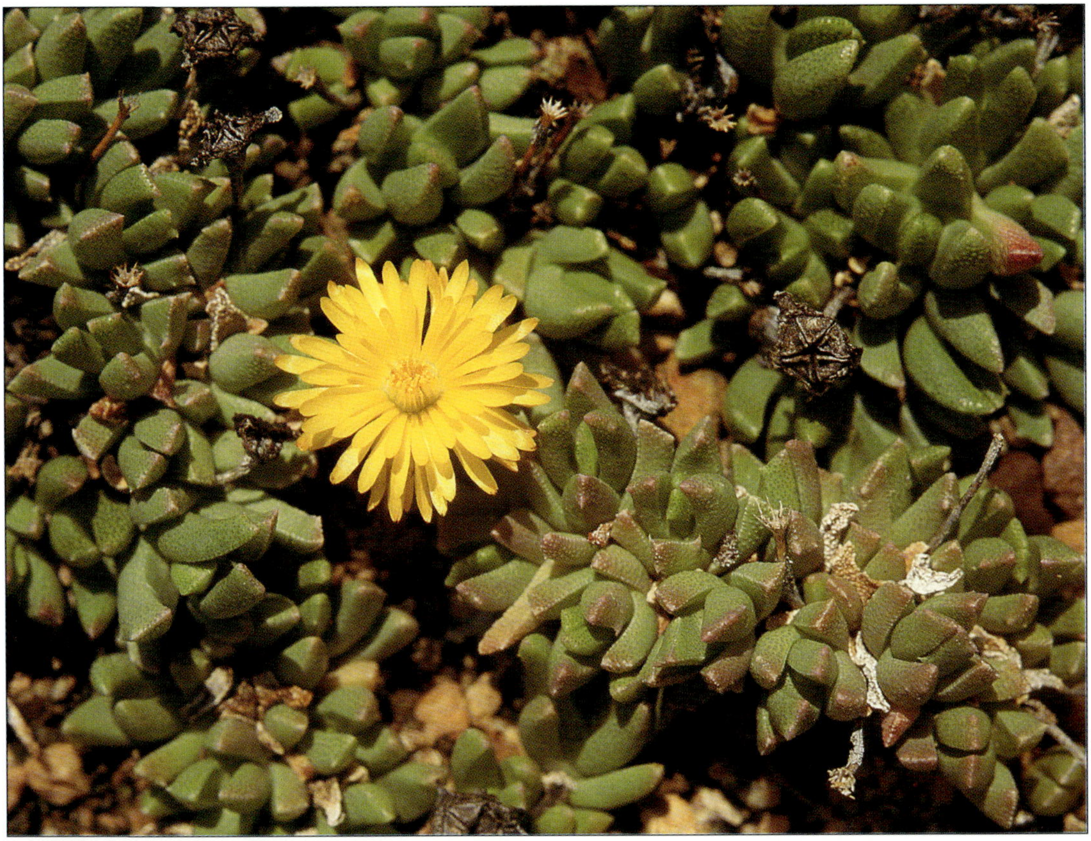

Ben-Erik van Wyk

C. musculinum

John Trager

C. braunsii

Peter Bruyns

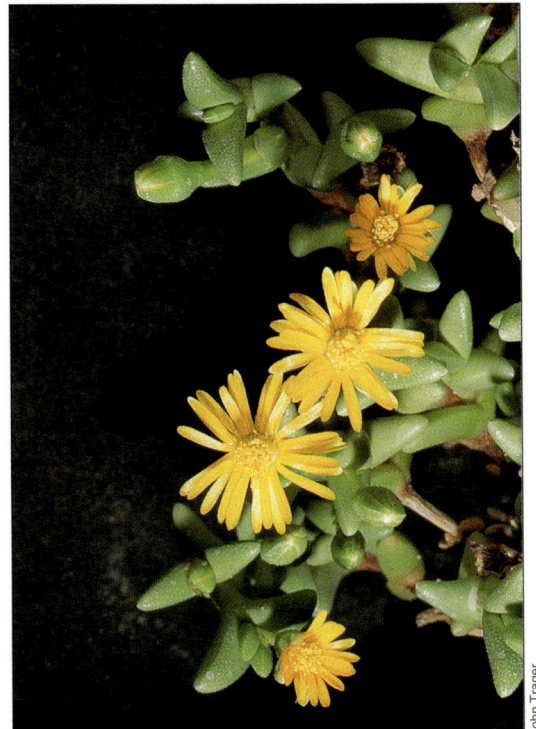

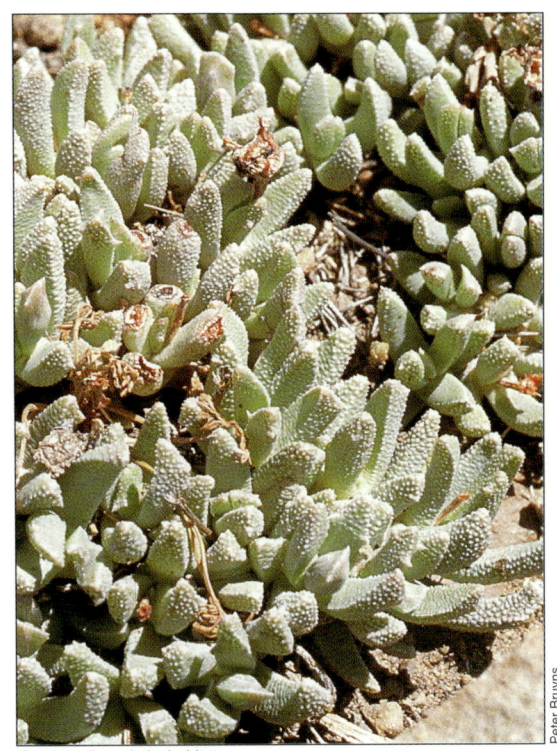

C. musculinum in its habitat

Faucaria

Derivation of genus name The genus name was derived from the Latin word *faux* (jaw). *Faucaria* refers to a collection of jaws, in allusion to the arrangement of the toothed leaves of the plants.

Common names Two species, *F. felina* and *F. tigrina*, are known as *tierbekvygie* or tiger-jaw(s).

Description Plants generally form small clumps of miniature aloe-like rosettes. The roots are only somewhat fleshy. Short stems are often formed with age, but are generally obscured by the clump-forming habit of the plants. The dark green, yellowish green, reddish purple or rarely bluish leaves are broadly triangular with smooth to variously tuberculed or marbled and ribbed surfaces. Leaves of all the species have distinct margins which are either smooth, white and cartilaginous or distinctly toothed. In some species the teeth are elongated and end in a thin thread. Flowers are yellow to a rusty orange red, especially when withered, or rarely white. Fruit capsules are usually five-locular, almost stalkless, deep, dark, tough and persistent. After a long time they are squeezed out by the growth of new leaves.

Distinguishing characters The genus is distinguished by its triangular, dark green leaves, generally with distinct margins which usually have rows of soft teeth that are curved towards the centre of the rosettes.

Flowering time Species of *Faucaria* generally flower in late summer or autumn (January to April in South Africa). The flowers open in the afternoon and close during the evening.

Distribution and ecology *Faucaria* species are components of the subtropical thickets of the Eastern Cape Province of South Africa. This type of vegetation was previously referred to as Valley Bushveld and typically consists of rather thorny, impenetrable thickets, interspersed with small open areas, where *Faucaria* species are likely to be found. In the western parts of the distribution range the genus extends into the Little and Great Karoos of the Western Cape Province. Tiger-jaws usually grow in the shade of surrounding vegetation or as beautifully coloured, but rather stunted plants in the open.

Cultivation These plants are generally very easy to cultivate. They make excellent subjects for growing in pots and flower freely in cultivation. Water sparingly during winter and summer. Plants can readily be propagated from seed.

Notes Tooth-leaved species of *Faucaria* do not look unlike fat-leaved venus fly traps. This rather exotic appearance has undoubtedly contributed to their popularity amongst collectors. It is likely that too many species are currently recognised in the genus; critical taxonomic revision may well reduce the number of species upheld. It is furthermore interesting to note that only *Faucaria tigrina*, the most commonly cultivated *tierbekvygie*, is regarded as threatened, probably as a result of urban expansion. *Faucaria* species have been in cultivation in Europe for more than 300 years.

Faucaria Schwantes

Number of species/subspecies/varieties (33/0/4)

Species list and conservation status
- *F. acutipetala* L.Bolus
- *F. albidens* N.E.Br.
- *F. boscheana* (A.Berger) Schwantes var. *boscheana*
- *F. boscheana* (A.Berger) Schwantes var. *haagei* (Tischer) Jacobsen
- *F. britteniae* L.Bolus
- *F. candida* L.Bolus [nt]
- *F. coronata* L.Bolus
- *F. cradockensis* L.Bolus
- *F. crassisepala* L.Bolus
- *F. duncanii* L.Bolus

Closed flowers of *F. candida*

Open flowers of *F. candida*

Open flowers of *F. tuberculosa*

Closed flowers of *F. tuberculosa*

Unusual form of *F. tuberculosa*

F. britteniae

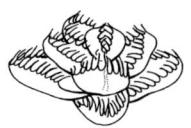

F. felina (Weston) Schwantes & Jacobsen var. felina [nt]
F. felina (Weston) Schwantes & Jacobsen var. jamesii L.Bolus
F. grandis L.Bolus
F. gratiae L.Bolus
F. kingiae L.Bolus
F. latipetala L.Bolus
F. laxipetala L.Bolus
F. longidens L.Bolus
F. longifolia L.Bolus
F. lupina (Haw.) Schwantes
F. militaris Tischer
F. montana L.Bolus
F. multidens L.Bolus var. multidens
F. multidens L.Bolus var. paardepoortensis L.Bolus
F. nemorosa L.Bolus ex L.E.Groen
F. paucidens N.E.Br.
F. peersii L.Bolus
F. plana L.Bolus
F. ryneveldiae L.Bolus
F. smithii L.Bolus
F. speciosa L.Bolus
F. subindurata L.Bolus
F. subintegra L.Bolus
F. tigrina (Haw.) Schwantes var. splendens Jacobsen [V]
F. tigrina (Haw.) Schwantes var. tigrina [V]
F. tuberculosa (Rolfe) Schwantes
F. uniondalensis L.Bolus

Literature

SCHWANTES, G. 1926. Zur Systematik der Mesembrianthemen. *Zeitschrift für Sukkulentenkunde* 2: 176.

BOLUS, H.M.L. 1937/1938. Note on *Faucaria* and *Orthopterum*. *Notes on Mesembryanthemum and allied genera.* University of Cape Town, Cape Town. Vol. 3: 109–115.

F. tuberculosa

F. gratiae

Red form of *F. tigrina*

Purple form of *F. tigrina*

Hammeria

Derivation of genus name The genus was named in honour of Steven Hammer, co-author of this book and an avid student of mesembs for the last 35 years.

Common names No vernacular names are known.

Description The plants are dwarf, creeping shrublets of up to 50 mm in height. Long shoots may spread from the central stem, rooting at the widely spaced nodes. Basal leaves are joined for a fifth to a third of their lengths. In the case of *H. salteri* basal leaves become membranous and persistent, forming sheaths around the young leaves which form the next leaf pair. The erect leaf pairs are 10 to 20 mm in length, D-shaped in cross section and are borne close together. The leaf margins are toothed at their tips and a layer of flaking wax covers the leaf surfaces. Leaf colour varies from grey-green in the rainy season to buff in the dry season, with dark green translucent dots. The flowers are about 20 mm in diameter and are borne upright on stalks of up to 30 mm in length. They are pale pink with white centres or dark pink and are subtended by bracts that resemble the leaves. The five equal sepals may have toothed tips. Petals are in two to three whorls and stamens bend inwards, forming a central cone. Short staminodes are present. Nectar glands form a ring around the five stigmas which are longer than the stamens. Fruit capsules have five locules, ample covering membranes, winged valves and diverging expanding keels. There are no closing bodies or rodlets. Seeds are smooth, honey-coloured and have darker tips.

Distinguishing characters The small creeping growth form, with long branches rooting at the nodes, together with the short, toothed leaves are characteristic and distinguish *Hammeria* from *Antimima*. The leaves are borne close together and therefore plants resemble some *Cheiridopsis* species, but the fruits are five-locular and not 10 to 20-locular as in *Cheiridopsis*.

Flowering time Flowers appear after good rains in the spring (August to September in South Africa) and are open from midday to mid-afternoon.

Distribution and ecology *Hammeria* is restricted to the Tanqua Karoo and parts of the Little Karoo in the Western and Northern Cape Provinces, South Africa. Plants grow in full sun on soils derived from dark coloured shales and quartzites of the Karoo sediments. Winter rainfall is predominant in this region and never exceeds 400 mm per annum.

Cultivation Plants can be grown from seed or from cuttings which root easily. They should be grown in well-drained soils in direct sunlight. Water sparingly in the summer months when the plants are dormant.

Notes One species of this newly described genus has long resided within *Ruschia*, but the capsules are quite different. They are shallow with complete covering membranes, ample valve wings and no closing bodies. The genus is included here because of the toothed leaves but it is closely related to genera of the BEAD-LEAVED MESEMBS.

Hammeria P.M. Burgoyne

Number of species/subspecies/varieties (2/0/0)

Species list and conservation status
H. salteri (L. Bolus) P.M. Burgoyne
H. gracilis P.M. Burgoyne & S.A. Hammer

Literature
BOLUS, L. 1932. *Notes on* Mesembrianthemum *and allied genera*. University of Cape Town, Cape Town, South Africa. Vol. 2: 371–372.

BURGOYNE, P.M., SMITH, G.F. & CHESSELET, P. 1998. *Hammeria* P.M. Burgoyne: a new genus of Aizoaceae from South Africa. *Cactus and Succulent Journal (US)* 70 (in press).

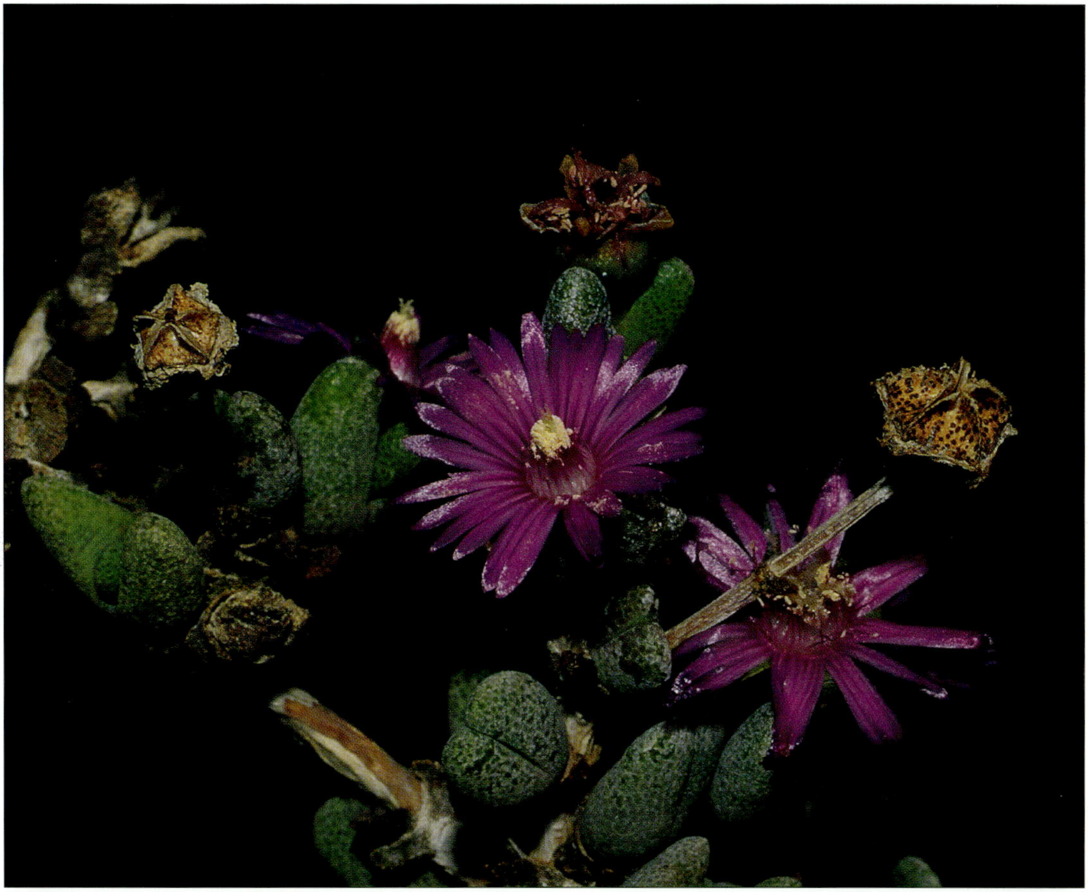

Chris Barnhill

H. salteri

Priscilla Burgoyne

H. salteri in its habitat

John Trager

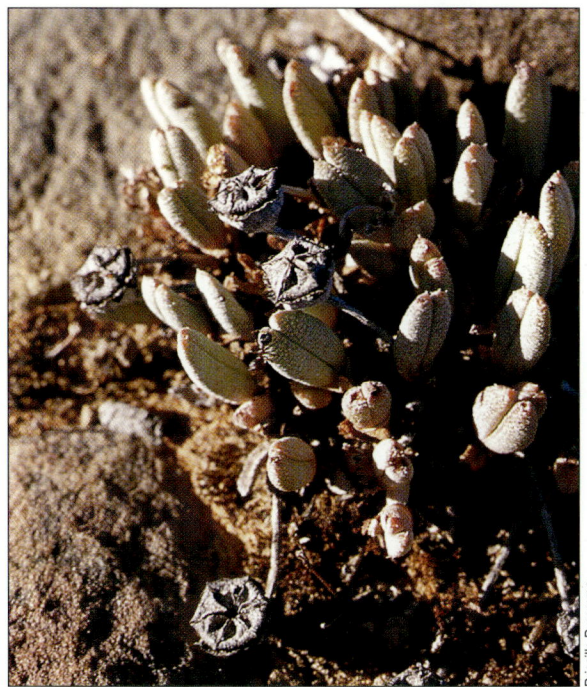

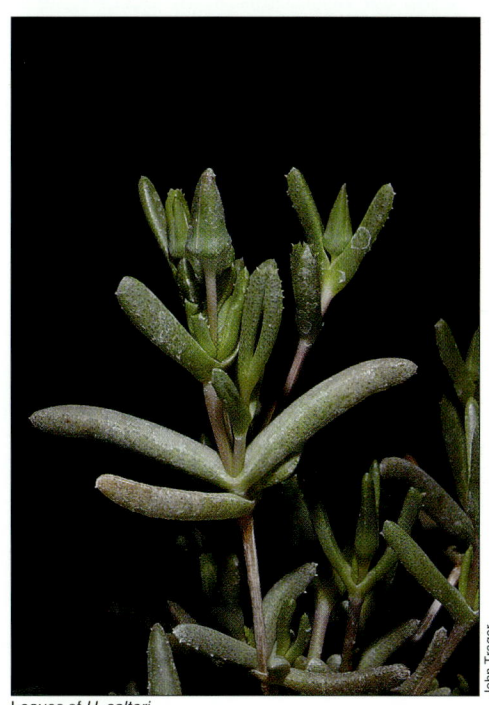

Leaves of *H. salteri*

ODONTOPHORUS

DERIVATION OF GENUS NAME The name is a combination of the Greek words *odus* (tooth) and *phorus* (bearer).

COMMON NAMES No vernacular names seem to have been recorded.

DESCRIPTION Plants, particularly of *Odontophorus nanus*, usually produce small clumps or mats, which will only in the case of one species, *O. angustifolius*, form large cushions of up to 500 mm in diameter. The rootstock is comparatively fleshy. Another species in the genus, *O. marlothii*, sends out trailing branches once the plant has flowered. The keeled leaves are greyish green, soft and fleshy and often fuzzy or hairy. Although the leaves can be spread out or placed compactly together, the typical tiger-jaw growth form is maintained throughout. The marginal teeth are usually fairly conspicuous and white, only rarely pinkish. Flower colour is mostly light yellow to deep yellowish orange; the occurrence of pink and white flowering forms is rare. The fruit capsules are eight to 11-locular, rounded and sometimes covered in soft hairs.

DISTINGUISHING CHARACTERS Species of *Odontophorus* can be distinguished by the dull greyish green, hairy appearance of their chunky, dentate leaves, the often short, erectly borne petals, and hairy fruits (in *O. nanus*).

FLOWERING TIME Species in the genus flower in late winter to early spring (July to August in South Africa). The strongly scented flowers open by midday and close by twilight.

DISTRIBUTION AND ECOLOGY The genus is found in the vicinity of Steinkopf in the Richtersveld, Northern Cape Province, South Africa.

CULTIVATION *Odontophorus* species thrive in containers outside of their habitat. They are best grown in sandy, well-drained soil. Seeds germinate readily and plants can also be grown from cuttings. Plants grow in winter.

NOTES *Odontophorus herrei*, which has been included in *O. nanus*, differs considerably from that species in its long pedicels and orange-yellow flowers. The leaves turn bright red when the plants go dormant. *O. nanus* may be distinguished by the creamy yellow colour of the flowers, as opposed to the bright yellow flowers of *O. angustifolius*. *O. angustifolius* subsp. *protoparcoides* differs in its fierce, curling black teeth and smaller, more compact habit. *O. marlothii* differs from *O. nanus* in its elongated side shoots.

Odontophorus N.E.Br.

NUMBER OF SPECIES/SUBSPECIES/VARIETIES (5/1/0)

SPECIES LIST AND CONSERVATION STATUS
O. angustifolius L.Bolus subsp. *angustifolius*
O. angustifolius L.Bolus subsp. *protoparcoides* S.A.Hammer
O. herrei L. Bolus
O. marlothii N.E.Br. [I]
O. nanus L.Bolus
O. pusillus S.A.Hammer

LITERATURE
HARTMANN, H.E.K. 1976. Monographie der Gattung *Odontophorus* N.E.Br. (Mesembryanthemaceae Fenzl). Monographien der Subtribus Leipoldtiinae Sch. I. *Botanische Jahrbücher* 97: 161–225.

HAMMER, S.A. 1993. Macro-observations on *Cheiridopsis* N.E.Br. and *Odontophorus* N.E.Br. (Aizoaceae). *Bradleya* 11: 75–85.

HAMMER, S.A. 1995. Up a sleeve: further observations on *Cheiridopsis* N.E.Br. and *Odontophorus* N.E.Br. (Aizoaceae). *Piante Grasse* Supplement 15: 59–87.

O. *nanus* in its habitat

Open flowers of *O. marlothii*

O. angustifolius subsp. *angustifolius*

Closed flowers of *O. marlothii*

ORTHOPTERUM

DERIVATION OF GENUS NAME The genus name was derived from the Greek words *orthos* (erect) and *pteron* (wing) on account of the erect wings projecting from the top of the open capsule.

COMMON NAMES No vernacular names seem to have been recorded.

DESCRIPTION Plants are dwarf, branched, compact succulents, forming small, cushion-like clumps. Branches have six to eight leaves. The slightly unequal, smooth, shiny green or reddish leaves are borne in pairs in small aloe-like rosettes. They are somewhat elongated and three angled, with few or no teeth on the margins. Old leaves are persistent. The solitary flowers are large and stalkless, with five subequal sepals. The petals are golden yellow within and reddish on the outside and are arranged in three or four whorls. Stamens are collected into a cone in the centre of a flower around the five greenish brown nectar glands which are quite distinct. There are five to six slender stigmas. Fruit capsules have five or six locules. The septa of the fruit are separated into two halves, their upper part not flat as in *Faucaria*, but arching on top of the capsule. The seeds are pear-shaped, with minute spines on their surfaces.

DISTINGUISHING CHARACTERS The smooth, soft, unmarked and usually greenish leaves and the relatively small winter flowers distinguish this genus from *Faucaria*. Other less obvious features are the presence of closing bodies and the peculiar arrangement of the septa of the fruits.

FLOWERING TIME Flowering is during late autumn and winter (May to July in South Africa). The flowers open in the afternoon and close by evening.

DISTRIBUTION AND ECOLOGY Both species occur in the subtropical thickets of the Eastern Cape Province of South Africa where they receive rain throughout the year, but with peaks in autumn and spring. They typically grow under the protection of low bushes and often co-occur with *Faucaria* species, which usually grow close by, often in the open.

CULTIVATION The two species of *Orthopterum* are easy to cultivate. They will tolerate large amounts of water if the soil mixture is well drained. Only in summer and winter should plants be watered sparingly. It is best to keep plants in a greenhouse. They can be propagated by seed or cuttings during summer.

NOTES *Orthopterum waltoniae*, which occurs near Fort Brown and Grahamstown, differs from the western species, *O. coegana*, in its longer, more narrowly tapering leaves. Both species occasionally have teeth on the leaves.

Orthopterum L.Bolus

NUMBER OF SPECIES/SUBSPECIES/VARIETIES
(2/0/0)

SPECIES LIST AND CONSERVATION STATUS
O. coegana L.Bolus [E]
O. waltoniae L.Bolus

LITERATURE
BOLUS, L. 1927. South African succulents: *Mesembrianthemum* and allied genera. *South African Gardening and Country Life* 17: 281.

BOLUS, H.M.L. 1937/8. Note on *Faucaria* and *Orthopterum*. Notes on *Mesembryanthemum* and allied genera. University of Cape Town, Cape Town. Vol. 3: 109–115.

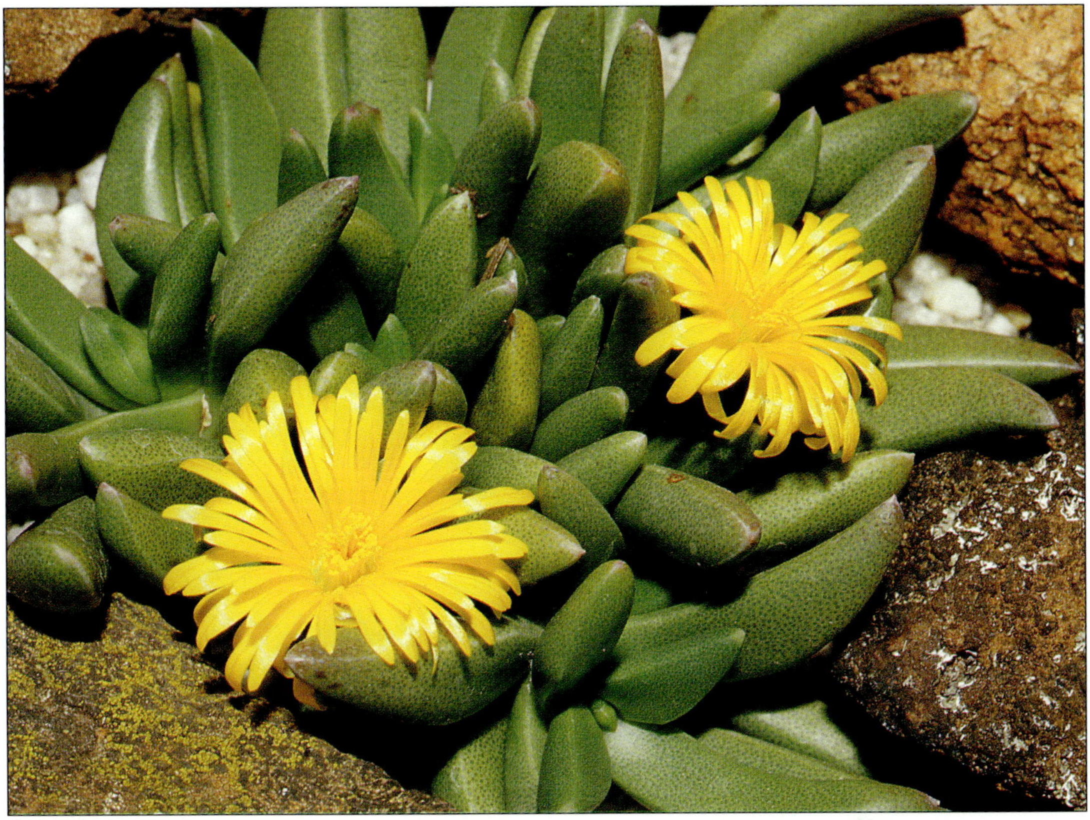

Bob Chinnock

O. coegana

Gerhard Marx

O. waltoniae

John Trager

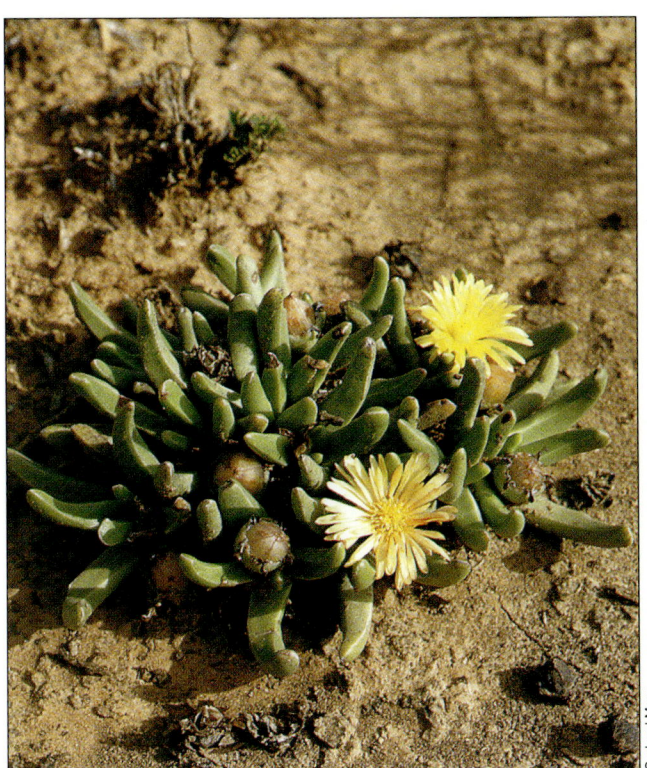

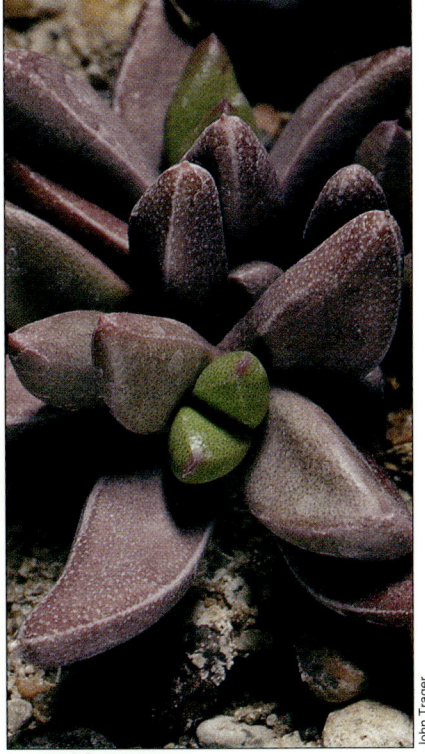

Leaves of *O. coegana*

STOMATIUM

DERIVATION OF GENUS NAME The genus name was derived from the Greek word *stomation* (open mouth), in reference to the toothed leaf pairs which resemble gaping mouths.

COMMON NAMES Two common names have been recorded: *tierbekvygie* (tiger mouth mesemb) for *Stomatium ermininum*, and *kussingvygie* (cushion mesemb) for *S. mustellinum*.

DESCRIPTION Plants form small to medium-sized clumps that tend to die down from the centre, gradually resulting in a small ring of rosettes. The leaves are narrowly to rather broadly triangular, olive green to drab grey, often purple when stressed but sometimes permanently reddish or purple. They are cylindrical towards the base, but usually keeled towards the tips, giving them a triangular appearance in cross-section. The leaf margins are usually armed with small, soft, white teeth. The scented, yellow or rarely white flowers are stalkless or shortly stalked, giving the impression that they arise directly from the leaves. The capsules are five-locular, rather soft and usually possess covering membranes; closing bodies are absent.

DISTINGUISHING CHARACTERS The numerous marginal teeth on the leaves are generally short and stiff, but quite harmless. The leaf surfaces of most species are minutely pimpled, unlike those of species of *Faucaria*, which are either perfectly smooth, rarely warty or ribbed.

FLOWERING TIME Flowering is in early spring to early summer (September to November in South Africa). Flowers generally open late in the afternoon, or in the early evening and stay open until about midnight.

DISTRIBUTION AND ECOLOGY Species of *Stomatium* occur in three east-west bands across the arid South African interior. These areas include parts of Namaqualand, Bushmanland and the Great Karoo in the Northern Cape Province. Further south, species grow in the arid Great Karoo which stretches across the Western, Northern and Eastern Cape Provinces. The third area is mostly confined to the climatically severe central Free State. Rocky outcrops are favoured habitats with the leaves of the plants often taking on the same drab colour as their surroundings.

CULTIVATION *Stomatium* species thrive in cultivation and can be grown in a greenhouse, or on a rockery. If the climate is suitably dry in winter, they can withstand considerable cold. Plants are mostly summer growers and should be watered sparingly during the winter months. The soil should be sandy, well drained and mineral rich. Propagation is by seeds or cuttings.

NOTES This is another of the few mesemb genera with obviously scented flowers. In the case of *Stomatium*, the flowers give off a rich sugary smell. The uninitiated can easily confuse *Stomatium* species with some *Faucaria* species, but the former generally have smaller leaves and are considerably untidier than the typical tiger jaw species. Like so many other mesemb genera, *Stomatium* will benefit greatly from a detailed taxonomic study. They are favourite plants for rockeries because of their tolerance of cold. The Namaqualand species, *S. alboroseum*, is the least hardy. Uniquely, its flowers turn from white to rosy pink.

Stomatium Schwantes

NUMBER OF SPECIES/SUBSPECIES/VARIETIES (39/0/1)

SPECIES LIST AND CONSERVATION STATUS
S. *acutifolium* L.Bolus
S. *agninum* (Haw.) Schwantes var. *agninum*

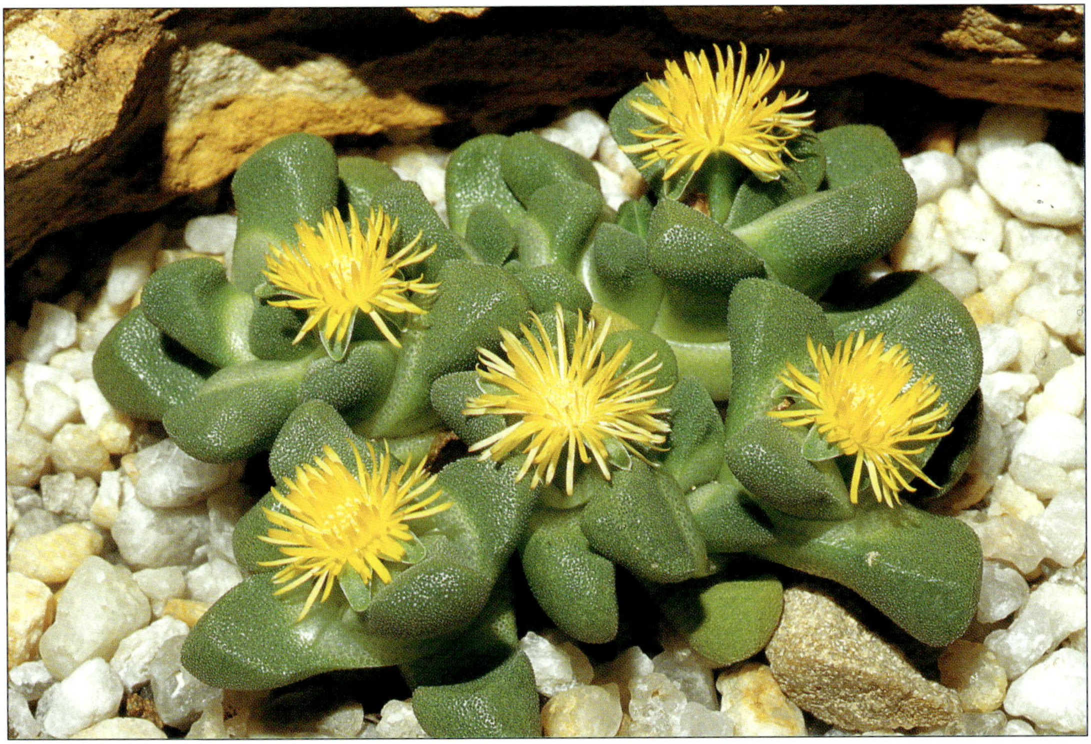
Bob Chinnock

S. suaveolens

Ben-Erik van Wyk

S. agninum

John Trager

Leaves of *S. suaveolens*

Bob Chinnock

S. niveum

Ernst van Jaarsveld

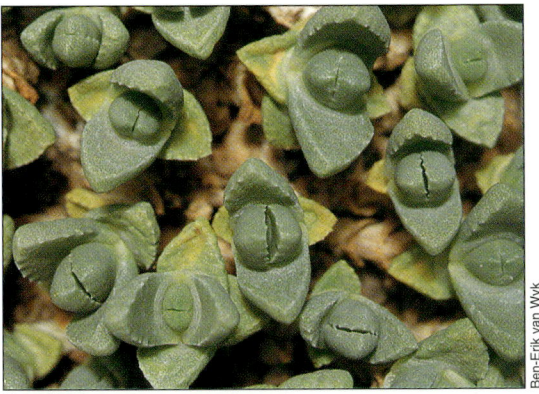

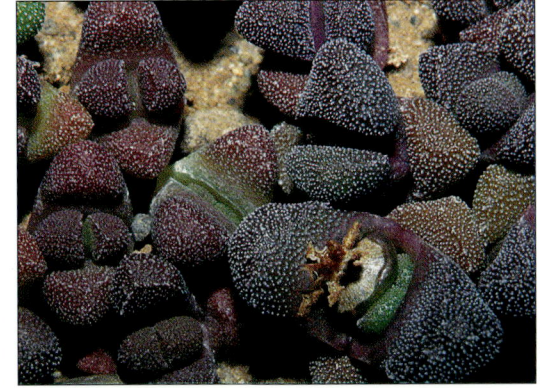

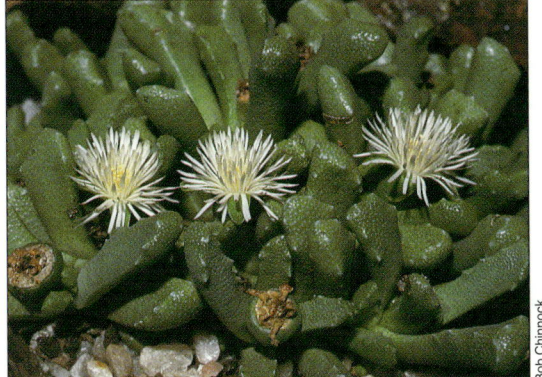

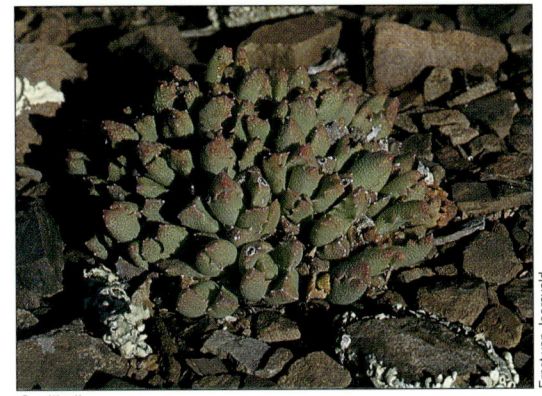

S. villetii

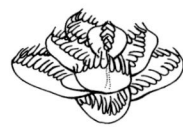

S. agninum (Haw.) Schwantes var. *integrifolium* (Salm-Dyck) Volk
S. alboroseum L.Bolus
S. angustifolium L.Bolus
S. beaufortense L.Bolus
S. braunsii L.Bolus
S. bryantii L.Bolus
S. deficiens L.Bolus
S. difforme L.Bolus
S. duthiae L.Bolus
S. ermininum (Haw.) Schwantes
S. fulleri L.Bolus
S. geoffreyi L.Bolus [I]
S. gerstneri L.Bolus
S. grandidens L.Bolus
S. integrum L.Bolus
S. jamesii L.Bolus
S. latifolium L.Bolus
S. lesliei (Schwantes) Volk
S. leve L.Bolus
S. loganii L.Bolus
S. middelburgense L.Bolus
S. murinum (Haw.) Schwantes
S. mustellinum (Salm-Dyck) Schwantes
S. niveum L. Bolus
S. patulum Jacobsen
S. paucidens L.Bolus
S. peersii L.Bolus
S. pluridens L.Bolus
S. pyrodorum (Diels) L.Bolus
S. resedolens L.Bolus
S. ronaldii L.Bolus [I]
S. rouxii L.Bolus
S. ryderae L.Bolus
S. suaveolens Schwantes
S. suricatinum L.Bolus
S. trifarium L.Bolus
S. villetii L.Bolus
S. viride L.Bolus

LITERATURE
SCHWANTES, G. 1926. Zur Systematik der Mesembrianthemen. *Zeitschrift für Sukkulentenkunde* 2: 176.

S. mustellinum

Flowers of S. mustellinum

Flowers of S. alboroseum

S. alboroseum

S. lesliei

S. suricatinum

VANHEERDEA

DERIVATION OF GENUS NAME The genus was named after Mr Pieter van Heerde (1893 to 1979) of Springbok, who was a school teacher and plant collector with a special interest in the succulents of Namaqualand.

COMMON NAMES No vernacular names seem to have been recorded.

DESCRIPTION Species of Vanheerdea form small, compact branched plants, each growth point bearing a pair of leaves. The very fleshy leaves are a characteristic light yellowish green colour (sometimes tinged with red or even entirely flushed with purple) and are triangular to semi-circular in diameter. They are often raised above ground level, especially in the rainy season, or could be sunken with the leaf tips flush with the ground. One of the two recognised species, V. primosii, has windows on the leaf tips. The margins of the leaves carry small, harmless, widely or narrowly spaced teeth. These structures also occur on the more or less central keel running for some distance down the back of the fat leaves. The leaf surface can either be felty (V. roodiae) or entirely smooth (V. primosii). The flowers are fairly large, up to 50 mm in diameter when fully open, and a pleasant bright yellow colour. The conical fruit capsules are seven to 15-locular and are borne on short or long stalks. Since the fruits do not close entirely after the seeds have been released, the terminal part is the shape of an inverted cone.

DISTINGUISHING CHARACTERS Vanheerdea species are characterised by very thick yellow-green leaves with short, stout, teeth on the margins and keels and also by the rich yellow-orange tinge of the petals. Each branch can bear up to three flowers, but these do not appear simultaneously. This distinguishes the genus from Cheiridopsis and Ihlenfeldtia, where the flowers are invariably single.

FLOWERING TIME Plants flower in late winter (July to August in South Africa). The strongly scented flowers open in the mid-afternoon and close at night.

DISTRIBUTION AND ECOLOGY Species of Vanheerdea occur in Bushmanland, Northern Cape Province, South Africa, on the border between the winter and summer rainfall regions. They usually grow fully exposed in barren areas covered by sharp gravel and seem to prefer soils derived from limestone.

CULTIVATION The species of Vanheerdea are rather uncommon in collections. The little that is known about their growing preferences indicates that they are not exceptionally difficult to grow if watered sparingly. Their occurrence on the border between the summer and winter rainfall regions implies their tolerance of watering during the hot summer months. Sow seeds in summer or autumn. V. primosii is notoriously prone to cracking when watered and rarely flowers in cultivation.

NOTES The two species of the genus Vanheerdea show affinities with a number of species included in other mesemb genera: similar fruits are found in some species of Aloinopsis, Ihlenfeldtia, Octopoma, Tanquana and Titanopsis. The leaves of some Aloinopsis species have the same felty surface as those of Vanheerdea roodiae.

Vanheerdea L.Bolus ex H.E.K.Hartmann

NUMBER OF SPECIES/SUBSPECIES/VARIETIES (2/0/0)

SPECIES LIST AND CONSERVATION STATUS
V. primosii (L.Bolus) L.Bolus ex H.E.K.Hartmann
V. roodiae (N.E.Br.) L.Bolus ex H.E.K.Hartmann

LITERATURE
HARTMANN, H.E.K. 1992. Revision of the genus Vanheerdea L.Bolus ex H.E.K.Hartmann (Aizoaceae). Bradleya 10: 5–16.

Flowers of *V. roodiae*

Leaves of *V. roodiae*

V. roodiae in its habitat

V. primosii

Ebracteola wilmaniae

Marlothistella stenophylla

Cheiridopsis meyeri

Cylindrophyllum dyeri

Tufted Mesembs

GROUP 7

The succulent, finger-like leaves are erect and crowded together; they may be cylindrical or three-sided; plants have short stems, often with thickened, fleshy or woody roots.

The tufted leaves usually have distinct dots on their surfaces, giving the leaf surfaces a textured appearance. The plants often form several tufts due to the branching of the thickened underground parts (e.g. *Hereroa*, *Khadia*). A number of genera included in the group have yellow flowers that open in the early afternoon, in the early evening or at night. Others have pink or white flowers and usually open in sunlight. The TUFTED MESEMBS have a wide variety of leaf shapes, ranging from the cylindrical, finger-like leaves of *Cylindrophyllum* to the laterally flattened leaves of *Rhombophyllum*, and the angular leaves of *Rabiea*.

TUFTED MESEMBS comprise 13 genera and 118 species.

Bergeranthus (10 species)	*Hereroa* (33 species)	*Psammophora* (4 species)
Calamophyllum (3 species)	*Khadia* (5 species)	*Rabiea* (6 species)
Cheiridopsis (33 species)	*Machairophyllum* (10 species)	*Rhombophyllum* (2 species)
Cylindrophyllum (6 species)	*Marlothistella* (1 species)	*Ruschianthus* (1 species)
Ebracteola (4 species)		

BERGERANTHUS

DERIVATION OF GENUS NAME The genus was named after Alwin Berger (1871 to 1931) of Germany, well-known student of South African succulent plants. The Latinised suffix *-anthus* was taken from the Greek word *anthos* which means flower.

COMMON NAMES No vernacular names seem to have been recorded.

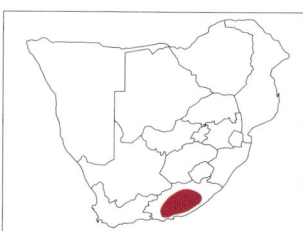

Reinet and Port Elizabeth in the west and East London and Queenstown in the east. The climate in the topographically diverse Eastern Cape is ameliorated by the adjacent warm Indian Ocean. The vegetation is diverse, but consists primarily of spiny, impenetrable subtropical thickets. Rain falls mainly in spring and autumn.

DESCRIPTION Plants are low-growing and clump-forming, with a thickened rootstock. The rather thin, smooth-edged, often somewhat elongated leaves are upright or droop slightly, and taper into a pointed tip. The usually dark green or yellowish green leaves (sometimes reddish purple, or even bluish green from wax) are more or less triangular in cross-section. The colour of the flowers ranges from bright yellow to a deep yellow-orange. They arise in groups of three to five on a branched stalk with several bracts. The stamens are numerous and the five thread-like to awl-shaped stigmas are longer than the stamens. Nectar glands are yellowish green, large or small. Fruit capsules are five-locular with rigid covering membranes and large closing bodies. They may open while still green. Seeds are more or less pear-shaped.

FLOWERING TIME Plants flower at almost any time of the year. Flowers open from late afternoon to early evening and close before midnight.

DISTINGUISHING CHARACTERS *Bergeranthus* plants are easy to spot by the long, grassy, triangular, smooth leaves and the branching flower clusters. The dots on the leaves are never raised, unlike those of *Hereroa* or *Rabiea*.

DISTRIBUTION AND ECOLOGY In contrast to so many other mesemb genera that are geographically centred in the winter rainfall area of South Africa, *Bergeranthus* occurs only in the Eastern Cape Province, between Graaff-

CULTIVATION Species of *Bergeranthus* do not present any difficulties in cultivation. They grow quite easily if some attention is paid to regular watering. Flowers are also produced quite readily. In cultivation plants are prone to red spider infestations.

NOTES The genus *Bergeranthus* represents yet another assemblage of mesemb species that is in dire need of taxonomic treatment because many of the species seem to merge and are difficult to distinguish. The widely accepted *B. glenensis*, recorded from the Free State, is nowadays correctly placed in the genus *Hereroa*.

Bergeranthus Schwantes

NUMBER OF SPECIES/SUBSPECIES/VARIETIES
(10/0/0)

SPECIES LIST AND CONSERVATION STATUS
B. addoensis L.Bolus [K]
B. artus L.Bolus
B. concavus L.Bolus
B. jamesii L.Bolus
B. katbergensis L.Bolus [K]
B. leightoniae L.Bolus
B. longisepalus L.Bolus
B. multiceps (Salm-Dyck) Schwantes
B. scapiger (Haw.) N.E.Br.
B. vespertinus (A.Berger) Schwantes

LITERATURE
SCHWANTES, G. 1926. Zur Systematik der Mesembrianthemen. *Zeitschrift für Sukkulentenkunde* 2: 179–181.

Bergeranthus sp.

B. artus

B. jamesii

B. scapiger

CALAMOPHYLLUM

DERIVATION OF THE GENUS NAME The genus name was derived from the Greek words *calamus* (reed) and *phyllon* (leaf).

DESCRIPTION Plants grow as small, branched, short-stemmed tufts of up to 200 mm in height. The roots are thin and fibrous. Leaves are somewhat blue-green and basally cylindrical, with blunt tips. Each leaf has three ridges running lengthwise from near the base to the tip. This gives the upper parts of the leaves an angular appearance. The flowers are comparatively large, ranging from 30 to 50 mm in diameter and are borne singly on short stalks which have two bracts. Flower colour apparently varies from uniformly red, to red with a white centre. The blunt-tipped petals, subtended by four sepals, are often slightly ragged at their tips and are arranged in several whorls. The stamens are white with red anthers and surround the outwardly bent stigmas which number between 10 and 12. Fruit capsules are 10 to 12-locular.

DISTINGUISHING CHARACTERS Plants seem to be distinguished by their curiously shaped, bluish green, angled leaves with blunt tips. However, a lack of living material makes it impossible to adequately characterise the genus. Flower features, in particular, are insufficiently known and cannot be used satisfactorily to establish the identity of the genus.

DISTRIBUTION AND ECOLOGY The natural distribution of *Calamophyllum* species is unknown. For this reason no map is included.

CULTIVATION Species of *Calamophyllum* are nowadays unknown in cultivation. Previous records indicate that propagation is easy from seed or cuttings and that plants grow easily and flower readily.

NOTES There is considerable confusion about this genus as it was not clearly described and the original habitats were not recorded, so that the species have not been found again for more than 200 years. To this day, there is considerable uncertainty about the true identity of these plants, with their angular, finger-like, tufted leaves. The colour of the flowers of *Calamophyllum* is shown as white in the benchmark book on the mesembs of Hans Herre. However, all the descriptive information that could be traced for the genus would seem to indicate that the flowers are deep red, or at best red with a white centre. In the original descriptions of two of the *Calamophyllum* species, their leaves are likened to those of *Mesembryanthemum fissoides*, which is nowadays treated as a species of *Antegibbaeum*.

Calamophyllum Schwantes

NUMBER OF SPECIES/SUBSPECIES/VARIETIES (3/0/0)

SPECIES LIST AND CONSERVATION STATUS
C. cylindricum (Haw.) Schwant.
C. teretifolium (Haw.) Schwant.
C. teretiusculum (Haw.) Schwant.

LITERATURE
JACOBSEN, H. 1977. *Lexicon of succulent plants.* Part II. Family Mesembryanthemaceae. Blandford, Poole.

Calamophyllum sp.

CHEIRIDOPSIS

DERIVATION OF GENUS NAME The name *Cheiridopsis* was derived from the Greek words *cheiris,* meaning sheath, and *opsis*, meaning resembling, in reference to the papery leaf sheaths that are formed during the resting period. These sheaths cover the next pair of leaves.

COMMON NAMES The following common names have been recorded: clock plant for *Cheiridopsis aurea* (now *C. robusta*) and *eseloor* (donkey ear) for *C. peculiaris*.

DESCRIPTION Plants range from dwarf to fairly large shrublets. Leaves of similar appearance are borne in pairs. Consecutive pairs are often dissimilar in shape, size and degree of fusion. Leaves may be small and almost triangular, to fairly long and slender. In sheath-forming species, the one pair of leaves dries to a prominent papery sheath enveloping the next pair. The leaf surfaces are often velvety, while leaf tips sometimes carry a few small prickles. The flowers are carried on short or long stalks. Petals are borne in several whorls, giving the flowers an almost brush-like appearance. Flower colour varies from the more commonly encountered yellow, to white and rarely even pinkish, red or purple. There are four or five sepals. Nectar glands form a ring surrounding the 10 to 20 feathery stigmas. The fruit capsules have 10 to 20 locules, covering membranes and large closing bodies. Seeds are whitish to brownish.

DISTINGUISHING CHARACTERS *Cheiridopsis* is such a variable genus that no obvious feature is common to all species. A graceful, curving flower stalk is found in many species, but this is not unique to *Cheiridopsis*. Instead of large valve wings (as in *Cephalophyllum*), the fruit capsules have awn-like points at the tips of the expanding keels. The leaves that make up a pair are often closely pressed together, resembling the beak of a bird. Leaf sheaths are found in only a third of the species but they are not always apparent in cultivated plants.

FLOWERING TIME Species of *Cheiridopsis* flower from autumn to early spring (April to October in southern Africa). The flowers are scented and open around midday, closing again at sunset.

DISTRIBUTION AND ECOLOGY The distribution range of the species of *Cheiridopsis* covers a broad band stretching from Lüderitz, in Namibia, southwards through the Richtersveld and Namaqualand in the Northern Cape Province, to the northern parts of the Western Cape Province, South Africa. The largest concentration of species is in the vicinity of Steinkopf and Platbakkies. Winter rainfall predominates across the entire distribution.

CULTIVATION Species of *Cheiridopsis* thrive in cultivation but should be grown in a greenhouse outside of their habitats. Soils should be sandy, well-drained and mineral rich. Plants respond well to an autumn and winter watering regime and water should be withheld during the hot summer months. In warm climates, plants are not shy to flower and some are very attractive.

NOTES Some 100 species were once recognised in the genus but only about a quarter of this number is accepted nowadays. *Cheiridopsis* species are reknowned for their diverse flower colours. Within one, *C. speciosa*, flower colour may range from butter yellow to magenta with a carmine centre, to the orangy and lilac colours of sunset. In two species (*C. glomerata* and *C. purpurea*), the petals do not unfold fully but rather hide the stamens (just as in *Namaquanthus*). A well-known species is *C. cigarettifera*, the old sheaths of which resemble cigarette holders. *C. peculiaris* is a good example of a species with two leaf types and two leaf placements – the one pair lying flat on the ground and the other like an

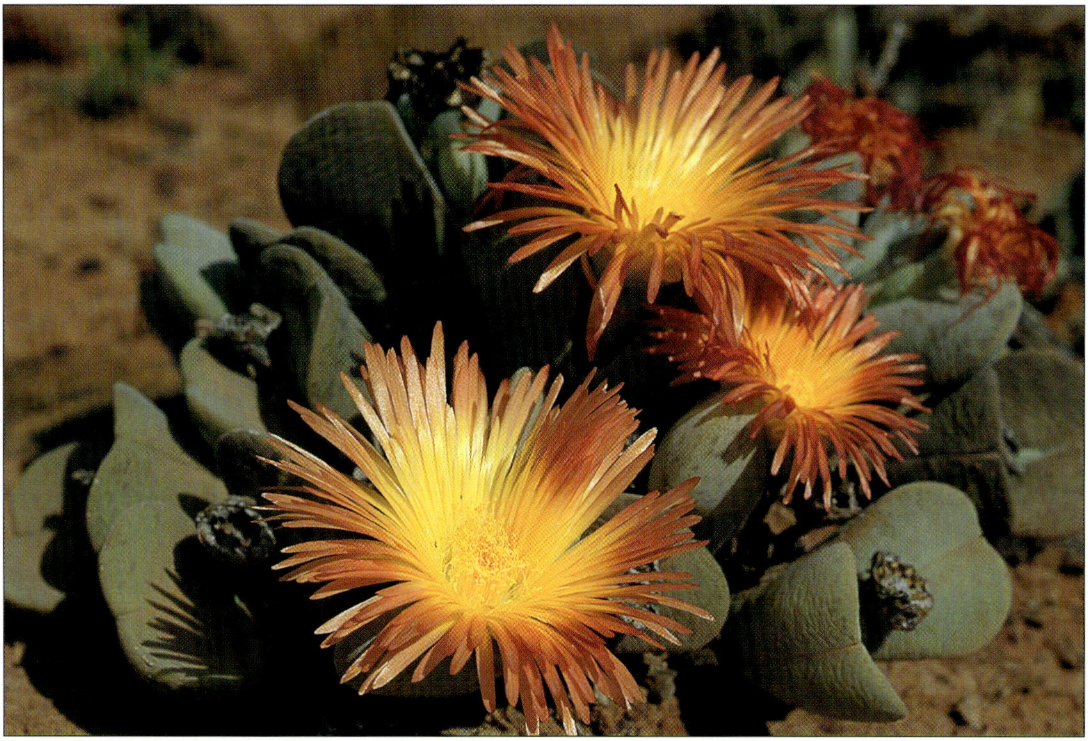

Graham Williamson

C. pillansii

Graham Williamson

C. imitans

Ernst van Jaarsveld

C. acuminata

Ernst van Jaarsveld

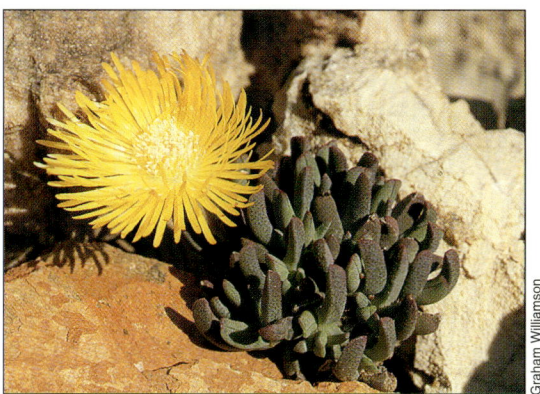

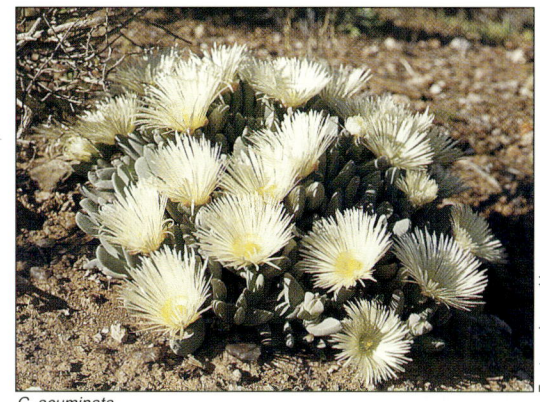

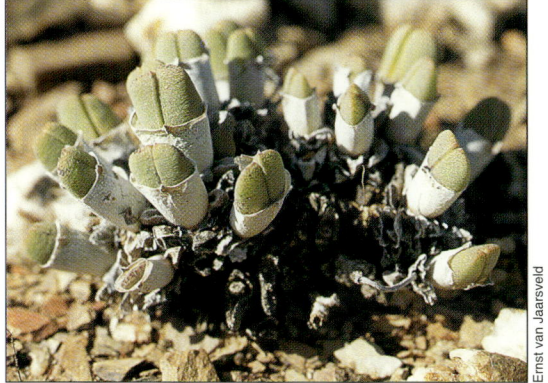

Dormant plant of *C. cigarettifera*

John Manning

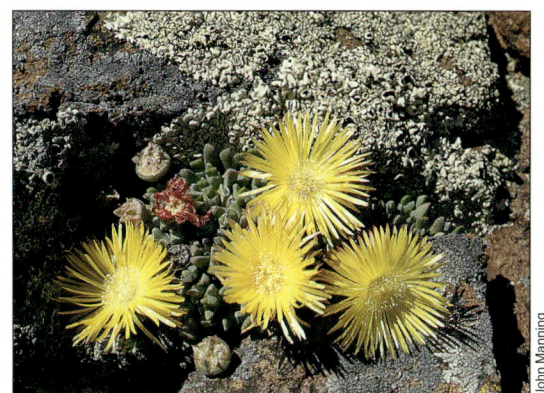

Flowering plant of *C. cigarettifera*

upright unopened clam. Both types of leaves often have the same brownish purple colour as the stones amongst which they grow. The very rare *C. umdausensis* has remarkable whitish warts on the leaves.

Cheiridopsis N.E.Br.

NUMBER OF SPECIES/SUBSPECIES/VARIETIES
(33/0/0)

SPECIES LIST AND CONSERVATION STATUS
   C. acuminata L.Bolus
   C. amabilis S.A.Hammer
   C. alata L.Bolus [I]
   C. aspera L.Bolus
   C. brownii Tischer
   C. caroli-schmidtii (Dinter & A.Berger) N.E.Br.
   C. cigarettifera (A.Berger) N.E.Br.
   C. delphinoides S.A.Hammer [R]
   C. denticulata (Haw.) N.E.Br.
   C. derenbergiana Schwantes
   C. gamoepensis S.A.Hammer
   C. glomerata S.A.Hammer
   C. herrei L.Bolus
   C. imitans L.Bolus
   C. meyeri N.E.Br.
   C. minima Tischer
   C. nelii Schwantes
   C. pearsonii N.E.Br. [V]
   C. peculiaris N.E.Br. [V]
   C. pillansii L.Bolus
   C. pilosula L.Bolus
   C. ponderosa S.A.Hammer
   C. purpurata L.Bolus
   C. purpurea L.Bolus
   C. robusta (Haw.) N.E.Br.
   C. rostrata (L.) N.E.Br.
   C. rudis L.Bolus [I]
   C. schlechteri Tischer
   C. speciosa L.Bolus
   C. turbinata L.Bolus
   C. umdausensis L.Bolus [V]
   C. velox S.A.Hammer [I]
   C. verrucosa L.Bolus

LITERATURE
HARTMANN, H.E.K. & DEHN, M. 1987. Monographie der Leipoldtiinae. VII. Monographie der Gattung *Cheiridopsis* (Mesembryanthemaceae). *Botanische Jahrbücher* 108: 567–663.

HAMMER, S.A. 1993. Macro-observations on *Cheiridopsis* N.E.Br. and *Odontophorus* N.E.Br. (Aizoaceae). *Bradleya* 11: 75–85.

HAMMER, S.A. 1995. Up a sleeve: further observations on *Cheiridopsis* N.E.Br. and *Odontophorus* N.E.Br. (Aizoaceae). *Piante Grasse* Supplement 15: 59–87.

C. robusta

C. glomerata

Fruiting plant of *C. rostrata*

Flowering plant of *C. rostrata*

C. peculiaris

C. umdausensis

CYLINDROPHYLLUM

DERIVATION OF GENUS NAME The name is derived from the Greek words *kylindros* (cylinder) and *phyllon* (leaf).

COMMON NAMES No vernacular names seem to have been recorded.

DESCRIPTION This genus comprises small or large tufted plants with short branches. The leaves are cylindrical, grey-green to silvery green, often reddish, finely dotted and up to about 100 mm in length. The solitary flowers are white to yellow or pinkish, often with darker colours in the centre. The numerous stamens are surrounded by orange or reddish filamentous staminodes that are collected into a cone. The awl-shaped stigmas are much shorter than the stamens. Fruit capsules are sharply conical, with five to 11 locules and long, narrow, awn-like valve wings (absent only in *C. tugwelliae*) and parallel, toothed expanding keels. Covering membranes are present and closing bodies very small. The seeds are dark brown.

DISTINGUISHING CHARACTERS The plants are very distinctive, being tufted, stemless succulents with cylindrical leaves and large, solitary flowers which are borne amongst the erect finger-like leaves. The fruit capsules are also very typical in that they have toothed, parallel expanding keels and usually long awn-like valve wings. Like *Faucaria* and *Neohenricia*, *Cylindrophyllum* has partially axile placentation (some seeds are attached to the central part of the capsule as in the WEEDY MESEMBS).

FLOWERING TIME Plants flower from spring to early summer (September to November in South Africa) and the flowers open at noon, closing by evening.

DISTRIBUTION AND ECOLOGY The genus occurs mainly in the Little Karoo in the Western Cape Province, South Africa. *Cylindrophyllum dyeri* has been recorded from Jansenville in the Eastern Cape Province and *C. hallii* on Brakfontein, near Loeriesfontein in the Northern Cape Province. The plants grow fully exposed on rough rocky hills.

CULTIVATION Species of *Cylindrophyllum* prefer fairly rich soils and are good rockery plants in dry areas. In wet climates, plants are best grown in a greenhouse. Water sparingly during winter and summer. Plants can be propagated easily from cuttings or seeds, which are best sown during the summer.

NOTES *Cylindrophyllum* plants often form large striking cushions which dominate dry hillsides, sometimes in the company of *Rhinephyllum* species. *C. hallii* is an outlying species recorded from Loeriesfontein, but the plant has not been found again.

Cylindrophyllum Schwantes

NUMBER OF SPECIES/SUBSPECIES/VARIETIES (6/0/0)

SPECIES LIST AND CONSERVATION STATUS
C. calamiforme (L.) Schwantes
C. comptonii L.Bolus
C. dyeri L.Bolus
C. hallii L.Bolus
C. obsubulatum (Haw.) Schwantes
C. tugwelliae L.Bolus

LITERATURE
SCHWANTES, G. 1927. Zur Systematik der Mesembrianthemen. *Zeitschrift für Sukkulentenkunde* 3: 15, 28.

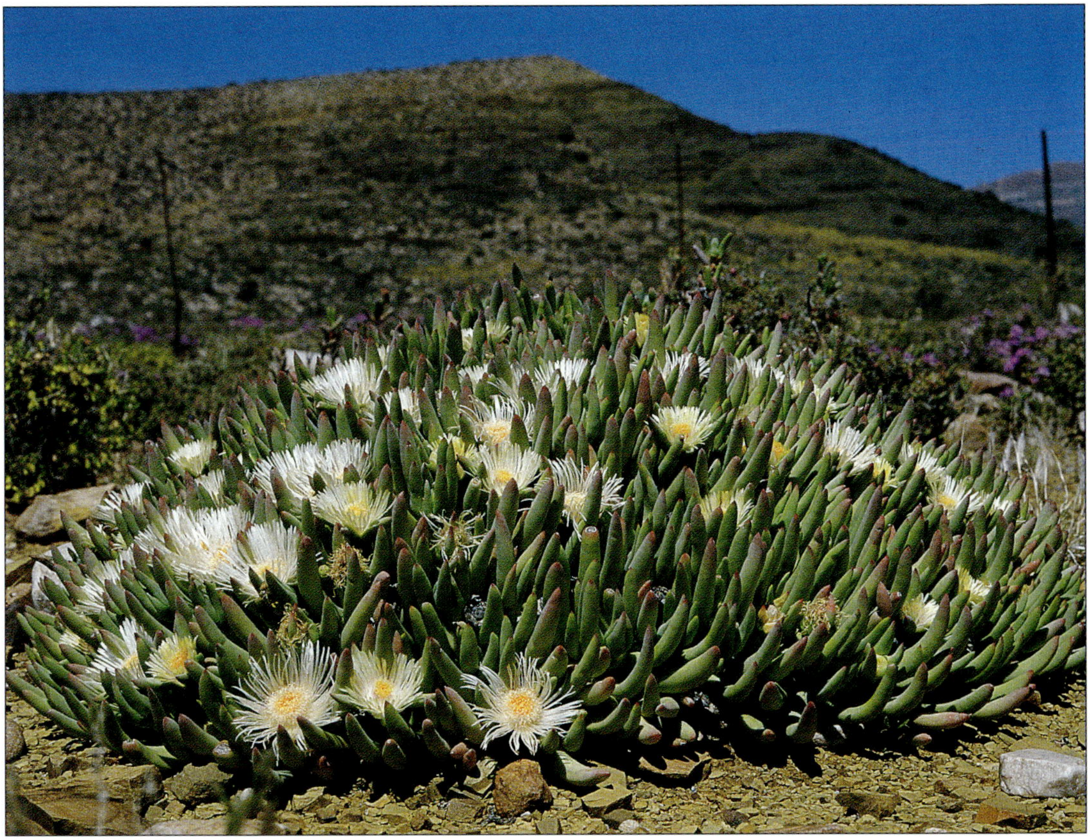
Ernst van Jaarsveld

C. tugwelliae

Ernst van Jaarsveld

C. comptonii

Ben-Erik van Wyk

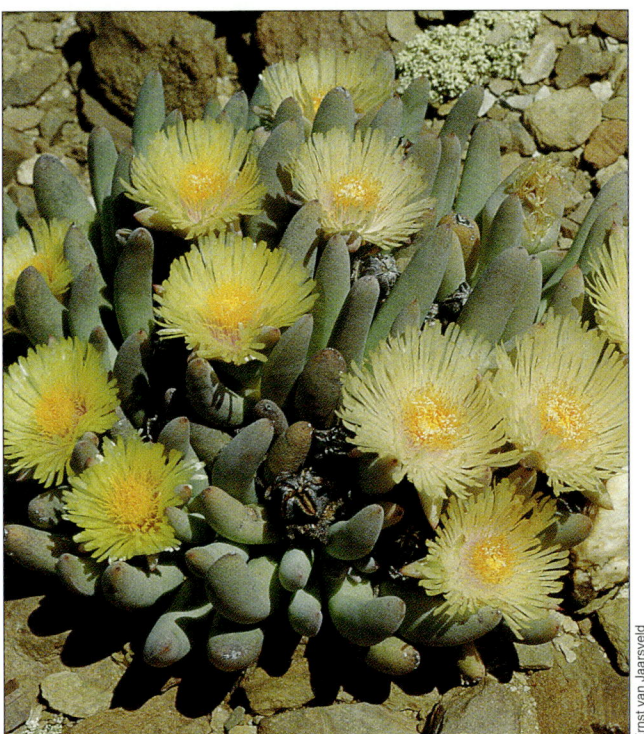

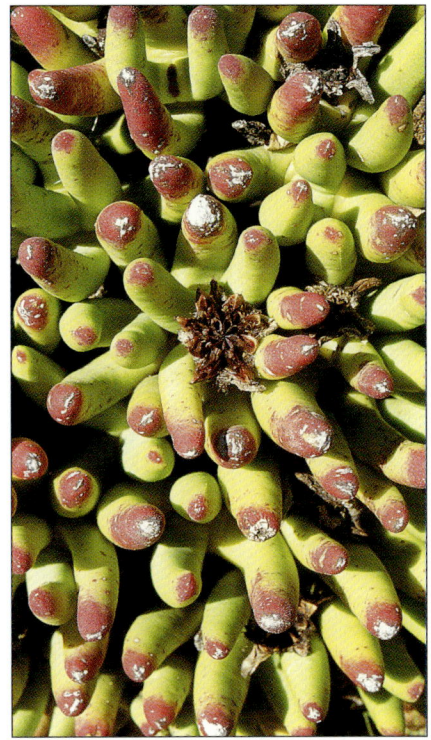

Leaves and fruit capsule of *C. dyeri*

EBRACTEOLA

DERIVATION OF GENUS NAME The genus name was derived from the Latin word *bracteola*, meaning small bract and the prefix *e-*, without, suggesting (wrongly) that bracteoles are absent.

COMMON NAMES No vernacular names are known.

DESCRIPTION Plants grow in small, much-branched tufts. The smooth, pointed leaves are comparatively short and stiff, sprouting from a thick, prominent rootstock. The leaves, which can be densely black-dotted or green-dotted, take on a beautiful blue-green colour in full sunlight, where they become tinged with red. The very shortly stalked flowers are light purple, with the colour more intense towards the tips of the petals, or pure white. The flower stalks are seemingly without bracts, but these are borne low down and are difficult to see when withered. The stamens and staminodes are arranged in a dense cone surrounding the ring-shaped nectar glands and five awl-shaped stigmas. The fruit capsules have five locules and persist for many years on the plants. The seeds are whitish.

DISTINGUISHING CHARACTERS The bicoloured flowers (bright purple fading into light purple to almost white) in combination with the short, stiff, blue-green leaves sprouting from a thickened rootstock, characterise the genus.

FLOWERING TIME Different species flower at different times of the year. The summer rainfall species, *E. wilmaniae*, flowers during spring and summer, while the others flower almost throughout the year. Flowers open during late morning and close again towards the evening.

DISTRIBUTION AND ECOLOGY *Ebracteola* has a horseshoe-shaped distribution, extending from central Namibia through the Northern Cape Province to central South Africa. Recent fieldwork has extended the distribution into the North-West Province, the Free State and Gauteng. In the grasslands of the southern African interior, the plants are inconspicuous. In Namibia, they are often found in deep, coarse sands and can be abundant in localised patches. They produce enormous tap roots.

CULTIVATION Plants are easily cultivated, either in pots, open beds or in a greenhouse. They are susceptible to root rot and red spider. They will flower freely once established.

NOTES The northernmost species, *Ebracteola montis-moltkei*, can be recognised by its extremely thin blade-like leaves and modest habit. *E. derenbergiana*, widespread in the southern Namib, is very robust, with bluish white waxy leaves. In habitat, white-flowered and pink-flowered plants are randomly interspersed. *E. fulleri*, from the western edge of Bushmanland (Steinkopf to Pofadder), has delicate, more or less cylindrical leaves which are also bluish white. The leaves of *E. wilmaniae* are darker grey-green to brownish green. When not in flower, this plant is easily confused with a yellow-flowered *Hereroa* by the same name (*H. wilmaniae*), amongst which it grows.

Ebracteola Dinter & Schwantes

NUMBER OF SPECIES/SUBSPECIES/VARIETIES (4/0/0)

SPECIES LIST AND CONSERVATION STATUS
  E. derenbergiana (Dinter) Dinter & Schwantes
  E. fulleri (L.Bolus) Glen
  E. montis-moltkei (Dinter) Dinter & Schwantes [K]
  E. wilmaniae (L.Bolus) Glen

LITERATURE
GLEN, H.F. 1986. Numerical taxonomic studies in the subtribe Ruschiinae (Mesembryanthemaceae) — *Astridia*, *Acrodon* and *Ebracteola*. *Bothalia* 16: 203–226.

HARTMANN, H.E.K. 1996. Miscellaneous taxonomic notes on Aizoaceae. *Bradleya* 14: 29–56.

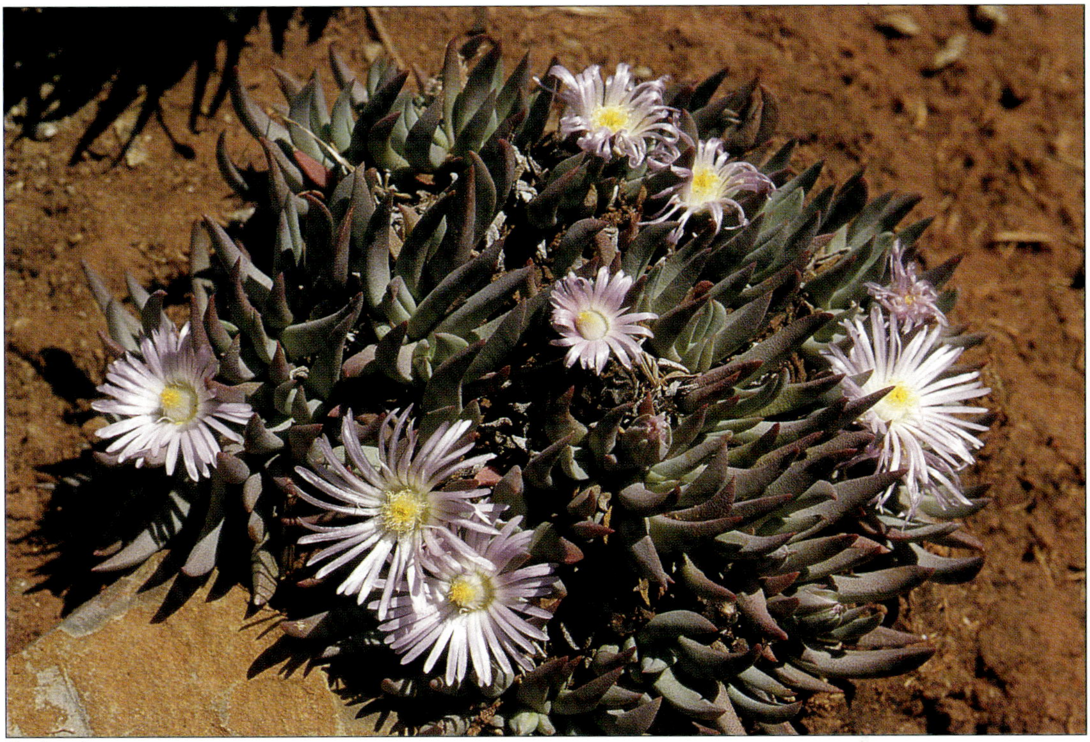

E. wilmaniae

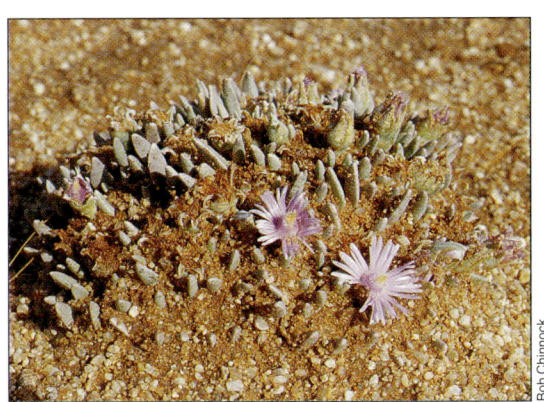

Ben-Erik van Wyk

E. derenbergiana

Bob Chinnock

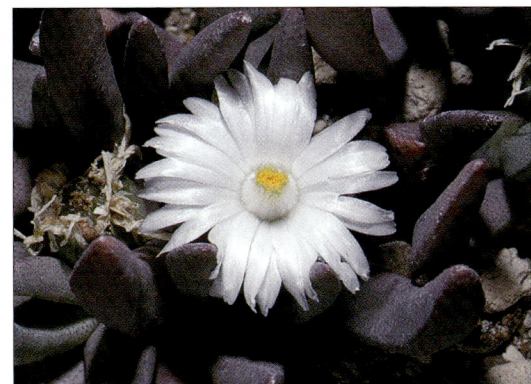

White form of *E. wilmaniae*

John Trager

E. fulleri

John Trager

E. montis-moltkei

Bob Chinnock

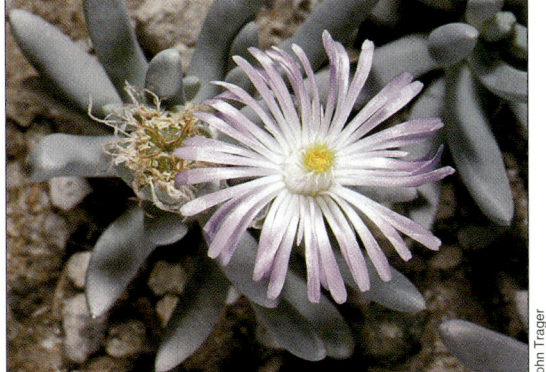

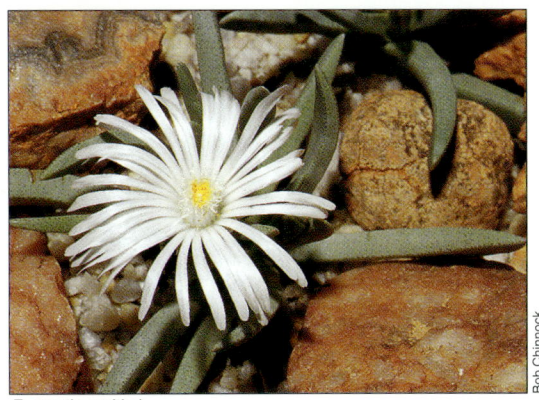

Hereroa

DERIVATION OF GENUS NAME The genus is named after the Herero people of Namibia.

COMMON NAMES The common names are *slaapvygie* (sleep mesemb) or clock plants because their flowers open in the late afternoon to evening when all else is preparing to retire. The Free State species, *Hereroa glenensis*, is known as *klein-s'keng-keng*.

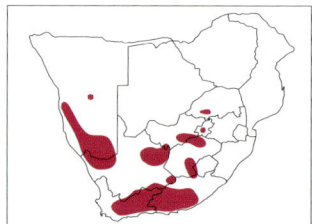

DESCRIPTION This genus comprises dwarf succulents that grow in tufts or form short shrubs of up to 200 mm high or rarely more. The leaves are generally finger-like, noodle-like or laterally compressed, narrowing to a blunt tip. Leaf colour varies from bright to dark green, often greyish or purplish green. The hairless surfaces are covered with dark dots which are raised when the leaves are not fully swollen. Flowers are usually rich yellow, rarely white. They are solitary or in the form of a cluster on a longish flower stalk with bracts. Five unequal sepals encircle the several whorls of petals. Nectar glands are free or fused into a ring. The five thread-like stigmas are longer than the stamens. Fruit capsules have five locules with tiny closing bodies, expanding keels without wings or membranous tips, and covering membranes. Seeds are small and pear-shaped.

DISTINGUISHING CHARACTERS This genus can be confused with *Khadia* at first glance, but differs from it by its yellow flowers, and by the absence of filamentous staminodes (staminodes are plentiful in *Khadia*). Some of the species of *Hereroa* may be confused with *Bergeranthus* but they differ in the presence of raised dots on partially dried out leaves and in the absence of sharp leaf tips.

FLOWERING TIME Flowers appear mainly in summer (December to February in southern Africa) and open in the afternoon or evening. The later they open, the more scented they seem to be.

DISTRIBUTION AND ECOLOGY The distribution of *Hereroa* is extensive yet discontinuous in southern Africa. It is widespread in southern Namibia and in South Africa where it is found in all the provinces except Mpumalanga and KwaZulu-Natal. Two recent South African collections of uncertain identity are reflected on the distribution map, one from the Waterberg in the North-West Province, and one from Walkerville in Gauteng. Plants favour rocky areas and full sunlight. They occur in winter and summer rainfall areas receiving less than 800 mm per year.

CULTIVATION Propagation is easy from seed or cuttings. Seedlings may flower in their first or second year. Plants grow well on rockeries in suitably dry climates or in the greenhouse. Soil should be sandy and mineral rich and it is recommended that watering be reduced in summer.

NOTES Plants of *Hereroa* vary from the tiny *H. brevifolia* from the Prince Albert district, to the imposing *H. fimbriata* from the same area. The petals of *Hereroa* usually have reddish undersides or tips, creating a halo-effect. There are many poorly defined species, so that identification is often puzzling. The well-known and commonly cultivated *Hereroa dyeri* is in fact not a *Hereroa*; the correct name for this plant is *Rhombophyllum dolabriforme*.

Hereroa (Schwantes) Dinter & Schwantes

NUMBER OF SPECIES/SUBSPECIES/VARIETIES (33/0/2)

SPECIES LIST AND CONSERVATION STATUS
- *H. acuminata* L.Bolus
- *H. albanensis* L.Bolus
- *H. angustifolia* L.Bolus
- *H. aspera* L.Bolus
- *H. brevifolia* L.Bolus
- *H. calycina* L.Bolus
- *H. carinans* (Haw.) L.Bolus

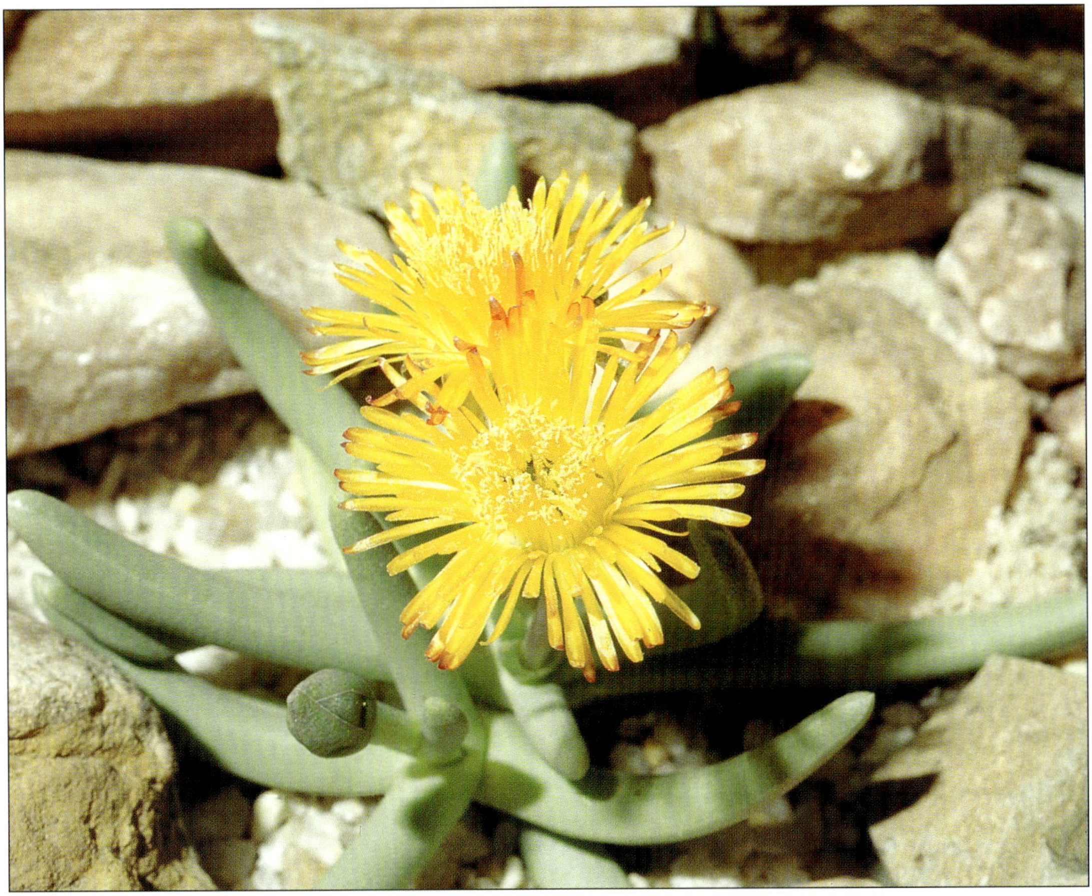
Bob Chinnock

H. pallens

Heidi Hartmann

H. pallens in its habitat

Ben-Erik van Wyk

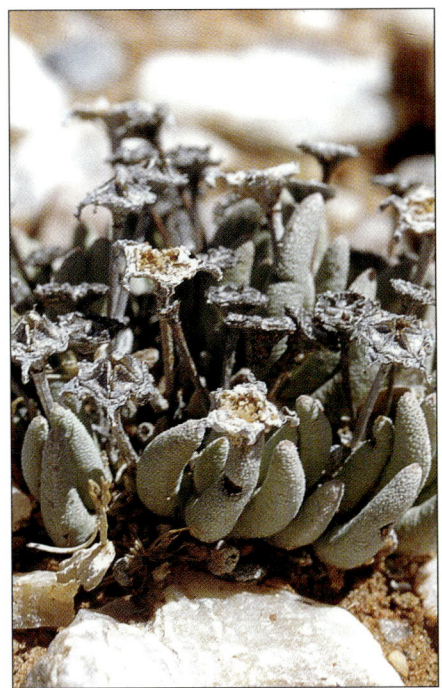

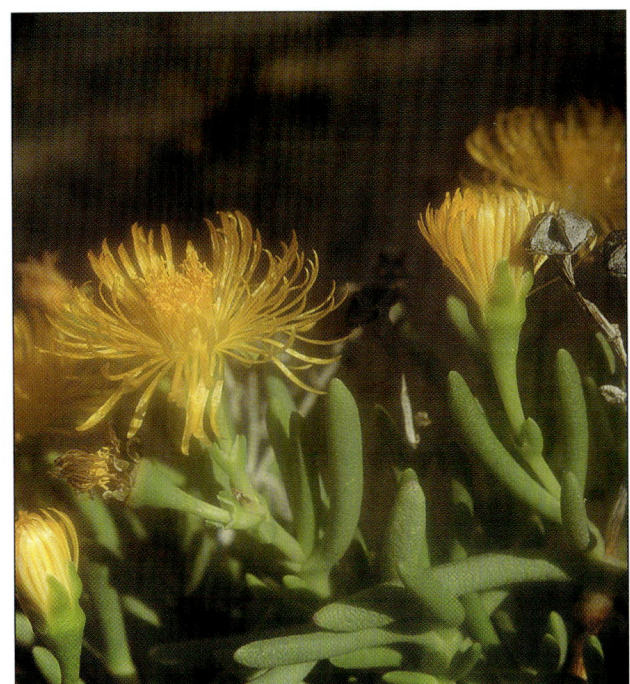

H. tenuifolia

H. concava L.Bolus
H. crassa L.Bolus
H. fimbriata L.Bolus
H. glenensis (N.E.Br.) L.Bolus
H. gracilis L.Bolus var. *compressa* L.Bolus
H. gracilis L.Bolus var. *gracilis*
H. granulata (N.E.Br.) Dinter & Schwantes
H. herrei Schwantes
H. hesperantha (Dinter) Dinter & Schwantes
H. incurva L.Bolus
H. joubertii L.Bolus
H. latipetala L.Bolus
H. muirii L.Bolus
H. nelii Schwantes
H. odorata (L.Bolus) L.Bolus
H. pallens L.Bolus
H. puttkameriana (A.Berger & Dinter) Dinter & Schwantes

H. rehneltiana (A.Berger) Dinter & Schwantes
H. stanfordiae L.Bolus
H. stanleyi (L.Bolus) L.Bolus
H. stenophylla L.Bolus
H. tenuifolia L.Bolus
H. teretifolia L.Bolus
H. tugwelliae (L.Bolus) L.Bolus
H. uncipetala (N.E.Br.) L.Bolus
H. willowmorensis L.Bolus
H. wilmaniae L.Bolus var. *langebergensis* L.Bolus
H. wilmaniae L.Bolus var. *wilmaniae*

LITERATURE
SCHWANTES, G. 1927. Zur Systematik der Mesembrianthemen. *Zeitschrift für Sukkulentenkunde* 3: 23.

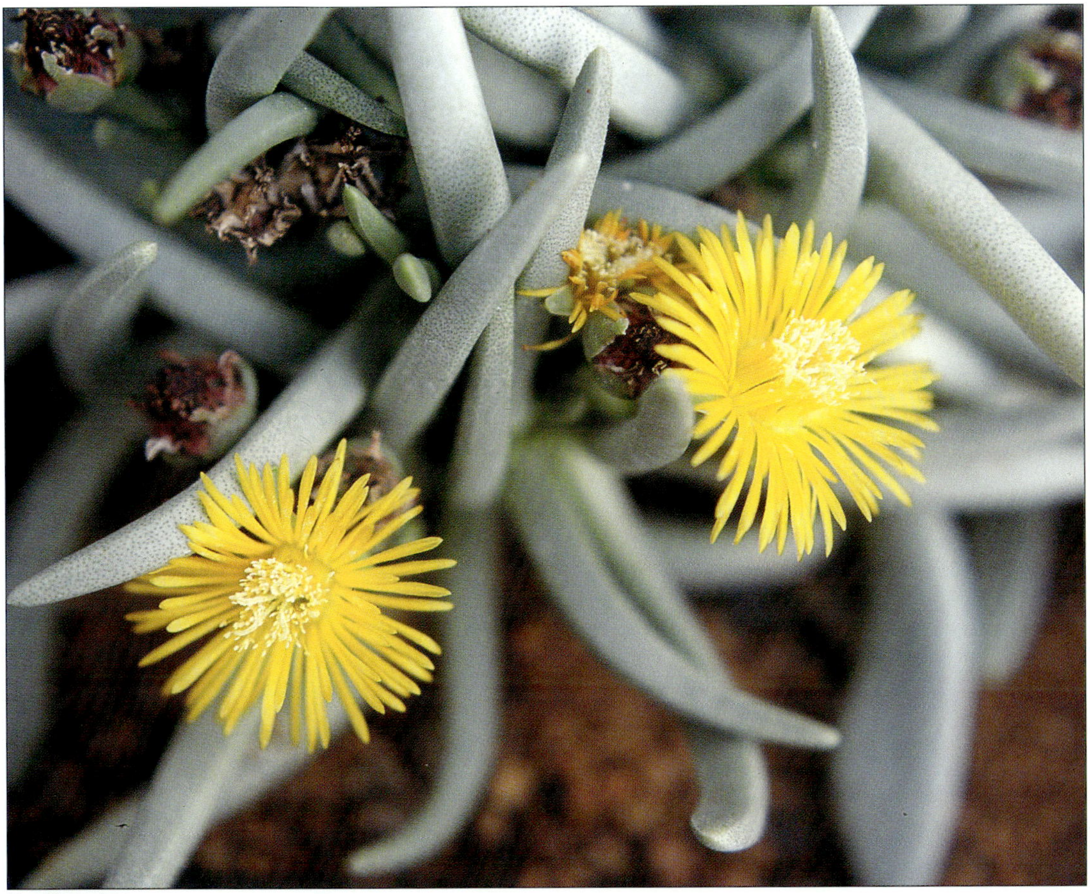

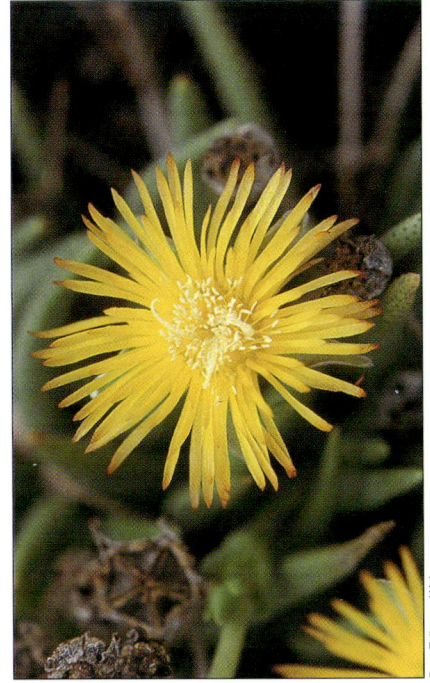

Ben-Erik van Wyk

H. puttkameriana

Ben-Erik van Wyk

H. angustifolia

Ernst van Jaarsveld

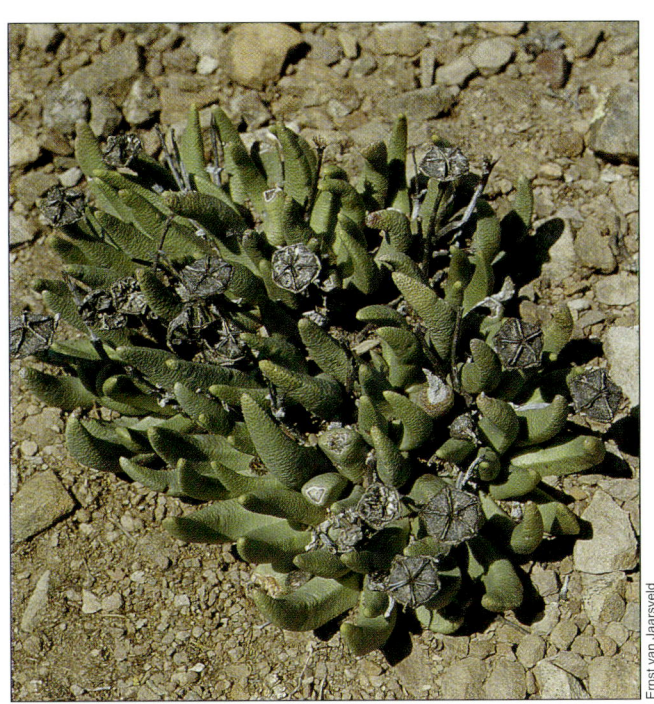

H. calycina in its habitat

KHADIA

Derivation of genus name The genus is named after the South Sotho word *khadi,* referring to the use of the starchy roots in brewing beer.

Common names *Khadia acutipetala* is known as *khadi, khadiwortel* or khadi root.

Description Plants are compact dwarf succulents with branches arising from a thickened rootstock resulting in a dense or sparse mat-like appearance. The dark green or grey-green leaves are fused and semi-cylindrical at the base with three-edged tips. Dots are concentrated along the keel and margins of the leaves. The white or pink to purple flowers occur singly on long or short flower stalks and are surrounded by five or six sepals. Staminodes form a distinct cone in newly opened flowers, spreading and bending backwards with time. Fruit capsules are woody, dark brown and reddish when young, with five to 11 locules, mostly with persistent flower stalks, otherwise breaking off and forming tumble fruit. The fruit has characteristic large valve rims and complete, rather firm covering membranes with low rims at their distal ends. The membrane itself is filled with a whitish, spongy tissue. Seeds are about 1 mm long, reddish orange or brown.

Distinguishing characters *Khadia* is distinguished by the large, woody root system, the sharp-edged and sharp-tipped grey-green to dark green leaves with visible translucent dots, and the unique capsule structure.

Flowering time Flowering occurs in summer (October to December in South Africa). Flowers mostly open in the late morning, but in *K. carolinensis* they open in the afternoon.

Distribution and ecology *Khadia* occurs in the northeastern parts of South Africa, in three distinct areas. These are the Northern Province; Gauteng and the North-West Province; and Mpumalanga extending into northern KwaZulu-Natal. All known species of *Khadia* occur at altitudes above 1 400 m in quartzitic soils, in grassland vegetation, or in humic pockets on and between rock plates.

Cultivation Propagation is by seeds, cuttings or division. Once established, and given a suitable climate and somewhat acidic, sandy soil, plants grow easily. They are susceptible to red spider mite. Growth occurs in summer and plants should preferably be kept dry during winter. Good results are obtained from plants grown in containers, but they also flourish in rockeries.

Notes *Khadia acutipetala* is a well-known species with large, fluffy purple or rarely white flowers. The white-flowered *K. borealis* deviates somewhat from the general tufted growth by its trailing branches. The recently described *K. alticola* has short stubby leaves and pale pink flowers and should be very popular in horticulture. The cohesion of this genus is doubtful and it is not unlikely that future investigations will lead to the recognition of at least two or three genera.

Khadia N.E.Br.

Number of species/subspecies/varieties (5/0/0)

Species list and conservation status
K. acutipetala (N.E.Br.) N.E.Br.
K. alticola Chesselet & H.E.K.Hartmann [nt]
K. beswickii (L.Bolus) N.E.Br. [E]
K. borealis L.Bolus
K. carolinensis (L.Bolus) L.Bolus

Literature
CHESSELET, P. & HARTMANN, H.E.K. 1995. *Khadia alticola* Chess. & H.E.K.Hartm. spec. nov. (Mesembryanthema, Aizoaceae). *Aloe* 32: 46–49.

CHESSELET, P., HARTMANN, H.E.K., HAHN, N., BURGOYNE, P. & SMITH, G.F. 1998. Taxonomic notes on the genus *Khadia* (Mesembryanthemaceae). *Bothalia* 28: 25–33.

K. carolinensis

K. acutipetala

K. alticola

K. borealis

MACHAIROPHYLLUM

DERIVATION OF GENUS NAME The name alludes to the sharply tapering shape and large size of the leaves. It is derived from the Greek *machaira* (sabre) and *phyllon* (leaf) and means dagger-leaf.

COMMON NAMES *Machairophyllum* species are sometimes referred to as dagger plants.

DESCRIPTION The plants are robust to dwarf succulents with either a compact or a branching habit of up to 1,2 m in diameter. They are highly variable in size, depending on exposure and nutrition. The leaves are opposite, crowded, up to about 200 mm in length and up to 20 mm broad, sharply triangular in cross-section and taper to sharp points. Old leaves persist for a long time. Leaf surfaces are invariably dotless, very smooth, pale green, sometimes flushed with red and may become whitish green. The large flowers are golden yellow to orange and are mostly borne on very long stalks, usually in threes but occasionally solitary. The stalks have large bracts, situated basally or halfway up (on the same plant!). The petals are densely arranged in three to seven whorls. They are copper-red on the outside, and yellow or golden orange within. Numerous stamens collectively resemble a powder-puff. Fruit capsules are five to 15-locular, with covering membranes and closing bodies. Seeds are egg-shaped and golden brown.

DISTINGUISHING CHARACTERS Leaves in this genus are sharply pointed, sometimes exceptionally large, pale milky green, dotless, with a waxen appearance. The flowers are large to very large, usually nocturnal, with very numerous petals and staminodes.

FLOWERING TIME *Machairophyllum* species flower in late winter to early summer. The flowers open after dark and close by daybreak or later in the following day. They are not strongly scented, unlike those of most other nocturnal species. A few species have flowers which open in the late afternoon, closing by dusk.

DISTRIBUTION AND ECOLOGY This is predominantly a Little Karoo genus, extending from Barrydale in the west to Willowmore in the east. It has also been recorded from the Zuurberg in the Eastern Cape Province and Lekkersing in the southern Richtersveld, a strange and unique disjunction. Plants occur on fire-prone slopes in fynbos or on rocky outcrops in mountainous areas. Two dwarf species occur on the crumbling silcrete hills around De Rust in the Little Karoo where their habitat is gradually disappearing and the mesembs with it!

CULTIVATION Plants are easily grown, but they need a bright sunny spot and abundant water to flower well; otherwise they sulk. Well-grown plants are remarkably showy.

NOTES The species vary in shape to a confusing degree; permanent runts occur near giant specimens. However, *Machairophyllum brevifolium* is always small, and its late-diurnal flowers suggest those of *Faucaria*. The best known species is the giant-leaved *M. albidum*.

Machairophyllum Schwantes

NUMBER OF SPECIES/SUBSPECIES/VARIETIES (10/0/0)

SPECIES LIST AND CONSERVATION STATUS
M. acuminatum L.Bolus
M. albidum (L.) Schwantes
M. baxteri L.Bolus
M. bijlii (N.E.Br.) L.Bolus
M. brevifolium L.Bolus
M. cookii (L.Bolus) Schwantes
M. latifolium L.Bolus
M. stayneri L.Bolus
M. stenopetalum L.Bolus
M. vanbredai L.Bolus

LITERATURE
HAMMER, S.A. 1991. Friendly daggers, or, notes on *Machairophyllum*. *Mesemb Study Group Bulletin*. 4: 77–78.

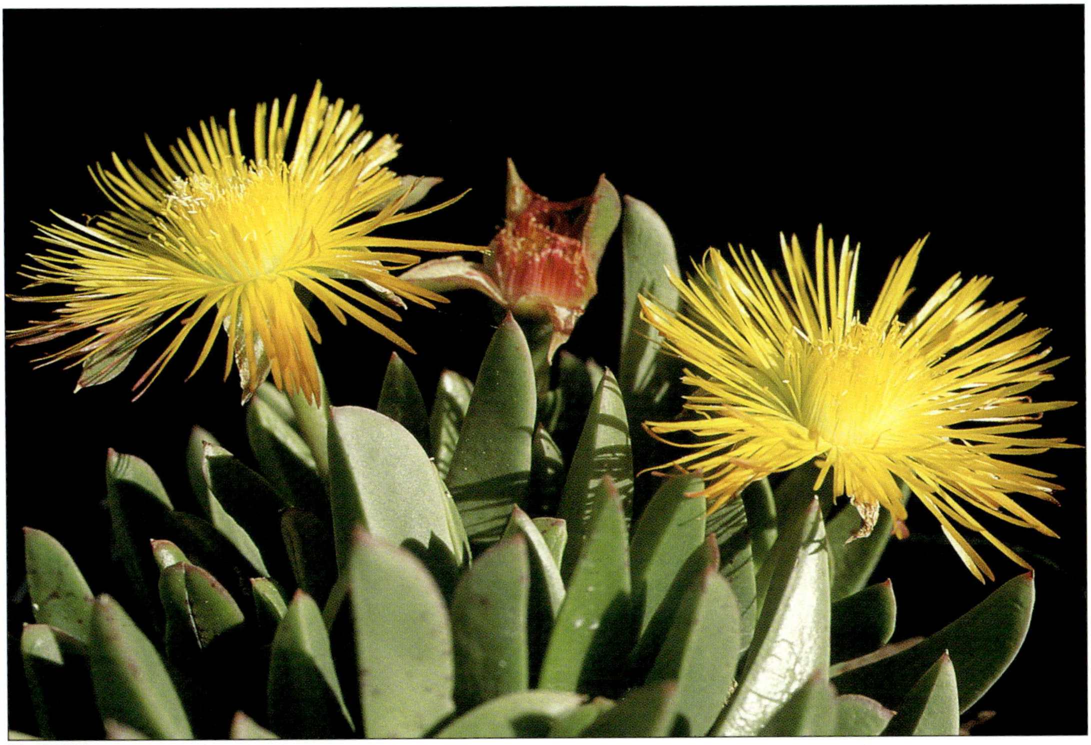

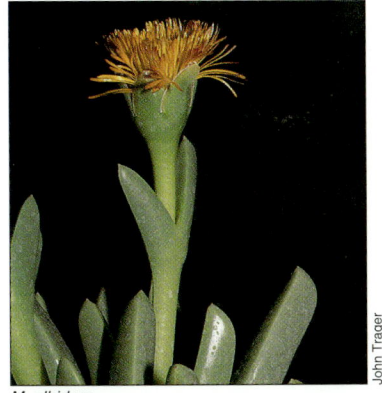

M. albidum

Ernst van Jaarsveld

M. albidum

John Trager

M. acuminatum

Heidi Hartmann

M. brevifolium

Ernst van Jaarsveld/NBI

M. latifolium

John Trager

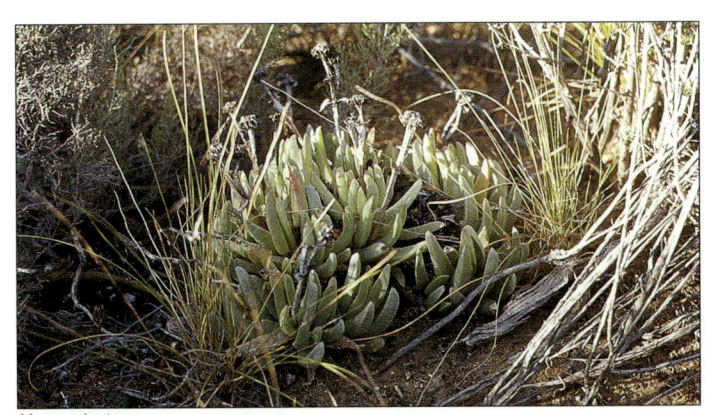

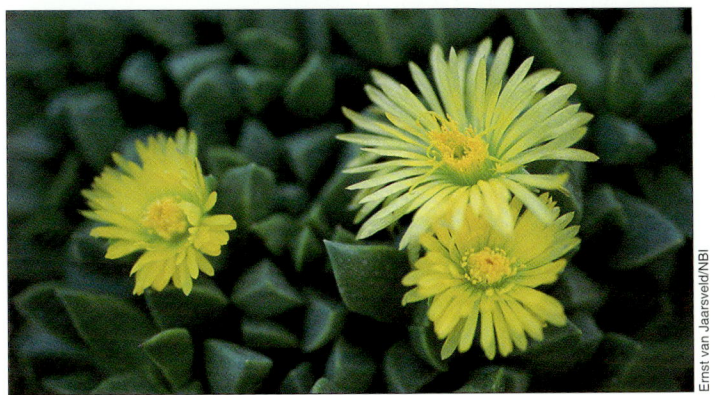

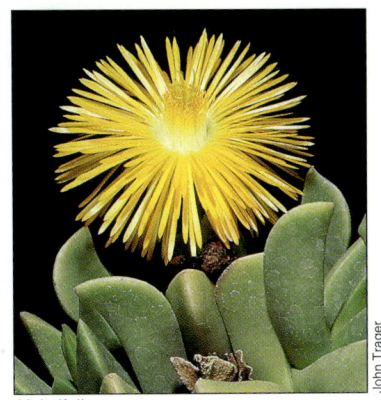

MARLOTHISTELLA

DERIVATION OF GENUS NAME The genus was named after Dr Hermann Wilhelm Rudolf Marloth (1855 to 1931), apothecary and botanist with a strong interest in the flora of South Africa.

COMMON NAMES No vernacular names seem to have been recorded.

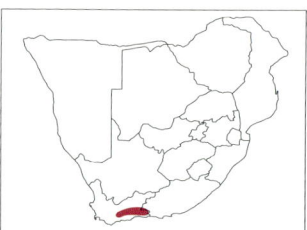

DESCRIPTION Plants have a large underground rootstock, resembling a series of carrots sprouting from a single growing point. Leaf shape is rather variable, ranging from short and stout to elongated and drooping. The most commonly encountered form is one where the leaves are comparatively thin, soft and needle-like. The leaves are more or less cylindrical in appearance and circular in cross-section. However, plants show some variation in these characters. The stalked flowers are a pleasant, uniform purple colour, with a slightly darker stripe of purple running down the centre of the petals. The anther-bearing filaments are typically bunched together in the centres of the flowers. Fruit capsules have five locules, stiff covering membranes and large closing bodies. Seeds are rounded and rough.

DISTINGUISHING CHARACTERS The single species is primarily characterised by the thickened roots, from which tufts of thin leaves sprout. Furthermore, the relatively large and showy purple flowers are borne singly, with a single darker purple line running down the length of the petals.

FLOWERING TIME *Marlothistella* flowers in winter (during July and August in South Africa). Flowers open in the morning and stay open all day.

DISTRIBUTION AND ECOLOGY *Marlothistella* is restricted to the Little Karoo in the Western Cape Province, South Africa. The species has been collected in the districts of Calitzdorp, De Rust, Joubertina, Prince Albert, Uniondale and Willowmore. The species often occurs among grasses or in the open on quartz flats. Some plants show a tendency to die back to the thickened rootstock.

CULTIVATION As with many mesembs that are grown too soft, i.e. with too much water or too little light, the leaves tend to become drooping and are not borne erectly as one would encounter in habitat. For a species with such a thick tap root it is surprisingly easy to cultivate. Although rather unassuming plants, the beautiful purple flowers of *Marlothistella* more than make up for what they otherwise lack in appearance. *Marlothistella* thrives in cultivation. It is best grown in containers, however. Plants require a sunny spot and sandy acid soil. Water mainly during summer. Propagation can be by division or seed.

NOTES *Marlothistella* is another genus of mesemb with one species only. This species is, however, quite variable and has in the past been included in the lucky packet genus *Ruschia*. When the plants are in flower they are covered by a showy purple carpet.

Marlothistella Schwantes

NUMBER OF SPECIES/SUBSPECIES/VARIETIES (1/0/0)

SPECIES LIST AND CONSERVATION STATUS
M. stenophylla (L.Bolus) S.A.Hammer

LITERATURE
HAMMER, S.A. 1995. *Marlothistella* Schwantes: a rising star. *Bradleya* 13: 82–87.

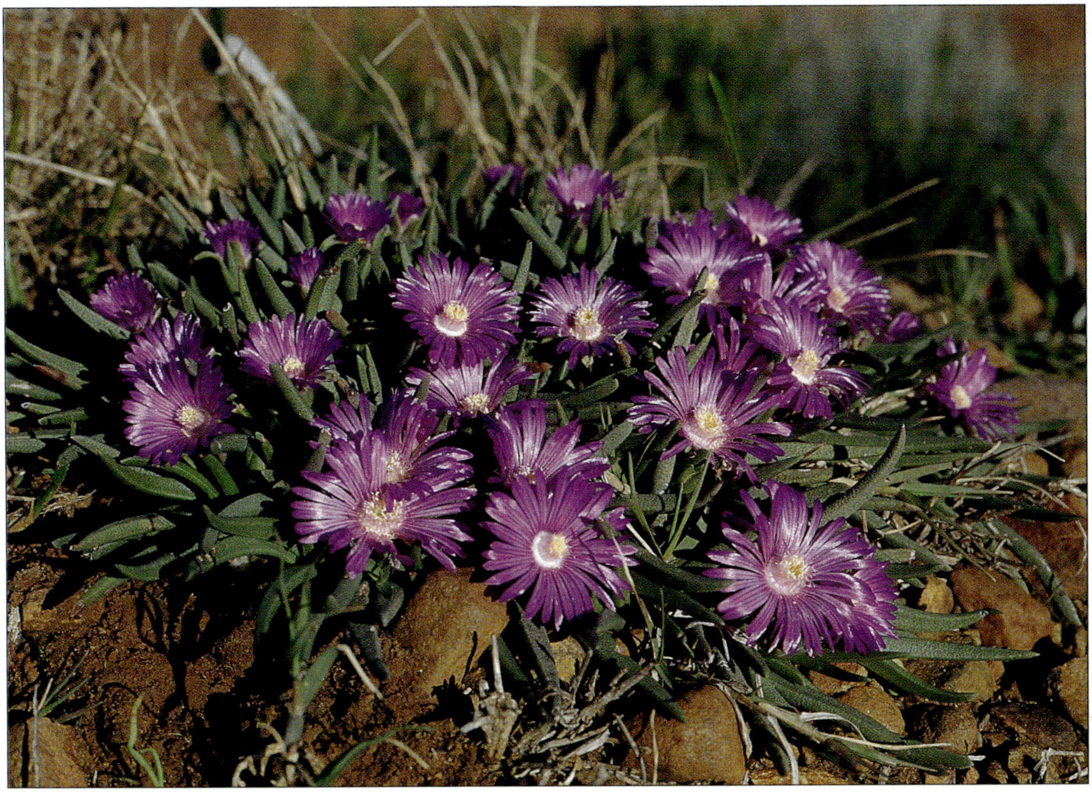

Ernst van Jaarsveld

M. stenophylla

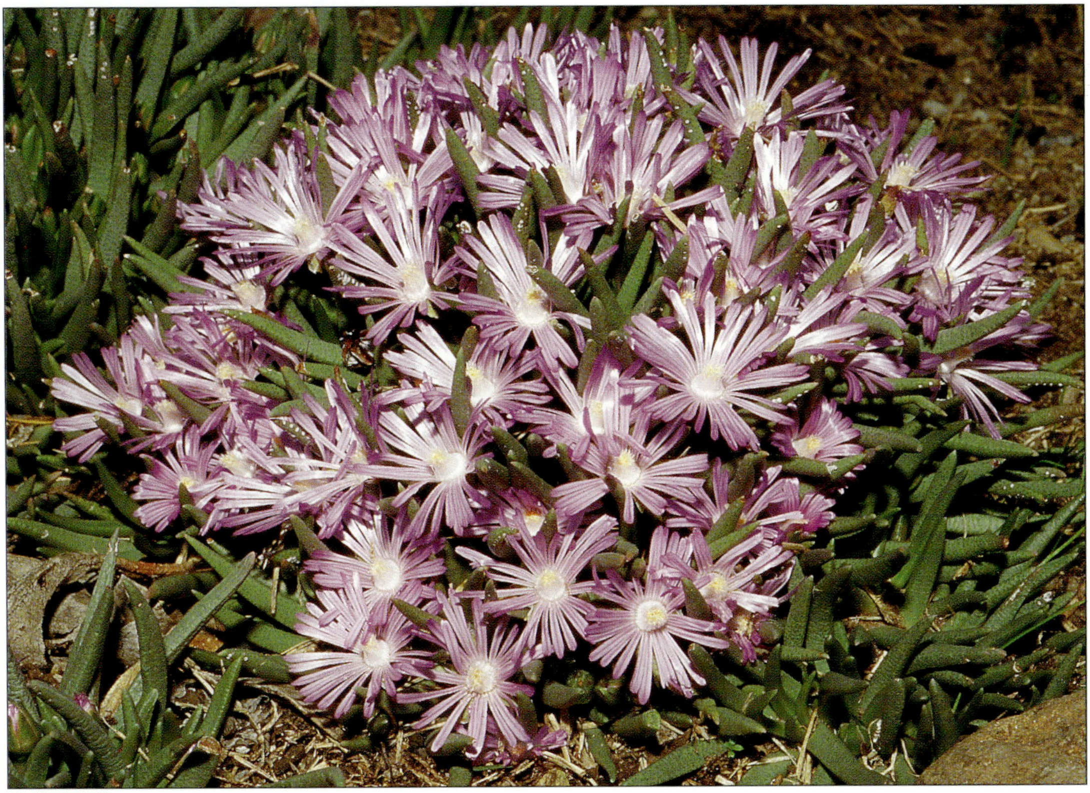

Bob Chinnock

M. stenophylla

PSAMMOPHORA

DERIVATION OF GENUS NAME The name is derived from the Greek words *psammos* (sand) and *phorein* (to bear) as the leaves of these plants exude a sticky substance to which sand adheres.

COMMON NAMES The common name *gomvy* (glue plant) is sometimes used for these sticky plants.

DESCRIPTION This genus comprises shrubs up to 600 mm in height, or compactly tufted dwarf shrubs of about 150 mm in height. The grey-green to bluish green leaves are fleshy, elongate, and are three-sided to almost cylindrical, with a rounded keel and often a pinkish base. Leaf surfaces produce a viscose substance to which grains of sand adhere, enhancing or obscuring the colour of the leaves. The solitary, medium-sized white to light violet flowers are borne terminally on flower stalks which have two pairs of bracts. Four or five unequal and sticky sepals surround the petals which are narrow and blunt at their tips. There are numerous stamens and staminodes. The nectar glands are in a ring, surrounding five to seven thread-like stigmas. Fruit capsules are five or six-locular, without closing bodies and with much reduced or absent covering membranes. The pear-shaped seeds are light brown with an orange tint.

DISTINGUISHING CHARACTERS The presence of sand grains on the leaves of *Psammophora* is diagnostic, as it is one of only two mesemb genera that have this characteristic, the other being *Arenifera*. It differs from *Arenifera* in that its fruit capsules have no closing bodies, and covering membranes are rudimentary or absent, whereas those of *Arenifera* are complete.

FLOWERING TIME *Psammophora* species flower in winter (May to July in southern Africa) and the scented flowers open at midday, closing by dusk.

DISTRIBUTION AND ECOLOGY *Psammophora* grows in full sun on rocky slopes and plains from Lüderitz, Namibia, to the Richtersveld, Northern Cape Province, South Africa. The areas where it is found receive sparse winter rainfall of less than 100 mm per year. There appears to be a higher calcium content in the soils of the northern localities.

CULTIVATION Propagation is not difficult from seed and is also possible from cuttings. Gravelly soils are recommended and watering must not be excessive. Keep dry during summer. Outside of its habitat, plants are best grown in a greenhouse.

NOTES The ability of *Psammophora* species to collect sand on their leaves is not peculiar to natural habitats; the same process occurs in cultivated plants. The sticky cover peels back like a scab, leaving behind a raw, shiny surface which then exudes more "glue" for the process to repeat itself. The genus can be sharply divided into two groups based on the growth form of the plants: *P. longifolia* and *P. nissenii* comprise the group of low-growing species, *P. nissenii* being the flatter and more compact of the two, while *P. modesta* and *P. axicola* comprise a shrubby group. It is this group which can easily be confused with species of *Arenifera*.

Psammophora Dinter & Schwantes

NUMBER OF SPECIES/SUBSPECIES/VARIETIES (4/0/0)

SPECIES LIST AND CONSERVATION STATUS
P. longifolia L.Bolus
P. modesta (Dinter & A.Berger) Dinter & Schwantes
P. nissenii (Dinter) Dinter & Schwantes
P. saxicola H.E.K. Hartmann

LITERATURE
HARTMANN, H.E.K. 1996. Miscellaneous taxonomic notes on Aizoaceae. *Bradleya* 14: 29–56.

P. modesta

P. longifolia

P. saxicola

P. nissenii

Rabiea

Derivation of genus name The genus name commemorates Reverend W. A. Rabie of the Free State, South Africa.

Common names *Rabiea albinota* and its varieties are known as *s'keng-keng* (*vygie*).

Description Plants form small, compact, aloe-like rosettes that arise from fleshy to thickly tuberous roots. The highly succulent, often sickle-shaped leaves are tightly packed on the stem and are dotted with small, but prominent white or blackish green spots. The leaves are more or less flat on their upper surfaces, whereas the lower surfaces are distinctly to indistinctly keeled. The flowers vary in colour from bright yellow to yellowish orange and arise on a short stalk which has two pairs of bracts. Numerous stamens surround the cup-shaped flower base which has nectar glands positioned at its rim. Fruit capsules have seven to 10 locules, are fairly rounded at the top and base, have valve wings and covering membranes. Seeds are pear-shaped and dark brown.

Distinguishing characters The thickly succulent, sickle-shaped leaves that are covered with closely packed white spots and the yellow flowers characterise *Rabiea*. The nearly stalkless, thick-walled, rather dark fruit capsules are quite unlike those of *Hereroa*, which otherwise resembles and could be confused with *Rabiea*.

Flowering time Plants flower in spring and summer in southern Africa. Flowers open at midday and close by evening.

Distribution and ecology The genus *Rabiea* occurs from the central Eastern Cape Province through the central parts of the Free State, South Africa. An outlier in the distribution has been recorded from the eastern Northern Cape Province. The winter climate of this primarily summer rainfall region can best be described as severe, as a result of regular frost, while summer temperatures can soar to over 40°C.

Cultivation Species of *Rabiea* are not difficult to cultivate. Plants adapt readily to pot culture, provided that good drainage is provided. As a result of their hardiness, they also do exceptionally well as rockery plants. Keep dry during winter. Red spider mite can sometimes be a problem. Propagation is by seed, division or cuttings.

Notes The genus is notorious for confounding taxonomists. The species are principally and optimistically distinguished by the thickness or thinness, and the position of their leaves. Plants tend to lose the white calcium-rich spots in cultivation.

Rabiea N.E.Br.

Number of species/subspecies/varieties (6/0/3)

Species list and conservation status
R. *albinota* (Haw.) N.E.Br. var. *albinota*
R. *albinota* (Haw.) N.E.Br. var. *longipetala* L.Bolus
R. *albinota* (Haw.) N.E.Br. var. *microstigma* L.Bolus
R. *albipuncta* (Haw.) N.E.Br. var. *albipuncta*
R. *albipuncta* (Haw.) N.E.Br. var. *major* L.Bolus
R. *comptonii* (L.Bolus) L.Bolus
R. *difformis* (L.Bolus) L.Bolus
R. *jamesii* (L.Bolus) L.Bolus [I]
R. *lesliei* N.E.Br.

Literature
BOLUS, H.M.L. 1958. *Rabiea* N.E.Br. *Notes on* Mesembryanthemum *and allied genera*. University of Cape Town, Cape Town. Vol. 3: 369–372.

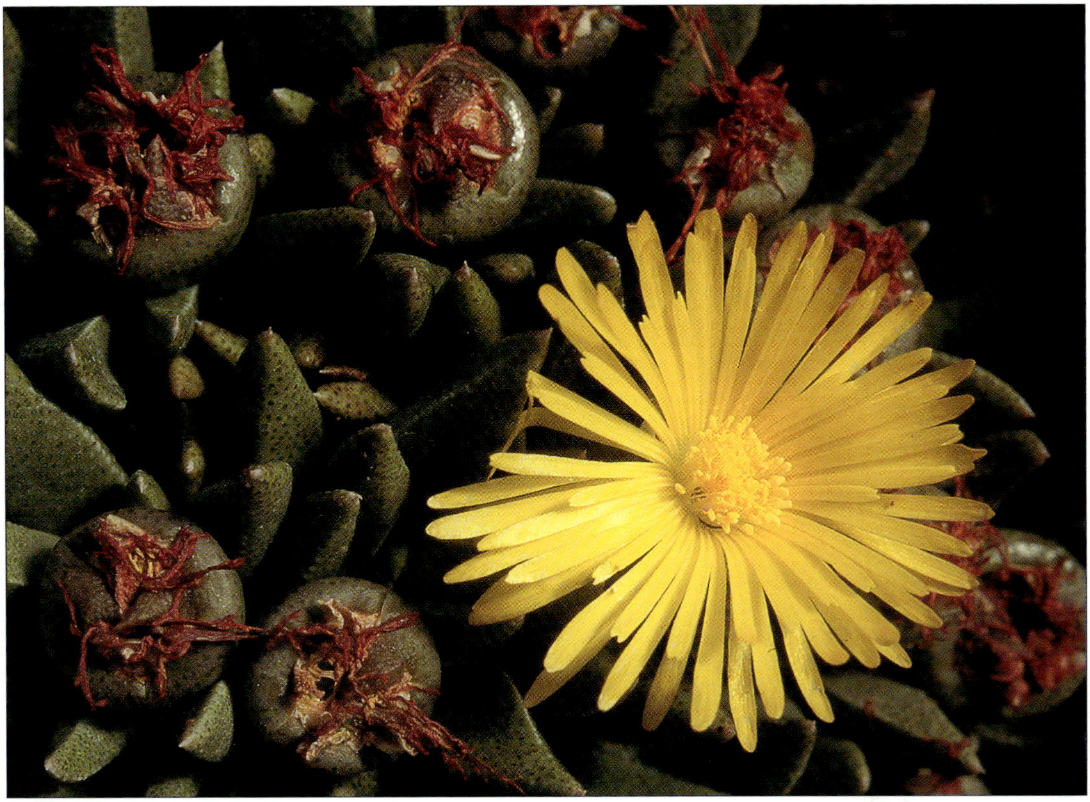

Ben-Erik van Wyk

R. albipuncta

Ernst van Jaarsveld

R. difformis in its habitat

NBI

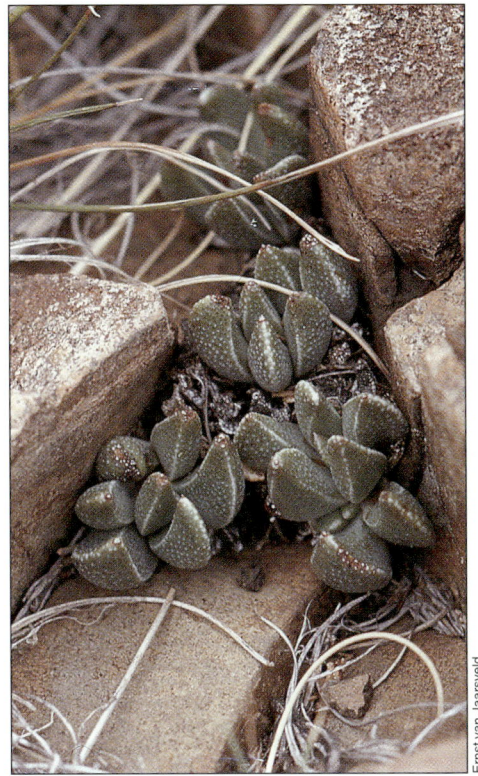

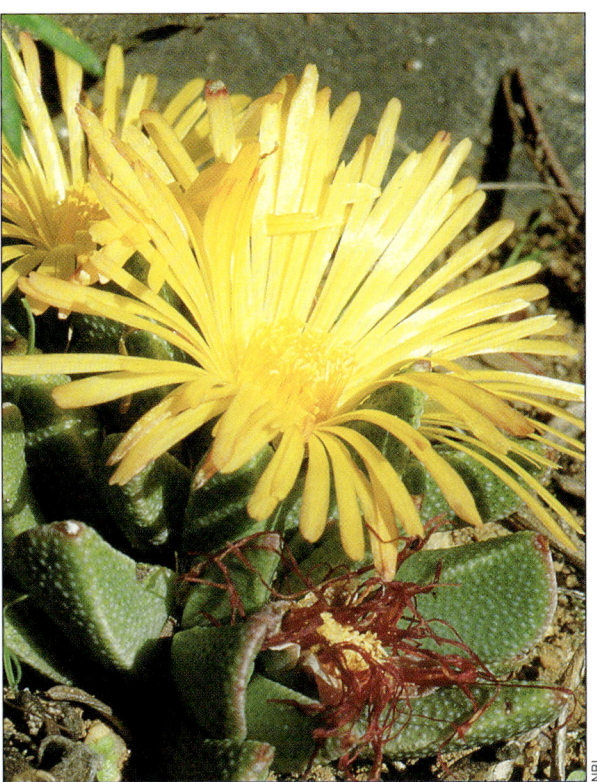

R. albinota

RHOMBOPHYLLUM

DERIVATION OF GENUS NAME The genus name was derived from the Greek words *rhombos,* meaning lozenge and *phyllon,* meaning leaf, in reference to the somewhat rhomboidally shaped leaves.

COMMON NAMES No vernacular names seem to have been recorded.

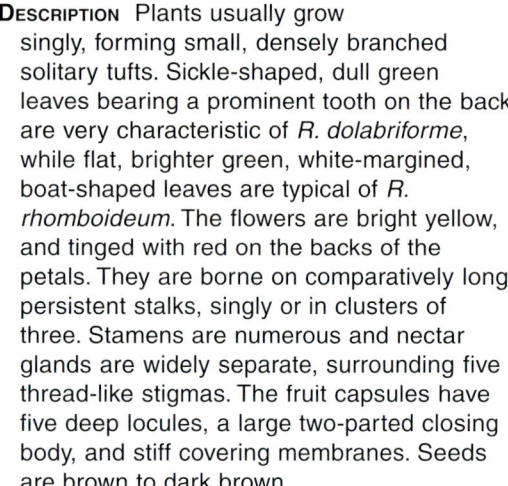

DESCRIPTION Plants usually grow singly, forming small, densely branched solitary tufts. Sickle-shaped, dull green leaves bearing a prominent tooth on the back are very characteristic of *R. dolabriforme,* while flat, brighter green, white-margined, boat-shaped leaves are typical of *R. rhomboideum.* The flowers are bright yellow, and tinged with red on the backs of the petals. They are borne on comparatively long persistent stalks, singly or in clusters of three. Stamens are numerous and nectar glands are widely separate, surrounding five thread-like stigmas. The fruit capsules have five deep locules, a large two-parted closing body, and stiff covering membranes. Seeds are brown to dark brown.

DISTINGUISHING CHARACTERS The irregular sickle or boat-shaped leaves are typical. The flower stalks tend to remain attached to the plants, giving them a somewhat spiny appearance. The large two-parted closing bodies are also characteristic of the genus.

FLOWERING TIME The plants flower in spring and early summer. Flowers open in the afternoon or evening.

DISTRIBUTION AND ECOLOGY This eastern Cape genus is practically confined to the subtropical thicket vegetation around Uitenhage and Port Elizabeth, and further afield near Graaff-Reinet in the Great Karoo. An outlying population has been recorded from the southeastern corner of the Northern Cape Province. *R. rhomboideum* loves limestone formations; *R. dolabriforme* prefers rocky shale hills and hybridises with *Faucaria boscheana* around Graaff-Reinet.

CULTIVATION Plants are easily propagated from cuttings and from seed. They grow readily in well-drained soil mixtures and do not seem to rot, even if grossly over-watered.

NOTES The rather strange and outlandish appearance of the species of *Rhombophyllum* make them some of the more desirable mesembs to grow. The leaves of certain forms of *R. dolabriforme* tend to be rough and almost crystalline, making them particularly attractive for cultivation. An example is shown in the first illustration.

Rhombophyllum (Schwantes) Schwantes

NUMBER OF SPECIES/SUBSPECIES/VARIETIES (2/0/0)

SPECIES LIST AND CONSERVATION STATUS
R. dolabriforme (L.) Schwantes
R. rhomboideum (Salm-Dyck) Schwantes

LITERATURE
SCHWANTES, G. 1927. Zur Systematik der Mesembrianthemen. *Zeitschrift für Sukkulentenkunde* 3: 16, 23.

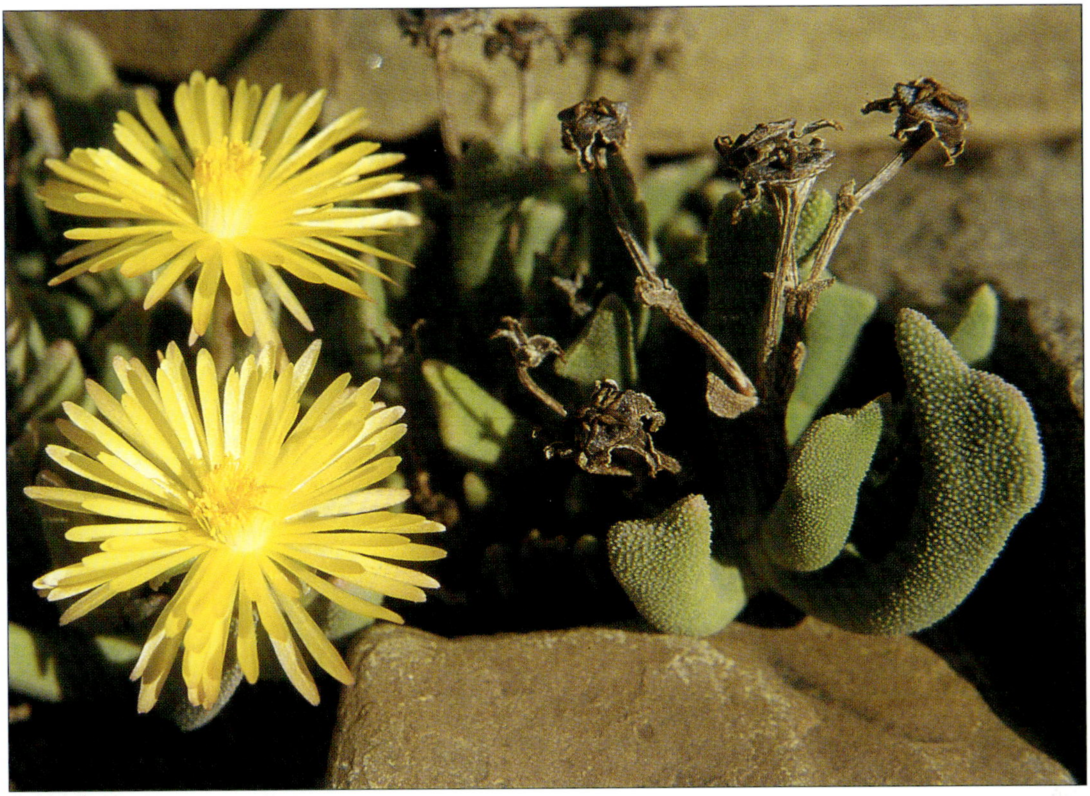
Ben-Erik van Wyk

R. dolabriforme

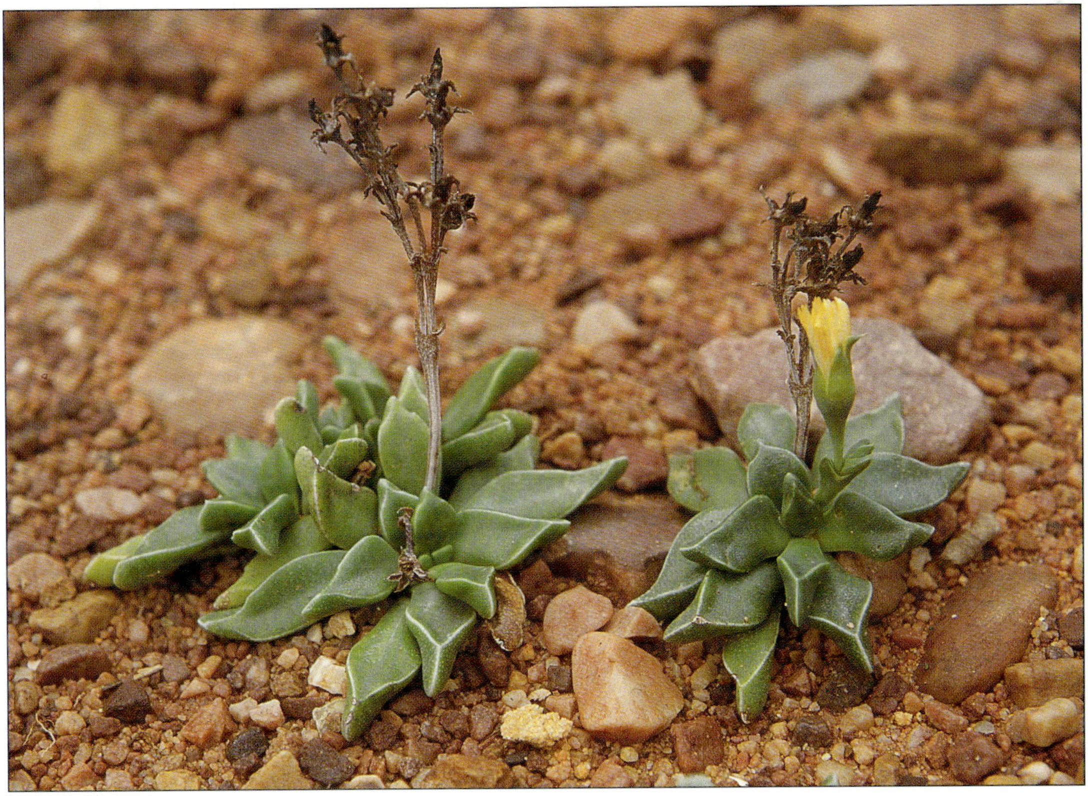
Ben-Erik van Wyk

R. rhomboideum

Ruschianthus

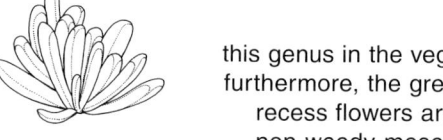

Derivation of the genus name The genus is named after the Namibian farmer Ernst Rusch; *anthos* is the Greek word for flower.

Common names No vernacular names seem to have been recorded.

Description The compact plants possess sickle-shaped, spreading leaves in a dense arrangement; the leaves are light grey in appearance due to a thick layer of crystals below the leaf surface and a thick layer of wax on the surface. Flowers are borne on short stalks enclosed by the leaves, the greenish yellow petals therefore never spread widely. The filamentous staminodes are bearded to the tips and are positioned, together with the petals and the stamens, on a thick, swollen ring-shaped cushion above the short cup-shaped flower base. Five nectar glands, only slightly separated from each other, surround the short stigmas. In this way, a recess (hypanthium) is formed in which copious amounts of nectar accumulate as an attraction for pollinators. The stalks elongate while the fruit is ripening. Fruit capsules have five locules. They are without covering membranes and have narrow valve wings. Seeds are light brown in colour.

Distinguishing characters The sickle-shaped, dotless leaves in a stemless tuft are typical of this genus in the vegetative state; furthermore, the greenish yellow recess flowers are unique in the non-weedy mesembs.

Flowering time The flowers appear in winter and open around midday, but after a few days they do not close again.

Distribution and ecology The single species in this genus occurs in a very restricted area in southern Namibia, at the edge of the Namib Desert.

Cultivation The plants grow easily from seed but are seldom cultivated. They are best grown in a bright position in a greenhouse.

Notes Few mesemb species develop greenish petals. The cup-shaped flower base is also rare. Both features may be adaptations to special pollinators, which are not known as yet.

Ruschianthus L.Bolus

Number of species/subspecies/varieties (1/0/0)

Species list and conservation status
R. falcatus L.Bolus

Literature
BOLUS, L. 1961. *Ruschianthus* gen. nov. *Journal of South African Botany* 27: 62–63.

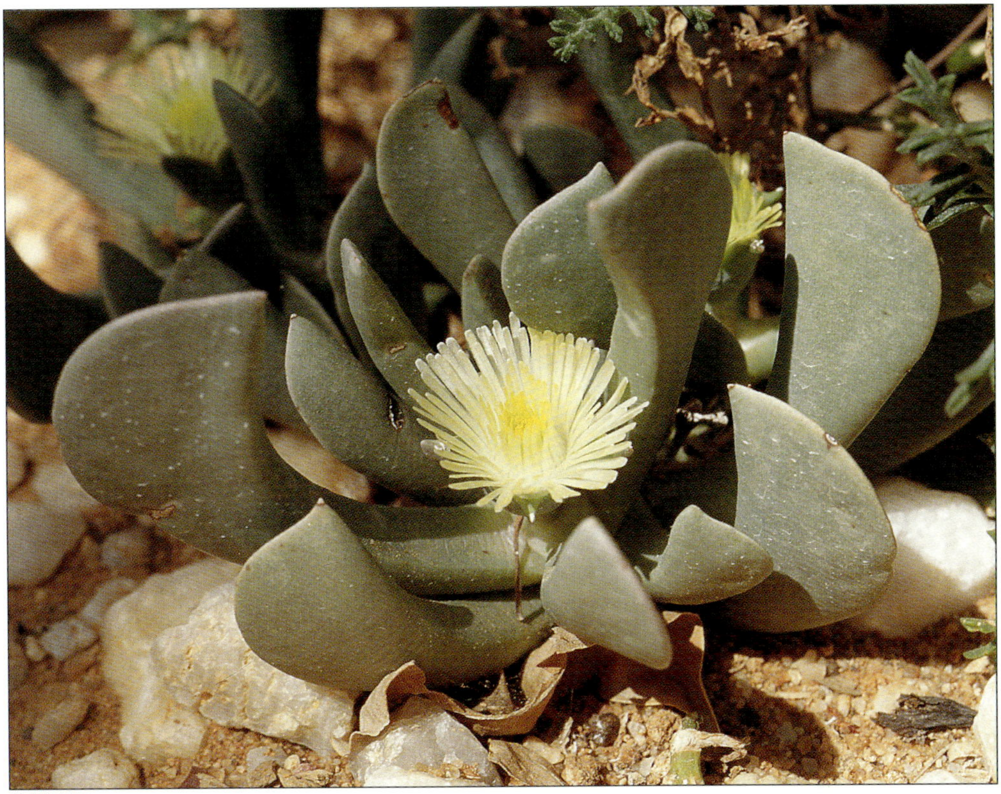
Ernst van Jaarsveld

R. falcatus

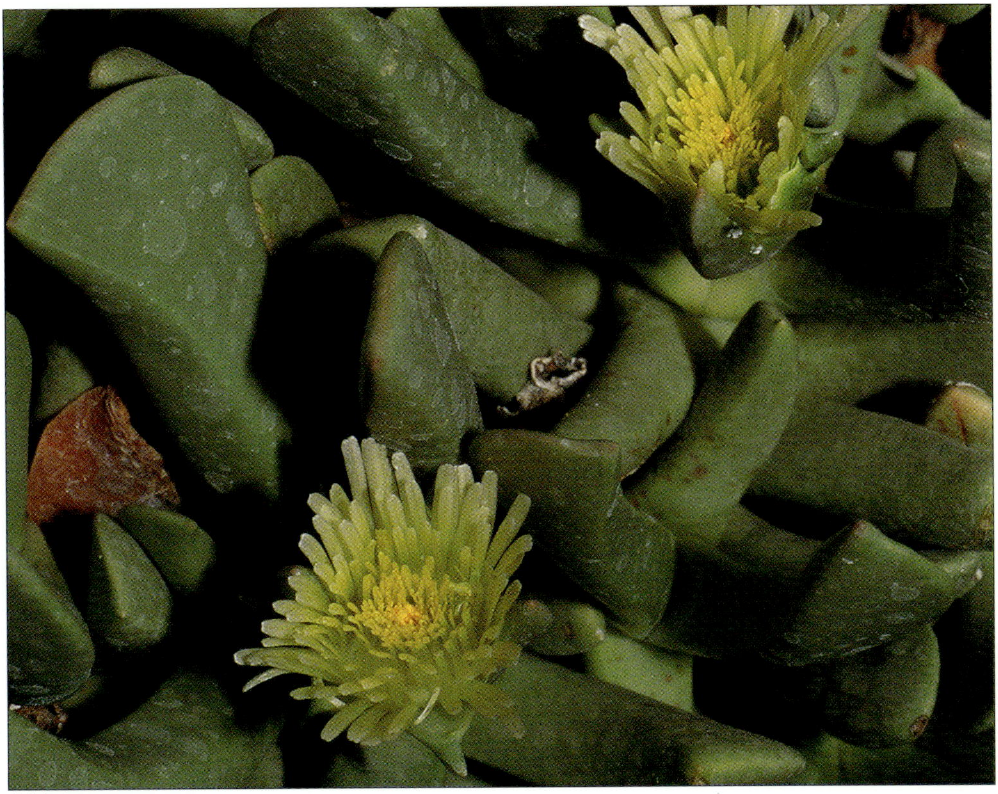
John Trager

R. falcatus

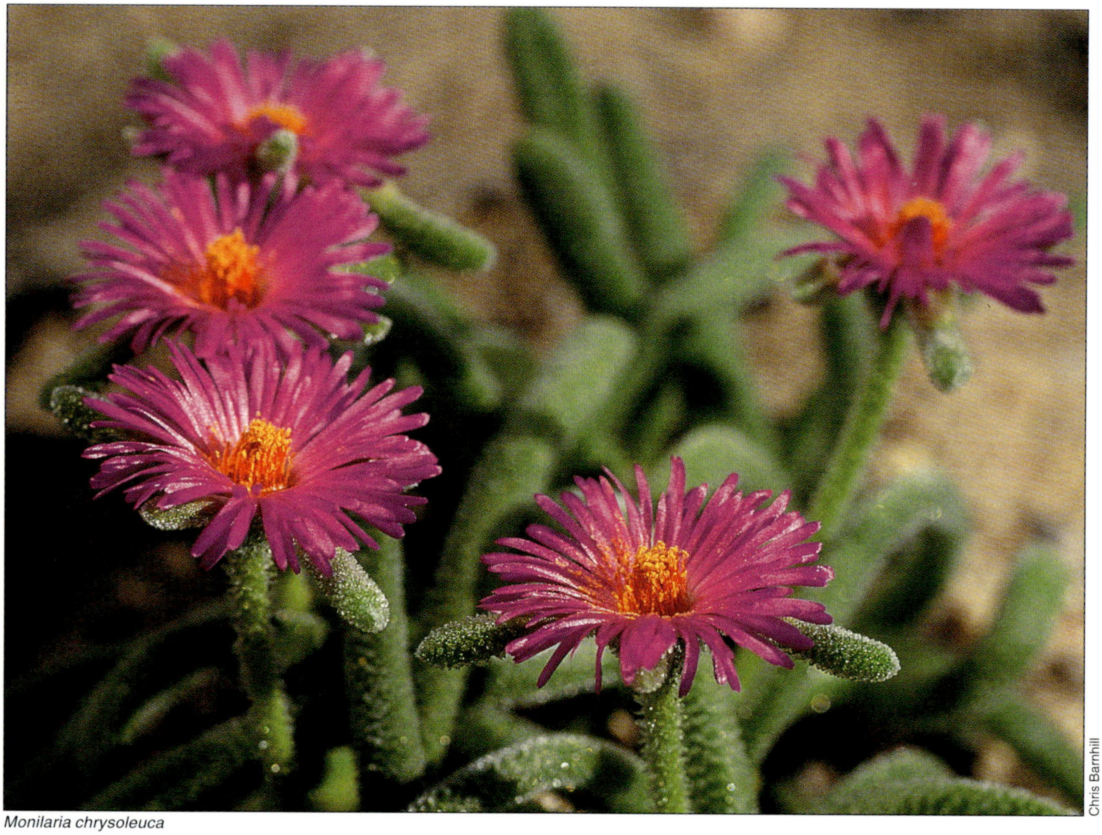
Chris Barnhill

Monilaria chrysoleuca

Ben-Erik van Wyk

Monilaria chrysoleuca

Ben-Erik van Wyk

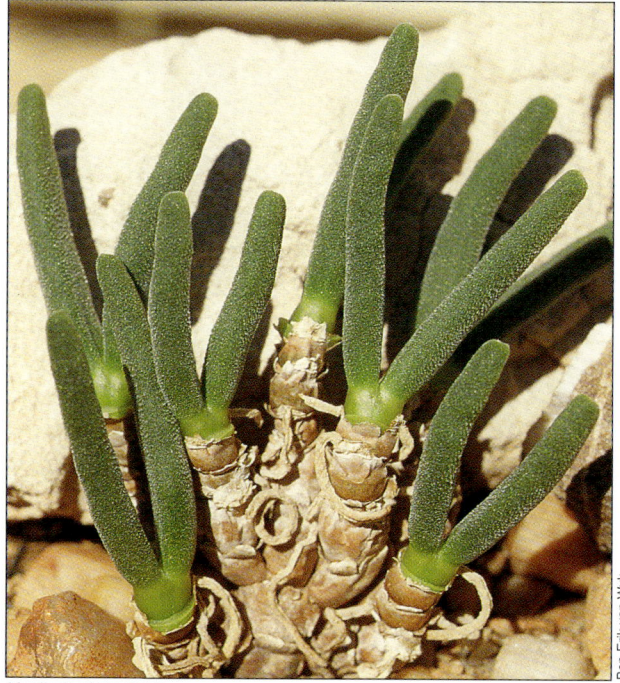

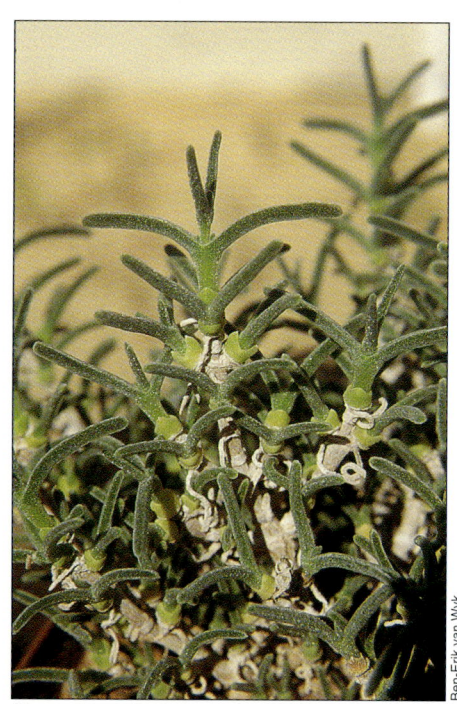

Dicrocaulon sp. (Doringrivier, Ceres Karoo)

Bead-leaved Mesembs

GROUP 8

The succulent leaves are arranged so that the newly formed leaf pair differs markedly from the older leaf pair from which it has emerged, thus resulting in a continually alternating series of different leaf pairs; the older globular leaf pairs dry out and protect the elongated new leaf pairs; this group does not include genera where the two leaves in each pair differ from one another.

The variously sized leaf bodies along the thin stems often resemble strings of beads, hence the name BEAD-LEAVED MESEMBS. The plants included here form a natural group of related genera.

The differences between leaf pairs are associated with interesting growth patterns. In the genus *Dicrocaulon* for example, the following unusual sequence of leaf formation can be observed, starting with autumn (the beginning of the growing season), when the topmost beady leaf pair swells and bursts from its white sheath. An oblong leaf pair with long free parts now emerges from it. Within the base of these long leaves, further long leaves may develop in some species, under favourable conditions. After these, a new beady leaf pair develops within the base of the uppermost long leaf pair. After the oblong leaves have been absorbed and have withered away, this bead then remains and will be reactivated the following autumn. A consequence of these growth patterns is an alteration of long and short internodes.

BEAD-LEAVED MESEMBS comprise 5 genera and 21 species.

Dicrocaulon (7 species)
Jacobsenia (2 species)
Meyerophytum (1 species)
Mitrophyllum (6 species)
Monilaria (5 species)

DICROCAULON

DERIVATION OF THE GENUS NAME The name is derived from the Greek *dicros* (fork) and *caulon* (stem), in reference to the branched stems (they fork much more than *Monilaria*, with which it is often confused).

COMMON NAMES The common name *opstandingsvygie* (resurrection mesemb) has been used for this genus.

DESCRIPTION The woody shrublets can form rounded or flat cushions, or can grow up to 500 mm in height, sometimes with marked differences between long shoots and dense short shoots and between oblong leaf pairs and beady leaf pairs (see description of the group). The flowers are usually white, but sometimes vivid magenta or white, turning rosy pink. In some species, they are embraced at their bases by a fused leaf pair forming a cup and the petals can be long, thread-like and tousled, the inner ones merging into filamentous staminodes. In other species, the petals are relatively wide, and open fully. The fruit capsules have four to nine locules (five or six being most common) and they are flat, with high rims on the valves. The expanding sheets merge into the expanding keels. Seeds are light brown and shiny.

DISTINGUISHING CHARACTERS The presence of two different kinds of leaf pairs on woody, hard, thin stems is typical of the genus. In summer, the once globular, now shrunken resting leaf pairs are not obvious because they are hidden by the sheaths which were formed by the previous leaf pairs. If the resting leaf pair is squeezed, there seems to be nothing inside, while in *Monilaria*, the squeeze test reveals a firm plant body inside.

FLOWERING TIME Flowers emerge in late winter to spring (August to October in South Africa), before or after the fresh leaves become visible. The flowers open either in the morning or at midday, closing at night. After a few days, some species keep their flowers open permanently.

DISTRIBUTION AND ECOLOGY *Dicrocaulon* occurs in Namaqualand, South Africa. Plants are generally found in quartzitic soils between Vanrhynsdorp and Hondeklipbaai and seldom at the coast. Rainfall in these regions hardly exceeds 125 mm per year.

CULTIVATION The plants grow easily from seed and need a very strict dry season in summer to maintain the development of two different kinds of leaves. The drying leaves make a bright spring and summer show, turning orange or red. During their resting state, the plants appear to be dead. These curious plants are best grown in a greenhouse. Soil should be sandy and mineral rich. Unfortunately, flowers are rarely formed in cultivation.

NOTES There are two distinct groups of resurrection mesembs: the first (*D. brevifolium* and its relatives) has small, tousled, non-showy flowers and small, delicate leaves; in one species, *D. humile*, the plant grows flat on the ground. The other group (*D. grandiflorum* and allies) has large, wide open flowers in a range of colours, and robust leaves.

Dicrocaulon N.E.Br.

NUMBER OF SPECIES/SUBSPECIES/VARIETIES (7/0/0)

SPECIES LIST AND CONSERVATION STATUS
D. brevifolium N.E.Br.
D. grandiflorum Ihlenf.
D. humile N.E.Br.
D. microstigma (L.Bolus) Ihlenf.
D. ramulosum (L.Bolus) Ihlenf.
D. spissum N.E.Br.
D. trichotomum (Thunb.) N.E.Br.

LITERATURE
IHLENFELDT, H.-D., HARTMANN, H. & POPPENDIECK, H.-H. 1978. Chromosomen-zahlen der Mitrophyllinae Schwantes (Mesembryanthemaceae). *Mitteilungen aus dem Institut für Allgemeine Botanik, Hamburg* 16: 171–182.

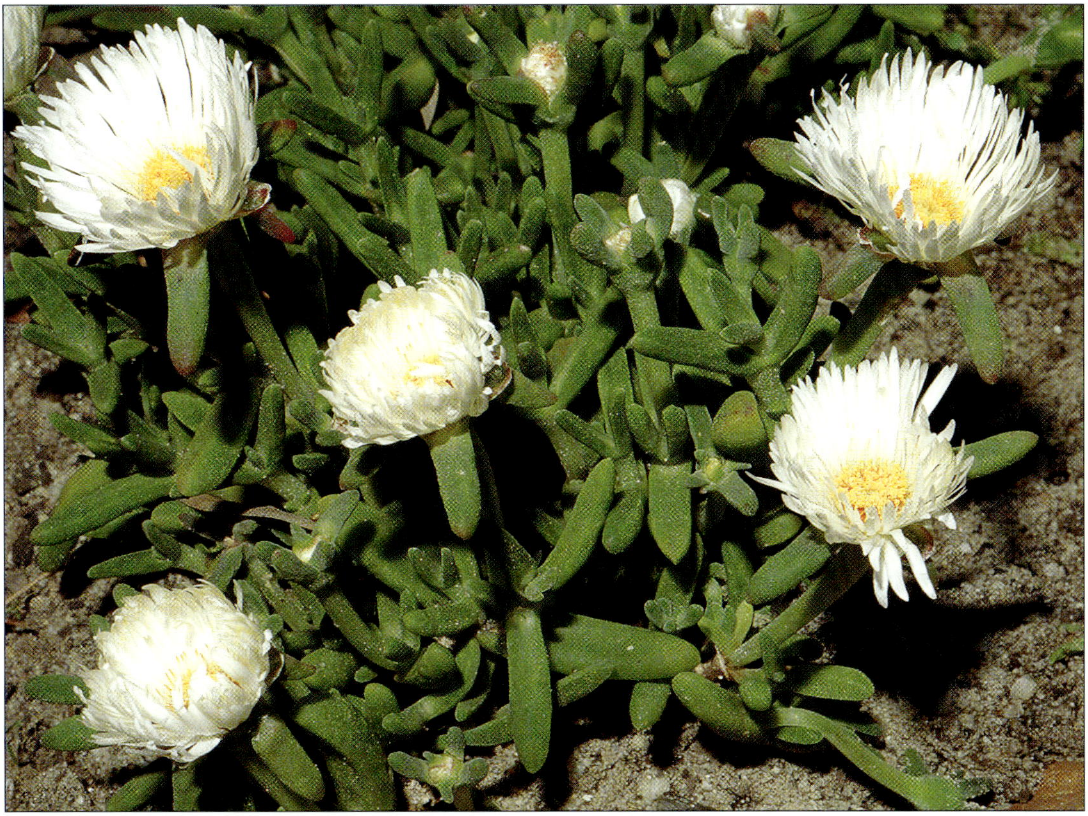
Bob Chinnock

D. grandiflorum

Ben-Erik van Wyk

D. brevifolium

John Trager

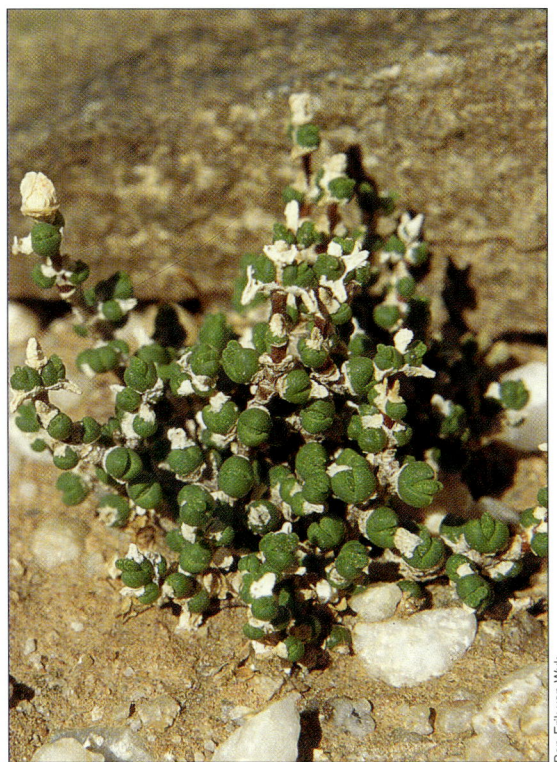

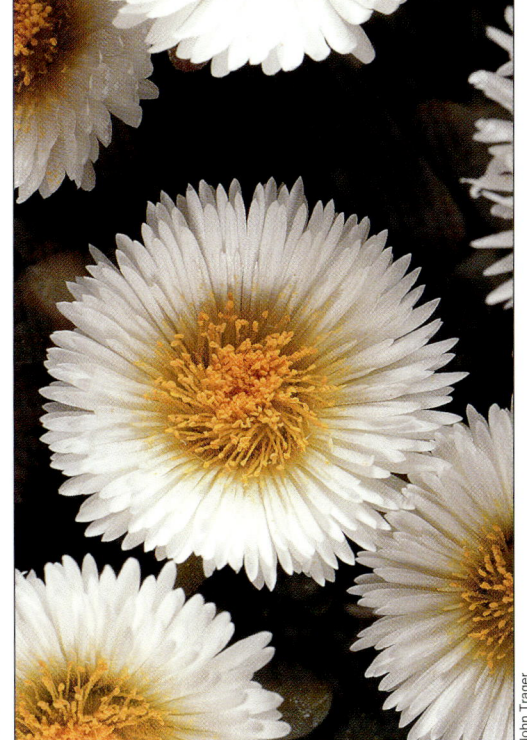

Flowers of *D. grandiflorum*

JACOBSENIA

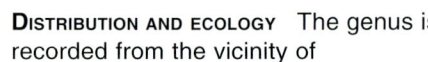

DERIVATION OF GENUS NAME The genus is named after Dr. H. Jacobsen, a past curator of Kiel Botanic Gardens, Germany.

COMMON NAMES No vernacular names seem to have been recorded.

DESCRIPTION The plants are succulent shrublets with long erect branches bearing the flower clusters. Their annual growth consists of two or more pairs of cylindrical, glistening or velvety leaves that are slightly flattened on their upper surfaces. The white to lemon yellow or rarely purplish flowers are solitary, large and completely lack staminodes. The five-locular ovary has a ring of fused nectar glands and five to seven awl-shaped stigmas. It develops into a fruit capsule which resembles that of *Mitrophyllum* with diverging expanding keels (here with toothed margins) and membranous wings. Covering membranes are present in *Jacobsenia* but closing bodies are not. The brown seeds are rounded and have minute nipples.

DISTINGUISHING CHARACTERS The large, usually white flowers and very long, red-brown bare stems in the only well-known species (*J. kolbei*) may help to distinguish this genus from other beady mesembs. In contrast to *Mitrophyllum*, where mature leaves can be fused into plant bodies, *Jacobsenia* has leaves forming plant bodies in the juvenile stages only.

FLOWERING TIME Plants flower in early summer (August to October in South Africa). The highly scented flowers open in the morning and close in the evening.

DISTRIBUTION AND ECOLOGY The genus is recorded from the vicinity of Vanrhynsdorp in the northern part of the Western Cape Province, straddling the border to the Northern Cape Province, South Africa. They grow on quartz outcrops or coarse granite hills.

CULTIVATION *Jacobsenia* species thrive in containers and are best grown in a greenhouse outside of their habitat. Plants should be kept dry in summer. Propagation is from seed sown in autumn, or by pruning old plants and rooting the stems.

NOTES While the leaves of a mature plant are only united at the very bases, the first leaf pair of the seedling is joined together almost to the top and so forms a small beady plant body. *Jacobsenia* is therefore included in the BEAD-LEAVED MESEMBS. However, the high degree of fusion in the seedling leaves gradually decreases as the plant grows older until finally the adult stage is reached. As a result, mature plants do not look like other beady mesembs. *J. hallii* has compact growth and velvety leaves; *J. kolbei* has lanky stems and glistening leaves.

Jacobsenia L.Bolus & Schwantes

NUMBER OF SPECIES/SUBSPECIES/VARIETIES (2/0/0)

SPECIES LIST AND CONSERVATION STATUS
J. hallii L.Bolus [R]
J. kolbei (L.Bolus) L.Bolus & Schwantes

LITERATURE
BOLUS, H.M.L. & SCHWANTES, G. 1954. *Notes on* Mesembryanthemum *and allied genera*. University of Cape Town, Cape Town. Vol. 3: 255.

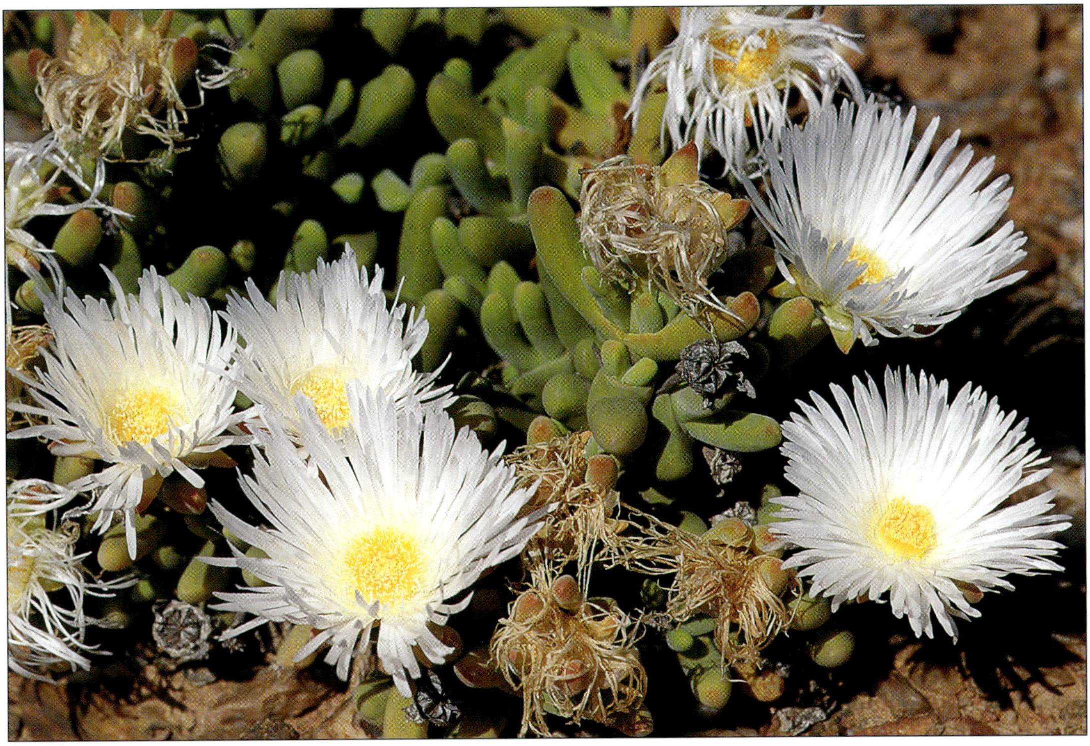

Peter Bruyns

J. hallii

Ben-Erik van Wyk

Flower of *J. kolbei*

Ernst van Jaarsveld

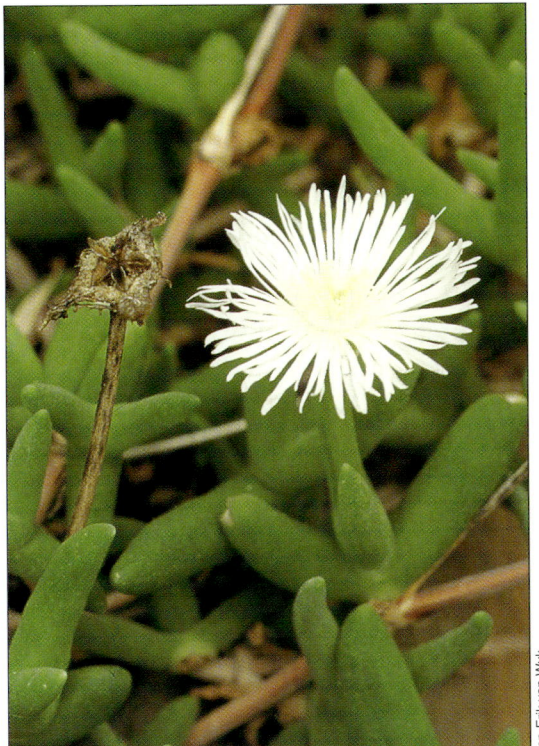

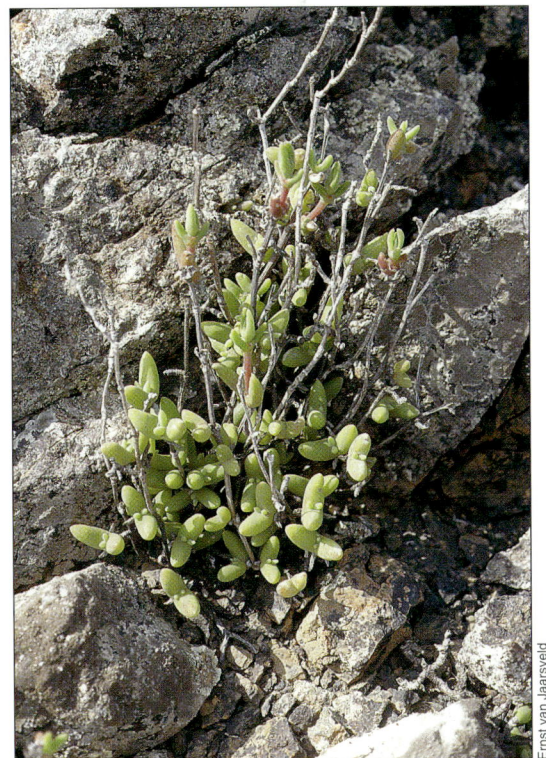

J. kolbei in its habitat

MEYEROPHYTUM

DERIVATION OF GENUS NAME This genus was named after reverend L.G. Meyer (1867 to 1958), formerly of Steinkopf, South Africa.

COMMON NAMES No vernacular names are known.

DESCRIPTION Plants are compact shrublets with two different kinds of leaf pairs. Branches are upright or low-lying, the lower parts often covered with persistent old dry leaf sheaths. The first leaf pair of the season is rounded and forms a body that is grey-green in colour. The second pair is longer, fused at the base for up to a third of its length. It is deeper green and has conspicuous water cells on the surfaces. Before the dry period commences, the base of the first pair of leaves develops into a papery sheath that enfolds the assimilating leaf pair of the next season. The violet-purple or white flowers are relatively large, solitary and have a long flower stalk that bears them well above the uppermost leaf pair. The five unequal sepals have membranous margins and the linear petals are arranged in three or four whorls. There are numerous upright stamens, the inner ones having bearded bases. Filaments are a deep rose-purple and anthers are purple-brown. The ovary is raised or is deeply concave with the nectar glands in a ring surrounding five short, broad, feathery stigmas. Fruit capsules have five locules, deeply concave covering membranes, amply winged valves and broad, flat expanding keels. Seeds are egg or pear-shaped and smooth.

DISTINGUISHING CHARACTERS The stems are weak and repeatedly branched in *Meyerophytum*, while in *Monilaria* they are constricted at intervals to give the appearance of beads. In *Meyerophytum*, the resting leaf bodies (beads) are conspicuous, whereas in *Dicrocaulon* they are shrunken and less conspicuous.

FLOWERING TIME Flowers appear after good rains in spring (August to September in South Africa) and they open at midday and close in the late afternoon.

DISTRIBUTION AND ECOLOGY *Meyerophytum* is restricted to the Northern Cape Province, South Africa, from the Richtersveld to southern Namaqualand. This area receives predominantly winter rainfall, never exceeding 200 mm per year. Plants grow fully exposed, or in partial shade between white, light-reflecting, quartzite rocks. The microhabitat (often on southern slopes) is relatively wet as a result of fog, dew and runoff.

CULTIVATION Plants are best propagated from seed sown in autumn, as cuttings do not root readily. Soil should be sandy or loamy, with good drainage properties. Plants should be grown in direct, bright sunlight in a greenhouse. Once the new leaves have absorbed water from the older leaves, watering is superfluous for several months. The plants should be kept dry in summer when they are dormant.

NOTES *Meyerophytum meyeri* var. *holgatense* differs from the typical form of the species by the white centres of the especially large flowers as seen in the illustration. Some populations (around Riethuis near Hondeklip Bay) have white flowers, which quickly turn pink and then persist for several days.

Meyerophytum Schwantes

NUMBER OF SPECIES/SUBSPECIES/VARIETIES (1/0/1)

SPECIES LIST AND CONSERVATION STATUS
  M. meyeri (Schwantes) Schwantes var. *holgatense* L. Bolus
  M. meyeri (Schwantes) Schwantes var. *meyeri*

LITERATURE
SCHWANTES, G. 1927. *Meyerophytum* Schwantes gen. nov. Möllers Deutsche Gärten-Zeitung 42: 436.

BOLUS, L. 1958. *Meyerophytum* Schwant. Notes on Mesembryanthemum *and allied genera.* University of Cape Town, Cape Town. Vol. 3: 345–346.

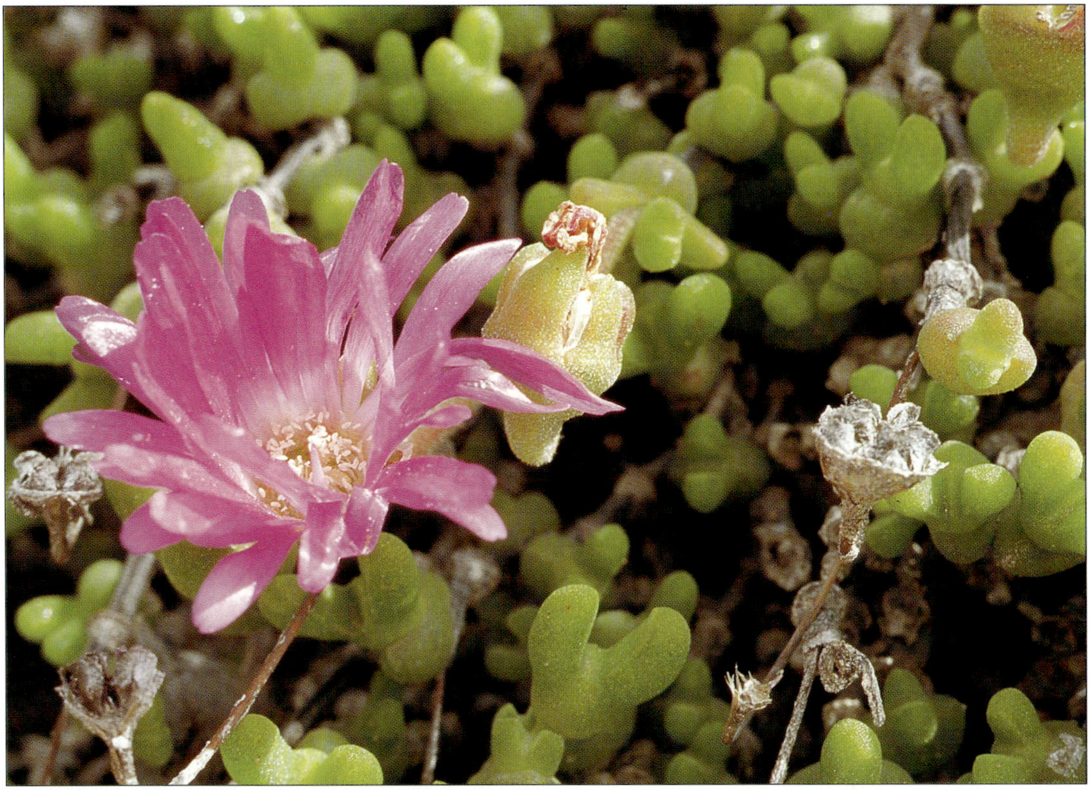

Bob Chinnock

M. meyeri var. *meyeri*

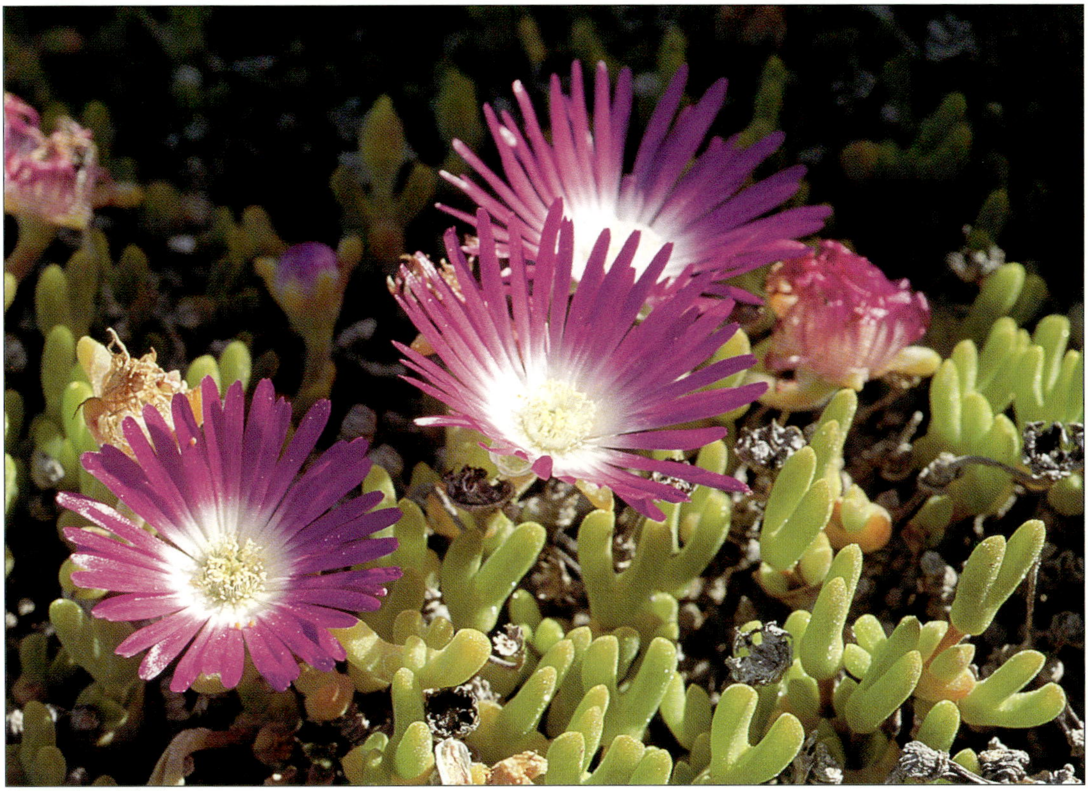

Graham Williamson

M. meyeri var. *holgatense*

Mitrophyllum

Derivation of genus name The name is derived from the Greek words *mitra* (bishop's mitre, bishop's cap), and *phyllon* (leaf).

Common names The common names clock plants or calendar plants have been used.

Description Species of *Mitrophyllum* are shrubs, often with articulated stems and nearly cylindrical leaves of two kinds. The two leaf-pairs of the same season differ in the length of the basal sheath: in the first pair the leaves are only slightly united at the base; in the second pair the leaves are extensively (to nearly completely) united and form a conical or cylindrical body which later develops into a papery, parchment-like skin protecting the next leaf-pair (which has the same form and function as the first pair) during the dry period. The white, yellow or pale violet-red flowers are solitary or borne in clusters. Stamens are numerous and are at first collected into a cone but spread later. Nectar glands are in a ring and surround five awl-shaped stigmas. The five-locular fruit capsules are with or without covering membranes and closing bodies. The seeds are egg or pear-shaped.

Distinguishing characters The enormous oblong leaf pairs, borne on erect woody stems, and the tattered papery sheaths, are highly characteristic for *Mitrophyllum*. The uppermost leaf pair looks remarkably similar to a bishop's cap.

Flowering time The flowering season is very long, from February to November in South Africa. The highly scented flowers (they smell like snapdragons) open at about noon and close in the late afternoon.

Distribution and ecology The genus is found in the Richtersveld in the Northern Cape Province of South Africa. The climate of this area is influenced by the Antarctic Benguela current and is characterised by the frequent occurrence of mist, which adds small amounts of moisture to an otherwise extremely dry area.

Cultivation Plants grow easily and rapidly from seeds sown in summer or autumn. Cuttings, however, do not root readily. A greenhouse is necessary in wet climates. Soil must be sandy and mineral rich. Although the plants grow rapidly, flowering can only be expected after several to many years.

Notes Plants of the same species may have radically different appearances at different times of the year. *M. dissitum* has thin, erect, elongated stems, resembling those of *Jacobsenia kolbei*, whereas the long thin stems of *M. roseum* are not erect but cascade over rocks. The latter species has indeterminate growth patterns so that an extended flowering period is induced under greenhouse conditions.

Mitrophyllum Schwantes (=*Conophyllum* Schwantes; =*Mimetophytum* L.Bolus)

Number of species/subspecies/varieties (6/0/0)

Species list and conservation status
M. *abbreviatum* L.Bolus [R]
M. *clivorum* (N.E.Br.) Schwantes
M. *dissitum* (N.E.Br.) Schwantes
M. *grande* N.E.Br.
M. *mitratum* (Marloth) Schwantes
M. *roseum* L.Bolus [R/V]

Literature
POPPENDIECK, H.–H. 1976. Untersuchungen zur Morphologie und Taxonomie der Gattung *Mitrophyllum* Schwantes s. lat. *Botanische Jahrbücher* 97: 339–413.

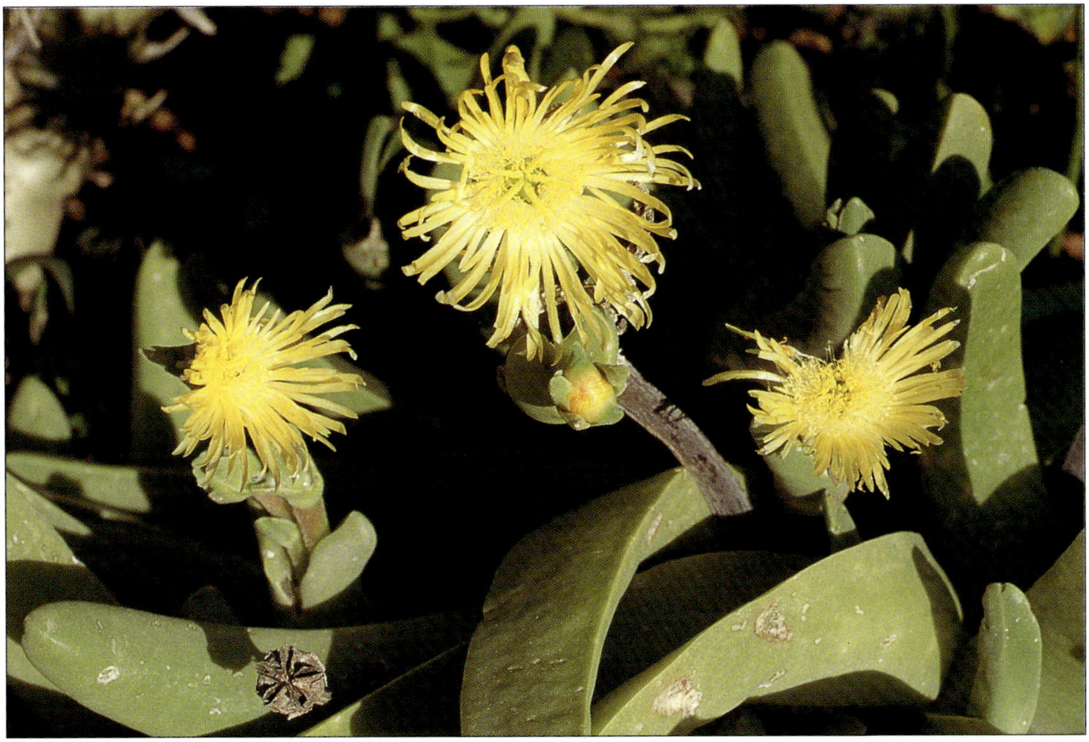

Ben-Erik van Wyk

M. grande

NBI

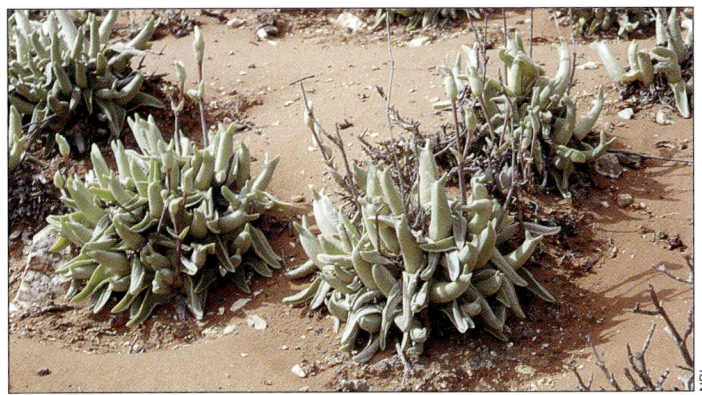

M. mitratum in its habitat

John Trager

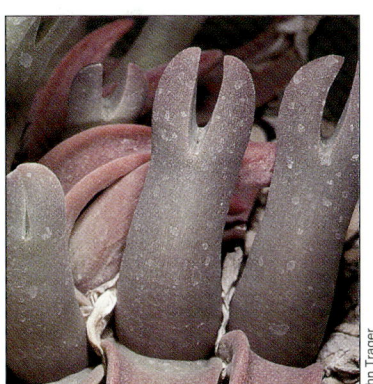

Leaves of *M. mitratum*

John Trager

M. abbreviatum x *M. roseum*

Graham Williamson

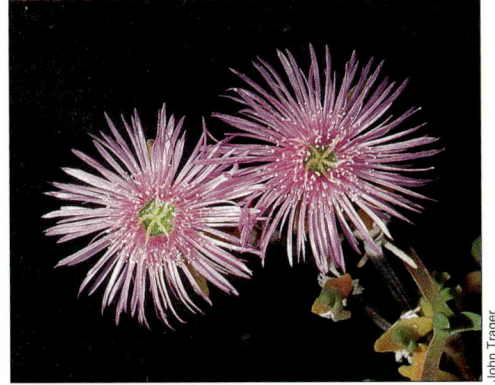

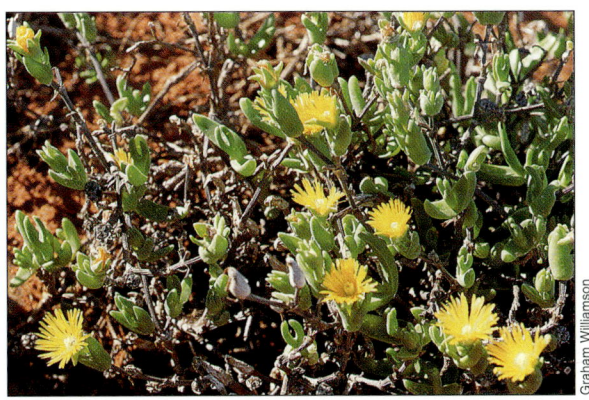

M. dissitum

Monilaria

Derivation of genus name The name is derived from the Latin word *monile* (string of pearls); *monilaria* means a collection of strings of pearls.

Common names *M. moniliformis* is known as *ertjievygie* (pea mesemb).

Description Species of *Monilaria* are small shrublets with branches constricted at the nodes into short, often bead- or button-like segments. A clear distinction between leaf types is visible here: plants form a short and a long leaf pair per season; the first pair is largely fused and appears as a small, fleshy, rounded body; the second pair emerges through the tip of this body and consists of cylindrical and elongated leaves which are only fused at their bases, and have glistening water cells. The base of the second leaf pair develops into a persistent, papery sheath. The white, pink to rarely yellow or salmon flowers are medium sized and occur singly on long stalks. The five to seven-locular fruit capsules have divergent expanding keels, narrow marginal wings, distinct covering membranes and no closing bodies. The seeds are egg-shaped and smooth.

Distinguishing characters The most distinct feature of this genus is the succulent, constricted stems and persistent leaf bases which have the appearance of strings of beads. The stems of *Monilaria* (and *Mitrophyllum*) are more succulent than those of *Dicrocaulon*.

Flowering time Species of *Monilaria* flower in autumn to winter (May to August in South Africa). In a given population, the flowering season is fairly short and concentrated to about three weeks. The highly scented flowers open at midday and close in the late afternoon.

Distribution and ecology This genus is found in Namaqualand, in the Western and Northern Cape Provinces of South Africa. Plants grow fully exposed (except for *M. obconica*), mostly on clay and quarzitic soils. Populations are often small.

Cultivation *Monilaria* species grow well in sandy soil in the greenhouse. Propagation is from seed (sown during autumn), since cuttings are difficult to root. Keep the plants dry in summer. The plants are exceptionally long-lived, possibly surviving for centuries.

Notes The most well-known species is *Monilaria moniliformis*, which can be identified by its large rounded beads. Flower colour varies tremendously; it is usually white, but even salmon shades are possible, contrasting beautifully with the often red stamens. The leaves of *M. chrysoleuca* var. *chrysoleuca* have such large and prominent water cells that they resemble pipe-cleaners. The magenta flower colour seen in the illustration is uncommon for this species, which usually has pale orange or golden flowers. The dormant plant of *M. moniliformis* in the illustration appears to be dead, but in fact it is about to commence active growth.

Monilaria (Schwantes) Schwantes

Number of species/subspecies/varieties (5/1/1)

Species list and conservation status
 M. chrysoleuca (Schltr.) Schwantes var. *chrysoleuca*
 M. chrysoleuca (Schltr.) Schwantes var. *polita* (L.Bolus) Ihlenf. & Jörg.
 M. moniliformis (Thunb.) Ihlenf. & Jörg.
 M. obconica Ihlenf. & Jörg.
 M. pisiformis (Haw.) Schwantes
 M. scutata (L.Bolus) Schwantes subsp. *obovata* Ihlenf. & Jörg.
 M. scutata (L.Bolus) Schwantes subsp. *scutata*

Literature
IHLENFELDT, H.-D. & JÖRGENSEN, S. 1973. Morphologie und Taxonomie der Gattung *Monilaria* (Schwantes) Schwantes s. str. (Mesembryanthemaceae). (Monographie der Mitrophyllinae Schwantes I). *Mitteilungen aus dem Institut für Allgemeine Botanik, Hamburg* 14: 49–94.

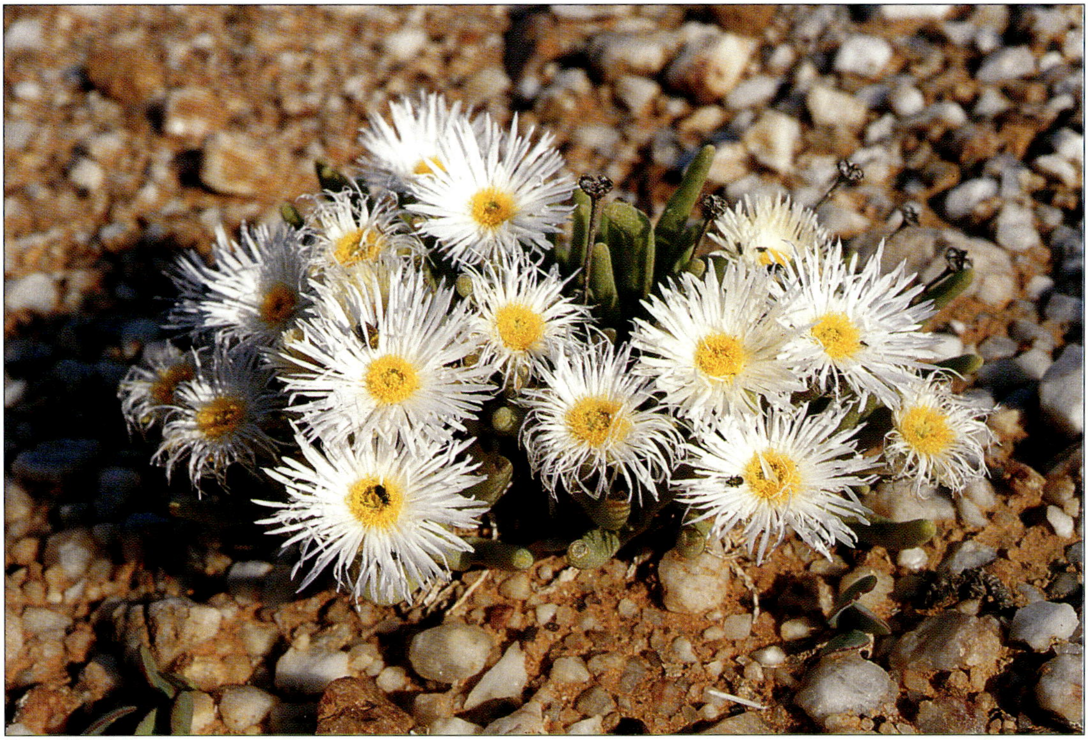
Ernst van Jaarsveld

M. moniliformis

NBI

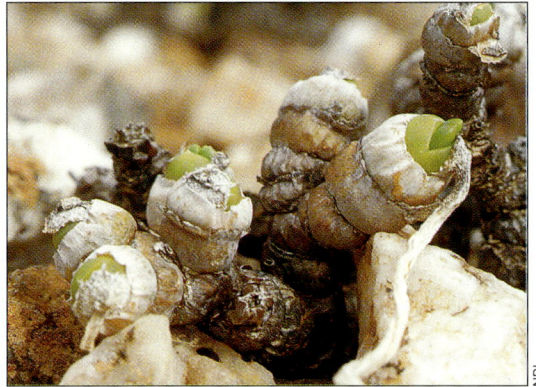

Dormant plant of *M. moniliformis*

John Trager

M. chrysoleuca

John Trager

M. obconica

Ben-Erik van Wyk

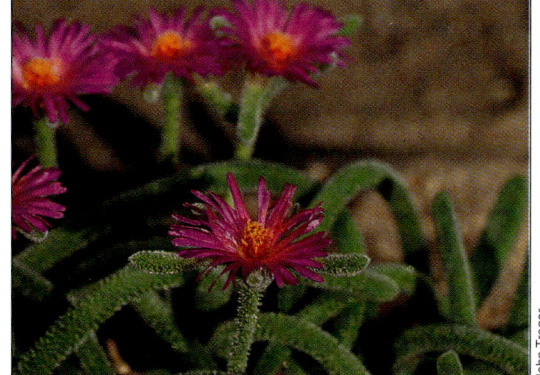

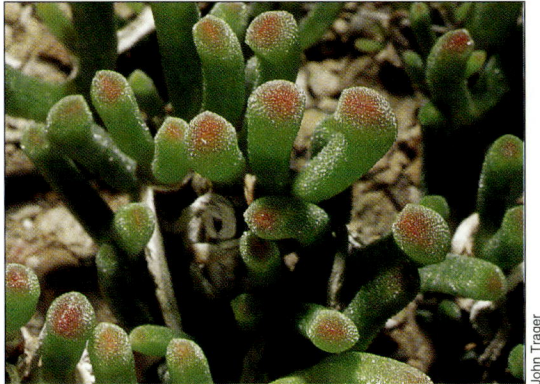

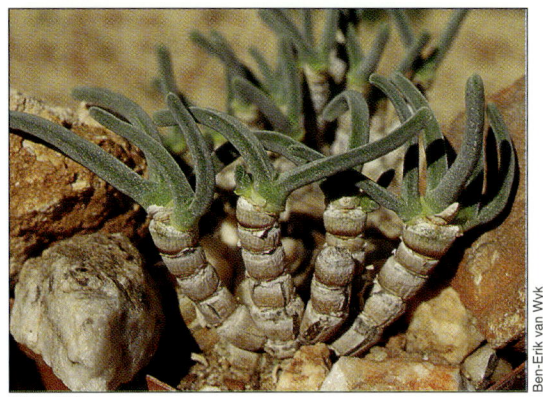

M. scutata

Carpobrotus acinaciformis

Jordaaniella dubia

Cephalophyllum ebracteatum

Disphyma crassifolium

Mat-forming Mesembs

GROUP 9

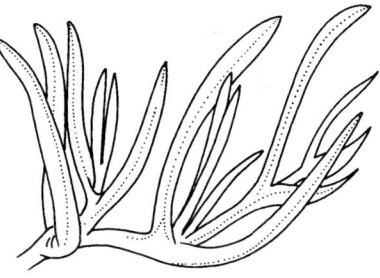

The succulent leaves are variable in size and shape but the plants all creep along the ground, forming dense or sparse mats.

Mesembs included in this artificial category have their creeping habit in common, and ease of identification is the only reason to group them together. Plants may form dense or sparse mats, usually rooting at the nodes. Leaves are very variable in size and shape and may be large (*Carpobrotus*) or small (*Mossia*). Fruit capsules in this group are highly diverse, ranging from the fleshy fruits of *Carpobrotus,* the tiny *Delosperma*-type capsules of *Mossia*, to the *Leipoldtia*-type capsules of *Antimima* and the multi-locular fruits with large closing bodies of *Cephalophyllum*.

MAT-FORMING MESEMBS comprise 8 genera and 168 species.

Antimima (99 species)
Carpobrotus (13 species)
Cephalophyllum (33 species)
Disphyma (4 species)
Jensenobotrya (1 species)
Jordaaniella (4 species)
Malephora (13 species)
Mossia (1 species)

Antimima

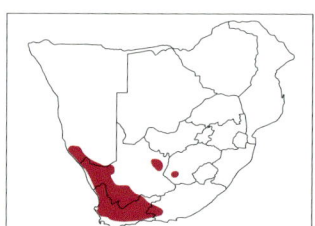

Derivation of the genus name The genus name was derived from the Greek word *antimimos* (imitating) because the first known species closely resembled *Argyroderma*.

Common names The name *kussingvygie* (cushion mesemb) has been used for some species, for example *Antimima lawsonii*. *A. saxicola* is commonly known as the *klipvygie* (rock mesemb).

Description The plants are compact, form mats or creep in rare cases, but they often possess a compact centre with long shoots developing from it. Leaves are either of one type only, or of two, rarely three, different types. In the latter case, one leaf pair forms a persistent sheath enveloping the subsequent leaf pair during the resting state. Different leaf types follow each other either alternatively, two pairs per season, or on a long shoot, the first leaf pair being a short, sheath-forming one. The smooth to minutely bumpy leaf surfaces are typically covered with wax. Flowers are pink to purple, rarely white, and occur singly or in clusters, with small bracts on the flower stalks. Petals are often gathered into five bunches per flower, as a result of the sepals pushing them into groups. The filamentous staminodes surround the central cone of stamens. Fruit capsules have five, rarely six locules and large, conspicuous closing bodies. The closing bodies are rarely small or absent. The valves expand widely with the help of broad, expanding keels, mostly reaching to the tip of the valve.

Distinguishing characters Fruit capsule characters, particularly the large closing bodies and the broad, narrowly and irregularly lobed expanding keels distinguish the genus from *Ruschia*, to which most species have been assigned in the past.

Flowering time Flowering occurs virtually throughout the year, but with a peak season in the winter. The sometimes strongly scented flowers open in the morning and close in the evening.

Distribution and ecology Most species occur in the winter rainfall area along the southwestern coast, from near Aus in Namibia to the south, extending into South Africa as far as the Eastern Cape Province. Only a few species are found in the Great Karoo of the Northern Cape Province and the Free State. The rainfall in the distribution area ranges from below 100 to about 500 mm per year. The species are in most cases adapted to particular soil conditions and inhabit rather restricted areas.

Cultivation The species with a single leaf type are easily grown from seed. The species with several types of leaves need a distinct dry summer period in order to maintain the leaf sequences typical of the species. Sow seed during autumn. Some species have considerable horticultural potential for dry and cold areas but so far very few have been introduced into gardens. Mat-forming species grow easily from cuttings.

Notes Most species of *Antimima* were until recently included in *Ruschia*. However, *Antimima* in its present delimitation probably includes more than one genus and further work is required. Although many species of *Antimima* have small flowers, the huge flowers of *A. ventricosa* make it one of the most spectacular sights on the west coast. One of the most extraordinary mesembs, *A. lawsonii*, a little-known species from the northern Karoo, forms dense, symmetrical, anthill-like cushions about 300 mm in diameter. The leaves of many species smell like old fish (for example *A. piscodora*). The sheath-forming species such as *A. pygmaea*, look dead during summer because the leaves are covered in a papery skin which opens up or disintegrates in autumn. *A. subtruncata* is a good example of those species with a

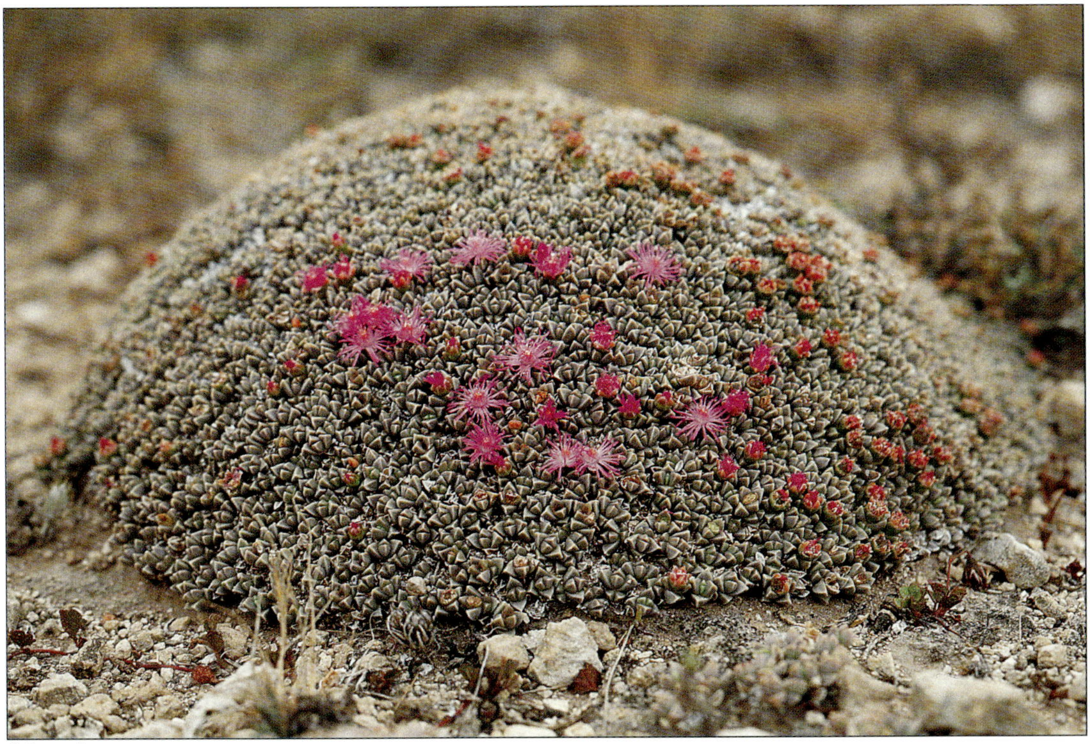
Ernst van Jaarsveld
A. lawsonii

Ernst van Jaarsveld
Flowers and leaves of *A. lawsonii*

John Manning
A. ventricosa

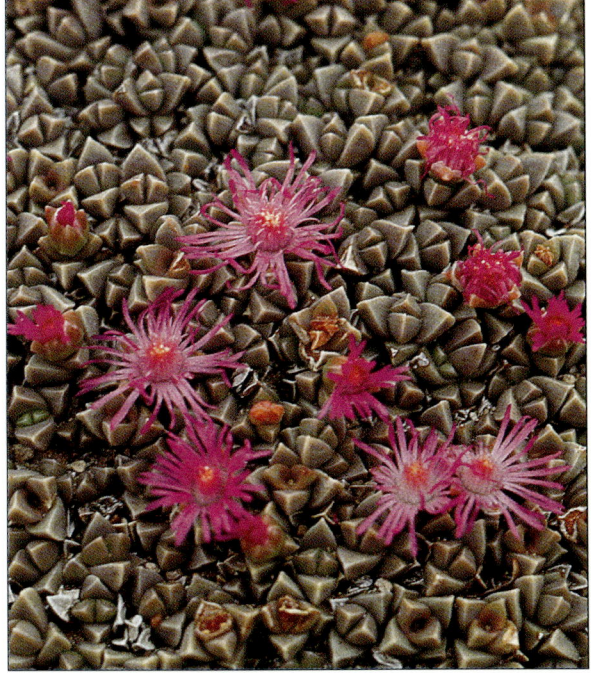

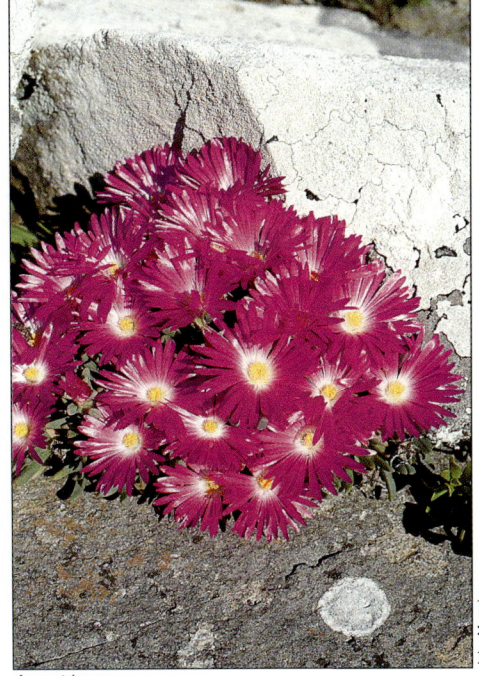

ANTIMIMA

compact centre and sporadic long shoots radiating from it.

Antimima N.E.Br.

NUMBER OF SPECIES/SUBSPECIES/VARIETIES (99/0/0)

SPECIES LIST AND CONSERVATION STATUS

- *A. addita* (L.Bolus) H.E.K.Hartmann
- *A. alborubra* (L.Bolus) Dehn
- *A. androsacea* (Marloth & Schwantes) H.E.K.Hartmann
- *A. argentea* (L.Bolus) H.E.K.Hartmann
- *A. aurasensis* H.E.K.Hartmann
- *A. biformis* (N.E.Br.) H.E.K.Hartmann
- *A. bina* (L.Bolus) H.E.K.Hartmann
- *A. bracteata* (L.Bolus) H.E.K.Hartmann
- *A. brevicarpa* (L.Bolus) H.E.K.Hartmann
- *A. brevicollis* (N.E.Br.) H.E.K.Hartmann
- *A. buchubergensis* (Dinter) H.E.K.Hartmann
- *A. compacta* (L.Bolus) H.E.K.Hartmann
- *A. compressa* (L.Bolus) H.E.K.Hartmann
- *A. concinna* (L.Bolus) H.E.K.Hartmann
- *A. condensa* (N.E.Br.) H.E.K.Hartmann
- *A. crassifolia* (L.Bolus) H.E.K.Hartmann
- *A. dasyphylla* (L.Bolus) H.E.K.Hartmann
- *A. defecta* (L.Bolus) H.E.K.Hartmann
- *A. dekenahi* (N.E.Br.) H.E.K.Hartmann
- *A. distans* (L.Bolus) H.E.K.Hartmann
- *A. dolomitica* (Dinter) H.E.K.Hartmann
- *A. dualis* (N.E.Br.) N.E.Br.
- *A. eendornensis* (Dinter) H.E.K.Hartmann
- *A. elevata* (L.Bolus) H.E.K.Hartmann
- *A. emarcescens* (L.Bolus) H.E.K.Hartmann
- *A. erosa* (L.Bolus) H.E.K.Hartmann
- *A. evoluta* (N.E.Br.) H.E.K.Hartmann
- *A. exsurgens* (L.Bolus) H.E.K.Hartmann
- *A. fenestrata* (L.Bolus) H.E.K.Hartmann
- *A. fergusoniae* (L.Bolus) H.E.K.Hartmann
- *A. gracillima* (L.Bolus) H.E.K.Hartmann
- *A. granitica* (L.Bolus) H.E.K.Hartmann
- *A. hallii* (L.Bolus) H.E.K.Hartmann
- *A. hamatilis* (L.Bolus) H.E.K.Hartmann
- *A. hantamensis* (Engler) H.E.K.Hartmann & Stüber
- *A. herrei* (Schwantes) H.E.K.Hartmann
- *A. intervallaris* (L.Bolus) H.E.K.Hartmann
- *A. ivori* (N.E.Br.) H.E.K.Hartmann
- *A. karroidea* (L.Bolus) H.E.K.Hartmann
- *A. klaverensis* (L.Bolus) H.E.K.Hartmann
- *A. koekenaapensis* (L.Bolus) H.E.K.Hartmann
- *A. komkansica* (L.Bolus) H.E.K.Hartmann
- *A. lawsonii* (L.Bolus) H.E.K.Hartmann [K]
- *A. leipoldtii* (L.Bolus) H.E.K.Hartmann [E]
- *A. leucanthera* (L.Bolus) H.E.K.Hartmann
- *A. limbata* (N.E.Br.) H.E.K.Hartmann
- *A. lodewykii* (L.Bolus) H.E.K.Hartmann
- *A. loganii* (L.Bolus) H.E.K.Hartmann
- *A. lokenbergensis* (L.Bolus) H.E.K.Hartmann
- *A. longipes* (L.Bolus) Dehn
- *A. luckhoffii* (L.Bolus) H.E.K.Hartmann
- *A. maleolens* (L.Bolus) H.E.K.Hartmann
- *A. maxwellii* (L.Bolus) H.E.K.Hartmann
- *A. menniei* (L.Bolus) H.E.K.Hartmann
- *A. mesklipensis* (L.Bolus) H.E.K.Hartmann
- *A. meyerae* (Schwantes) H.E.K.Hartmann
- *A. microphylla* (Haw.) Dehn
- *A. minima* (Tischer) H.E.K.Hartmann
- *A. minutifolia* (L.Bolus) H.E.K.Hartmann
- *A. modesta* (L.Bolus) H.E.K.Hartmann
- *A. mucronata* (Haw.) H.E.K.Hartmann
- *A. mutica* (L.Bolus) H.E.K.Hartmann
- *A. nobilis* (Schwantes) H.E.K.Hartmann
- *A. nordenstamii* (L.Bolus) H.E.K.Hartmann
- *A. oviformis* (L.Bolus) H.E.K.Hartmann
- *A. papillata* (L.Bolus) H.E.K.Hartmann
- *A. paucifolia* (L.Bolus) H.E.K.Hartmann
- *A. pauper* (L.Bolus) H.E.K.Hartmann
- *A. peersii* (L.Bolus) H.E.K.Hartmann
- *A. perforata* (L.Bolus) H.E.K.Hartmann
- *A. persistens* (L.Bolus) H.E.K.Hartmann
- *A. pilosula* (L.Bolus) H.E.K.Hartmann
- *A. piscodora* (L.Bolus) H.E.K.Hartmann
- *A. prolongata* (L.Bolus) H.E.K.Hartmann
- *A. propinqua* (N.E.Br.) H.E.K.Hartmann
- *A. prostrata* (L.Bolus) H.E.K.Hartmann
- *A. pumila* (Fedde & Schuster) H.E.K.Hartmann
- *A. pusilla* (Schwantes) H.E.K.Hartmann
- *A. pygmaea* (Haw.) H.E.K.Hartmann
- *A. quarzitica* (Dinter) H.E.K.Hartmann
- *A. roseola* (N.E.Br.) H.E.K.Hartmann
- *A. saturata* (L.Bolus) H.E.K.Hartmann
- *A. saxicola* (L.Bolus) H.E.K.Hartmann
- *A. schlechteri* (Schwantes) H.E.K.Hartmann
- *A. simulans* (L.Bolus) H.E.K.Hartmann
- *A. sobrina* (N.E.Br.) H.E.K.Hartmann
- *A. solida* (L.Bolus) H.E.K.Hartmann
- *A. stayneri* (L.Bolus) H.E.K.Hartmann

Dormant plant of *A. pygmaea*

Flowering plant of *A. pygmaea*

A. leipoldtii

A. fenestrata

A. stokoei (L.Bolus) H.E.K.Hartmann
A. subtruncata (L.Bolus) H.E.K.Hartmann
A. triquetra (L.Bolus) H.E.K.Hartmann
A. tuberculosa (L.Bolus) H.E.K.Hartmann
A. turneriana (L.Bolus) H.E.K.Hartmann
A. vanzylii (L.Bolus) H.E.K.Hartmann
A. varians (L.Bolus) H.E.K.Hartmann
A. ventricosa (L.Bolus) H.E.K.Hartmann
A. verruculosa (L.Bolus) H.E.K.Hartmann
A. watermeyeri (L.Bolus) H.E.K.Hartmann
A. wittebergensis (L.Bolus) H.E.K.Hartmann

Literature

DEHN, M. 1988. Untersuchungen zum Merkmalsbestand und zur Stellung der Gattung *Antimima* N.E.Br. emend. Dehn (Mesembryanthemaceae Fenzl). *Mitteilungen aus dem Institut für Allgemeine Botanik Hamburg* 22: 189–215.

HARTMANN, H.E.K. 1996. *Antimima aurasensis* H.E.K.Hartmann (Aizoaceae), eine neue Art aus Namibia. *Kakteen und andere Sukkulenten* 47: 229–234.

HARTMANN, H.E.K. 1998. New combinations in *Antimima* (Ruschioideae, Aizoaceae). *Bothalia* 28: 67–82.

A. evoluta

A. meyerae

CARPOBROTUS

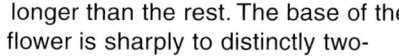

DERIVATION OF GENUS NAME The name is derived from the Greek words *karpos* (fruit) and *brota* (edible things), referring to the edible fruits.

COMMON NAMES The common names *suurvy* and sour fig are widely used, less commonly *perdevy* (horse fig) or *vyerank* (creeping fig). The vernacular name *gaukum* has been recorded for *C. deliciosus*. In Australia, *C. aequilateris* is known as the angular pigface, while *C. modestus* is known as the inland pigface.

DESCRIPTION Plants are robust succulents with trailing, angular, sometimes narrowly winged, smooth stems and short, more or less upright flowering branches. The leaves are somewhat fused at their bases, sharply three-angled, variously curved and sharply pointed, sometimes almost straight. They display various shades of green or grey-green and are often tinged with a reddish hue. The exceptionally large flowers occur in shades of purple, pink, white or yellow. They are borne singly on flower stalks, each with a pair of toothed bracts near its midpoint. There are five leaf-like sepals, often with membranous margins. In some species the sepals are uniform in size and shape; in others they are unequal, often with two much longer than the rest. The base of the flower is sharply to distinctly two-ridged and tapers into the flower stalk. In some species it is separated from the stalk by a more or less distinct constriction. Nectar glands are arranged in a ring. Fruit capsules have fleshy walls and lack valves. They gradually dry out with age but never open up as in most other mesembs. Seeds are relatively large and glossy brown, and are imbedded in a slimy pulp.

DISTINGUISHING CHARACTERS *Carpobrotus* flowers are the largest in the family and can reach a diameter of up to 150 mm. The fruits are fleshy, pulpy and do not break apart.

FLOWERING TIME Species of *Carpobrotus* flower abundantly from early spring to summer (August to January in the southern hemisphere). Flowers open in the morning and close at night.

DISTRIBUTION AND ECOLOGY *Carpobrotus* occurs in various parts of the world. In South Africa, most species are found along the coast at low altitudes but a few occur inland. The South African species are *C. acinaciformis*, *C. deliciosus*, *C. dimidiatus*, *C. edulis* subsp. *edulis*, *C. edulis* subsp. *parviflorus*, *C. mellei*,

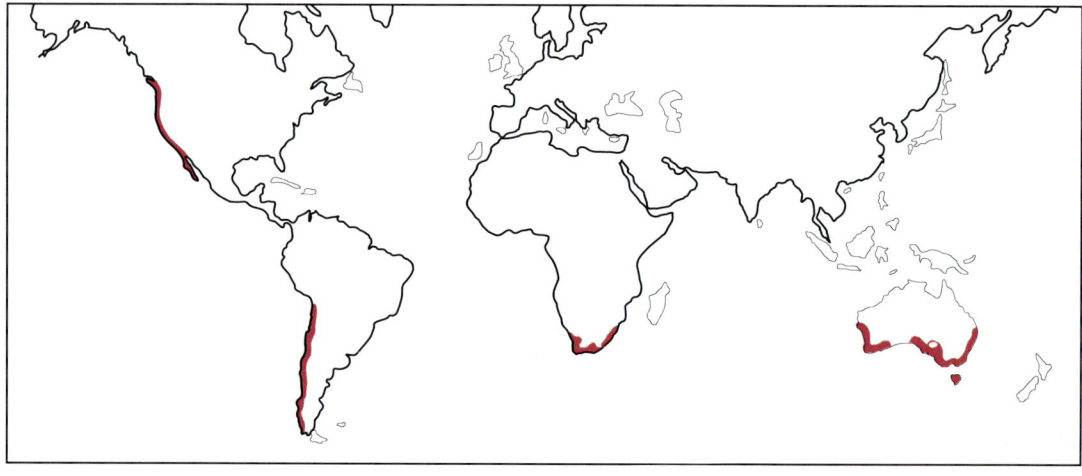

C. aequilaterus

C. chilensis

C. glaucescens

C. virescens

C. rossii

C. quadrifidus

Carpobrotus N.E.Br.

NUMBER OF SPECIES/SUBSPECIES/VARIETIES (13/1/0)

SPECIES LIST AND CONSERVATION STATUS
C. acinaciformis (L.) L.Bolus
C. aequilaterus (Haw.) N.E.Br.
C. chilensis (Mol.) N.E.Br.
C. deliciosus (L.Bolus) L.Bolus [nt]
C. dimidiatus (Haw.) L.Bolus
C. edulis (L.) L.Bolus subsp. edulis
C. edulis (L). L.Bolus subsp. parviflorus Wisura & Glen
C. glaucescens (Haw.) Schwantes
C. mellei (L.Bolus) L.Bolus
C. modestus S.T.Blake
C. muirii (L.Bolus) L.Bolus
C. quadrifidus L.Bolus
C. rossii (Haw.) Schwantes
C. virescens (Haw.) Schwantes

C. muirii and C. quadrifidus. The species which occur in Australia, New Zealand and Tasmania are C. aequilaterus, C. glaucescens, C. modestus, C. rossii and C. virescens. One species, C. chilensis, is found in Chile, Baja California (Mexico) and California (USA). C. edulis and C. chilensis are naturalised in many parts of the world, particularly in regions with winter rain and hot, dry summers, such as Australia and California.

CULTIVATION Plants grow easily from cuttings or seeds and are tolerant of a wide range of soils. They are popular in gardens and for landscaping, particularly to stabilise sandy banks and road edges, because rooted or unrooted cuttings planted out in spring thickly cover considerable areas in a short time. The fruits are widely eaten and are also used to make *suurvy* jam.

NOTES The Australian genus *Sarcozona* is sometimes considered to be closely related to *Carpobrotus*, but since the species are erect shrublets they have been included under DWARF SHRUBBY MESEMBS. Amongst the South African members of *Carpobrotus*, C. edulis is the best known and most widely harvested species. It is easily recognised by the pale yellow flowers which turn pink with age, and the two long sepals.

LITERATURE

BLAKE, S.T. 1969. A revision of *Carpobrotus* and *Sarcozona* in Australia, genera allied to *Mesembryanthemum* (Aizoaceae). *Contributions from the Queensland Herbarium* 7: 1–65.

WISURA, W. & GLEN, H.F. 1993. The South African species of *Carpobrotus* (Mesembryanthema – Aizoaceae). *Contributions from the Bolus Herbarium* 15: 76–107.

C. mellei

C. deliciosus

C. edulis subsp. *edulis*

Flowers of *C. edulis* subsp. *edulis*

C. dimidiatus

C. acinaciformis

CEPHALOPHYLLUM

DERIVATION OF THE GENUS NAME The name is derived from the Greek words *cephalos* (head) and *phyllon* (leaf), in allusion to the compact heads of leaves in several species.

COMMON NAMES The common name *rankvygie* (creeping mesemb) has been used for some species.

DESCRIPTION Plants are compact, cushion-forming, creeping or, in rare cases, erect; some species develop vegetative heads or rosettes with annual or perennial long shoots bearing flowers; the stems vary in texture from woody to spongy in different species. The leaves are three-angled, spindle-shaped, quill-shaped, or club-shaped, ranging from light to dark green in colour, with a smooth surface throughout. Flowers are most commonly borne in rather large flower clusters; petals can be white, yellow, pink to purple, red, copper, or orange, also in different combinations in one flower. Filamentous staminodes are found in one species only. The stamens normally spread widely, bearing anthers that are often strikingly different in colour, emphasising the centre of the flower prominently. Five sepals enclose the flowers in the bud. Stigmas are feathery and surrounded by a ring of nectar glands. The fruit capsules have between 10 and 24 locules, with large closing bodies, very broad valve wings and covering membranes with thick closing bulges at their distal opening. Seeds are small and brown.

DISTINGUISHING CHARACTERS Species of *Cephalophyllum* have brilliant multi-coloured flowers and are distinguished by the 10 to 24-locular capsules with large closing bodies, broad valve wings and closing bulges at the covering membranes.

FLOWERING TIME The flowers appear in winter, and open mainly over midday and in the afternoon.

DISTRIBUTION AND ECOLOGY Species of the genus occur in a broad band from Lüderitz to Aus in Namibia, stretching southwards through the Northern Cape Province and into the Western Cape Province, South Africa, occasionally reaching the sea, and extending as far east as Oudtshoorn. Most species receive 100 to 350 mm rainfall per year, mainly in winter; in other areas, peaks are experienced in autumn (March) and early summer (November).

CULTIVATION Most species with a creeping habit spread easily by suckers, the mother plant dying off after some years. In this way, the formation of ground covers on disturbed places such as roadsides or after floods occur within a few years, stabilising the exposed ground. These species are tardy in germinating their seeds, in contrast to species with compact growth, which can also establish rather large populations after floods. Flowering is much more spectacular in compact plants. Sow seed in autumn.

NOTES The genus can be divided into two groups of species: those with persisting vegetative heads or rosettes with long leaves and mostly short-lived flower-bearing branches with distinctly shorter leaves (*Cephalophyllum* subg. *Cephalophyllum*), and those with leaves of nearly equal size (*Cephalophyllum* subg. *Homophyllum*). *C. alstonii* is an example of the first group, and is famous for its spectacular anemone-red flowers. From the second group *C. pillansii* is a commonly cultivated creeper whose yellow flowers have cherry red centres. Flower colour in *Cephalophyllum* is marvellously variable. Even the capsules are beautiful, resembling starry brown flowers when they open.

Cephalophyllum N.E.Brown

NUMBER OF SPECIES/SUBSPECIES/VARIETIES (33/0/0)

C. rigidum

C. alstonii

Open fruit capsules of *C. diversiphyllum*

C. spissum

C. fulleri

C. caespitosum

Cephalophyllum

Species list and conservation status
 C. alstonii Marloth ex L.Bolus
 C. caespitosum H.E.K.Hartmann
 C. compressum L. Bolus
 C. confusum (Dinter) Dinter & Schwantes [R]
 C. corniculatum (L.) Schwantes
 C. curtophyllum (L.Bolus) Schwantes
 C. diversiphyllum (Haw.) N.E.Br.
 C. ebracteatum (Schltr. & Diels) Dinter & Schwantes
 C. framesii L.Bolus
 C. fulleri L.Bolus [V]
 C. goodii L.Bolus
 C. hallii L.Bolus
 C. herrei L.Bolus
 C. inaequale L.Bolus
 C. loreum (L.) Schwantes
 C. niveum L.Bolus
 C. numeesense H.E.K.Hartmann
 C. parvibracteatum (L.Bolus) H.E.K.Hartmann
 C. parviflorum L.Bolus
 C. parvulum (Schltr.) H.E.K.Hartmann [Ex]
 C. pillansii L.Bolus
 C. pulchellum L.Bolus [R]
 C. pulchrum L.Bolus [R/V]
 C. purpureo-album (Haw.) Schwantes
 C. regale L.Bolus
 C. rigidum L.Bolus
 C. rostellum (L.Bolus) H.E.K.Hartmann [K]
 C. spongiosum (L. Bolus) L. Bolus
 C. spissum H.E.K.Hartmann
 C. staminodiosum L.Bolus [R]

 C. subulatoides (Haw.) N.E.Br
 C. tetrastichum H.E.K.Hartmann [V]
 C. tricolorum (Haw.) Schwantes

Literature

HARTMANN, H.E.K. 1978. Zur Kenntnis der Gattung Cephalophyllum. *Botanische Jahrbücher* 99: 264–302.

HARTMANN, H.E.K. 1983. Interaction of ecology, taxonomy and distribution in some Mesembryanthemaceae. *Bothalia* 14: 653–659.

HARTMANN, H.E.K. 1984. Zur Biologie und Taxonomie des *Cephalophyllum curtophyllum*–Komplexes. *Mitteilungen aus dem Institut für Allgemeine Botanik, Hamburg* 19: 141–163.

HARTMANN, H.E.K. 1986. Chromosome numbers in the genus *Cephalophyllum* N.E.Br. (Mesembryanthemaceae). Notes on *Cephalophyllum* V. *Cactus and Succulent Journal (US)* 58: 263–266.

HARTMANN, H.E.K. 1988. Monographien der Subtribus Leipoldtiinae. IV. Monographie der Gattung *Cephalophyllum* (Mesembryanthemaceae). *Mitteilungen aus dem Institut für Allgemeine Botanik, Hamburg* 22: 93–187.

C. pillansii

C. numeesense

C. loreum

C. spongiosum

C. tricolorum

DISPHYMA

DERIVATION OF GENUS NAME N.E. Brown named the genus *Disphyma* from the Greek words *dis* (two, double), and *phyma* (tubercle). The name is an allusion to the bilobed closing bodies (structures found at the entrance of each locule in the capsule) of *D. crassifolium*, *D. dunsdonii* and *D. papillatum*.

COMMON NAMES The common name wishbone plant has been used.

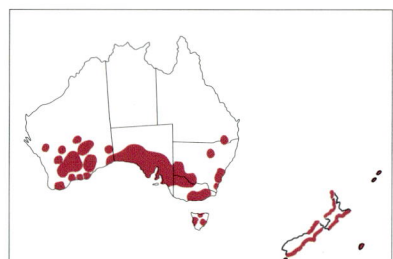

DESCRIPTION Plants lie flat on the ground and form dense mats which root at the nodes. The leaves are three-edged, almost rounded and sometimes minutely fringed at the base, in small rosettes at the nodes, and are translucently dotted when seen against light. They grow to about 30 mm in length, are bright green, later becoming reddish or yellowish. The medium-sized white, pink to purple flowers are borne at the ends of shoots, occurring singly or in threes or rarely in clusters of two to three on tufted, short side branches. The flower stalks are up to 30 mm in length. There are five unequal sepals. The inner ones have membranous margins. The stamens are numerous. The inner ones are bearded at the base. Nectar glands are pronounced, and sometimes arranged in a ring surrounding five awl-shaped, feathery stigmas. Fruit capsules have five locules, with (rarely without) large forked, spongy closing bodies. They are pale yellow and spongy in texture. The valves become reflexed and the expanding keels are diverging, with few teeth. The valve wings are more or less oblong and as long as the valves. Covering membranes are present. The seeds are more or less egg-shaped.

DISTINGUISHING CHARACTERS The spongy consistency of the capsule, the expanding keels which are widely separated at the base and diverging, as well as the two-lobed closing body at the opening to each seed chamber, serve to distinguish the genus.

FLOWERING TIME Flowering occurs during early summer in Australia (October to December) or in winter and early spring in South Africa (July to September). Flowers open at noon and close by evening.

DISTRIBUTION AND ECOLOGY This coastal genus occurs in Australia, Tasmania and New Zealand and also in the Western and Eastern Cape Provinces of South Africa. *D. australe* occurs in New Zealand, the Kermadec and Chatham Islands, *D. crassifolium* subsp. *crassifolium* is found in South Africa, and also in southern Australia and Tasmania. It is naturalised on the coast of Spain. *D. crassifolium* subsp. *clavellatum* is restricted to southern Australia, while *D. dunsdonii* has a limited distribution on the southern coast of South Africa. *D. papillatum* occurs in New Zealand east of Christchurch and on the Chatham Islands. Plants grow in sandy, slightly brackish soils.

CULTIVATION Plants thrive in cultivation, especially close to the seashore in coastal gardens, where they can be used as a ground cover. They are easily propagated from cuttings.

NOTES Along with *Carpobrotus*, *Disphyma* is the most widely distributed of all mesemb genera. Long distance dispersal (perhaps by seabirds?) may account for the presence of the same species (*D. crassifolium*) on two distant continents, but overall the South African and Australian species do not seem particularly closely related. The small and little known *D. dunsdonii* has short reddish leaves and is locally

D. crassifolium subsp. *clavellatum*

Fruit capsules of *D. crassifolium* subsp. *clavellatum*

Two colour forms of *D. australe*

abundant on the limestones around Bredasdorp.

Disphyma N.E.Br.

Number of species/subspecies/varieties
(4/1/0)

Species list and conservation status
  D. australe (Aiton) J.M.Black
  D. crassifolium (L.) L.Bolus subsp. *crassifolium*
  D. crassifolium (L.) L.Bolus subsp. *clavellatum* (Haw.) Chinnock
  D. dunsdonii L.Bolus
  D. papillatum Chinnock

Literature
CHINNOCK, R.J. 1971. Studies in *Disphyma* — a genus related to *Mesembryanthemum*. 1. A revision of *Disphyma australe* (Ait.) J.M.Black. *New Zealand Journal of Botany* 9: 331–344.

CHINNOCK, R.J. 1986. Studies in *Disphyma* 2. Infraspecific subdivision of *Disphyma australe* and notes on the Australian species of *Disphyma*. *New Zealand Journal Botany* 14: 77–78.

CHINNOCK, R.J. 1996. To the limits of *Disphyma* (Aizoaceae: Ruschioideae) and beyond. *Aloe* 33: 59–61.

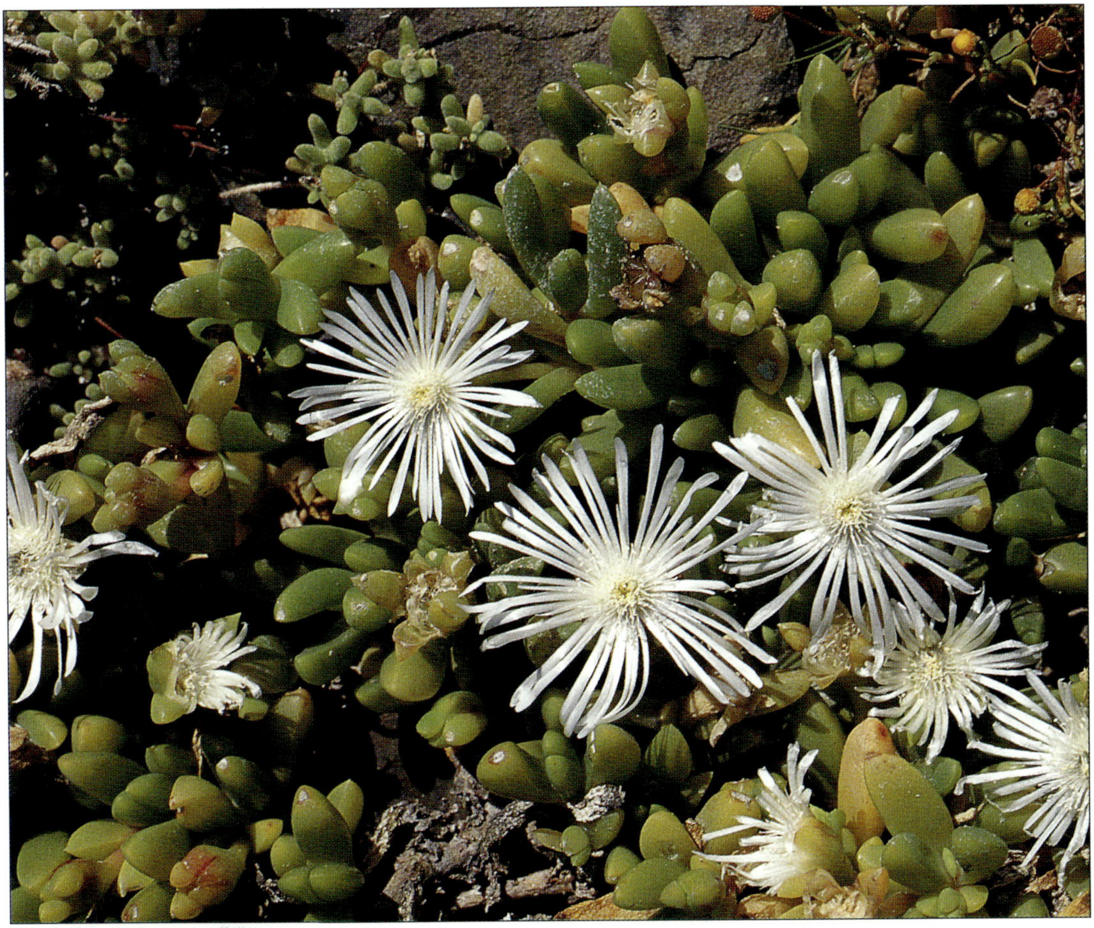

John Manning

D. crassifolium subsp. *crassifolium*

Chris Barnhill

D. dunsdonii in dormant state

Chris Barnhill

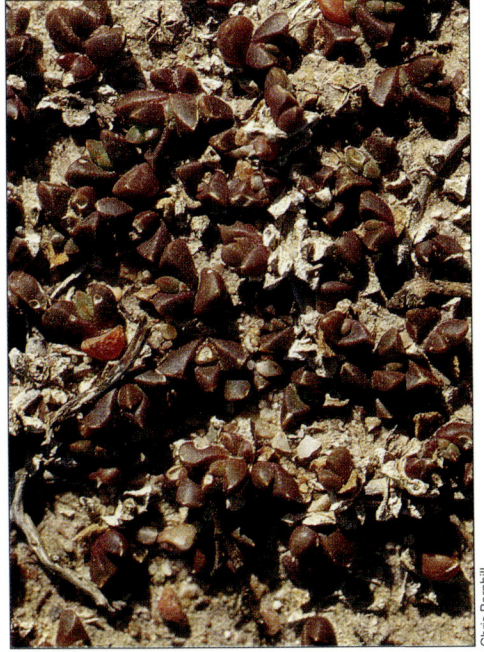

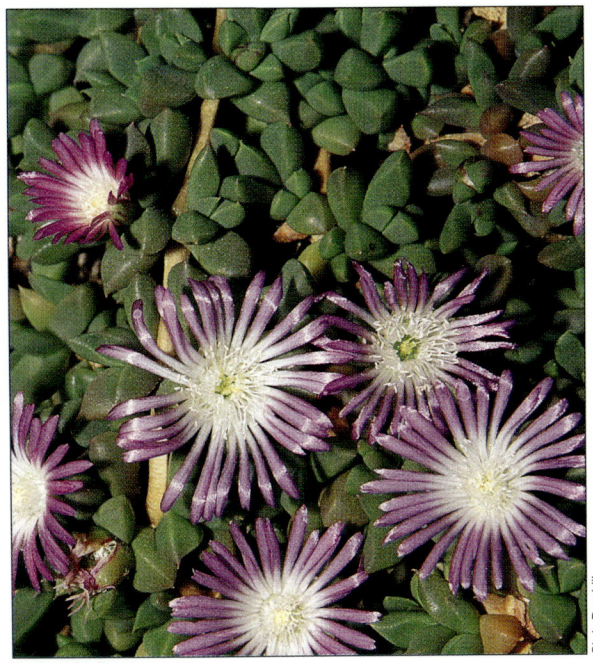

D. dunsdonii

JENSENOBOTRYA

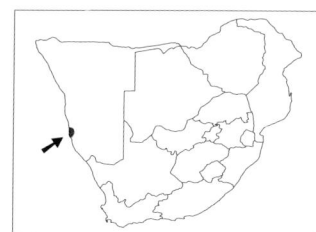

DERIVATION OF GENUS NAME This unusual genus was named after Emil Jensen, a succulent enthusiast and resident of Lüderitz and Walvis Bay, and is further derived from the Greek word *botrya* (bunch of grapes), describing the likeness of the leaves of *Jensenobotrya* to a bunch of grapes.

COMMON NAMES The vernacular name for *Jensenobotrya* is *druiwetrosvygie* (bunch-of-grapes mesemb).

DESCRIPTION The plants are low growing, often hanging woody shrubs with thick stems, forming large dense mats. The leaves are large, egg-shaped and sharply narrowing towards the base, very fleshy, smooth, grey, often purplish in colour. The small flowers are pink with white at the base, or they are pure white. Flowers occur solitary on short stalks. Sepals are unequal. The three inner ones have membranous margins but not the outer two. Stamens are numerous and basally bearded. The top of the flower is flat or somewhat concave with dark green nectar glands in a ring, surrounding five awl-shaped stigmas. Fruit capsules have five locules, without covering membranes, valve wings or closing bodies. The stigmas persist on the ripening capsule for a considerable period. The fruit capsules are more or less flat and up to 12 mm in diameter. Seeds are shiny and brown.

DISTINGUISHING CHARACTERS *Jensenobotrya* is easily recognised by its huge old stems with grape-shaped leaves hanging in bunches across the rock surfaces at the very specific localities where it occurs. Cultivated plants can still be recognised by the uniquely shaped, very smooth leaves which wrinkle like raisins.

FLOWERING TIME Plants flower in autumn and winter. The flowers open around noon and close by dusk. They are self-fertile, a condition possibly linked to the foggy, saline habitat.

DISTRIBUTION AND ECOLOGY *Jensenobotrya* is endemic to Namibia, and occurs at Dolphin Head in Spencer Bay, in the coastal desert between Lüderitz and Walvis Bay. This rare plant grows on cliffs exposed to strongly saline moisture, out of direct light in rock crevices. Although the rainfall in its habitat is less than 50 mm per year, the plant is adapted to regular fog and cool weather.

CULTIVATION Plants are readily propagated from cuttings or seed and flourish in well-drained sandy soil. Plants are fast growing and should regularly be regrown from fresh cuttings as they become scruffy in time. They are sensitive to strong light and burn easily; the strong light conditions of their natural environment are softened by mists.

NOTES In habitat, the stems of *Jensenobotrya* become very thick. Some plants are estimated to be 200 years old or more.

Jensenobotrya A.G.J.Herre

NUMBER OF SPECIES/SUBSPECIES/VARIETIES (1/0/0)

SPECIES LIST AND CONSERVATION STATUS *J. lossowiana* A.G.J.Herre [R]

LITERATURE
HERRE, H. 1970. Notes on South African succulents. *Cactus and Succulent Journal (US)* 42: 17–18.

John Trager

J. lossowiana

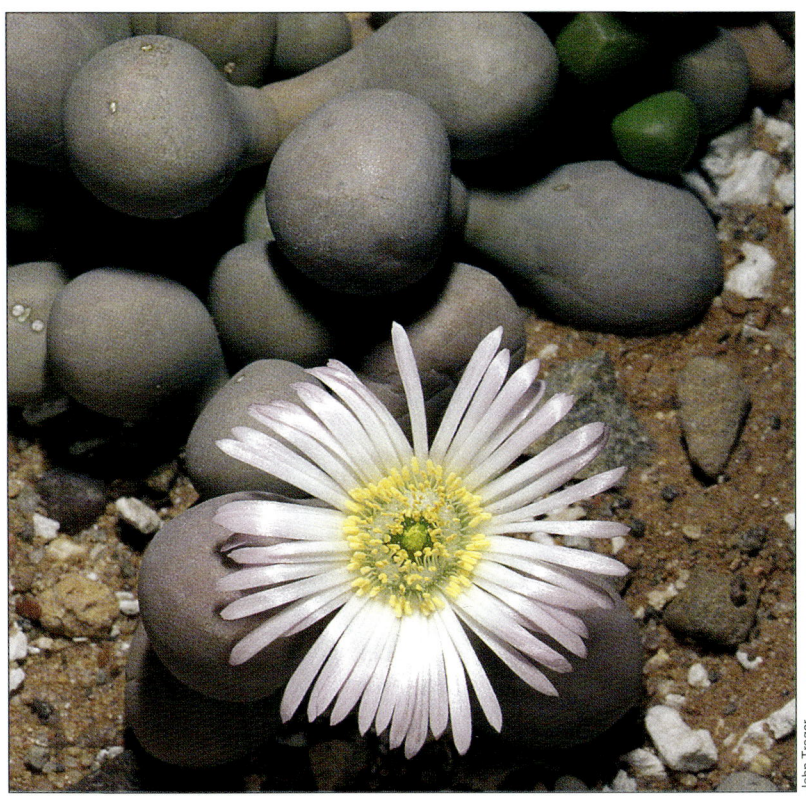

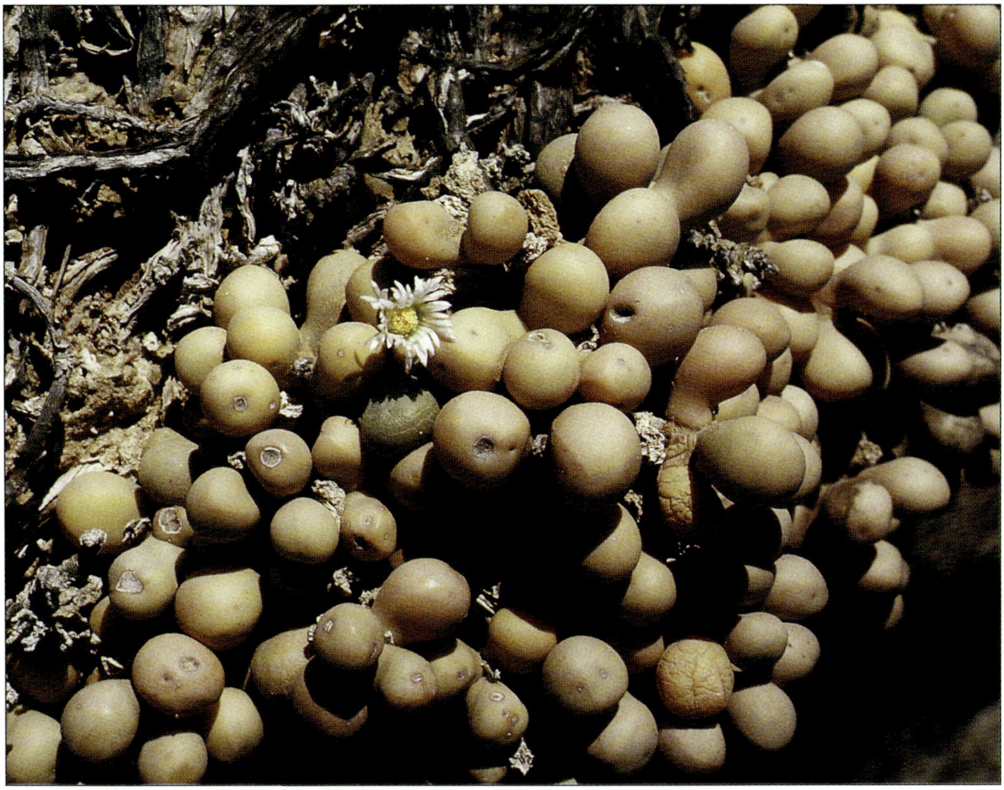

Derek Clark

J. lossowiana in its habitat

JORDAANIELLA

DERIVATION OF GENUS NAME The genus was named after Professor Pieter Gerhardus Jordaan (1913 to 1987), professor of Botany at the University of Stellenbosch.

COMMON NAMES The vernacular names *strandvygie* (beach mesemb) and *rankvygie* (creeping mesemb) are used for this genus.

DESCRIPTION Plants are creeping succulents, with long, orange or yellow internodes formed from a spongy persistent tissue. They have numerous roots arising from the nodes of the trailing branches. The leaves are broadest towards the middle, tapering towards both ends. Flowers are different shades of yellow or purple, rarely white. They occur singly on short side branches and have bracts on the flower stalks. Fruit capsules have 10 to 25 locules and are similar to those of *Leipoldtia* but have small closing bodies.

DISTINGUISHING CHARACTERS The genus is characterised by the trailing branches bearing leaves that taper at both ends and also by the 10 to 25-locular fruit capsules which are square on top.

FLOWERING TIME Plants flower in winter and spring. Flowers open around noon and close in the late afternoon.

DISTRIBUTION AND ECOLOGY *Jordaaniella* occurs mainly on coastal sands in strandveld and in the succulent karoo of the winter rainfall area. The distribution area is restricted to southern Namibia and the Northern and Western Cape Provinces of South Africa. Rainfall is mainly in winter and ranges from 50 to 400 mm per year.

CULTIVATION Plants are easily propagated from cuttings or seeds and they will thrive even in windy, sandy coastal gardens.

NOTES The flowers of *Jordaaniella* are large and attractive. Plants are rapid growers and form colourful mats. It may become necessary to repropagate them if they become untidy.

Jordaaniella H.E.K. Hartmann

NUMBER OF SPECIES/SUBSPECIES/VARIETIES (4/0/0)

SPECIES LIST AND CONSERVATION STATUS
J. clavifolia (L.Bolus) H.E.K.Hartmann [R]
J. cuprea (L.Bolus) H.E.K.Hartmann
J. dubia (Haw.) H.E.K.Hartmann [nt]
J. uniflora (L.Bolus) H.E.K.Hartmann [R/V]

LITERATURE
HARTMANN, H. 1983. Untersuchungen zum Merkmalsbestand und zur Taxonomie der Subtribus Leipoldtiinae (Mesembryanthemaceae). *Bibliotheca Botanica* 136: 1–67.

HARTMANN, H. 1984. Monographien der Subtribus Leipoldtiinae. VI. Monographie der Gattung *Jordaaniella* (Mesembryanthemaceae). *Botanische Jahrbücher* 104: 321–360.

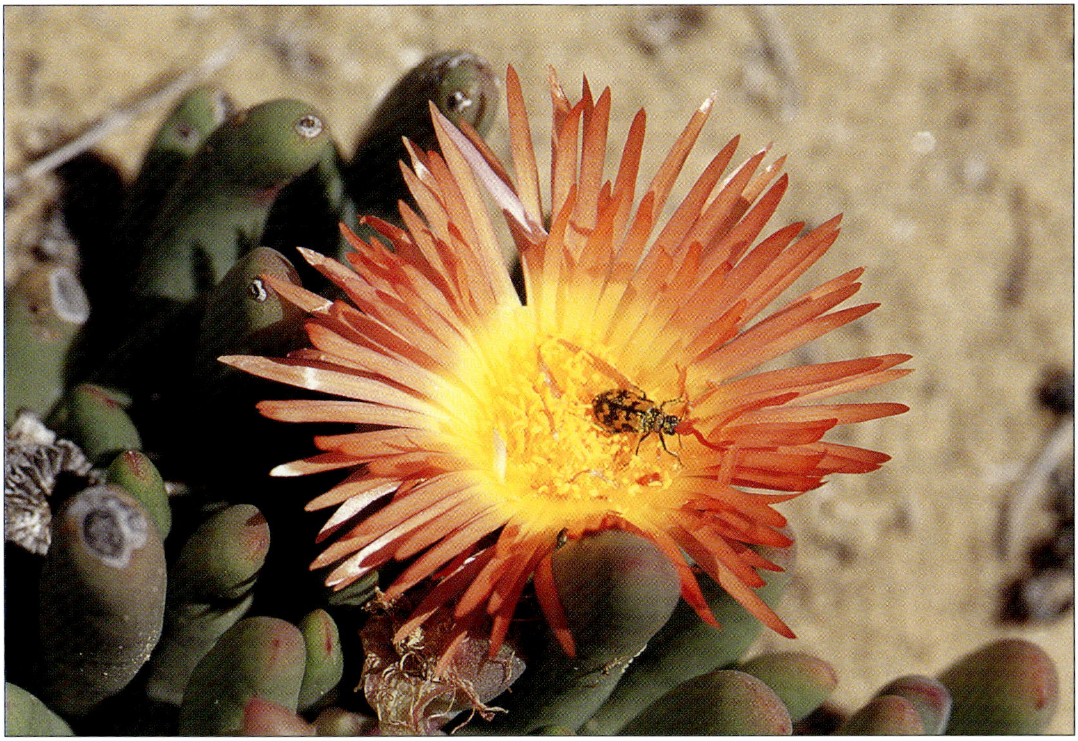
Graham Williamson

J. cuprea

Ernst van Jaarsveld

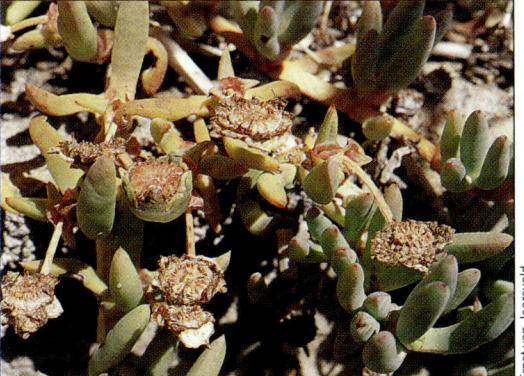

Fruit capsules of *J. clavifolia*

Ernst van Jaarsveld

White form of *J. dubia*

John Manning

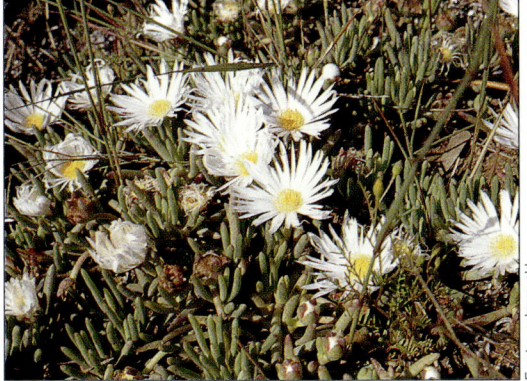

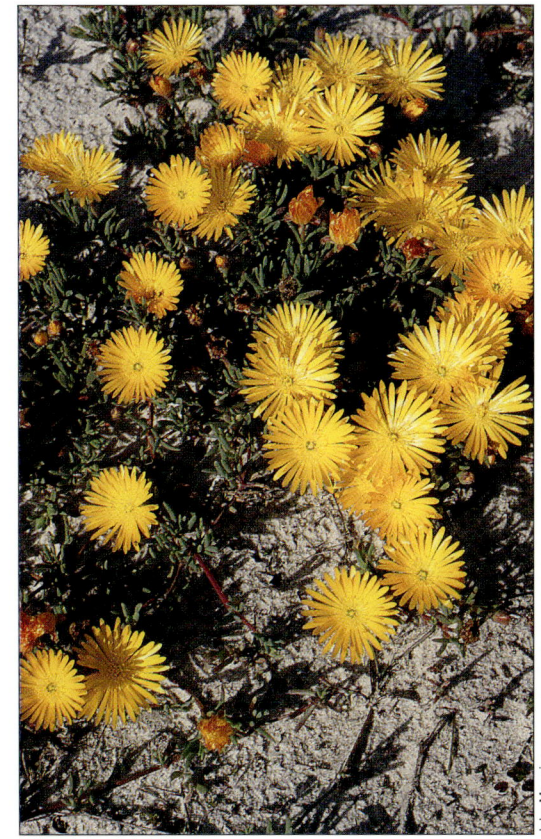

Yellow form of *J. dubia*

MALEPHORA

DERIVATION OF GENUS NAME The name *Malephora* is derived from the Greek words *male* (arm-hole) and *pherein* (to bear).

COMMON NAMES Vernacular names include *geelvingerkanna,* or *geelvingervygie* (yellow finger mesemb), referring to the fat leaves resembling fingers and *vingerkanna* (finger mesemb).

DESCRIPTION Plants are creeping, sometimes upright shrubs or subshrubs with greyish to yellow stems and distinct internodes. The opposite leaves are slightly fused, cylindrical to three-angled, abruptly pointed or semi-cylindrical and fleshy. They are up to 50 mm in length with a smooth surface and often with a bluish, waxy coating. Flowers are golden, yellow, pink or reddish purple and are medium to large, occurring singly or in clusters at the tips, or in the axils of leaves. The backs of the petals are often a rich purple colour, while the inside is golden. Four to five unequal sepals enclose the buds. Numerous stamens, often bearded towards their bases, surround the finely scalloped nectar glands, which are often united to form a ring around eight to 12 feathery stigmas. The fruit capsules have eight to 12 locules. The base of each locule is divided by a thin wall which forms pockets containing one or two seeds. These units are then dispersed. Seeds are flattish and rough-textured.

DISTINGUISHING CHARACTERS *Malephora* has extremely waxy leaves (the wax layer can easily be rubbed off). It may be confused with *Disphyma* (which has similar finger-like waxy leaves) but differs in the yellow to red flowers (white to pink in *Disphyma*). *Disphyma* and *Malephora* both have a tendency to root at the nodes. The bilobed closing body of *Malephora crocea* is a characteristic shared with *Disphyma crassifolium* and some species of *Drosanthemum*.

FLOWERING TIME Some species flower from June to August (winter) and others from December to March (summer in South Africa). The flowers are open at midday.

DISTRIBUTION AND ECOLOGY Species of *Malephora* grow from southern Namibia to Namaqualand in the Northern Cape Province, South Africa, through the Great and Little Karoos to the Eastern Cape Province. Some species are known from the southern Free State. Plants prefer direct sunlight and sandy, well-drained soils. They grow in winter and summer rainfall areas that receive less than 500 mm per year.

CULTIVATION Plants of this genus can be propagated with great ease, either from seed, or from cuttings, which flower in their first year. A few species are popular in rockeries and at least one of these, *M. herrei*, is extremely tolerant of frost because it comes from Fauresmith, a very cold place in South Africa.

NOTES Species of *Malephora* are difficult to identify because there have been no recent studies. The orange and yellow-flowered *M. crocea* is cultivated in many parts of the world and has been known for two centuries. The fattest of the species, *M. crassa*, is a more recent discovery from the Ceres Karoo. In contrast, another yellow-flowered species, *M. lutea*, has thin leaves. *M. uitenhagensis* has only a thin wax layer on its bright green leaves.

Malephora N.E.Br. (=*Crocanthus* L.Bolus; *Hymenocyclus*. Dinter & Schwantes)

NUMBER OF SPECIES/SUBSPECIES/VARIETIES (13/0/1)

SPECIES LIST AND CONSERVATION STATUS
  M. crassa (L.Bolus) H.Jacobsen & Schwantes
  M. crocea (Jacq.) Schwantes var. *crocea*
  M. crocea (Jacq.) Schwantes var. *purpureo-crocea* (Haw.) H.Jacobsen

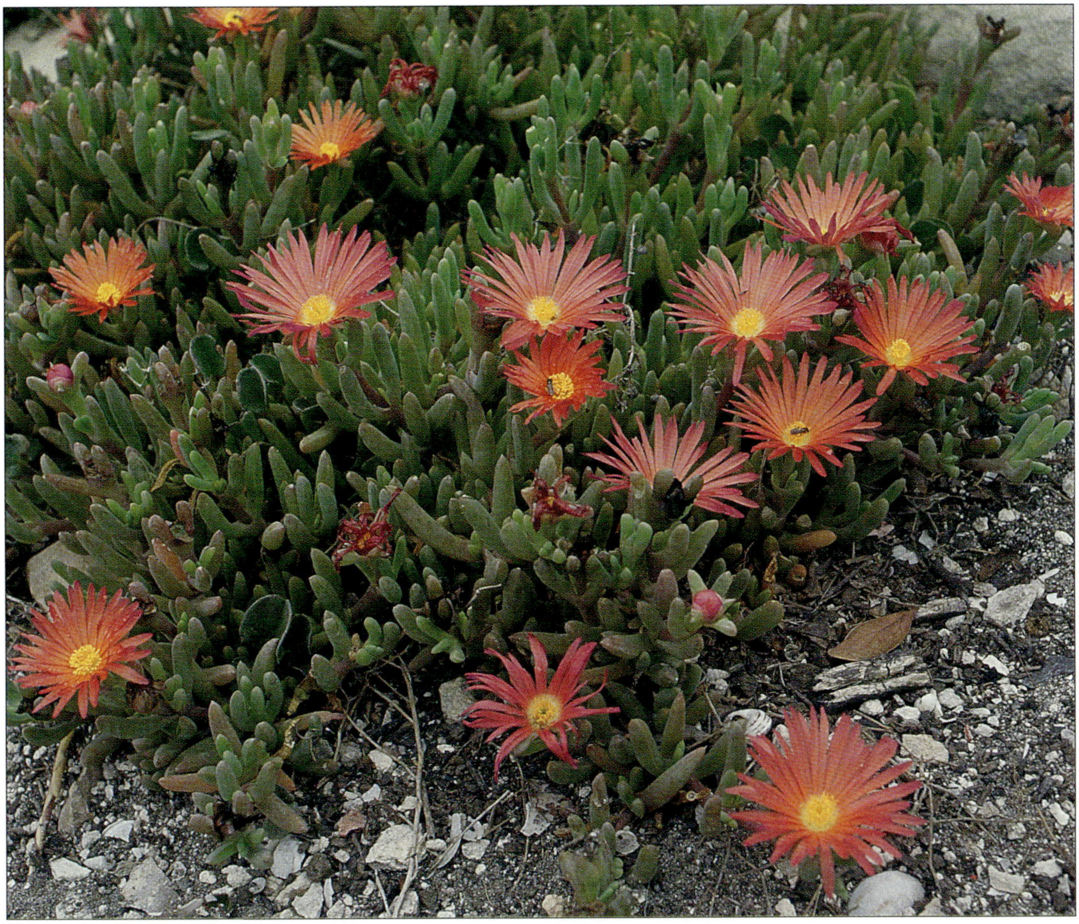

Ernst van Jaarsveld

M. crocea var. *crocea*

Ernst van Jaarsveld

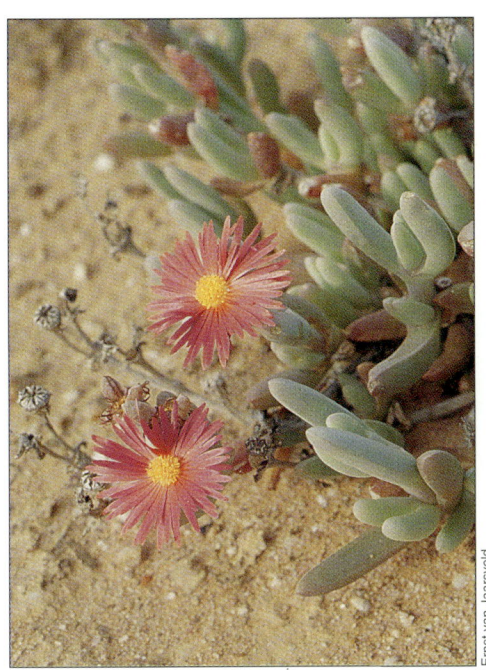

M. crocea var. *purpureo-crocea*

Ernst van Jaarsveld

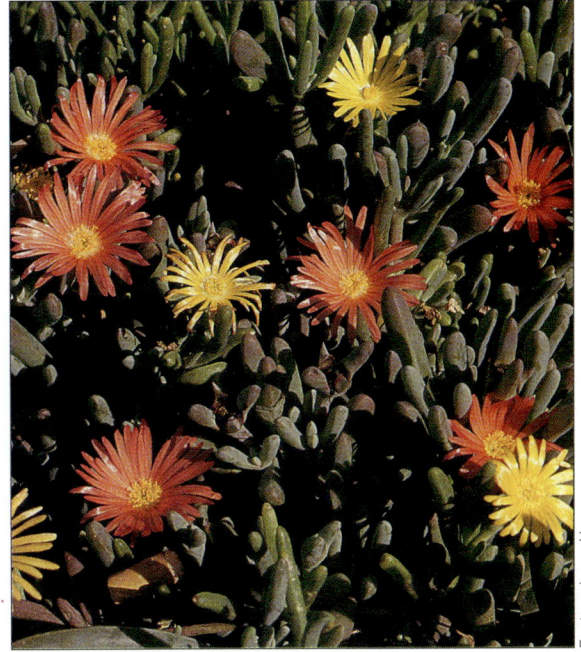

Yellow and orange forms of *M. crocea* var. *crocea*

Malephora

M. engleriana (Dinter & A.Berger) Schwantes
M. flavo-crocea (Haw.) H.Jacobsen & Schwantes
M. framesii (L.Bolus) H.Jacobsen & Schwantes
M. herrei (Schwantes) Schwantes
M. latipetala (L.Bolus) H.Jacobsen & Schwantes
M. lutea (Haw.) Schwantes
M. luteola (Haw.) Schwantes
M. mollis (Aiton) N.E.Br.
M. thunbergii (Haw.) Schwantes

M. uitenhagensis (L.Bolus) H.Jacobsen & Schwantes
M. verruculoides (Sond.) Schwantes

Literature

BROWN, N.E. 1927. *Mesembryanthemum* and some new genera separated from it. *Gardener's Chronicle* 81: 12.

JACOBSEN, H. & ROWLEY, G.D. 1958. Some name changes in succulent plants, Part IV. *National Cactus and Succulent Journal* 13: 77–78.

M. lutea

M. crassa

Mossia

Derivation of genus name The genus was named after Professor Charles Edward Moss (1870 to 1930), first professor of Botany at the South African School of Mines and Technology, now the University of the Witwatersrand.

Common names No vernacular names seem to have been recorded for this genus.

Description Plants have a typical creeping growth form, the long stems forming a tangled mass. The reddish brown stems are thin and wiry and carry the leaves in clusters which are far apart; the leaves are erect and rather short, usually no longer than about 10 mm. The light green to dull blue-green leaves have faint black spots on both surfaces, but are smooth to the touch and terminate in a rather soft, pointed tip. The flowers are white with the petals slightly incurved at their tips. The five sepals are almost of equal size, longer than the petals and united into a short tube. There are numerous stamens surrounding the nectar glands which are dark green and occur in more or less distinct groups corresponding with the sepals. The flower base has a shallow depression at the centre of which are five or six awl-shaped stigmas. The fruit capsules have five or seldom six locules, are shallow and flat on top, without covering membranes, valve wings or closing bodies.

Distinguishing characters *Mossia* is characterised by its creeping growth habit; white flowers that open at night, which have a distinct depression (hypanthium) at their centre; and by its short, dull blue-green leaves which are smooth and roundly triangular in cross-section.

Flowering time In South Africa *Mossia intervallaris* flowers from early to mid-summer (from late October) through November to early December in South Africa, depending on the first seasonal rain. The smoke-scented flowers open at night, more or less between 21:00 and 22:00.

Distribution and ecology *Mossia* occurs in a narrow band more or less to the west of the Drakensberg massif, from Barkly East in the Eastern Cape Province, South Africa to the eastern Free State and northwestern Lesotho. Isolated populations have been recorded from the Free State, Gauteng and Mpumalanga. It is a typical summer rainfall mesemb, restricted to high altitudes on the climatically severe inland escarpment where snowfalls and frost may occur. The frequent fires that occur in the highveld grasslands are avoided by the plants because they usually grow in shallow depressions on rocky outcrops, often on sheet rock in very thin layers of seasonally flooded soil.

Cultivation The species presents few problems in cultivation and it even tolerates winter rain. Plants will grow well in open beds, bench containers or hanging pots. It is not a very striking plant in cultivation and only the most dedicated of collectors will maintain it for any length of time.

Notes This is one of the more inconspicuous mesembs: the flowers are white and open at night, and the leaves are comparatively short and dull-coloured. It is therefore likely that thorough field studies will reveal a distribution wider than is currently known.

Mossia N.E.Br.

Number of species/subspecies/varieties (1/0/0)

Species list and conservation status
  M. intervallaris (L.Bolus) N.E.Br. [R]

Literature
SMITH, G.F., HARTZER, P. & VAN WYK, A.E. 1997. Plate 310. *Mossia intervallaris*. Curtis's Botanical Magazine 14: 16–22.

M. intervallaris

Growth form of M. intervallaris

Flower of M. intervallaris

Growth form of M. intervallaris

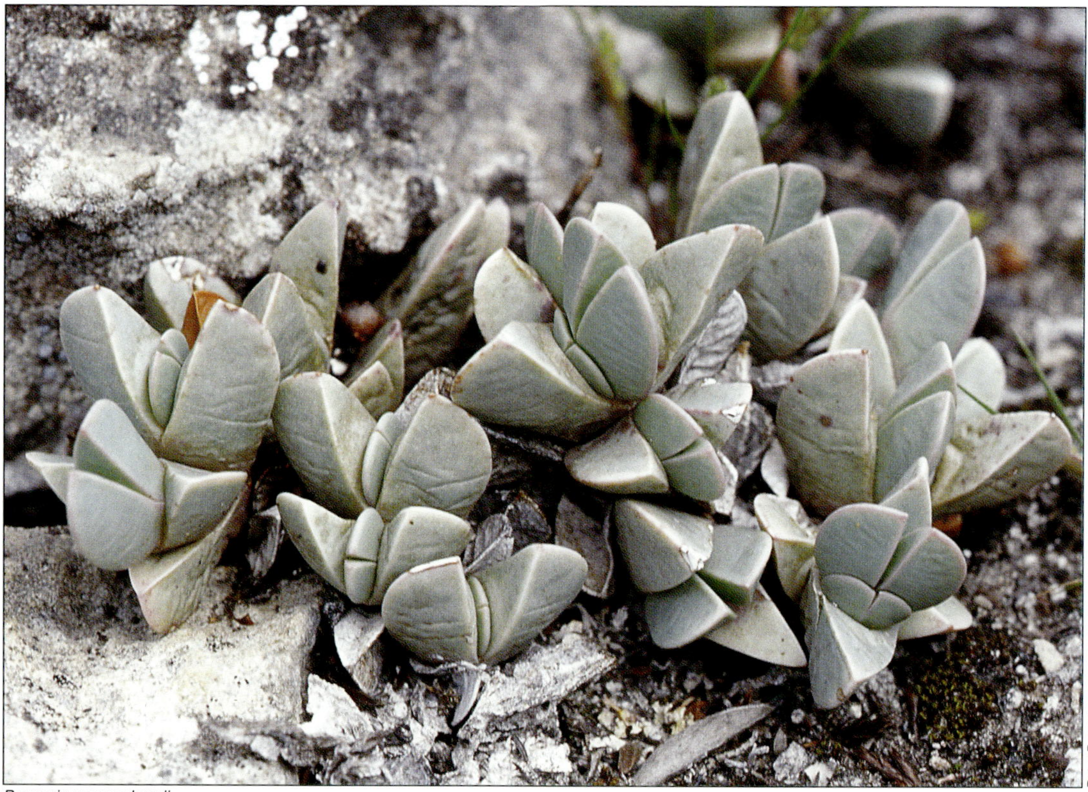

Peter Bruyns

Braunsia vanrensburgii

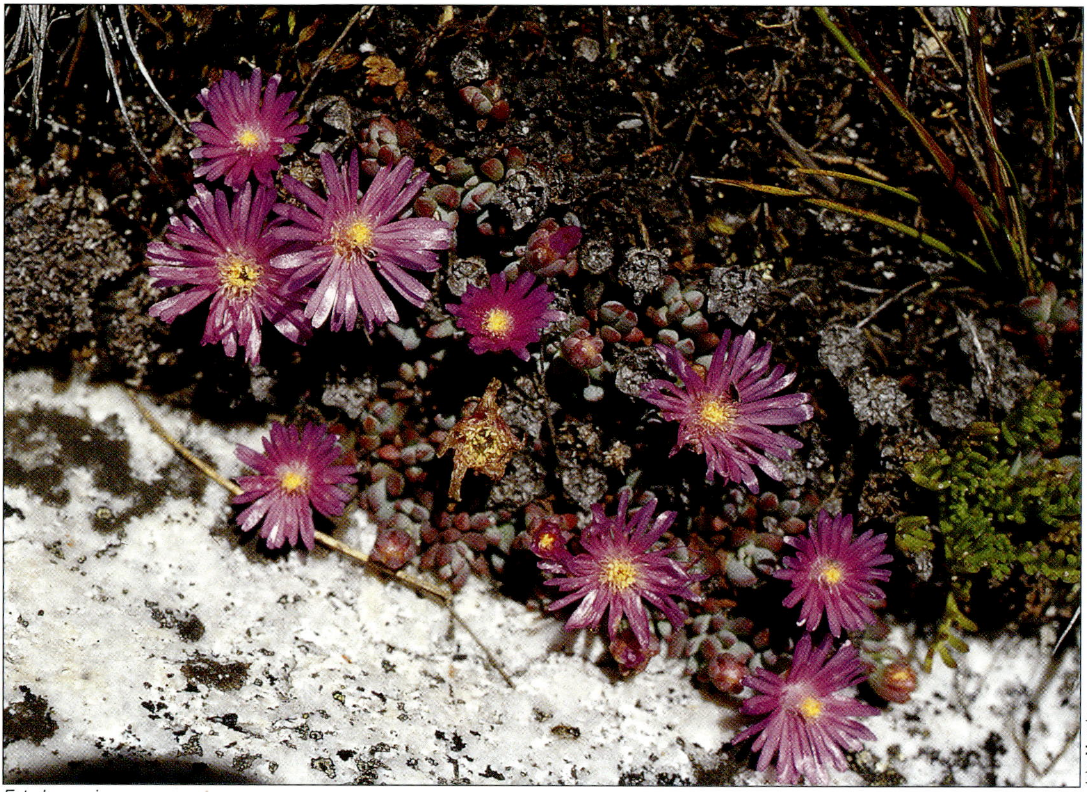

John Manning

Esterhuysenia sp.

Dwarf shrubby Mesembs

GROUP 10

The plants are all dwarf shrubs with visible, more or less erect stems; the succulent leaves are variable in size and shape.

The majority of the genera in this group are closely related and share *Lampranthus*-type fruit capsules (*Braunsia, Esterhuysenia, Zeuktophyllum*). *Hartmanthus*, which is thought to be related to *Jensenobotrya, Schlechteranthus* with its *Leipoldtia*-type capsules, *Corpuscularia* and *Sarcozona*, are included simply because they have dwarf shrubby growth forms.

DWARF SHRUBBY MESEMBS comprise 7 genera and 15 species.

Braunsia (5 species)	*Sarcozona* (2 species)
Corpuscularia (2 species)	*Schlechteranthus* (2 species)
Esterhuysenia (1 species)	*Zeuktophyllum* (1 species)
Hartmanthus (2 species)	

BRAUNSIA

DERIVATION OF GENUS NAME This genus was named in honour of Dr H. Brauns of Willowmore, South Africa.

COMMON NAMES No vernacular names seems to have been recorded.

DESCRIPTION Members of this genus are low-growing shrublets with woody stems which root at the nodes. Leaves are succulent, dull-green, and three-angled, with a velvety or smooth, waxy surface. They are fused for up to half of their length and leaf margins are minutely toothed, hairy or cartilaginous. The pink or white flowers are medium sized, solitary or appear in few-flowered clusters on short flower stalks. There are five sepals. The stamens are collected into an upright bundle and filamentous staminodes are present. Nectar glands form an inconspicuous series of small teeth in a continuous ring. The woody fruit capsules have five locules with valve wings. Covering membranes are well developed or reduced and a bunch of funicular hairs is present at each seed exit. Seeds are brown, globose and minutely spiny (echinate), hence the previous name for this genus – *Echinus*, meaning hedgehog.

DISTINGUISHING CHARACTERS The cartilaginous or serrated edges of the leaves, as well as their fusion to half their length or more, are quite characteristic of this genus. The seeds are echinate, which distinguishes them from the seeds of other genera allied to *Lampranthus*.

FLOWERING TIME This genus can be seen flowering from midsummer to early winter (January to July in South Africa). Flowers open in the middle of the day to late in the afternoon.

DISTRIBUTION AND ECOLOGY *Braunsia* is found in the southwestern part of South Africa, mostly in arid areas of the Western Cape Province including the Little Karoo, and at Matjiesfontein and Laingsburg in the Great Karoo. Plants are found growing mainly in gravelly areas, on sandstones, shales or limestones, in areas receiving more or less 200 mm of rain annually with peaks in March and November.

CULTIVATION *Braunsia* thrives in cultivation. Seeds should be sown in summer or autumn. Plants are also easily propagated from cuttings.

NOTES The fuzzy leaves of one species in this genus (*B. apiculata*) are sometimes tipped with a prominent point which hardens into a small brownish spine. Leaves of the other species are smooth, with a waxy surface. The leaves of the rare *B. vanrensburgii* (see illustration opposite the introductory page of DWARF SHRUBBY MESEMBS) are robust, with pink margins. Other species have showy flowers in shades of pink, salmon or white.

Braunsia Schwantes (=*Echinus* L. Bolus)

NUMBER OF SPECIES/SUBSPECIES/VARIETIES (5/0/0)

SPECIES LIST AND CONSERVATION STATUS
B. *apiculata* (Kensit) L.Bolus
B. *geminata* (Haw.) L.Bolus
B. *nelii* Schwantes [K]
B. *stayneri* (L.Bolus) L.Bolus [R]
B. *vanrensburgii* (L.Bolus) L.Bolus [R]

LITERATURE
SCHWANTES, G. 1928. Mesembriaceen-Studien. *Gartenwelt* 32: 644.

BOLUS, L. 1927. *Echinus apiculatus*. Flowering Plants of South Africa 7: 266.

BOLUS, L. 1967. Notes on *Mesembryanthemum* and allied genera. *Journal of South African Botany* 33: 306.

B. apiculata

B. stayneri

Flower of B. geminata

B. geminata

CORPUSCULARIA

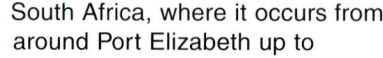

DERIVATION OF GENUS NAME The name is derived from the Latin words *corpusculum* (a body) and *aria* (a collection), thus "a group of bodies", in reference to the leaves which form bodies by their close fusion, and their aggregation into clusters.

COMMON NAMES No vernacular names seem to have been recorded.

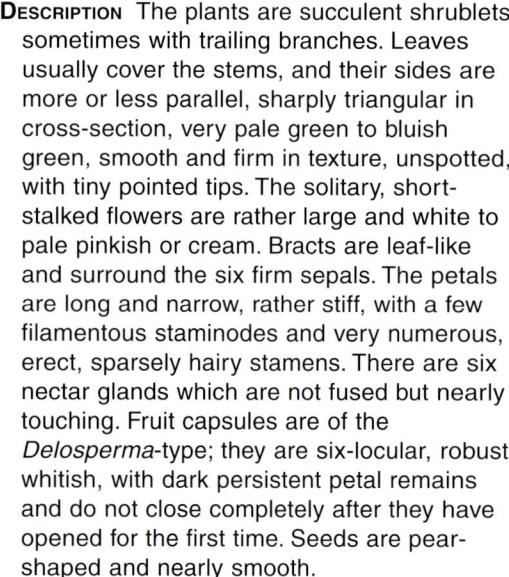

DESCRIPTION The plants are succulent shrublets, sometimes with trailing branches. Leaves usually cover the stems, and their sides are more or less parallel, sharply triangular in cross-section, very pale green to bluish green, smooth and firm in texture, unspotted, with tiny pointed tips. The solitary, short-stalked flowers are rather large and white to pale pinkish or cream. Bracts are leaf-like and surround the six firm sepals. The petals are long and narrow, rather stiff, with a few filamentous staminodes and very numerous, erect, sparsely hairy stamens. There are six nectar glands which are not fused but nearly touching. Fruit capsules are of the *Delosperma*-type; they are six-locular, robust, whitish, with dark persistent petal remains and do not close completely after they have opened for the first time. Seeds are pear-shaped and nearly smooth.

DISTINGUISHING CHARACTERS *Corpuscularia* species are recognised by their pale grey-green, severely three-angled, dotless leaves, their pale yellowish white flowers with stiff petals and six leaf-like sepals. The old petals typically persist as black wads atop the fruits.

FLOWERING TIME The plants flower mainly in the spring and summer months. The almost scentless flowers open in the morning and close in the evening.

DISTRIBUTION AND ECOLOGY The genus is restricted to the Eastern Cape Province, South Africa, where it occurs from around Port Elizabeth up to Grahamstown. Plants grow on rocky slopes where they are often shaded by the surrounding vegetation.

CULTIVATION These plants are indestructible. They are most active in summer, but are responsive to any season and adaptable to any ambience. They prefer slight shade but can be planted in open rockeries or containers. Propagation is from cuttings. Water sparingly in summer.

NOTES This is a poorly known genus, recently revived and recognised as distinct from *Delosperma*, but originally founded on a hyper-heterogenous bundle of species from the genera *Nelia*, *Antimima*, and *Malephora*. A further group of species with hairy yet firm leaf surfaces and self-fertile flowers may belong here. *Corpuscularia lehmannii* is a well-known garden plant with compact growth and creamy white flowers. *C. taylorii* is a scraggly shrublet with longer internodes. It has been propagated, mainly from one clone, for a century.

Corpuscularia Schwantes (=*Schonlandia* L.Bolus)

NUMBER OF SPECIES/SUBSPECIES/VARIETIES (2/0/0) and possibly more.

SPECIES LIST AND CONSERVATION STATUS
C. *lehmannii* (Ecklon & Zeyher) Schwantes
C. *taylorii* (N.E.Br.) Schwantes

LITERATURE
BOLUS H.M.L. 1927. *Schonlandia lehmannii*. Flowering Plants of South Africa 7: t. 259.

SCHWANTES G. 1926. Einige neue Mesembrianthemaceen. *Zeitschrift für Sukkulentenkunde* 2: 186.

C. taylorii

Heidi Hartmann

C. lehmannii

Ben-Erik van Wyk

C. lehmannii

Ernst van Jaarsveld

ESTERHUYSENIA

DERIVATION OF GENUS NAME
Esterhuysenia is named in honour of Ms Elsie Esterhuysen who was born in Cape Town in 1912, one of the most prolific plant collectors in South Africa. She is an authority on the Cape flora and has concentrated her amazing efforts on collecting from the highest mountain peaks in the Cape.

COMMON NAMES No vernacular names seem to have been recorded.

DESCRIPTION The plants are dwarf compact shrublets with fibrous roots and smooth, red stems from which a reddish powder may flake off. Leaves are minutely textured, always equal and similar, upright or spreading with pointed tips, the lower side rounded and the upper side flattened, up to 12 mm in length, often tinged with red. The medium-sized, pinkish purple flowers occur singly and are positioned on a flower stalk. They are enclosed by five sepals. The nectar glands are arranged in a continuous ring. Fruit capsules have five locules. They are woody with the expanding keels parallel at their bases and divergent at the tips. Covering membranes are flexible and do not cover the locules entirely. Seeds are brown and rough-textured.

DISTINGUISHING CHARACTERS The cinnamon red fruit capsules lack both closing bodies and valve wings. In addition, the keels are parallel at their bases and diverge widely at the tips. The stems are reddish and may be powdery if rubbed lightly.

FLOWERING TIME Plants flower between spring and summer (September to December in South Africa). Flowers open during the day.

DISTRIBUTION AND ECOLOGY
Esterhuysenia occurs on high mountain peaks north of Worcester in the winter rainfall region of the Western Cape Province of South Africa. Recent collections indicate that the distribution range may be wider than originally thought. Species grow on rock ledges, often in shady conditions, on upper slopes of sandstone mountains at altitudes from 500 to 2 000 metres.

CULTIVATION Few plants occur in cultivation. *Esterhuysenia* prefers porous, slightly acidic soil and ample water. Plants are difficult to maintain in a warm climate as they require cool conditions.

NOTES There are a number of undescribed species in *Esterhuysenia*, a genus in need of revision. The genus shows similarities with *Oscularia*, *Ruschia* and *Lampranthus*, so its relationships are still uncertain. The only currently recognised species, *E. alpina*, reminds one of a thick-leaved species of *Crassula*.

Esterhuysenia L.Bolus

NUMBER OF SPECIES/SUBSPECIES/VARIETIES (1/0/0)
More species will be added in future.

SPECIES LIST AND CONSERVATION STATUS
E. alpina L.Bolus [R]

LITERATURE
BOLUS, H.M.L. 1967. Notes on *Mesembryanthemum* and allied genera. *South African Journal of Botany* 33: 308.

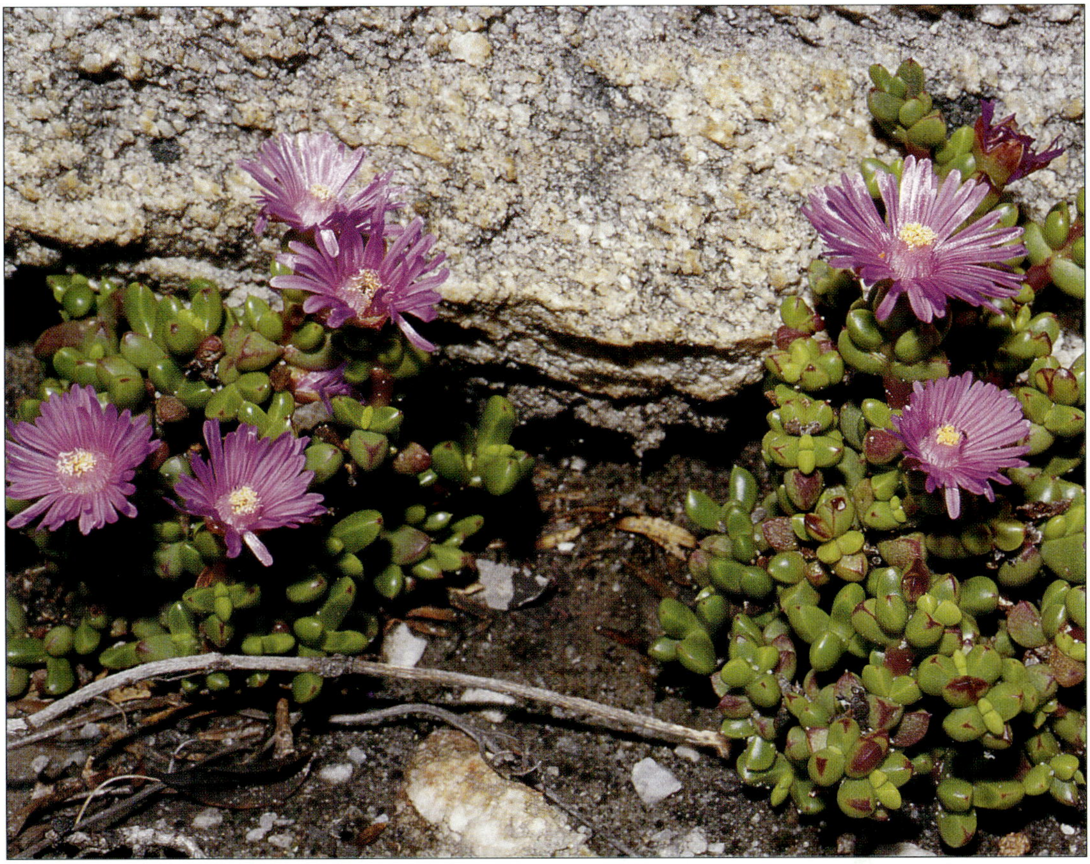
Bob Chinnock

E. alpina

Peter Bruyns

Esterhuysenia sp.

Ulrich Meve

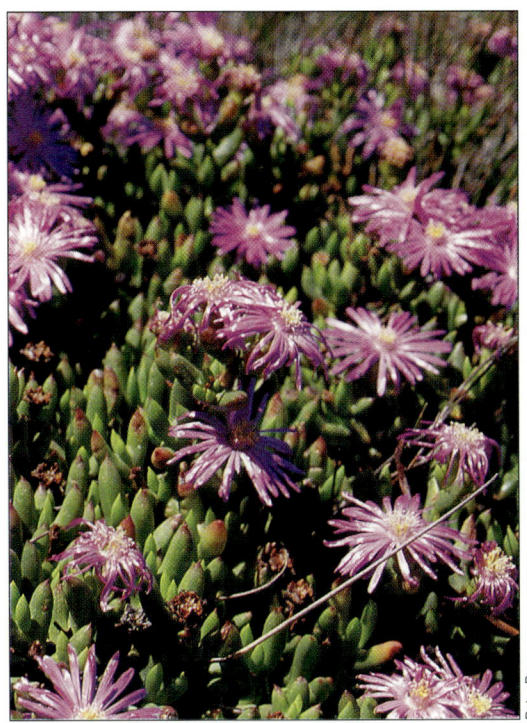

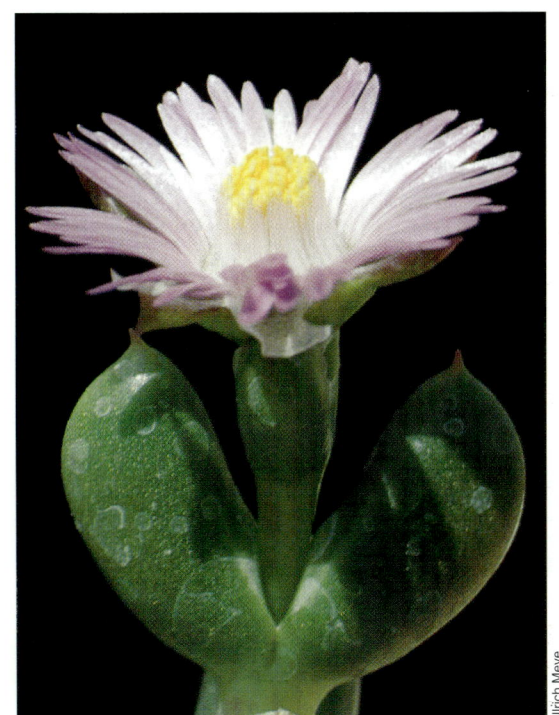

Flower of *E. alpina*

Hartmanthus

Derivation of genus name The genus was named after Dr Heidrun Elsbeth Klara Hartmann (born in 1942), botanist at the University of Hamburg, Germany, who has contributed much to our knowledge of mesembs of the world.

Common names No vernacular names seem to have been recorded.

Description Plants form small to medium-sized shrublets with erect or somewhat spreading, woody branches. Their growth form is often more compact and they generally have a "neater" appearance than many other mesemb shrubs. The greenish grey to silvery leaves are strongly compressed lengthwise and keeled on the back. The leaf margins are not very sharp and take on a reddish tinge in full sun. Leaf surfaces are rough to the touch. The flowering stalks are long, flattened and long-lived. Flowers are faintly scented, with the petal colour ranging from almost white to light or distinctly pink. The capsules are six-locular and whitish, with no covering membranes and no closing bodies, hence the former inclusion of this genus in *Delosperma*. The seeds are medium sized, pear-shaped and rather glossy.

Distinguishing characters The genus may be known by its keeled, greenish grey to silvery green leaves that are laterally flattened. The leaf surfaces are rough to the touch in one species (*H. hallii*), somewhat resembling the texture of fine sandpaper. The leaf margins are usually a reddish purple colour if exposed to sufficient light intensity. The flowering stalk is strongly flattened. Another interesting characteristic of the leaves is that they do not shrink evenly over the whole surface, but rather in patches.

Flowering time Although some variability exists, most plants flower during late winter and early spring. The flowers open during mid to late afternoon and close by dusk.

Distribution and ecology The two species of *Hartmanthus* occur in very arid regions in the northern Richtersveld, Northern Cape Province, South Africa, and in the Sperrgebiet in southern Namibia where they grow fully exposed. Plants growing at some of the localities benefit from evening mists rolling in from the cold Atlantic Ocean but rainfall is really scarce here. *H. pergamentaceum* occurs with *Juttadinteria albata* and looks very much like it.

Cultivation Amazingly these species do not present undue difficulties in cultivation. Protection against excessive moisture should, however, be provided if the plants are to be retained for any length of time. *Hartmanthus hallii* is the more desirable and colourful of the two species. Sow seed during autumn in well-drained mineral-rich soil.

Notes *Hartmanthus pergamentaceus* is a gangly, usually white-flowered (rarely pink) shrublet with smooth thin leaves. In habitat, and even in cultivation, it often appears half dead (see illustration). *H. hallii* has shorter internodes and fatter leaves, which are sometimes as rough as sandpaper. Its flowers are always pink. The two species have until recently been classified in the diverse and rather unwieldy genus *Delosperma*. The transfer of these species to their own genus is part of an attempt to more closely reflect relationships in this interesting group of plants.

Hartmanthus S.A.Hammer

Number of species/subspecies/varieties (2/0/0)

Species list and conservation status
  H. hallii (L.Bolus) S.A.Hammer [R]
  H. pergamentaceus (L.Bolus) S.A.Hammer

Literature
  HAMMER, S.A. 1995. *Hartmanthus*, a new genus in Aizoaceae. *Haseltonia* 3: 77–82.

Leaves of *H. pergamentaceus*

Flower of *H. hallii*

H. pergamentaceus in its habitat

Flowers of *H. pergamentaceus*

SARCOZONA

DERIVATION OF THE GENUS NAME The name was derived from the Greek words *sarkos* (flesh) and *zone* (involucre), which refers to the fleshy base of the fruit.

COMMON NAMES The genus name, *Sarcozona*, has become entrenched as its common name.

DESCRIPTION Plants are erect shrublets of up to 300 mm in height, and only occasionally do the stems tend to become somewhat creeping. The dull or bluish green leaves clasp the stems and are three-angled, the keels of the leaves being smooth or faintly toothed. Leaf surfaces are smooth (*Sarcozona bicarinata*), or rough to the touch (*S. praecox*) as a result of the presence of a dense covering of tiny warts. The white to pale pinkish flowers are borne singly or in groups of three. Flowers are stalkless or may have very short stalks. There are five sepals of which two are larger than the rest. The fruit capsules have four or five locules, each having a fleshy wall. They do not split open as is the case with most other mesembs. The flat seeds are egg or halfmoon-shaped and light to dark brown.

DISTINGUISHING CHARACTERS Plants are similar to *Carpobrotus* but they are smaller, erect, and in *S. praecox*, the leaves are warty. The flowers are less showy than those of *Carpobrotus*.

FLOWERING TIME The plants flower in spring (August to October in Australia). The flowers open in the morning and close in the evening.

DISTRIBUTION AND ECOLOGY The two species of *Sarcozona* occur in the southern parts of Australia. *S. praecox* grows in scattered populations along the south coast, and also further inland. It has been recorded from Western Australia, South Australia and Victoria. The other species, *S. bicarinata*, has a more restricted distribution on the Yorke and Eyre peninsulas in South Australia. Both species favour sandy soils. This genus is restricted to Australia and is therefore the only genus of mesemb that does not occur naturally in southern Africa.

CULTIVATION Cultivation is easy from cuttings or seed. Plants grow best in well-drained soils.

NOTES The two species of *Sarcozona* can be separated on the basis of their differing leaf textures as discussed above. In addition, the flowers of *S. bicarinata* have two keels running lengthwise down the tube formed by the fused sepals. These keels are absent in the case of *S. praecox*. Another character that can be used to distinguish between the species, albeit less useful due to its variability, is that *S. praecox* usually has four styles and only rarely five. The opposite is true for *S. bicarinata*. It has been proposed that the genus *Sarcozona* should be included in *Carpobrotus*, a genus here grouped with the MAT-FORMING MESEMBS, but there is no convincing evidence that the two genera should be combined.

Sarcozona J.M.Black

NUMBER OF SPECIES/SUBSPECIES/VARIETIES (2/0/0)

SPECIES LIST AND CONSERVATION STATUS
S. bicarinata S.T.Blake
S. praecox (F.Muell.) S.T.Blake ex H. Eichler

LITERATURE
VENNING, J. 1984. *Sarcozona*. In Flora of Australia vol. 4. Phytolacaceae to Chenopodiaceae, pp. 28, 30. Australian Government Publishing Service, Canberra.

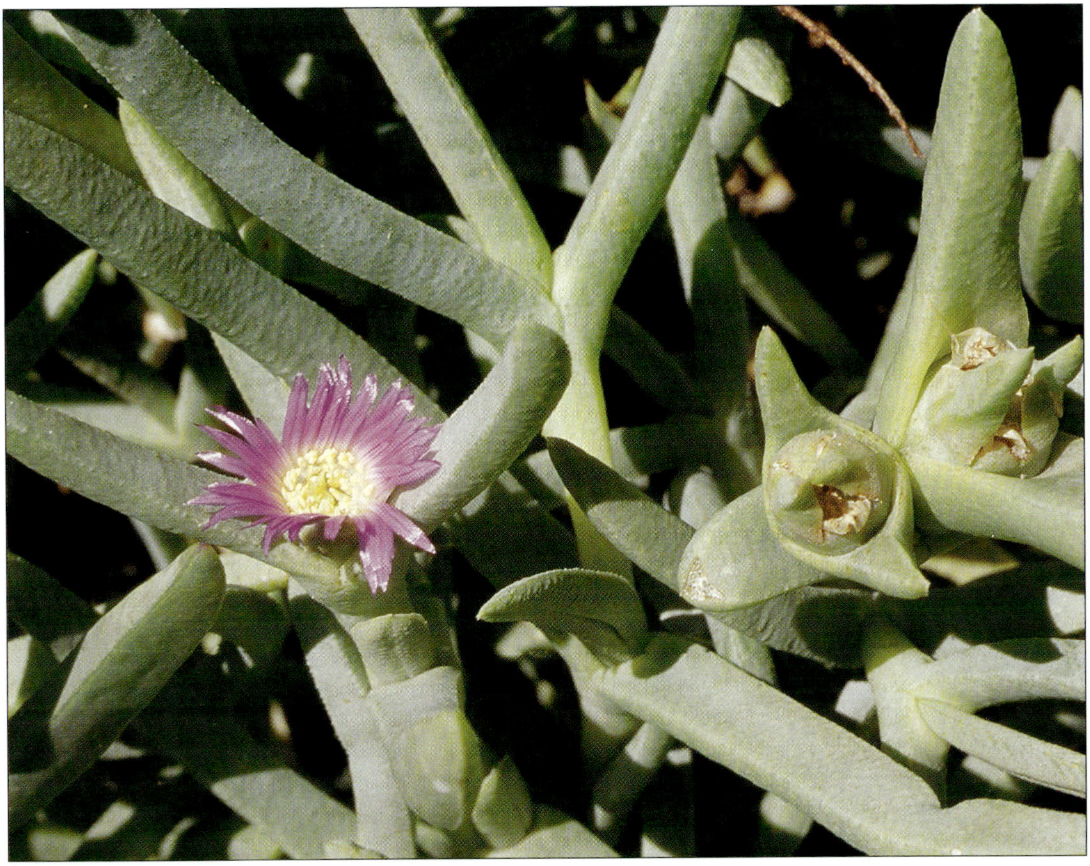
Hellmut Toelken

S. praecox

Hellmut Toelken

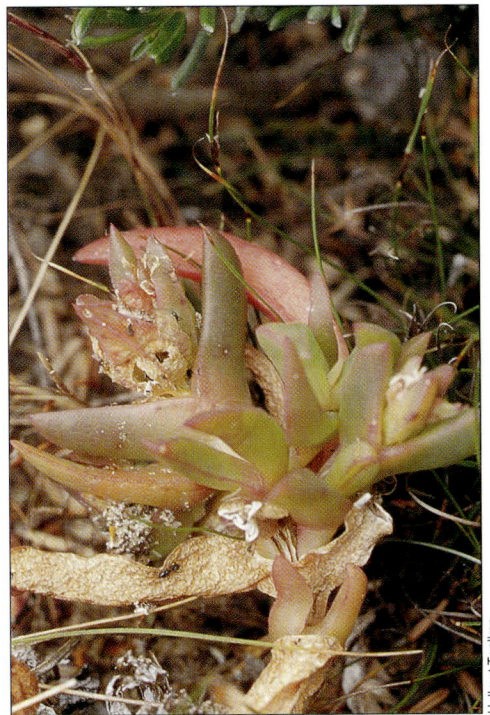

Fruit of *S. bicarinata* disintegrating to release seeds

Hellmut Toelken

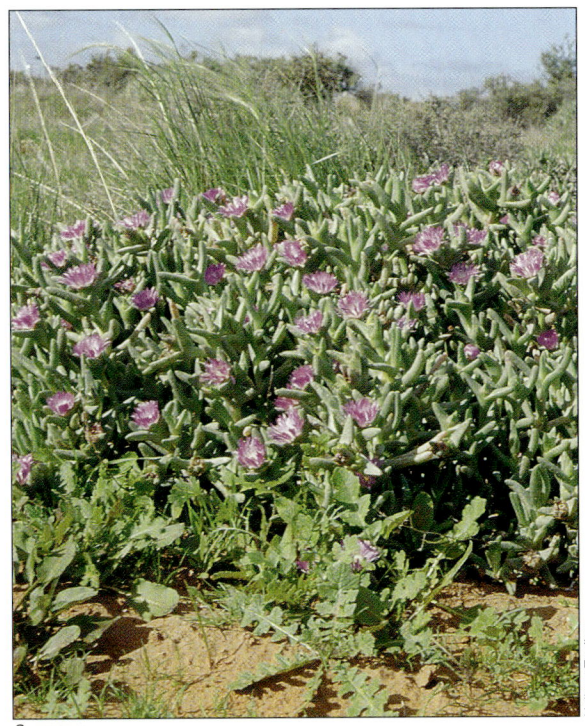

S. praecox

SCHLECHTERANTHUS

DERIVATION OF GENUS NAME The genus is named after the German trader, explorer and succulent plant collector, especially of Mesembryanthemaceae, Mr Max Schlechter (1874 to 1960), who settled in the Port Nolloth district for a while.

COMMON NAMES The vernacular name for both species of *Schlechteranthus* is *vleisbos* (meat bush).

DESCRIPTION Plants are dwarf shrubs with crowded branchlets, up to 400 mm in height. Leaves are smooth, glossy with a greyish, bluish or reddish colour, united for up to half their length into bodies, with flat upper surfaces and keeled lower surfaces, with one or two short teeth on the keels. Flowers are white, or pink to purple, solitary, with a very short stalk and subtended by two pairs of bracts. There are six unequal sepals. The numerous bearded stamens are collected into a central cone, surrounded by filamentous staminodes. The flower base is convex and nectar glands are arranged in a ring around long, awl-shaped stigmas. Fruit capsules have 10 to 12 locules, close to the *Leipoldtia*-type but with a variable and sometimes complete recurving of the covering membranes, with spur-like processes which nearly close the seed exits and parallel expanding keels which diverge towards the tips. Seeds are reddish yellow, pear-shaped and somewhat flattened and rough.

DISTINGUISHING CHARACTERS *Schlechteranthus* is recognised by its compact growth and grey-green short leaves. Flowers are white, or pink to purple, with staminodes collected into a cone. Fruit capsules are rounded on top and have 10 to 12 locules.

FLOWERING TIME Flowering takes place during the winter months. The flowers open from midday untill evening.

DISTRIBUTION AND ECOLOGY *Schlechteranthus* occurs in the Richtersveld, Northern Cape Province, South Africa. This extremely arid area receives rain in winter. Plants occur on rocky slopes. *S. maximiliani* occurs closer to the coast than *S. hallii* and receives more moisture in the form of fog.

CULTIVATION Plants are very attractive, compact with silver-green foliage. They are best grown in a greenhouse outside of their habitat and are easily cultivated in containers, provided they are kept dry during summer. Seeds sown during autumn germinate readily. The plants are slow growing, long-lived perennials.

NOTES *Schlechteranthus hallii* is a bushy shrublet with large pink to purple flowers, while *S. maximiliani* has white or pale yellow flowers, is more woody and taller, with smaller leaves.

Schlechteranthus Schwantes

NUMBER OF SPECIES/SUBSPECIES/VARIETIES (2/0/0)

SPECIES LIST AND CONSERVATION STATUS
S. hallii L.Bolus
S. maximiliani Schwantes

LITERATURE
SCHWANTES, G. 1929. Neue Mesembriaceen IV. *Monatschrift der Deutschen Kakteen Gesellschaft* 1: 16.

Flower of *S. maximiliani*

S. maximiliani in its habitat

Leaves and fruit capsules of *S. hallii*

Flowers of *S. hallii*

ZEUKTOPHYLLUM

DERIVATION OF GENUS NAME The name *Zeuktophyllum* is derived from the Greek words *zeuktos* (joint) and *phyllon* (leaf).

COMMON NAMES The common name *spookvygie* (ghost mesemb) has been used.

DESCRIPTION Plants are dwarf, sturdy shrublets, up to 100 mm in height with woody stems and fibrous roots. Leaves are erect, fused towards their bases, three-angled, tapering, with smooth surfaces, and up to 30 mm in length. The solitary white flowers are borne on short stalks. There are six sepals and numerous filamentous staminodes. The nectar glands are arranged in a continuous ring. Fruit capsules are woody, flat-topped and have 10 locules. The expanding keels end in a small stiff wing and covering membranes cover the seed chambers. Seeds are smooth.

DISTINGUISHING CHARACTERS The plants are compact, short shrublets with partially fused leaves, solitary, short-stalked white flowers and 10-locular, flat-topped fruit capsules without closing bodies. However, the genus is distinguished mostly by its obscurity.

FLOWERING TIME Plants flower in summer (January to March in South Africa). The flowers open in the morning and close in the evening.

DISTRIBUTION AND ECOLOGY *Zeuktophyllum* is found in the Little Karoo in the vicinity of Barrydale and Bakoven in the central Western Cape Province, South Africa. It grows on slopes of low, stony, arid hills or flats on glacial deposits or shales.

CULTIVATION Plants of the genus *Zeuktophyllum* are rarely cultivated. They thrive in containers and are best kept in a greenhouse.

NOTES This genus apparently comprises only one species but it is very poorly known and needs further study.

Zeuktophyllum N.E.Br.

NUMBER OF SPECIES/SUBSPECIES/VARIETIES (1/0/0)

SPECIES LIST AND CONSERVATION STATUS
Z. suppositum (L.Bolus) N.E.Br. [R]

LITERATURE
BROWN, N.E. 1927. *Mesembryanthemum* and some new genera separated from it. *Gardener's Chronicle* 81: 12.

Z. suppositum

Open fruit capsules of Z. suppositum

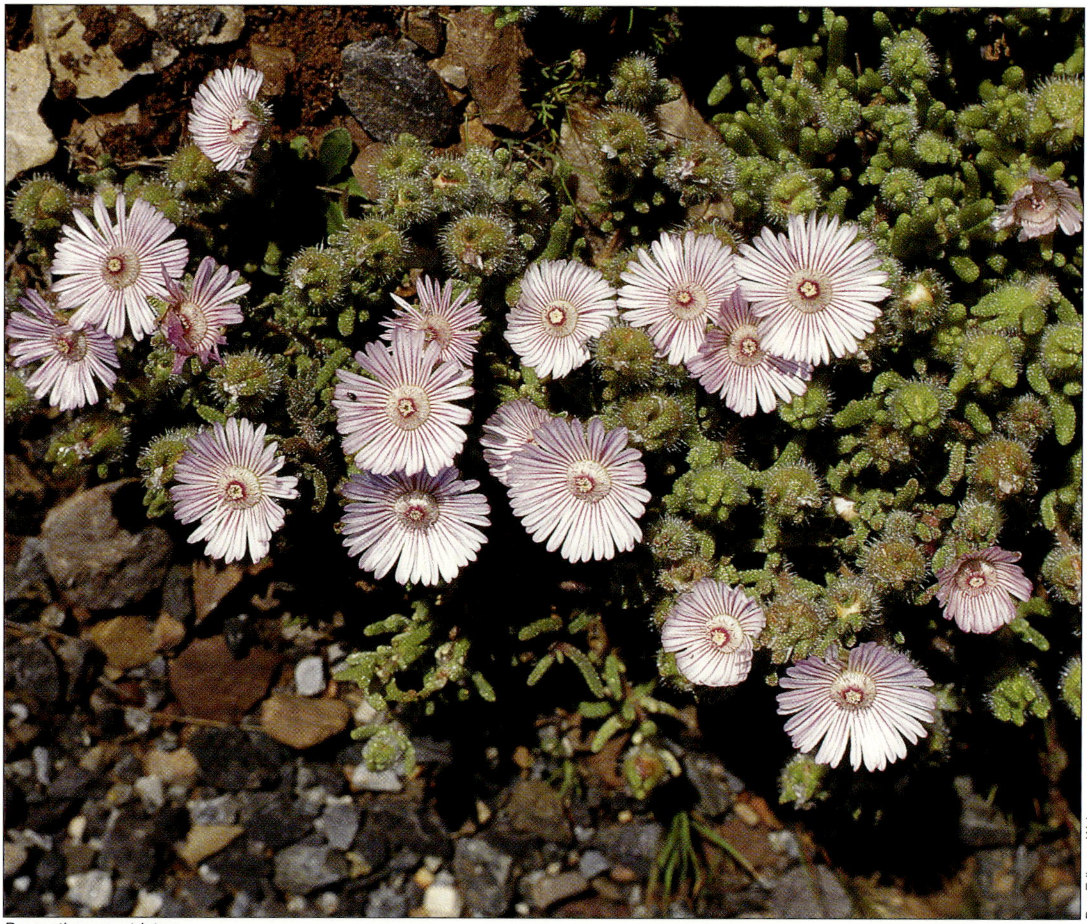

Ben-Erik van Wyk

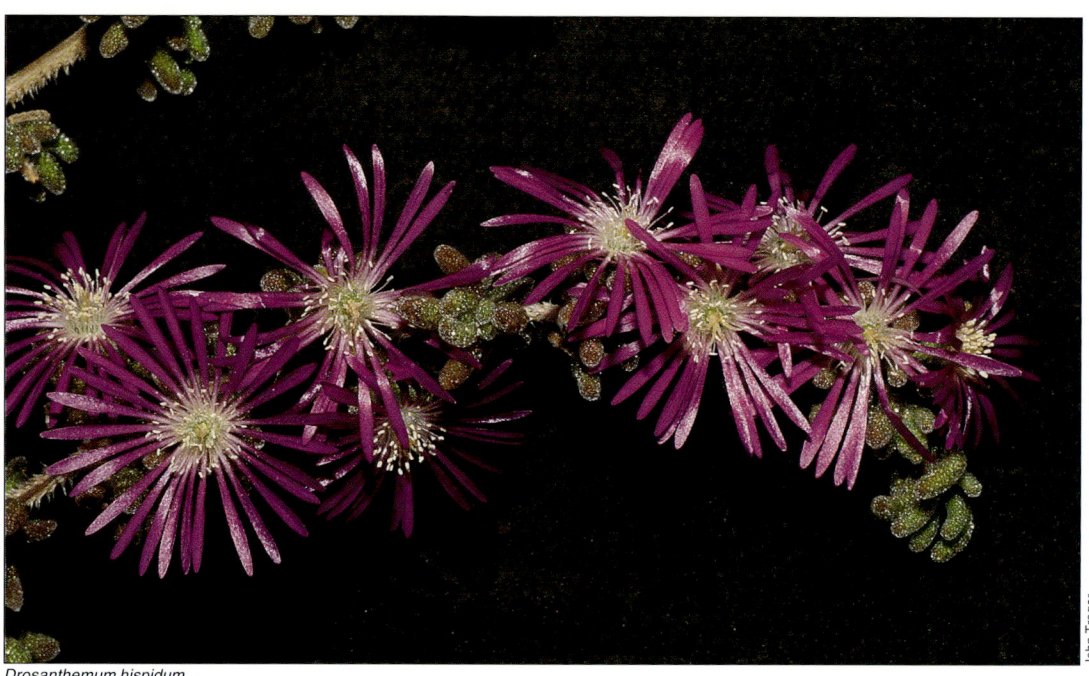

Drosanthemum striatum

John Trager

Drosanthemum hispidum

Glittering shrubby Mesembs

GROUP 11

The leaves have tiny cells storing water, giving the leaf surfaces a glittering appearance; the plants are all shrubby with visible internodes and creeping or erect stems; fruit capsules are of the *Drosanthemum* or *Delosperma*-type.

Despite differences in fruit capsule type, the four genera listed here are all related. Growth form in the large genera of shrubby mesembs (*Delosperma*, *Drosanthemum* and *Trichodiadema*) varies considerably amongst species. As a result, some tiny plants are included here but they nevertheless have glittering leaf surfaces.

GLITTERING SHRUBBY MESEMBS comprise 5 genera and 324 species.

Delosperma (163 species)
Drosanthemum (120 species)
Ectotropis (1 species)
Mestoklema (6 species)
Trichodiadema (34 species)

Delosperma

DERIVATION OF GENUS NAME The name *Delosperma* is derived from the Greek words *delos* (visible) and *sperma* (seed), in allusion to the fact that the capsules have no covering membranes. The seeds are therefore exposed when capsules are open.

COMMON NAMES The following vernacular names have been recorded: *klipvygie* or *rotsvygie* (rock mesemb) for *D. carolinense* and *D. vogtsii*; *skaapvygie* (sheep mesemb) for *D. floribundum*; *witbergvygie* (white mountain mesemb) for *D. herbeum*; *kalkklipvygie* (calcrete mesemb) for *D. litorale*; *rooibergvygie* (red mountain mesemb) for *D. macellum*; and *skaapvygie* or *slypvygie* (honing mesemb) for *D. ornatulum*.

DESCRIPTION Plants are small to medium-sized shrubs or herbs, many with deciduous or annual branches. They often grow flat on the ground, sometimes with tips curving upwards, or rarely scrambling, sometimes with long red stems. In taller, stronger shrubs the stems are often thickened with peeling bark. The rootstock may sometimes be tuberous or the lower roots may have multiple small tubers. Leaves are slightly fused basally or they can be free. Leaf colour ranges from bluish green to yellowish green or red. Leaf surfaces are sometimes finely to conspicuously hairy or prickly, usually bearing visible external water storage cells, giving the leaves a glistening appearance. The leaves are highly variable in shape, broadly triangular to cylindric, mostly subcylindric with a grooved or indented upper surface, rarely flat, usually soft and tender. Flower colour varies from white to pink, purple, yellow, salmon, orange, rust-red, rarely scarlet, carmine or wine-coloured. They are solitary or borne in three-flowered clusters which may form many-flowered branches. Flower stalks are variable in length, bearing leaf-like bracts. The five sepals are strongly unequal, the outer two being far larger, often horn-like. Petals are narrow to spoon-shaped and occur in one to four whorls. They are rarely hairy. The erect stamens are usually surrounded by short thread-like inner petals. There are five large, separate nectar glands; these are rarely united into a ring. There are four to six stigmas. Fruit capsules are four to six-locular, soft, usually pale in colour, with valve wings, but mostly without covering membranes and invariably without closing bodies. Seeds are globose, pale brown, and smooth to slightly textured.

DISTINGUISHING CHARACTERS *Delosperma* species are creeping or clump-forming shrubs with soft glistening leaves and, often, not particularly succulent leaves. Flower colour is often of many hues in one population. The fruit capsules are the simplest of all the mesembs fruit capsule types and are found in a number of other genera.

FLOWERING TIME Flowering occurs in spring to high summer, but in the case of many species, plants may flower sporadically. The flowers open in the morning and close at night. Individual flowers are not particularly strongly scented, but when in full flower the collective effect may be noticeable.

DISTRIBUTION AND ECOLOGY The genus occurs mostly in South Africa, where it is widespread in all the provinces except the climatically severe central Northern Cape Province. It also occurs in Lesotho, Swaziland, Botswana and further north to eastern Africa through Zimbabwe, Kenya and Tanzania to Ethiopia as well as Arabia and the Yemen. The Indian Ocean islands of Madagascar and Reunion also have a few indigenous species. Within South Africa, the genus is found mostly (and abundantly) in the summer rainfall region. The upland species often inhabit areas of severe

Delosperma sp.

D. asperulum

D. echinatum

D. rileyi

D. rogersii

D. repens

winter frost and have special adaptations to it (autumnal dehydrations, red pigments, perennial underground parts and annual shoots). Many species inhabit locally moist zones near watercourses, the drip-lines of rocks or under trees. The few Namaqualand species are tougher and grow well exposed.

CULTIVATION Plants are easily cultivated if one forgets that they are succulents and treat them like strawberry plants, that is, with abundant summer water and rich but well-drained, somewhat sandy soil. The tuberous species must be kept dry in winter.

NOTES *Delosperma* species are diverse and show many growth forms, from the tiny, compact *D. esterhuyseniae* to the large untidy creepers such as *D. litorale* and *D. repens*. The diversity in habit, leaves and flower colour is shown in the illustrations. *Delosperma* is a large heterogenous assemblage, colliding in part with *Trichodiadema*, *Mestoklema* and especially *Drosanthemum*. The species with xeromorphic (hard) leaves have been removed to *Hartmanthus* and *Corpuscularia*.

Delosperma N.E.Br.

NUMBER OF SPECIES/SUBSPECIES/VARIETIES (163/0/18)

SPECIES LIST AND CONSERVATION STATUS
 D. aberdeenense (L.Bolus) L.Bolus
 D. abyssinicum (Regel) Schwantes
 D. acocksii L.Bolus var. acocksii
 D. acocksii L.Bolus var. luxurians L.Bolus
 D. acuminatum L.Bolus
 D. adelaidense Lavis
 D. aereum (L.Bolus) L.Bolus var. aereum
 D. aereum (L.Bolus) L.Bolus var. album (L.Bolus) L.Bolus
 D. affine Lavis
 D. algoense L.Bolus
 D. aliwalense L.Bolus
 D. alticolum L.Bolus
 D. angustifolium L.Bolus
 D. angustipetalum Lavis
 D. annulare L.Bolus

 D. appressum L.Bolus
 D. ashtonii L.Bolus [nt]
 D. asperulum (Salm-Dyck) L.Bolus
 D. ausense L.Bolus
D. basuticum L.Bolus
D. bosseranum Marais
D. brevipetalum L.Bolus
D. brevisepalum L.Bolus var. brevisepalum
D. brevisepalum L.Bolus var. majus L.Bolus
D. brunnthaleri (A.Berger) Schwantes
D. burtoniae L.Bolus
D. caespitosum L.Bolus forma caespitosum
D. caespitosum L.Bolus forma roseum (L.Bolus) L.Bolus
D. calitzdorpense L.Bolus
D. calycinum L.Bolus
D. carolinense N.E.Br. var. carolinense
D. carolinense N.E.Br. var. compacta L.Bolus
D. carterae L.Bolus
D. clavipes Lavis [K]
D. cloeteae Lavis
D. concavum L.Bolus
D. congestum L.Bolus
D. cooperi (Hook.f.) L.Bolus forma bicolor (L.Bolus) G.D.Rowley
D. cooperi (Hook.f.) L.Bolus forma cooperi
D. crassuloides (Haw.) L.Bolus
D. crassum L.Bolus
D. cronemeyerianum (A.Berger) Jacobsen
D. davyi N.E.Br.
D. deilanthoides S.A.Hammer
D. deleeuwiae Lavis
D. dunense L.Bolus
D. dyeri L.Bolus var. dyeri
D. dyeri L.Bolus var. laxum L.Bolus
D. echinatum (Aiton) Schwantes
D. ecklonis (Salm-Dyck) Schwantes var. ecklonis
D. ecklonis (Salm-Dyck) Schwantes var. latifolia L.Bolus
D. edwardsiae L.Bolus
D. erectum L.Bolus
D. esterhuyseniae L.Bolus
D. exspersum (N.E.Br.) L.Bolus var. decumbens L.Bolus
D. exspersum (N.E.Br.) L.Bolus var. exspersum
D. ficksburgense Lavis
D. floribundum L.Bolus
D. framesii L.Bolus
D. fredericii Lavis

D. herbeum (white form)

D. herbeum (pink form)

D. cooperi

D. floribundum

Fruit capsules of *D. litorale*

D. litorale

D. frutescens L.Bolus
D. galpinii L.Bolus var. galpinii
D. galpinii L.Bolus var. minus L.Bolus
D. gerstneri L.Bolus
D. giffenii Lavis
D. gracile L.Bolus
D. gracillimum L.Bolus
D. gramineum L.Bolus
D. grandiflorum L.Bolus
D. grantiae L.Bolus
D. gratiae L.Bolus
D. guthriei Lavis [V]
D. harazianum (Deflers) Poppend. & Ihlenf.
D. herbeum (N.E.Br.) N.E.Br.
D. hirtum N.E.Br. var. bicolor L.Bolus
D. hirtum (N.E.Br.) Schwantes var. hirtum
D. hollandii L.Bolus
D. imbricatum L.Bolus
D. inaequale L.Bolus [K]
D. incomptum (Haw.) L.Bolus var. ecklonis (Salm-Dyck) Jacobsen
D. incomptum (Haw.) L.Bolus var. gracile L.Bolus
D. incomptum (Haw.) L.Bolus var. incomptum
D. inconspicuum L.Bolus
D. intonsum L.Bolus
D. jansei N.E.Br.
D. karroicum L.Bolus
D. katbergense L.Bolus var. amatolense L.Bolus
D. katbergense L.Bolus var. angustifolium L.Bolus
D. katbergense L.Bolus var. katbergense
D. klinghardtianum (Dinter) Schwantes
D. knox-daviesii Lavis
D. kofleri Lavis
D. lavisiae L.Bolus var. lavisiae
D. lavisiae L.Bolus var. parisepalum L.Bolus
D. laxipetalum L.Bolus
D. lebomboense (L.Bolus) Lavis
D. leendertziae N.E.Br. [K]
D. leightoniae Lavis
D. liebenbergii L.Bolus
D. lineare L.Bolus var. lineare
D. lineare L.Bolus var. tenuifolium L.Bolus
D. litorale (Kensit) L.Bolus
D. longipes L.Bolus
D. lootsbergense Lavis
D. luckhoffii L.Bolus
D. luteum L.Bolus

D. lydenburgense L.Bolus var. acutipetalum L.Bolus
D. lydenburgense L.Bolus var. lydenburgense
D. macellum (N.E.Br.) N.E.Br.
D. macrostigma L.Bolus [K]
D. mahonii (N.E.Br.) N.E.Br.
D. mariae L.Bolus [K]
D. maxwelliae L.Bolus
D. minimum Lavis
D. monanthemum Lavis
D. muirii L.Bolus
D. multiflora L.Bolus
D. nakurense (Engl.) Herre
D. napiforme Schwantes
D. neethlingiae (L.Bolus) Schwantes
D. nelii L.Bolus
D. nubigenum (Schltr.) L.Bolus
D. obtusum L.Bolus
D. oehleri (Engl.) Herre [R]
D. ornatulum N.E.Br.
D. pachyrhizum L.Bolus var. pachyrhizum
D. pachyrhizum L.Bolus var. pubescens L.Bolus
D. pageanum (L.Bolus) L.Bolus
D. pallidum L.Bolus
D. papillatum (L.Bolus) L.Bolus
D. parviflorum L.Bolus
D. patersoniae (L.Bolus) L.Bolus
D. peersii Lavis
D. peglerae L.Bolus
D. pilosulum L.Bolus
D. platysepalum L.Bolus
D. pondoense L.Bolus [I]
D. pontii L.Bolus
D. pottsii (L.Bolus) L.Bolus
D. prasinum L.Bolus
D. pruinosum (Thunb.) J.W.Ingram
D. pubipetalum L.Bolus
D. repens L.Bolus
D. reynoldsii Lavis
D. rileyi L.Bolus
D. robustum L.Bolus
D. rogersii (Schonland & A.Berger) L.Bolus var. glabrescens L.Bolus
D. rogersii (Schonand & A.Berger) L.Bolus var. rogersii
D. roseopurpureum Lavis
D. saturatum L.Bolus
D. saxicolom Lavis
D. scabripes L.Bolus

D. esterhuyseniae

D. obtusum

D. leendertziae

D. vinaceum

D. sutherlandii

D. schimperi (Engl.) H.E.K.Hartmann & Niesler
D. smythae L.Bolus
D. sphalmanthoides S.A.Hammer
D. stenandrum L.Bolus
D. steytlerae L.Bolus [I]
D. subclavatum L.Bolus
D. subincanum (Haw.) Schwantes
D. subpetiolatum L.Bolus [I]
D. sulcatum L.Bolus
D. sutherlandii (Hook.f.) N.E.Br.
D. suttoniae Lavis [I]
D. testaceum (Haw.) Schwantes
D. tradescantioides (A.Berger) L.Bolus
D. truteri Lavis
D. uitenhagense L.Bolus
D. uncinatum L.Bolus
D. uniflorum L.Bolus
D. vandermerwei L.Bolus

D. velutinum L.Bolus [I]
D. verecundum L.Bolus
D. vernicolor L.Bolus
D. versicolor L.Bolus
D. vinaceum (L.Bolus) L.Bolus
D. virens L.Bolus
D. vogtsii L.Bolus
D. waterbergense L.Bolus
D. wethamae L.Bolus
D. wilmaniae Lavis
D. wiunii Lavis
D. zeederbergii L.Bolus
D. zoeae L.Bolus
D. zoutpansbergense L.Bolus

Literature

LAVIS, N. 1966. Notes on the genus *Delosperma* (Mesembrieae). *Journal of South African Botany* 32: 209.

D. pottsii

D. prasinum

Drosanthemum

Derivation of genus name The name is derived from the Greek words *drosos* (dew) and *anthos* (flower), describing the glittering water cells on the leaves of many species which resemble drops of dew.

Common names The names *porseleinbos* (porcelain bush) and *vleisbos* (meat bush) have been recorded for *D. striatum* and the general name *bergvygie* (mountain mesemb) is sometimes used for *Drosanthemum* species.

Description The low to erect shrubs possess maroon to ochre stems with or without short or long hairs. The mostly short-lived, easily dropping leaves are covered by large water-filled cells which glisten in the sunlight. The leaf tips are mostly rounded, the bases of a pair are rarely fused. One or two species shed all their leaves during the dry season. The white, yellow, orange, scarlet or purple flowers occur in clusters or, more rarely, solitarily. Filamentous staminodes are absent or present. When present they are white, pink or deep black. Black staminodes are associated with yellow or red flowers. When filamentous staminodes are present, the stamens stand in a central cone. When they are absent, the stamens spread widely. Fruit capsules are mostly borne on long, erect stalks. The bases of the capsules often have long hairs and a light-coloured portion of stem below each, clearly separated from the maroon stalk. The covering membranes are straight and in rare cases laterally reduced. The valves possess broad wings. Closing bodies are absent in most species, but in a few species bulges derived from expanding tissue can be seen at the seed exits. Overall, the fruits are light-coloured and rather short-lived.

Distinguishing characters The combination of covering membranes (absent in *Delosperma*) and broad valve wings (present in *Delosperma* and *Trichodiadema*) in the fruit, and the absence of elongate cells at the leaf tips (present in *Trichodiadema*) are typical of the genus. The flower is distinguished by its elongated, almost thread-like stigmas and nectar glands in five parts.

Flowering time Plants flower mostly in spring and summer. The flowers open around midday and close by evening.

Distribution and ecology The genus has a wide distribution range in the western parts of southern Africa. It stretches in a broad sweeping band from near the Angolan border in the north through the western parts of Namibia across southwestern South Africa, including the entire Western Cape Province and much of the Northern Cape Province, the Free State and the Eastern Cape Province. There are records from Lesotho. Most species are found in rather restricted areas in the winter rainfall region on gravelly soils and often in disturbed places. *D. candens* is grown as an ornamental plant in southeastern Australia. It is naturalised in parts of Portugal, England and the Azores.

Cultivation *Drosanthemum* species are perhaps the most popular and spectacular of all garden mesembs. Few other plants can match the glowing intensity of a stand of these mesembs in full flower. They grow easily from seed and can be used freely in gardening. Suitable species will form attractive, marvellously colourful cushions. It should be noted, however, that the plants die after some years and care should be taken to keep seedlings for replacement. Seed can be sown during autumn or even in summer. Cuttings usually root readily and are best taken after fruiting, from midsummer to autumn (December to April in South Africa).

Notes *Drosanthemum speciosum* is found in Mediterranean gardens throughout the world. This species has many colour forms. It shares dramatic black stamens (a character

D. speciosum in its habitat

D. speciosum

Unusual form of *D. speciosum* from Uniondale

D. splendens

D. collinum

otherwise unknown in the family) with its relatives *D. bicolor* and *D. micans*, two other spectacular and popular garden plants (see illustrations). Other garden favourites are the *vleisbos*, *D. striatum*, with its attractively striped petals, *D. collinum*, with its bright golden flowers, *D. floribundum*, with its pink, self-fertile flowers, and the ubiquitous *D. hispidum*, with its cascading purplish pink flowering branches. A particularly beautiful dwarf species with tuberous roots, usually known as *Trichodiadema fergusoniae*, is actually a *Drosanthemum*.

Drosanthemum Schwantes

NUMBER OF SPECIES/SUBSPECIES/VARIETIES (120/0/2)

SPECIES LIST AND CONSERVATION STATUS
 D. acuminatum L.Bolus
 D. acutifolium (L.Bolus) L.Bolus
 D. albens L.Bolus
 D. albiflorum (L.Bolus) Schwantes
 D. ambiguum L.Bolus
 D. anomalum L.Bolus
 D. archeri L.Bolus
 D. attenuatum (Haw.) Schwantes
 D. aureopurpureum (L.Bolus) L.Bolus
 D. austricolum L.Bolus [E]
 D. autumnale L.Bolus
 D. barkerae L.Bolus
 D. barwickii L.Bolus
 D. bellum L.Bolus [R]
 D. bicolor L.Bolus
 D. breve L.Bolus
 D. brevifolium (Aiton) Schwantes
 D. calycinum (Haw.) Schwantes
 D. candens (Haw.) Schwantes
 D. capillare (Thunb.) Schwantes
 D. cereale L.Bolus
 D. chrysum L.Bolus
 D. collinum (Sond.) Schwantes
 D. comptonii L.Bolus
 D. concavum L.Bolus
 D. crassum L.Bolus
 D. croceum L.Bolus
 D. curtophyllum L.Bolus
 D. cymiferum L.Bolus
 D. dejagerae L.Bolus
 D. delicatulum (L.Bolus) Schwantes
 D. diversifolium L.Bolus
 D. duplessiae L.Bolus
 D. eburneum L.Bolus
 D. edwardsiae L.Bolus
 D. erigeriflorum (Jacq.) Stearn
 D. exspersum (N.E.Br.) Schwantes
 D. filiforme L.Bolus
 D. flammeum L.Bolus
 D. flavum (Haw.) Schwantes
 D. floribundum (Haw.) Schwantes
 D. fourcadei (L.Bolus) Schwantes
 D. framesii L.Bolus
 D. fulleri L.Bolus
 D. giffenii (L.Bolus) Schwantes
 D. glabrescens L.Bolus
 D. globosum L.Bolus
 D. godmaniae L.Bolus
 D. gracillimum L.Bolus
 D. hallii L.Bolus [R]
 D. hermannii (Pax ex Engl.) Schwantes
 D. hirtellum (Haw.) Schwantes
 D. hispidum (L.) Schwantes
 D. hispifolium (Haw.) Schwantes
 D. inornatum (L.Bolus) L.Bolus
 D. insolitum L.Bolus
 D. intermedium (L.Bolus) L.Bolus
 D. jamesii L.Bolus
 D. karrooense L.Bolus
 D. latipetalum L.Bolus
 D. lavisii L.Bolus
 D. laxum L.Bolus
 D. leipoldtii L.Bolus
 D. leptum L.Bolus
 D. lignosum L.Bolus
 D. lique (N.E.Br.) Schwantes
 D. littlewoodii L.Bolus
 D. luederitzii (Engl.) Schwantes
 D. macrocalyx L.Bolus
 D. maculatum (Haw.) Schwantes
 D. marinum L.Bolus
 D. martinii L.Bolus
 D. mathewsii L.Bolus
 D. micans (L.) Schwantes [R]
 D. montaguense L.Bolus
 D. muirii L.Bolus
 D. nordenstamii L.Bolus [K]
 D. oculatum L.Bolus
 D. opacum L.Bolus
 D. otzenianum (Dinter) Friedrich

D. hispidum

D. latipetalum

Drosanthemum sp., presently known as *Trichodiadema fergusoniae*

D. albiflorum

D. calycinum

D. pallens (Haw.) Schwantes
D. parvifolium (Haw.) Schwantes
D. paxianum (Schltr. & Diels) Schwantes
D. pauper (Dinter) Dinter & Schwantes
D. pickhardii L.Bolus
D. praecultum (N.E.Br.) Schwantes
D. prostratum L.Bolus
D. pulchellum L.Bolus
D. pulchrum L.Bolus
D. pulverulentum (Haw.) Schwantes
D. ramosissimum (Schltr.) L.Bolus
D. robustum L.Bolus
D. roridum L.Bolus
D. roseatum L.Bolus
D. salicolum L.Bolus
D. schoenlandianum (Schltr.) L.Bolus
D. semiglobosum L.Bolus
D. sessile (Thunb.) Schwantes
D. speciosum (Haw.) Schwantes
D. splendens L.Bolus
D. stokoei L.Bolus
D. striatum (Haw.) Schwantes var. *hispifolium* (Haw.) G.D.Rowley
D. striatum (Haw.) Schwantes var. *pallens*

(Haw.) G.D.Rowley
D. striatum (Haw.) Schwantes var. *striatum*
D. strictifolium L.Bolus
D. subalbum L.Bolus
D. subclausum L.Bolus
D. subcompressum (Haw.) Schwantes
D. subglobosum (Haw.) Schwantes
D. subplanum L.Bolus
D. tardum L.Bolus
D. thudichumii L.Bolus var. *gracilius* L. Bolus [R]
D. thudichumii L.Bolus var. *thudichumii*
D. torquatum (Haw.) Schwantes
D. tuberculiferum L.Bolus
D. uniflorum (L. Bolus) Friedrich
D. vaginatum L.Bolus
D. vandermerwei L.Bolus
D. vespertinum L.Bolus
D. wittebergense L.Bolus
D. worcesterense L.Bolus
D. zygophylloides (L.Bolus) L.Bolus

LITERATURE
SCHWANTES, G. 1927. Zur Systematik der Mesembrianthemen. *Zeitschrift für Sukkulentenkunde* 3: 14, 29.

D. bicolor

D. micans

D. bellum

Colour variant of *D. speciosum*

D. diversifolium

D. floribundum

ECTOTROPIS

DERIVATION OF GENUS NAME The name is derived from the Greek words *ectos* (outside) and *tropis* (keel of a ship), possibly in reference to the extreme bending back of the keels on the open fruit capsules.

COMMON NAMES No vernacular names seem to have been recorded for this poorly known and inconspicuous species.

DESCRIPTION Plants are tiny perennial tufts, scarcely more than 15 mm in height and 30 mm wide, with small carrot-like tubers. The minute glistening leaves, about 4 mm in length, are oblong, more or less cylindrical, minutely textured and reddish in colour. They are arranged on slender prostrate rooting branches. The flowers are white, solitary, very small, with few petals, no staminodes and about 25 stamens. Fruit stalks are about 10 mm long. The fruit capsules are five-locular, small, and extremely fragile, with reduced wings and without covering membranes or closing bodies. Seeds are brownish and minutely textured.

DISTINGUISHING CHARACTERS This distinctive miniscule plant is unlikely to be confused with other mesembs. It is very rare and occurs on moist rocks.

FLOWERING TIME Flowering time is in summer (November and December in South Africa). The short-lived flowers open in the morning and close in the evening.

DISTRIBUTION AND ECOLOGY *Ectotropis* has a restricted distribution at Hogsback in the Amatola Mountains and the Katberg, Eastern Cape Province, South Africa. It occurs at high altitudes on large boulders near streams. By all accounts this is a very rare genus.

CULTIVATION Hardly any plants have ever been cultivated. Its horticultural requirements are obscure, but it probably prefers moist, cool conditions.

NOTES The original description of *Ectotropis* is puzzlingly meager, but a much better description appeared later. The plant is known mainly from a few herbarium specimens. Despite the fragility of its fruits, this poorly known plant is perhaps better placed in *Delosperma*.

Ectotropis N.E. Br.

NUMBER OF SPECIES/SUBSPECIES/VARIETIES (1/0/0)

E. alpina N.E.Br. [I]

LITERATURE
BOLUS, H.M.L. 1927. Novitates Africanae. *Annals of the Bolus Herbarium* 4: 99–100.

BROWN, N.E. 1927. *Mesembryanthemum* and some genera recently separated from it. *Gardener's Chronicle* 81: 12.

HAMMER, S.A. 1997. Mesembs from A-Z: *Ectotropis alpina*. *Aloe* 33: 96–97.

E. alpina

Seedlings of *E. alpina*

Mestoklema

Derivation of genus name The name *Mestoklema* is derived from the Greek words *mestos* (full) and *klema* (a small branch) referring to the dense branching found in this genus.

Common names *Mestoklema* plants are commonly known as *donkievygie*, (donkey mesemb – eaten by donkeys), *vybossie*, (fig bush), *lidjiesganna* or *hongerdoring* (hungry thorn, in reference to its value as a famine food for stock).

Description Plants are densely branched shrubs of more than a metre in height, with tuberous storage roots. The young branches are covered with small, water-filled cells, becoming pallid and slightly rough when dry. The flower stalks are persistent, becoming spiny with age. Leaves are three-angled to almost cylindrical. When young they are covered with fine water-filled cells. The base of the leaf remains on the stem at leaf fall. The tiny flowers are orange, pink or salmon pink (rarely white), and flowers are borne on flower stalks in very rich clusters. Five unequal sepals with membranous margins are present. Petals are narrow and are arranged in a single whorl surrounding numerous stamens, of which the inner ones are bearded. Nectar glands are separate, encircling five awl-shaped stigmas. Fruit capsules are small, have five locules and resemble those of *Drosanthemum*. Valves are recurved when expanded, with the expanding keels nearly touching. They diverge toward the tips and have narrow, membranous, pointed wings. Covering membranes are present but are sometimes reduced by half. Closing bodies are absent. Seeds are smooth and brown.

Distinguishing characters *Mestoklema* is characterised by its huge roots and tiny flowers in coppery shades. It can be confused with *Delosperma*, but differs from that genus in that covering membranes are present in the fruit capsules and whorls of filamentous staminodes are absent.

Flowering time Plants can be found in flower from spring to mid-summer (September to December in southern Africa). Flowers open at midday.

Distribution and ecology This genus occurs in Namibia and South Africa and is widely distributed in a broad north-south band in South Africa, stretching southward from the central North-West Province through the Free State and Northern Cape Province to the coast of the Eastern and Western Cape Provinces. It extends into the Little Karoo in the west and occurs at two outlying localities in eastern and central Namibia. Drier areas of southern Africa are favoured, where species receive summer or winter rainfall of less than 500 mm per year. Plants grow in full sun in a variety of soil types.

Cultivation Propagation can be from seed or cuttings, which root easily. Porous, well-drained soils are recommended. Plants are summer growers which must be kept dry in winter.

Notes The enormous tubers of *Mestoklema*, with their shiny bronze-coloured bark, are perhaps the largest of all mesembs.

Mestoklema N.E.Br. ex Glen

Number of species/subspecies/varieties (6/0/1)

Species list and conservation status
   M. albanicum N.E.Br. ex Glen [K]
   M. arboriforme (Burch.) N.E.Br. ex Glen
   M. copiosum N.E.Br. ex Glen
   M. elatum N.E.Br. ex Glen
   M. illepidum N.E.Br. ex Glen
   M. tuberosum (L.) N.E.Br. ex Glen var. *macrorrhizum* (Haw.) N.E.Br. ex Glen
   M. tuberosum (L.) N.E.Br. ex Glen var. *tuberosum*

Literature
GLEN, H.F. 1981. Mesembryanthemaceae. Nomenclature in the genus *Mestoklema*. *Bothalia* 13: 454.

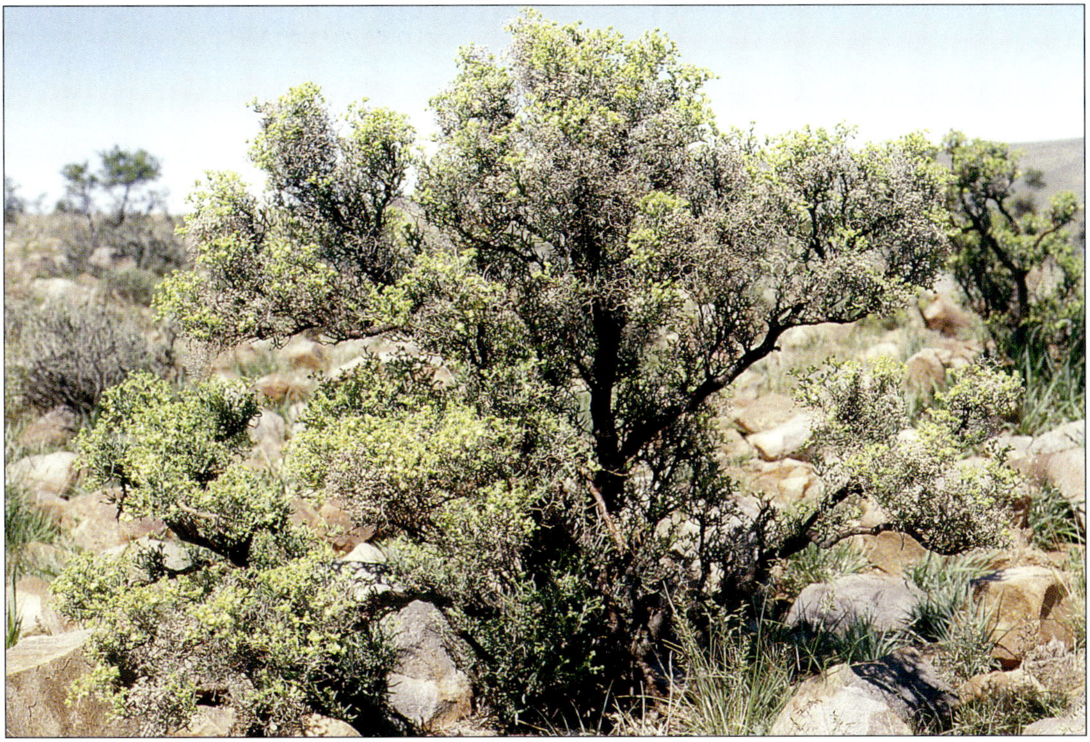

M. arboriforme

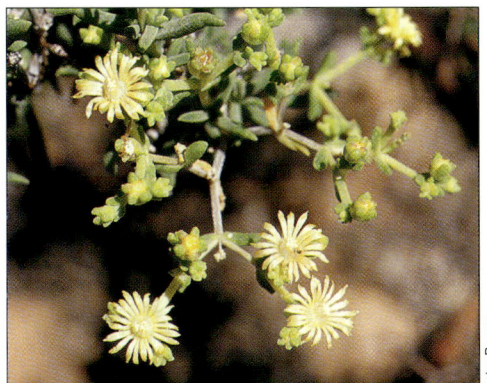

Peter Bruyns

Leaves and flowers of *M. arboriforme*

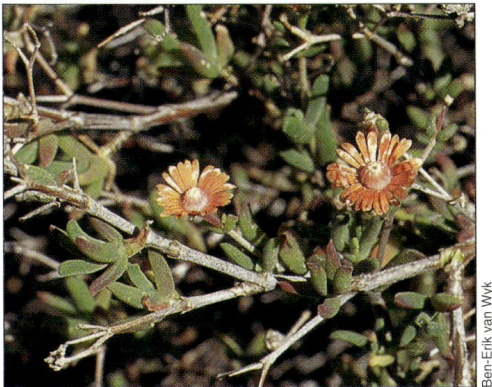

Peter Bruyns

Leaves and flowers of *M. tuberosum*

Ben-Erik van Wyk

M. tuberosum showing tuberous growth form

Ben-Erik van Wyk

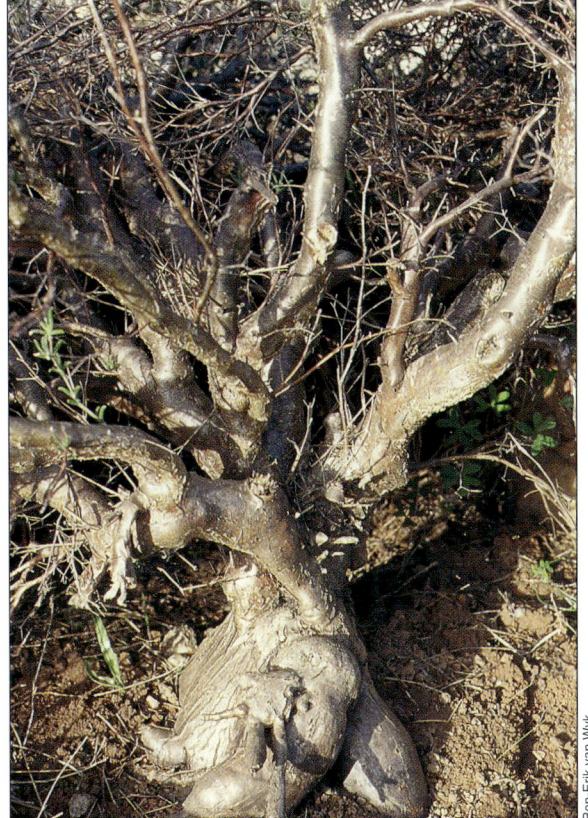

TRICHODIADEMA

DERIVATION OF GENUS NAME This genus name is derived from the Greek words *trix* (hair) and *diadema* (crown) referring to the characteristic tuft of bristles borne on the tips of the leaves.

COMMON NAMES Common names include *kareemoervygie* (karee potato vygie — underground tubers), *kierievygie* (walking-stick vygie — tubers resemble the upper knob of a walking stick), *perdevygie* (horse vygie), *donkievygie* (donkey vygie — as donkeys and horses have a tendency to eat these plants in times of drought), *duinevygie* (dune vygie) and *soetaartappel* (sweet potato — underground tubers resemble those of a sweet potato).

DESCRIPTION Plants are either shrubs with long, slender, arched branches or they are short-stemmed subshrubs, ranging in height from 30 to 600 mm. Roots are woody or may be tuberous, and clusters of tubers may be present. Leaves are small, fused, semi-cylindrical and are tipped with a cluster of bristles forming the "diadem". The leaf surfaces glisten with specialised water-storing cells that are elevated and pointed at one or both ends. The small, white, cream, yellow, light pink to dark pink flowers are solitary and borne on short flower stalks. Five to eight bristle-tipped sepals surround the flowers. Numerous narrow to lance-shaped petals are arranged in one whorl. Filamentous staminodes are present. Stamens are collected into a cone and are sometimes hairy at their bases. Nectar glands are separate and surround five to eight stigmas that are widest at their bases and much shorter than the stamens. The fruit capsules have four to seven, mostly five or six locules. Expanding keels lie close to each other and diverge somewhat towards the tips; the valve wings are broad and rounded. Covering membranes are well developed or somewhat reduced. The brown or yellowish seeds are pear-shaped and minutely textured.

DISTINGUISHING CHARACTERS *Trichodiadema* is unique due to the presence of a crown of radiating hairs on the leaf tips and sepals. This "diadem" is similar in appearance to the spines of some of the members of the Cactus family but the spines are of a different origin: those of the cacti being derived from stem structures while those of *Trichodiadema* originate from leaf structures. This genus is closely allied to *Delosperma* but differs from it in the presence of well-developed covering membranes in the fruit capsule. Fruit capsule features to distinguish the species are the shapes of the expanding keels and covering membranes.

FLOWERING TIME Flowers appear in winter to early summer (August to January in southern Africa) after good rains and the flowers are fully open in the mid-afternoon.

DISTRIBUTION AND ECOLOGY This genus is widespread in the more arid areas of southern Africa. Species grow in southern Namibia, the Richtersveld, Bushmanland and the Great and Little Karoos. In South Africa the distribution of *Trichodiadema* covers extensive areas in the Northern, Western and Eastern Cape Provinces as well as the western Free State. These areas receive winter and summer rainfall of less than 600 mm per year. Species of *Trichodiadema* prefer full sun and are often found growing in rocky areas.

CULTIVATION Plants are easily propagated by means of cuttings which will flower in their first year. Reproduction by seed is also worthwhile, as the small seedlings with their densely packed leaves, each tipped with a circlet of hairs, are very attractive. Sow seed during autumn. Outside of their habitat plants are best grown in a greenhouse.

NOTES *Trichodiadema densum*, a commonly cultivated species with conspicuously ornate diadems, is covered with large, very long-

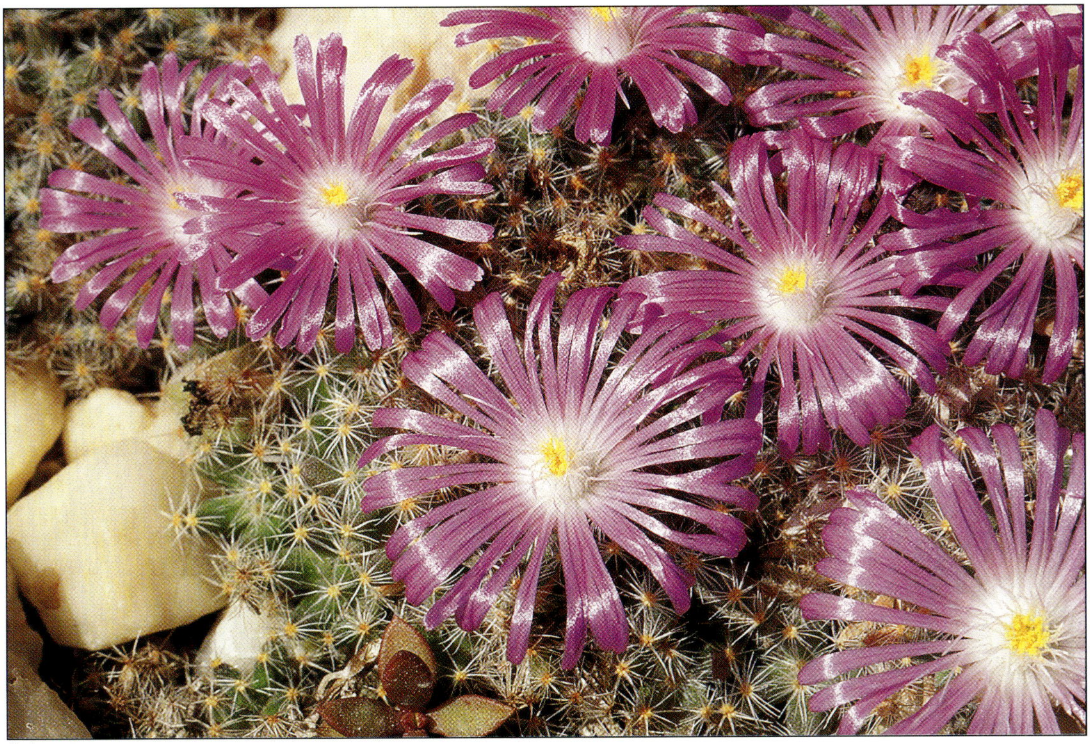
Ernst van Jaarsveld

T. densum

Ernst van Jaarsveld

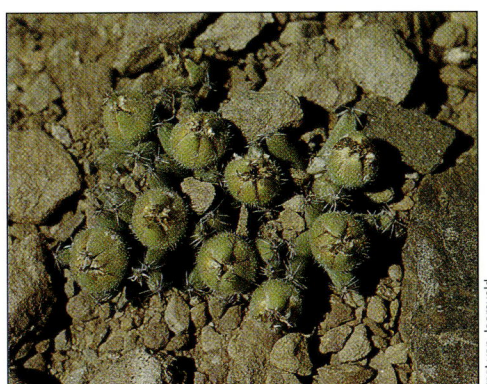

Fruiting plant of *T. densum* in its habitat

Bob Chinnock

T. decorum

Ben-Erik van Wyk

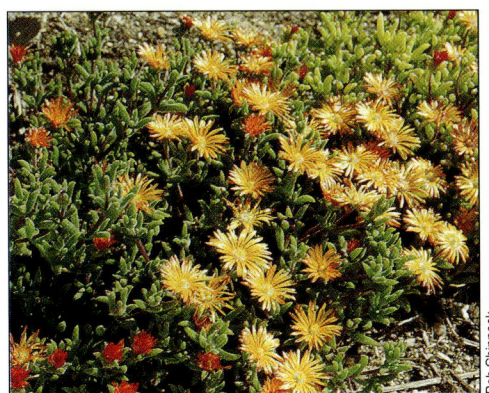

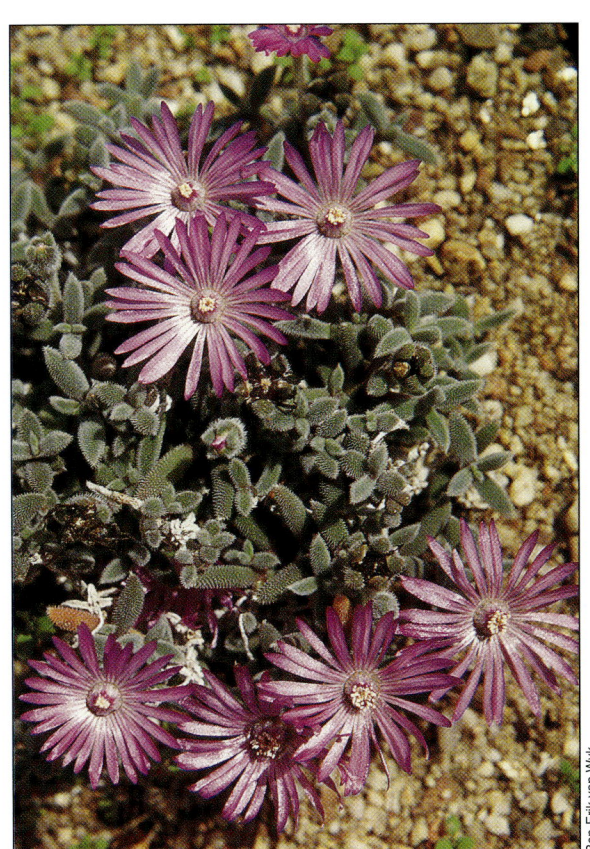

T. pygmaeum

lasting flowers in the late winter to early spring. In contrast, *T. pygmaeum* is very rare and has no proper diadems. Although most *Trichodiadema* species have purple flowers, yellow and orange shades are also possible. It is thought that the radiating tufts of hairs crowning each leaf of *Trichodiadema* species may serve to absorb water or dew. It has been noted that in cultivation, particularly if there is an over-abundance of water, the characteristic circle of hairs is not formed, thus substantiating their postulated role in water absorption. A careful taxonomic assessment of the genus would probably reveal fewer valid species than is currently the case.

Trichodiadema Schwantes

NUMBER OF SPECIES/SUBSPECIES/VARIETIES (34/0/4)

SPECIES LIST AND CONSERVATION STATUS
T. *attonsum* (L.Bolus) Schwantes
T. *aureum* L.Bolus [I]
T. *barbatum* (L.) Schwantes
T. *bulbosum* (Haw.) Schwantes
T. *burgeri* L.Bolus [R]
T. *calvatum* L.Bolus
T. *concinnum* L.Bolus
T. *decorum* (N.E.Br.) Stearn
T. *densum* (Haw.) Schwantes [nt]
T. *emarginatum* L.Bolus
T. *fergusoniae* L.Bolus
T. *fourcadei* L.Bolus
T. *gracile* L.Bolus var. *gracile*
T. *gracile* L.Bolus var. *piliferum* L.Bolus
T. *gracile* L.Bolus var. *setiferum* L.Bolus
T. *hallii* L.Bolus [R]
T. *hirsutum* (Haw.) Stearn
T. *imitans* L.Bolus
T. *inornatum* L.Bolus
T. *intonsum* (Haw.) Schwantes
T. *littlewoodii* L.Bolus forma *alba* L.Bolus
T. *littlewoodii* L.Bolus forma *littlewoodii*
T. *marlothii* L.Bolus
T. *mirabile* (N.E.Br.) Schwantes var. *leptum* L.Bolus
T. *mirabile* (N.E.Br.) Schwantes var. *mirabile*
T. *obliquum* L.Bolus [I]
T. *occidentalis* L.Bolus
T. *olivaceum* L.Bolus
T. *orientale* L.Bolus
T. *peersii* L.Bolus [I]
T. *pomeridianum* L.Bolus
T. *pygmaeum* L.Bolus [R]
T. *rogersiae* L.Bolus [I]
T. *rupicolum* L.Bolus [I]
T. *ryderae* L.Bolus
T. *setuliferum* (N.E.Br.) Schwantes var. *niveum* L.Bolus
T. *setuliferum* (N.E.Br.) Schwantes var. *setuliferum*
T. *stayneri* L.Bolus
T. *strumosum* (Haw.) L.Bolus

LITERATURE
IHLENFELDT, H.-D. 1980. Der Haarapparat ("Diadem") der Gattung *Trichodiadema* Schwant. (Mesembryanthemaceae). *Mitteilungen aus dem Institut für Allgemeine Botanik, Hamburg* 17: 145–163.

NIESLER, I.M. 1997. Is *Trichodiadema* Schwantes one genus only? — Problems in delimitation or "what is a diadem?" *Bradleya* 15: 13–27.

T. bulbosum

T. mirabile

T. barbatum

Lampranthus amoenus

Lampranthus-like shrubby Mesembs

GROUP 12

Shrubby plants with creeping or erect stems with characteristic fruit capsules of the *Lampranthus*-type.

The most striking features of the *Lampranthus*-type of fruit are the bunches of funicular hairs at the exits of the locules, i.e. in lieu of closing bodies. Covering membranes are quite rigid and persistently convex in shape. They have a distinct recurved rim above and closing ledge on the distal undersurface of the membrane. Valve wings may be present or absent in this capsule type. This type of fruit capsule usually has five but may have up to 10 locules.

LAMPRANTHUS-LIKE SHRUBBY MESEMBS comprise 10 genera and 270 species.

Amphibolia (6 species)
Circandra (1 species)
Enarganthe (1 species)
Erepsia (27 species)

Lampranthus (227 species)
Namaquanthus (1 species)
Oscularia (3 species)

Scopelogena (2 species)
Smicrostigma (1 species)
Wooleya (1 species)

Amphibolia

Derivation of genus name The name is taken directly from the Greek word *amphibolia* (uncertainty or doubt) because it was doubtful whether this genus could indeed be separated from the genus *Stoeberia*.

Common names No vernacular names are known.

Description The shrubs are sprawling, often broader than they are tall and up to 2 m wide, with almost white or light yellow-coloured stems. The leaves are sharp-pointed or slightly blunt and blue-green in colour. Flowers occur solitarily or in clusters of three on short flower stalks, which have pink bracts below the flowers. The blunt or somewhat sharp-pointed petals occur in two whorls and have a distinct colour pattern: there is a dark area at the tip with a dark line running down to the base, where it broadens and darkens to form a clearly demarcated area which is surrounded by a whitish region. The stamens are arranged in a central cone and filamentous staminodes are mostly absent or, when present, they are bent and occur in a single row. The fruit capsules have five locules. Once open they do not close again. The valves have very broad, almost rectangular wings and the expanding keels are erect and irregularly torn. The covering membranes sometimes have two spur-like processes. Closing bodies are small, and closing rodlets are present on the distal surfaces of the covering membranes. The glossy brown seeds are roundish.

Distinguishing characters The main features distinguishing this genus are the fruit capsules: they have winged valves as in *Lampranthus* and closing bodies as in *Ruschia*. The leaf surfaces are always smooth and regularly covered by a fine layer of sharp-edged wax platelets of irregular shape (as seen under the microscope). The flower clusters often have a one-sided appearance due to solitary flowers which are arranged in a sequence on the shoots. The whole flower has a distinctive star-like colour pattern.

Flowering time The plants flower in summer and the flowers open during the day.

Distribution and ecology In South Africa, *Amphibolia* species occur in the Western Cape Province and Namaqualand. In Namibia they occur on the coastal plain between Lüderitz and Oranjemund.

Cultivation *Amphibolia* species are easily propagated from seed sown in the autumn. However, the genus is seldom cultivated.

Notes This genus is poorly known and is difficult to recognise when the plants are not in flower.

Amphibolia L.Bolus ex A.G.J.Herre

Number of species/subspecies/varieties (6/0/0)

Species list and conservation status
  A. gydouwensis (L.Bolus) L.Bolus ex Toelken & Jessop
  A. hallii (L.Bolus) L.Bolus ex Toelken & Jessop
  A. littlewoodii (L.Bolus) L.Bolus ex Toelken & Jessop
  A. maritima L.Bolus ex Toelken & Jessop
  A. rupis-arcuatae (Dinter) H.E.K.Hartmann
  A. stayneri L.Bolus ex Toelken & Jessop

Literature
BOLUS, H.M.L. 1965. Notes on *Mesembryanthemum* and allied genera. *Journal of South African Botany* 31: 169–174.

TÖLKEN, H.R. & JESSOP, J.P. 1976. Mesembryanthemaceae. Nomenclature of the genus *Amphibolia*. *Bothalia* 12: 64.

HARTMANN, H.E.K. & DEHN, M. 1989. A re-examination of the genus *Amphibolia* (Mesembryanthemaceae). *Bothalia* 19: 179–182.

HARTMANN, H.E.K. 1996. Miscellaneous taxonomic notes on Aizoaceae. *Bradleya* 14: 29–56.

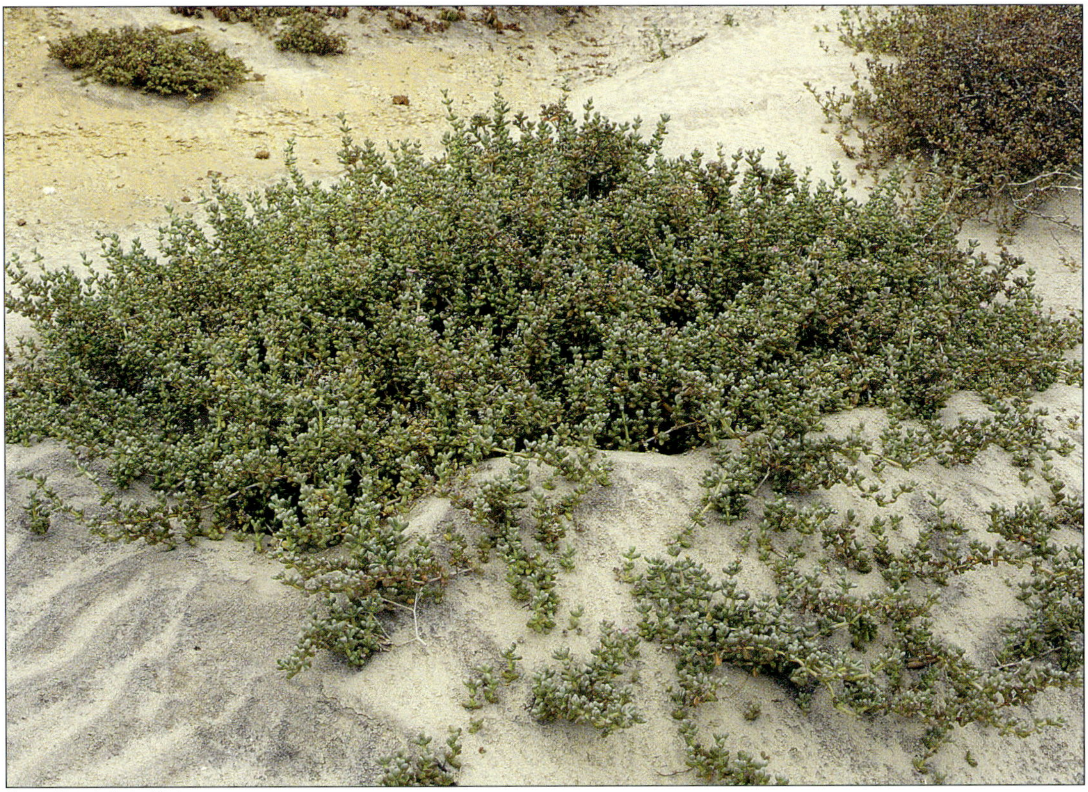

Ben-Erik van Wyk

A. maritima in its habitat

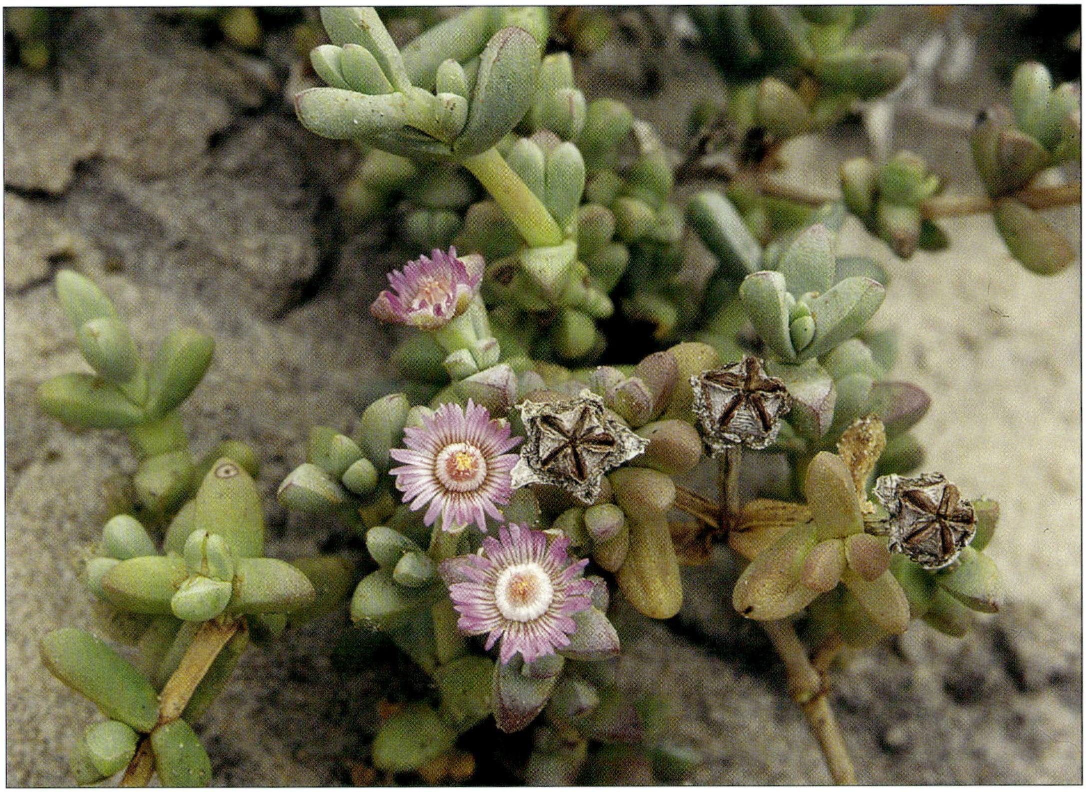

Ben-Erik van Wyk

A. maritima

CIRCANDRA

DERIVATION OF GENUS NAME The name is derived from the Greek words *kirkos* (a ring) and *aner* (male) in allusion to the remarkable manner in which the very short stamens are arranged in a ring around the top of the ovary, clearly exposing the latter to view.

COMMON NAMES No vernacular names seem to have been recorded.

DESCRIPTION This plant is a sparsely branched succulent shrub of up to 600 mm in height with fibrous roots. The stems are smooth and up to 5 mm in diameter. The slender, sharply three-sided leaves are up to 35 mm in length, with a sharp tip and toothed margins. The flowers are borne solitarily on the tips of the branches and are 50 mm in diameter. The five sepals are unequal in length. Bright yellow petals surround the very short stamens. There are no filamentous staminodes. Fruit capsules are 12 mm in diameter, with five locules. The expanding keels are diverging and extend into an awn with broad wings. Covering membranes cover the seed locules, but closing bodies are absent. The seeds are large with rough surfaces.

DISTINGUISHING CHARACTERS *Circandra serrata* is a conspicuous, sparsely branched shrub of up to 600 mm in height, with finely toothed (serrated) leaves and solitary yellow flowers of about 50 mm in diameter.

FLOWERING TIME Early records show that the plants flower in early summer (October and November in South Africa).

DISTRIBUTION AND ECOLOGY *Circandra* has been recorded from a small area in the Western Cape Province of South Africa and has been collected from the surroundings of Ceres, Tulbagh and Villiersdorp, at low altitudes of up to 500 m.

CULTIVATION There are no plants in cultivation.

NOTES Although *Circandra* was already known to the famous Swedish botanist Linnaeus in the 1750s, it is only known today from very few herbarium collections and has not been collected since 1913. In the southwestern Cape, there has been extensive destruction of the natural vegetation for agriculture and it is highly likely that the species is extinct.

Circandra N.E.Br.

NUMBER OF SPECIES/SUBSPECIES/VARIETIES (1/0/0)

SPECIES LIST AND CONSERVATION STATUS
 C. serrata (L.) N.E.Br. [Ex]

LITERATURE
 LIEDE, S. 1989. Untersuchungen zum Merkmalsbestand und zur Taxonomie der "Erepsiinae" (Mesembryanthemaceae). *Beiträge zur Biologie der Pflanzen* 64: 391–479.

Dried herbarium specimen of *Circandra serrata*

ENARGANTHE

DERIVATION OF GENUS NAME The name is derived from the Greek words *enarges* (shining) and *anthos* (flower), in allusion to the large conspicuous flowers.

COMMON NAMES No vernacular names seem to have been recorded.

DESCRIPTION The plant is an erect shrub with woody branches, smooth stems and fibrous roots. The stout, smooth, three-sided leaves are up to 15 mm in length. They occur in pairs and are not fused at their bases. The flowers occur singly or in few-flowered clusters on the branch tips. They are up to 60 mm in diameter and are borne on short stalks. The four sepals are unequal in size – two are much larger and fleshy. The bright pink petals are all of the same length and filamentous staminodes are absent. Nectar glands are present as a series of small teeth in a continuous ring. The woody fruit capsules have eight locules. The expanding keels are parallel at the base but diverge towards the ends, ending in short spines. Covering membranes cover the seed cavities, the valves have wings and closing bodies are present as small tubercles. The smooth, kidney-shaped seeds are ochre to light brown.

DISTINGUISHING CHARACTERS *Enarganthe octonaria* has very large bright pink flowers. The fruit capsules have eight locules and small closing bodies are present.

FLOWERING TIME The plants flower in midwinter (July to August in South Africa). Flowers open around noon and close in the late afternoon.

DISTRIBUTION AND ECOLOGY *Enarganthe* occurs in the Richtersveld, between the Augrabies Mountain and Brakfontein in the Northern Cape Province, South Africa. It grows on quartzite and schist slopes in the dry winter rainfall region.

CULTIVATION *Enarganthe* thrives in cultivation. Plants can be grown from seed sown in autumn or from cuttings rooted in sand. Keep dry in summer. Outside of its habitat it is best grown in a greenhouse.

NOTES There is only one species in this genus. *Enarganthe* appears to be closely allied to *Namaquanthus,* from which it differs by the presence of closing bodies in the capsules (lacking in *Namaquanthus*); the fruit capsules with eight locules (eight to 16 locules in *Namaquanthus*); and the smooth seeds (the seed surface has minute spines in *Namaquanthus*).

Enarganthe N.E.Br.

NUMBER OF SPECIES/SUBSPECIES/VARIETIES (1/0/0)

SPECIES LIST AND CONSERVATION STATUS
E. *octonaria* (L.Bolus) N.E.Br.

LITERATURE
BROWN, N.E. 1930. *Mesembryanthemum* and some new genera separated from it. *Gardener's Chronicle* 87: 71.

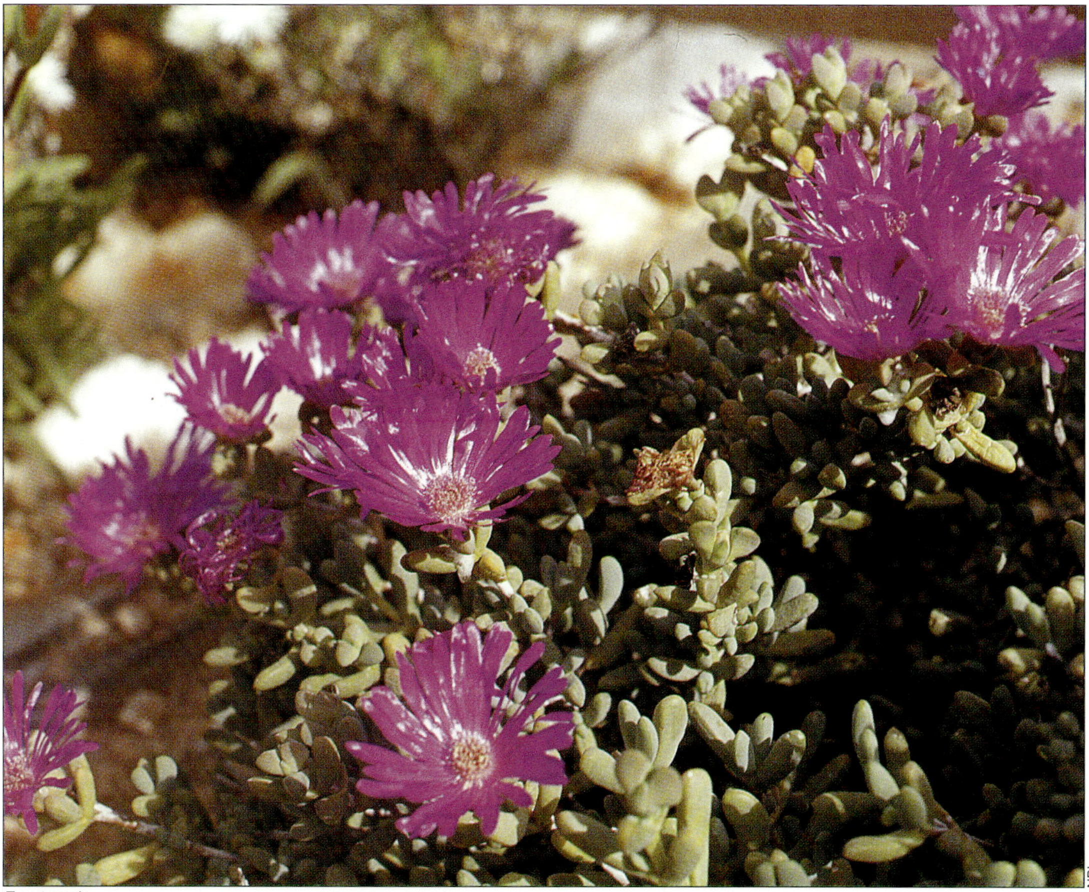

E. octonaria

NBI

E. octonaria

Ben-Erik van Wyk

E. octonaria

Ernst van Jaarsveld

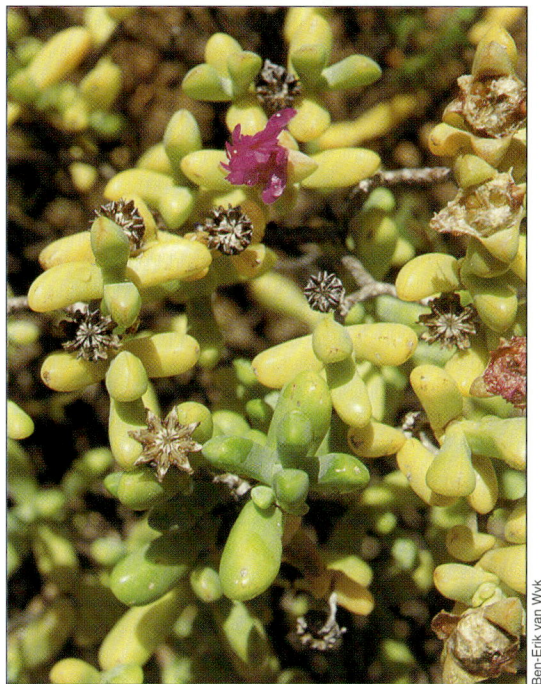

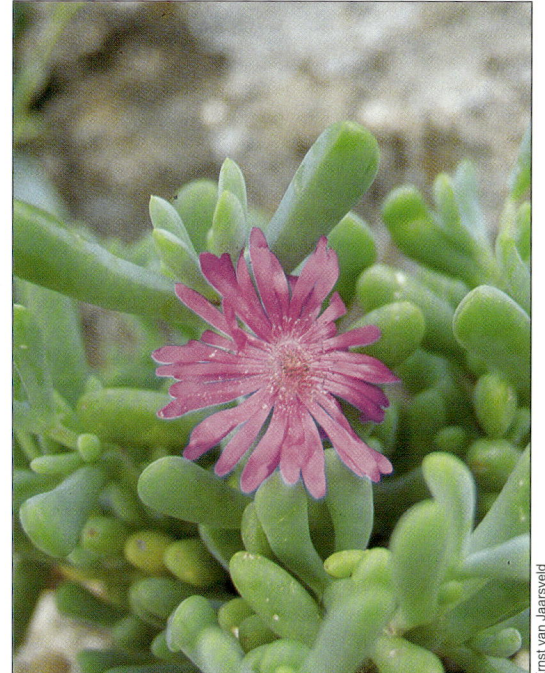

EREPSIA

DERIVATION OF GENUS NAME The name is derived from the Greek word *erepsis* ("I hide myself"), since the reproductive parts of the flower are completely hidden by the staminodes.

COMMON NAMES *Erepsia inclaudens* (and other species) are commonly known as *altydvygie(s)*. The names *Piketbergvygie* for *Erepsia pillansii* and *Paarlvygie* for *Erepsia lacera* have also been recorded.

DESCRIPTION These shrubs or shrublets have smooth stems of variable thickness which often branch from the base of the plant. The leaves are three-sided and sharply angled. In one species, *E. lacera*, the leaf edges have sharp horny serrations. The leaves always have a sharp, somewhat thorny tip and the surface is more or less smooth, with a thin wax layer. Flowers are borne in clusters or singly and are sometimes almost without stalks. There are five sepals and numerous white, pink or puce petals, arranged in several whorls. In one species, *E. pillansii*, the petals are spoon-shaped. Filamentous staminodes either partly or completely conceal the stamens. The surface of the ovary is hollow, so that the flower forms a short tube from which the stamens bend down towards the ovary. Nectar glands are present as a series of small teeth in a continuous ring, or in rare cases in groups of five. The woody fruit capsules are funnel-shaped at their bases and they usually have five locules, but rarely eight to 13 locules. The expanding keels end in sharp tips and the valve wings are variable in shape. Covering membranes are present and they have lateral to marginal closing ledges. There are no closing bodies. The dark brown, egg-shaped seeds are large, with rough surfaces.

DISTINGUISHING CHARACTERS The sharp tips of the leaves are characteristic of the genus. In addition, *Erepsia* has a peculiar flower structure, which is easily seen in longitudinal section: the surface of the ovary is hollow (concave), forming a short tube (hypanthium). As a result, the thread-like staminodes bend down and usually conceal the stamens below. The fruits are woody and lack closing bodies.

FLOWERING TIME In South Africa, flowering occurs mainly between November and April, but a few species flower at other times of the year. The flowers of some species stay open day and night.

DISTRIBUTION AND ECOLOGY The genus is mainly found in the winter rainfall area of the Western Cape Province, South Africa, with an outlier around Port Elizabeth in the Eastern Cape Province. Plants usually occur on mineral-poor quartzitic sandstone soils in fynbos or sometimes in richer shale or granite soils in renosterveld. The habitat varies from coastal flats to mountain tops. Rainfall in this region is mainly in winter and ranges from 500 to 2 000 mm per year. Many species are highly localised and occur only in small areas. Fire is an essential element in the ecology of *Erepsia* species and in the case of some species, successful seedling recruitment is dependent on regular fires. Most species do not resprout but regenerate from seed after a fire. Pollination is also specific and with many species the flowers remain open day and night, hence the name *altydvygie* (*altyd* means always).

CULTIVATION Cuttings are easily rooted in sand and can be taken at any time of the year. Seed should preferably be sown during the autumn or winter months but germination is sometimes slow. Most species thrive in cultivation and are useful for fynbos gardens.

NOTES Since its description, *Erepsia promontorii* has never been found again and is thought to be extinct. Two genera, both with only a single species, were shown to be part of *Erepsia*. These are *Semnanthe* (now *E. lacera*), a tooth-leaved plant resembling an

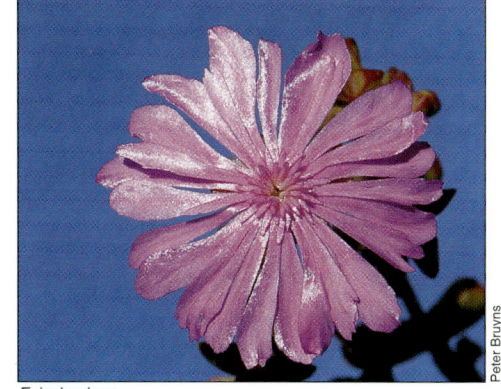

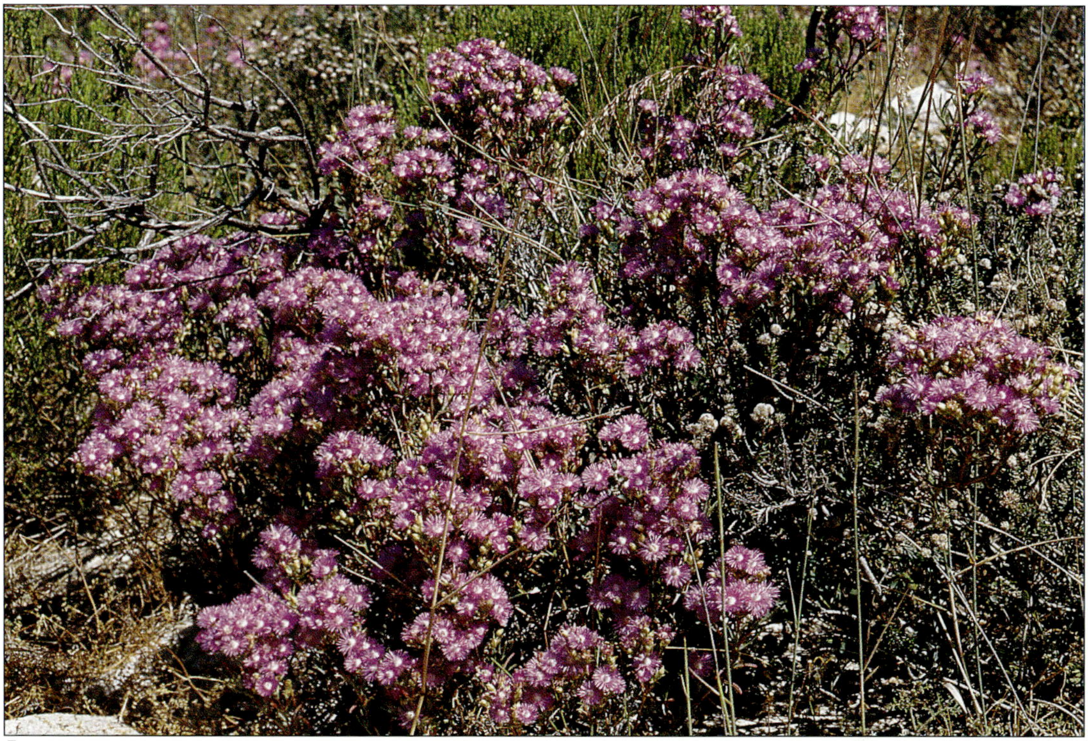
E. aspera

Ernst van Jaarsveld

E. mutabilis

Bob Chinnock

Flowers of E. aspera

Ernst van Jaarsveld

E. inclaudens

Peter Bruyns

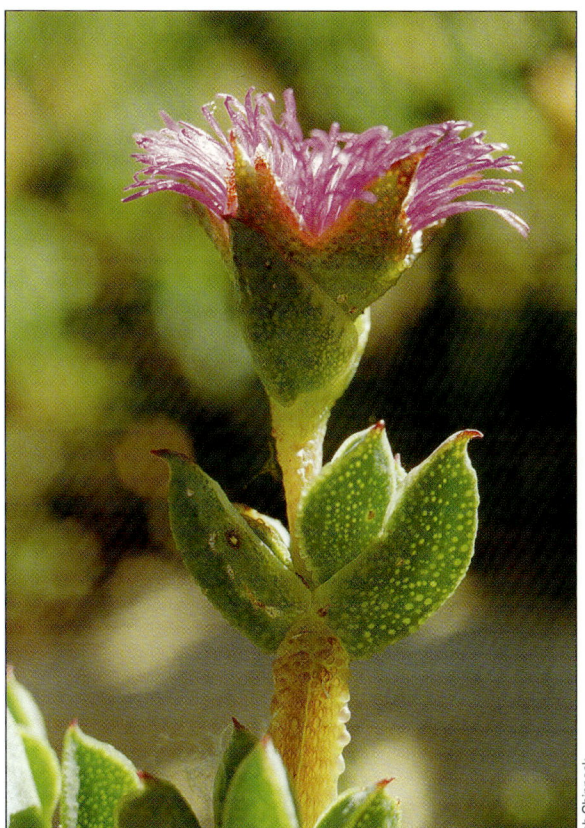

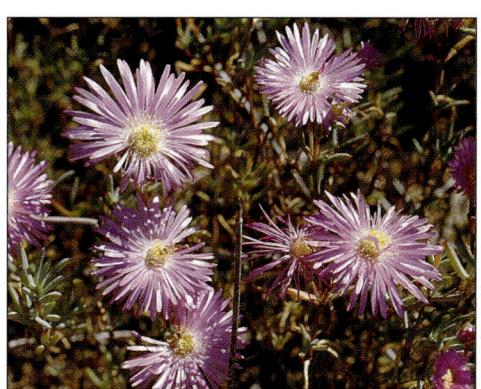

erect *Carpobrotus;* and *Kensitia* (now *E. pillansii*), with its distinctive spoon-shaped petals. Some species of *Erepsia* are hardly succulent and not very long-lived, lasting for about five years.

Erepsia N.E.Br. (= *Kensitia* Fedde; = *Semnanthe* N.E.Br.)

NUMBER OF SPECIES/SUBSPECIES/VARIETIES (27/0/0)

SPECIES LIST AND CONSERVATION STATUS
- *E. anceps* (Haw.) Schwantes
- *E. aperta* L.Bolus
- *E. aspera* (Haw.) L.Bolus
- *E. babiloniae* Liede
- *E. bracteata* (Aiton) Schwantes
- *E. bracteata* (Aiton) Schwantes x *E. anceps* (Haw.) Schwantes
- *E. bracteata* (Aiton) Schwantes x *E. ramosa* L.Bolus
- *E. brevipetala* L.Bolus [E]
- *E. distans* L.Bolus
- *E. dubia* Liede [R]
- *E. esterhuyseniae* L.Bolus
- *E. forficata* (L.) Schwantes
- *E. gracilis* (Haw.) L.Bolus
- *E. hallii* L.Bolus [I]
- *E. heteropetala* (Haw.) Schwantes
- *E. inclaudens* (Haw.) Schwantes
- *E. insignis* (Schltr.) Schwantes [R]
- *E. lacera* (Haw.) Liede
- *E. oxysepala* (Schltr.) L.Bolus
- *E. patula* (Haw.) Schwantes [R]
- *E. pentagona* (L.Bolus) L.Bolus [V]
- *E. pillansii* (Kensit) Liede [I]
- *E. polita* (L.Bolus) L.Bolus [E]
- *E. polypetala* (A.Berger & Schltr.) L.Bolus [R]
- *E. promontorii* L.Bolus [Ex]
- *E. ramosa* L.Bolus
- *E. saturata* L.Bolus
- *E. steytlerae* L.Bolus [I]
- *E. villiersii* L.Bolus [V]

LITERATURE
LIEDE, S. 1989. Untersuchungen zum Merkmalsbestand und zur Taxonomie der "Erepsiinae" (Mesembryanthemaceae). *Beiträge zur Biologie der Pflanzen* 64: 391–479.

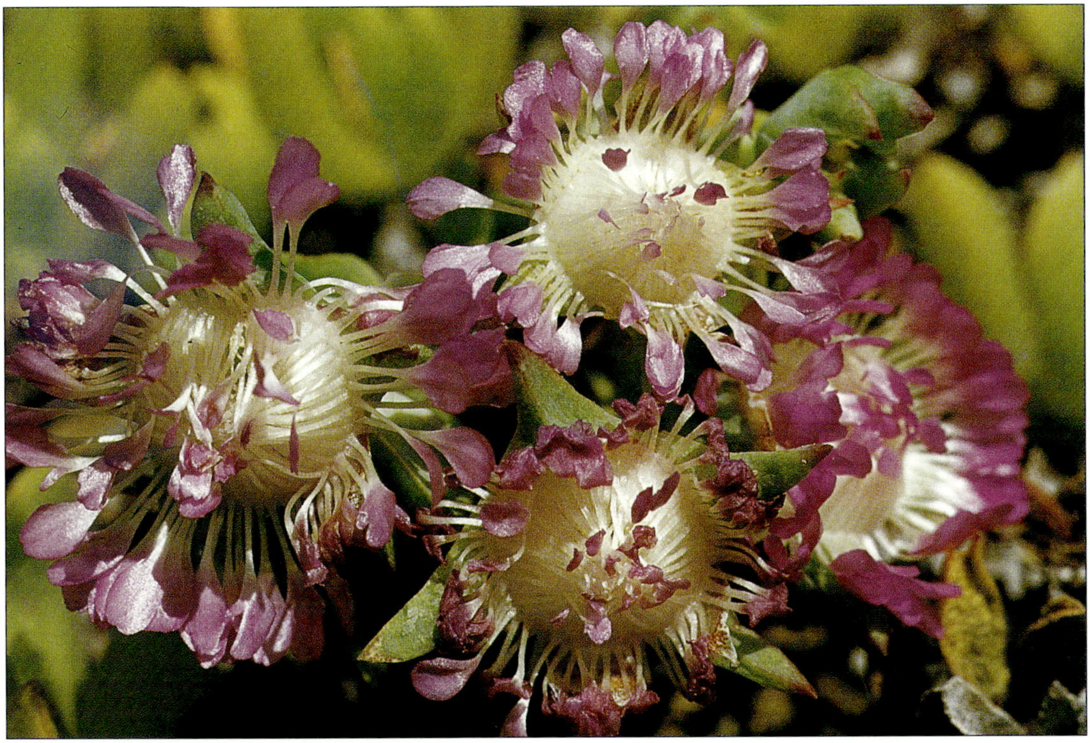

NBI

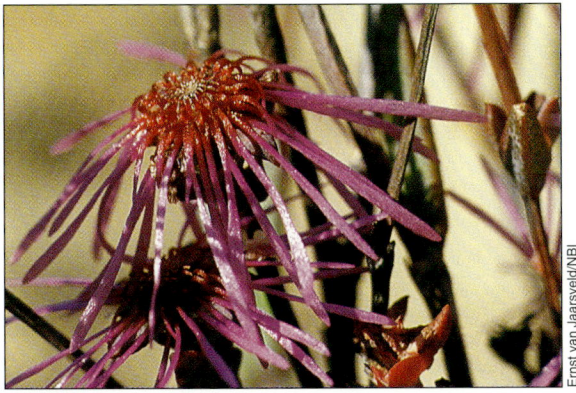

Flowers of *E. pillansii*

Ernst van Jaarsveld/NBI

E. saturata

Ben-Erik van Wyk

E. pillansii

NBI

Flower and leaves of *E. lacera*

Ben-Erik van Wyk

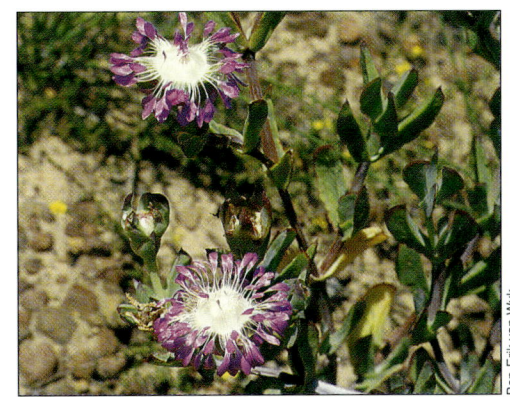

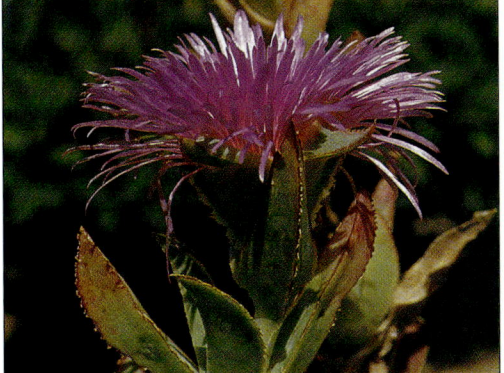

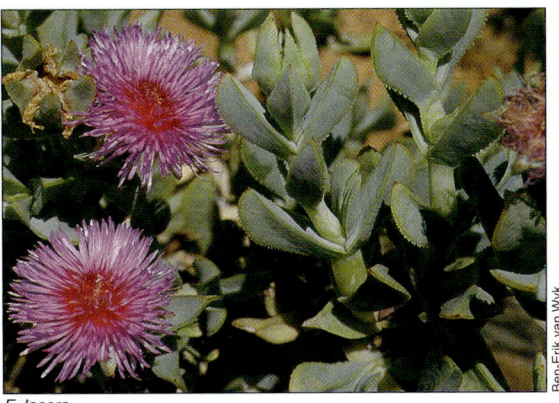

E. lacera

LAMPRANTHUS

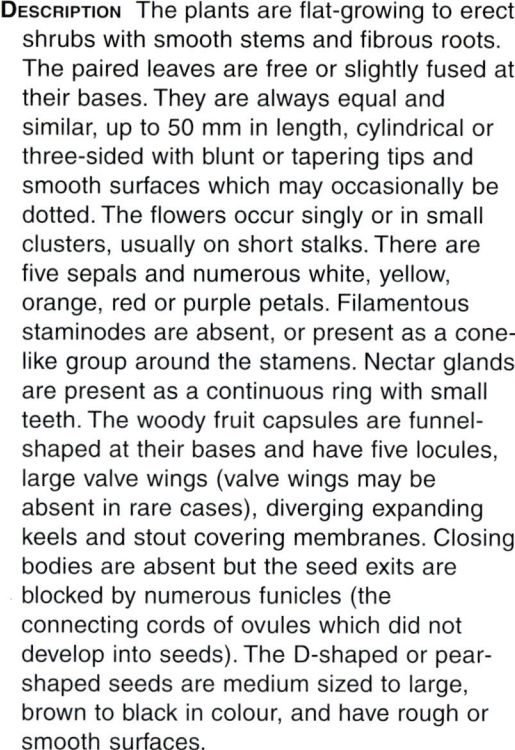

DERIVATION OF GENUS NAME The name is derived from the Greek words *lampros* (bright) and *anthos* (flower), referring to the large showy flowers of many species.

COMMON NAMES *Lampranthus arbuthnotiae* is known as *rankvygie* (creeping mesemb). *Vleisbos* (meat bush) refers to *L. lunatus*. *Rooivygie* (red mesemb) is a common name for *L. ornatus*.

DESCRIPTION The plants are flat-growing to erect shrubs with smooth stems and fibrous roots. The paired leaves are free or slightly fused at their bases. They are always equal and similar, up to 50 mm in length, cylindrical or three-sided with blunt or tapering tips and smooth surfaces which may occasionally be dotted. The flowers occur singly or in small clusters, usually on short stalks. There are five sepals and numerous white, yellow, orange, red or purple petals. Filamentous staminodes are absent, or present as a cone-like group around the stamens. Nectar glands are present as a continuous ring with small teeth. The woody fruit capsules are funnel-shaped at their bases and have five locules, large valve wings (valve wings may be absent in rare cases), diverging expanding keels and stout covering membranes. Closing bodies are absent but the seed exits are blocked by numerous funicles (the connecting cords of ovules which did not develop into seeds). The D-shaped or pear-shaped seeds are medium sized to large, brown to black in colour, and have rough or smooth surfaces.

DISTINGUISHING CHARACTERS *Lampranthus* is characterised by smooth, cylindrical, often waxy leaves and stems, in contrast to *Drosanthemum*, which has numerous water cells on the stems and leaves. In contrast to *Ruschia*, which has only white or pink flowers, *Lampranthus* exhibits a large variety of flower colours. In addition, the fruits of *Lampranthus* always lack closing bodies.

FLOWERING TIME Flowering occurs at all times of the year but the most commonly cultivated species usually flower in spring or early summer. The flowers open by mid-morning and close again in the late afternoon. The flowers of some species have a fragrance reminiscent of ripe strawberries.

DISTRIBUTION AND ECOLOGY The majority of species is found in a broad band along the western and southern coasts of the Northern, Western and Eastern Cape Provinces, South Africa. Further towards the east it grows near the Mthamvuna River mouth on the border between the Eastern Cape Province and KwaZulu-Natal. *Lampranthus* also extends as far north as southern Namibia. Species grow in a variety of habitats: they are found in sandy places along the coast as well as in shaly or loamy flats. Species are also common in rocky areas and on sandstone soils at higher altitudes.

CULTIVATION *Lampranthus* species are commonly cultivated and many of them are popular rockery plants, especially amongst water-wise gardeners. They can withstand drought well but do not thrive in areas which are subject to prolonged periods of frost. Plants grow easily from cuttings or seeds. Sow seed during the summer or winter months. Cuttings are best rooted after fruiting in mid-June.

NOTES The illustrations included here can hardly do justice to the tremendous variation in colour and form of *Lampranthus* flowers. All of them are shining and many are bicoloured, capturing light in different ways. The majority of species have purple to pink flowers, but yellow, golden orange and red are not uncommon. Some of the most commonly cultivated species include *L. haworthii*, with its purple flowers; *L. aureus*, with its wide, gilded petals; *L. blandus*, with its soft, pale pink flowers, and *L. roseus*, which hides its leaves completely in the

L. vernalis

L. roseus

L. multiradiatus

Salmon form of L. roseus

L. coralliflorus

L. tegens

flowering season. Studies have shown that *Lampranthus* is very heterogenous and that it should perhaps be subdivided. There is evidence that only the large-flowered species should remain in this genus, while the small-flowered species may be moved to other genera. This process has already started with the transfer of species to *Oscularia*.

Lampranthus N.E.Br.

NUMBER OF SPECIES/SUBSPECIES/VARIETIES
(227/0/13)

SPECIES LIST AND CONSERVATION STATUS
- *L. acrosepalus* (L.Bolus) L.Bolus
- *L. acutifolius* (L.Bolus) N.E.Br.
- *L. aduncus* (Haw.) N.E.Br.
- *L. aestivus* (L.Bolus) L.Bolus
- *L. affinis* L.Bolus
- *L. albus* (L.Bolus) L.Bolus
- *L. algoensis* L.Bolus [R]
- *L. altistylus* N.E.Br. [E]
- *L. amabilis* L.Bolus
- *L. amoenus* (Salm-Dyck) N.E.Br. [E]
- *L. antemeridianus* (L.Bolus) L.Bolus
- *L. antonii* L.Bolus
- *L. arbuthnotiae* (L.Bolus) L.Bolus [E]
- *L. arenicolus* L.Bolus
- *L. arenosus* L.Bolus
- *L. argenteus* (L.Bolus) L.Bolus
- *L. argillosus* L.Bolus
- *L. aurantiacus* (DC.) Schwantes
- *L. aureus* (L.) N.E.Br.
- *L. austricolus* (L.Bolus) L.Bolus
- *L. baylissii* L.Bolus
- *L. berghiae* (L.Bolus) L.Bolus
- *L. bicolor* (L.) N.E.Br.
- *L. blandus* (Haw.) Schwantes
- *L. borealis* L.Bolus
- *L. brachyandrus* (L.Bolus) N.E.Br.
- *L. brevistamineus* (L.Bolus) L.Bolus
- *L. brownii* (Hook.f.) N.E.Br.
- *L. caespitosus* (L.Bolus) N.E.Br. var. *caespitosus*
- *L. caespitosus* (L.Bolus) N.E.Br. var. *luxurians* (L.Bolus) H.Jacobsen
- *L. calcaratus* (Wolley-Dod) N.E.Br. [E]
- *L. candidus* L.Bolus
- *L. capillaceus* (L.Bolus) L.Bolus
- *L. caudatus* L.Bolus
- *L. cedarbergensis* (L.Bolus) L.Bolus
- *L. ceriseus* (L.Bolus) L.Bolus
- *L. citrinus* (L.Bolus) L.Bolus
- *L. coccineus* (Haw.) N.E.Br.
- *L. comptonii* (L.Bolus) N.E.Br. var. *comptonii* forma *comptonii*
- *L. comptonii* (L.Bolus) N.E.Br. var. *comptonii* forma *roseus* (L.Bolus) G.D.Rowley
- *L. comptonii* (L.Bolus) N.E.Br. var. *comptonii* forma *angustifolius* (L.Bolus) L.Bolus
- *L. compressus* L.Bolus
- *L. conspicuus* (Haw.) N.E.Br.
- *L. convexus* (L.Bolus) L.Bolus
- *L. copiosus* (L.Bolus) L.Bolus
- *L. coralliflorus* (Salm-Dyck) N.E.Br.
- *L. creber* L.Bolus
- *L. curvifolius* (Haw.) N.E.Br. var. *curvifolius*
- *L. curvifolius* (Haw.) N.E.Br. var. *minor* (Salm-Dyck) G.D.Rowley
- *L. cyathiformis* (L.Bol.) N.E.Br.
- *L. debilis* (Haw.) N.E.Br.
- *L. deflexus* (Aiton) N.E.Br.
- *L. densifolius* (L.Bolus) L.Bolus
- *L. densipetalus* L.Bolus
- *L. dependens* (L.Bolus) L.Bolus
- *L. diffusus* (L.Bolus) N.E.Br.
- *L. dilutus* N.E.Br.
- *L. diutinus* (L.Bolus) N.E.Br.
- *L. dregeanus* (Sond.) N.E.Br.
- *L. dulcis* (L.Bolus) L.Bolus
- *L. dunensis* (Sond.) L.Bolus [E]
- *L. edwardsiae* (L.Bolus) L.Bolus
- *L. egregius* (L.Bolus) L.Bolus
- *L. elegans* (Jacq.) Schwantes
- *L. emarginatoides* (Haw.) N.E.Br.
- *L. emarginatus* (L.) N.E.Br. var. *emarginatus*
- *L. emarginatus* (L.) N.E.Br. var. *puniceus* (Jacq.) Schwantes
- *L. ernestii* (L.Bolus) L.Bolus
- *L. esterhuyseniae* L.Bolus
- *L. excedens* (L.Bolus) L.Bolus
- *L. eximius* L.Bolus
- *L. explanatus* (L.Bolus) N.E.Br. [V]
- *L. falcatus* (L.) N.E.Br. var. *falcatus*
- *L. falcatus* (L.) N.E.Br. var. *galpiniae* (L.Bolus) L.Bolus
- *L. falciformis* (Haw.) N.E.Br. var. *falciformis*
- *L. falciformis* (Haw.) N.E.Br. var. *maritimus*

L. falciformis

L. blandus

L. emarginatus

L. saturatus

Pink form of *L. haworthii*

L. austricolus

(L.Bolus) L.Bolus
L. fergusoniae (L.Bolus) L.Bolus var. *fergusoniae*
L. fergusoniae (L.Bolus) L.Bolus var. *crassistigma* L.Bolus
L. filicaulis (Haw.) N.E.Br. [V]
L. flexifolius (Haw.) N.E.Br.
L. foliosus L.Bolus
L. formosus (Haw.) N.E.Br.
L. framesii (L.Bolus) N.E.Br.
L. francisci L.Bolus
L. fugitans L.Bolus [K]
L. furvus (L.Bolus) N.E.Br.
L. galpiniae (L.Bolus) L.Bolus [R]
L. glaucoides (Haw.) N.E.Br.
L. glaucus (L.) N.E.Br. var. *glaucus*
L. glaucus (L.) N.E.Br. var. *tortuosus* (Haw.) Schwantes
L. globosus (L.Bolus) L.Bolus
L. glomeratus (L.) N.E.Br.
L. godmaniae (L.Bolus) L.Bolus var. *godmaniae*
L. godmaniae (L.Bolus) L.Bolus var. *grandiflorus* (L.Bolus) L.Bolus
L. gracilipes (L.Bolus) N.E.Br. forma *gracilipes*
L. gracilipes (L.Bolus) N.E.Br. forma *luxurians* L.Bolus
L. guthriae (L.Bolus) N.E.Br.
L. hallii (L.Bolus) N.E.Br.
L. haworthii (Donn) N.E.Br.
L. henricii (L.Bolus) N.E.Br.
L. hiemalis (L.Bolus) L.Bolus
L. hoerleinianus (Dinter) Friedrich
L. holensis L.Bolus
L. hollandii (L.Bolus) L.Bolus
L. hurlingii (L.Bolus) L.Bolus
L. imbricans (Haw.) N.E.Br.
L. immelmaniae (L.Bolus) N.E.Br.
L. inaequalis (Haw.) N.E.Br.
L. inconspicuus (Haw.) Schwantes
L. incurvus (Haw.) Schwantes
L. intervallaris L.Bolus
L. laetus (L.Bolus) L.Bolus
L. lavisii (L.Bolus) L.Bolus var. *lavisii*
L. lavisii (L.Bolus) L.Bolus var. *concinnus* L.Bolus
L. laxifolius (L.Bolus) N.E.Br.
L. leightonii (L.Bolus) L.Bolus
L. leipoldtii (L.Bolus) L.Bolus
L. leptaleon (Haw.) N.E.Br.
L. leptosepalus (L.Bolus) L.Bolus
L. lewisiae (L.Bolus) L.Bolus
L. liberalis (L.Bolus) L.Bolus
L. littlewoodii L.Bolus
L. longisepalus (L.Bolus) L.Bolus
L. longistamineus (L.Bolus) L.Bolus
L. lunatus (Willd.) N.E.Br.
L. lunulatus (A.Berger) L.Bolus
L. macrocarpus (A.Berger) N.E.Br.
L. macrosepalus (L.Bolus) L.Bolus
L. macrostigma L.Bolus
L. magnificus (L.Bolus) N.E.Br.
L. marcidulus N.E.Br.
L. mariae (L.Bolus) L.Bolus
L. martleyi (L.Bolus) L.Bolus
L. maturus N.E.Br.
L. matutinus (L.Bolus) N.E.Br.
L. maximilianii (Schltr. & A.Berger) L.Bolus
L. meleagris L.Bolus
L. microsepalus L.Bolus
L. microstigma (L.Bolus) N.E.Br.
L. middlemostii (L.Bolus) L.Bolus
L. montaguensis (L.Bolus) L.Bolus
L. monticolus (L.Bolus) L.Bolus
L. mucronatus L.Bolus
L. multiradiatus (Jacq.) N.E.Br.
L. multiseriatus (L.Bolus) N.E.Br.
L. mutans (L.Bolus) N.E.Br.
L. nardouwensis (L.Bolus) L.Bolus
L. nelii L.Bolus
L. neostayneri L.Bolus
L. obconicus (L.Bolus) L.Bolus
L. occultans L.Bolus
L. ornatus L.Bolus
L. paarlensis L.Bolus
L. pakhuisensis (L.Bolus) L.Bolus
L. palustris (L.Bolus) L.Bolus
L. parcus N.E.Br.
L. paardebergensis (L.Bolus) L.Bolus
L. pauciflorus (L.Bolus) N.E.Br.
L. paucifolius (L.Bolus) N.E.Br.
L. peacockiae (L.Bolus) L.Bolus [E]
L. peersii (L.Bolus) N.E.Br.
L. perreptans L.Bolus
L. persistens (L.Bolus) L.Bolus
L. piquetbergensis (L.Bolus) L.Bolus
L. plautus N.E.Br.
L. pleniflorus L.Bolus
L. plenus (L.Bolus) L.Bolus
L. pocockiae (L.Bolus) N.E.Br.
L. polyanthon (Haw.) N.E.Br.

L. sauerae

L. variabilis

Orange form of L. aureus

Yellow form of L. aureus

L. explanatus

L. aurantiacus

L. praecipitatus (L.Bolus) L.Bolus
L. prasinus L.Bolus
L. primivernus (L.Bolus) L.Bolus
L. productus (Haw.) N.E.Br. var. *productus*
L. productus (Haw.) N.E.Br. var. *lepidus* (Haw.) Schwantes
L. productus (Haw.) N.E.Br. var. *purpureus* (L.Bolus) L.Bolus
L. prominulus (L.Bolus) L.Bolus
L. promontorii (L.Bolus) N.E.Br. [V]
L. proximus L.Bolus
L. purpureus L.Bolus
L. rabiesbergensis (L.Bolus) L.Bolus
L. recurvus (L.Bolus) Schwantes
L. reptans (Aiton) N.E.Br. [E]
L. roseus (Willd.) Schwantes
L. rubroluteus (L.Bolus) L.Bolus
L. rupestris (L.Bolus) N.E.Br.
L. rustii (A.Berger) N.E.Br. [R]
L. salicolus (L.Bolus) L.Bolus
L. salteri (L.Bolus) L.Bolus
L. saturatus (L.Bolus) N.E.Br.
L. sauerae (L.Bolus) L.Bolus
L. scaber (L.) N.E.Br. [V]
L. schlechteri (Zahlbr.) N.E.Br. [Ex]
L. serpens (L.Bolus) L.Bolus [V]
L. simulans L.Bolus
L. sociorum (L.Bolus) N.E.Br. [E]
L. sparsiflorus L.Bolus
L. spectabilis (Haw.) N.E.Br.
L. spiniformis (Haw.) N.E.Br.
L. staminodiosus (L.Bolus) Schwantes
L. stanfordiae L.Bolus
L. stayneri (L.Bolus) N.E.Br.
L. steenbergensis (L.Bolus) L.Bolus
L. stenopetalus (L.Bolus) N.E.Br.
L. stenus (Haw.) N.E.Br. [R]
L. stephanii (Schwantes) Schwantes
L. sternens L.Bolus
L. stipulaceus (L.) N.E.Br.
L. stoloniferus L.Bolus
L. suavissimus (L.Bolus) L.Bolus var. *suavissimus*
L. suavissimus (L.Bolus) L.Bolus var. *suavissimus* forma *fera* (L.Bolus) L.Bolus
L. suavissimus (L.Bolus) L.Bolus var. *oculatus* (L.Bolus) L.Bolus
L. subaequalis (L.Bolus) L.Bolus
L. sublaxus L.Bolus

L. subrotundus L.Bolus
L. subtruncatus L.Bolus var. *subtruncatus*
L. subtruncatus L.Bolus var. *wupperthalensis* L.Bolus
L. superans (L.Bolus) L.Bolus
L. swartbergensis (L.Bolus) N.E.Br.
L. tegens (F. Muell.) N.E.Br. [E]
L. tenuifolius (L.) N.E.Br. [V]
L. tenuis L.Bolus
L. thermarum (L.Bolus) L.Bolus
L. tulbaghensis (A.Berger) L.Bolus
L. turbinatus (Jacq.) N.E.Br.
L. uncus (L.Bolus) Schwantes
L. vallis-gratiae (Schlechter & A.Berger) N.E.Br.
L. vanheerdei L.Bolus
L. vanputtenii L.Bolus
L. vanzijliae (L.Bolus) N.E.Br. [Ex]
L. variabilis (Haw.) N.E.Br.
L. verecundus (L.Bolus) L.Bolus
L. vernalis (L.Bolus) L.Bolus
L. vernicolor (L.Bolus) L.Bolus
L. versicolor (Haw.) L.Bolus
L. viatorus (L.Bolus) N.E.Br.
L. villiersii (L.Bolus) L.Bolus
L. violaceus (DC.) Schwantes
L. virgatus L.Bolus
L. vredenburgensis L.Bolus
L. walgateae L.Bolus
L. watermeyeri (L.Bolus) N.E.Br.
L. woodburniae (L.Bolus) N.E.Br.
L. wordsworthiae (L.Bolus) N.E.Br.
L. zeyheri (Salm-Dyck) N.E.Br.

LITERATURE

JACOBSEN, H. 1977. *Lexicon of succulent plants*. Part II. Family Mesembryanthemaceae. Pp. 489-501. Blandford, Poole.

GLEN, H.F. 1978. A taxonomic monograph of *Lampranthus* and allied genera (Mesembryanthemaceae). Unpublished Ph.D. dissertation, University of Cape Town.

DEHN, M. 1992. Untersuchungen zum Verwandtschaftskreis der Ruschiinae (Mesembryanthemaceae Fenzl). *Mitteilungen aus dem Institut für Allgemeine Botanik, Hamburg* 24: 91–198.

L. bicolor in its habitat

Flower and fruit capsules of *L. bicolor*

L. reptans

White form of *L. haworthii*

L. filicaulis

Lampranthus sp. (Karoo Gardens, Worcester)

Namaquanthus

DERIVATION OF GENUS NAME The name is derived from the word Namaqualand, where this plant grows, and the Greek word *anthos* (flower).

COMMON NAMES Plants are commonly known as *namakwavygie*.

DESCRIPTION The plants are shrublets of up to about 300 mm in height, with woody stems and short internodes and are covered by persistent, dead leaves. The leaves are almost cylindrical, fused at their bases into a sheath and up to 60 mm in length. Leaf surfaces are smooth. The large flowers are pink to purple, terminal on the flower stalk, with two bracts subtending each solitary flower. There are four sepals of unequal sizes. The outer petals are spoon-shaped and the inner ones become shorter towards the centre of the flower, where the purple stamens are crowded into a group. The nectar glands are pink and arranged in a continuous ring. Fruit capsules have eight to 16 locules. They are woody and rounded, with valve wings and covering membranes. Seeds are large, brown and coarsely hairy.

DISTINGUISHING CHARACTERS *Namaquanthus* is recognised by its compact growth and the pairs of partially fused, finger-like leaves; its brilliant pink to magenta flowers which do not open fully, with spoon-shaped outer petals which become shorter towards the centre of the flower; and its hairy seeds, a rare character otherwise only found in *Braunsia*, *Astridia* and *Antegibbaeum*. Fruits with nine to 10 locules are common.

FLOWERING TIME Flowering takes place during winter (July to August in South Africa). Flowers open in the morning and close in the evening, but the centre portions never open completely.

DISTRIBUTION AND ECOLOGY *Namaquanthus* is restricted to a small area northwest of Springbok in Namaqualand, Northern Cape Province, South Africa. Plants occur on a rocky hill in succulent karoo among sandstone rocks, where it is locally abundant. In habitat leaves are partly covered with black lichens. Rainfall is mainly in winter and ranges from 150 to 200 mm per year.

CULTIVATION Plants thrive in containers in a sandy-loam soil with ample compost. They are easily cultivated from seed which should preferably be older than a year. Sowing should take place in autumn. Outside of the habitat it is best grown in a greenhouse. Keep dry in summer. Plants are not often cultivated but the flowers are very showy. It is a slow grower and a long-lived perennial.

NOTES The blackness of the leaves was mentioned in the original description. It is now known that the colour is caused by minute lichens which also grow on other smooth-leaved species of *Nelia*, *Cephalophyllum* and *Jordaaniella*.

Namaquanthus L.Bolus

NUMBER OF SPECIES/SUBSPECIES/VARIETIES (1/0/0)

SPECIES LIST AND CONSERVATION STATUS
N. vanheerdei L.Bolus

LITERATURE
BOLUS, H.M.L. 1954. *Notes on Mesembryanthemum and allied genera*. University of Cape Town, Cape Town. Vol. 3: 257.

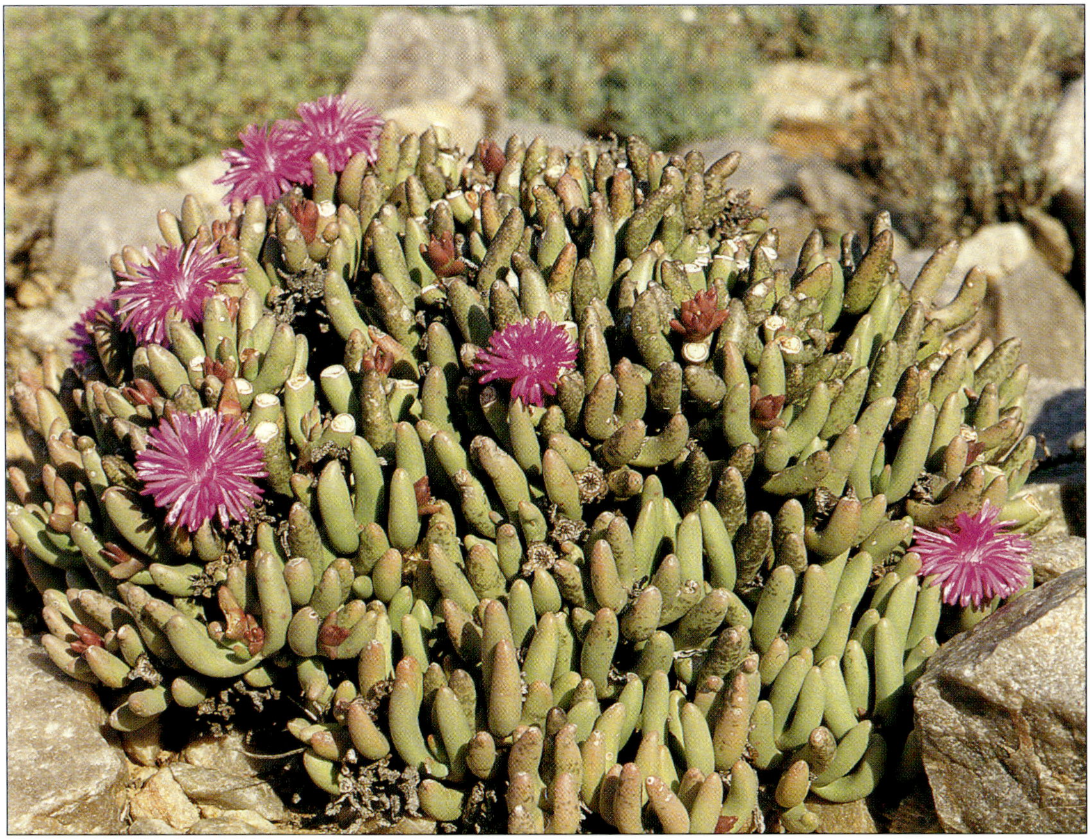
Peter Bruyns

N. vanheerdei in its habitat

NBI

Leaves of *N. vanheerdei*

Graham Williamson

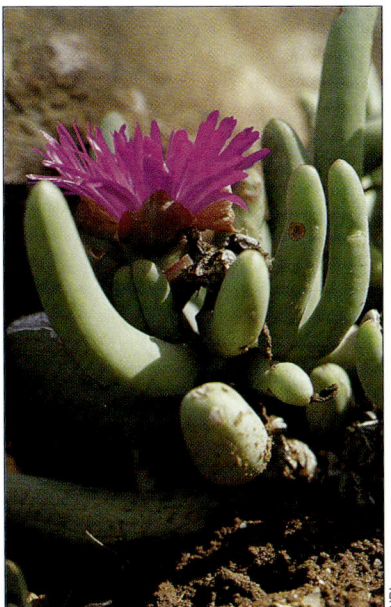

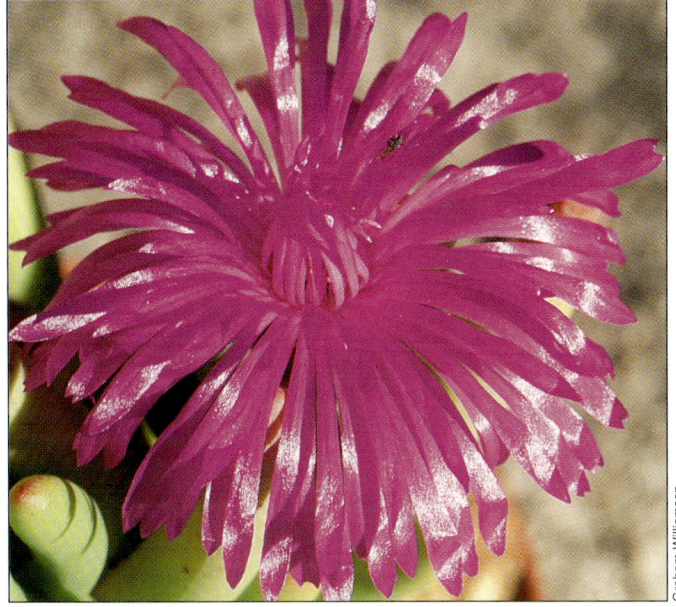

Flower of *N. vanheerdei*

OSCULARIA

DERIVATION OF GENUS NAME The name is derived from the Latin word *osculum* (small mouth). *Oscularia* means a group of small mouths and pertains to the toothed leaves of *O. deltoides*.

COMMON NAMES The name *sandsteenvygie* (sandstone mesemb) has been recorded.

DESCRIPTION Plants have reddish stems and are small shrubs with erect or spreading branches. The short, greyish green to pale blue waxy leaves are three-angled and vary from sickle to club-shaped. They are borne in opposite pairs, the leaves in a pair being somewhat dissimilar in size and united at their bases. The leaf margins and keels are prominently red-toothed in one species. The smallish white to pink flowers are borne singly or in multi-flowered clusters, mostly on stalks, each with two bracts. Numerous stamens and staminodes are collected into a cone. All these structures are hairy near their bases. The nectar glands are arranged in a circle and are dark green. There are between four and seven (mostly five) stigmas. Fruit capsules also have between four and seven locules, with the expanding keels diverging from their bases. Covering membranes are present, but closing bodies are lacking. The seeds are brown, egg-shaped and have rough surfaces.

DISTINGUISHING CHARACTERS The genus is distinguished by its greyish green, heavily wax-covered, three-angled, club-shaped to broadly sickle-shaped leaves. The flowers are generally borne in dense clusters and they are characteristically almond-scented.

FLOWERING TIME Flowering is from midwinter to midsummer (June to December in South Africa). The flowers open in the morning and close at night.

DISTRIBUTION AND ECOLOGY Species of *Oscularia* typically grow in fynbos vegetation of the Western Cape Province, South Africa, extending into the Northern Cape Province. A single species, *O. deltoides*, has been collected just east of Albertinia in the Western Cape Province. *Oscularia* species favour rocky outcrops of quartzitic sandstone and grow in acidic soil. Rainfall is mainly in winter and ranges from 300 to 1 500 mm per year.

CULTIVATION Plants are easily cultivated and are particularly suitable for sunny rockeries and steep embankments in fynbos gardens. They produce large numbers of beautiful white to pink flowers and are suitable as a ground cover among rocks, but will also grow happily in containers. Plants should be renewed every two to five years when they become untidy. Cuttings can be rooted in pure sand or in the required beds and they grow rapidly. Seed should be sown during autumn in sandy, acid soil.

NOTES *Oscularia deltoides* is widely used in horticulture, especially for its striking reddish teeth on the grey leaves. Most species have toothless, halfmoon or club-shaped leaves, always with a white wax layer.

Oscularia Schwantes

NUMBER OF SPECIES/SUBSPECIES/VARIETIES (3/0/1)

SPECIES LIST AND CONSERVATION STATUS
  O. caulescens (Mill.) Schwantes
  O. deltoides (L.) Schwantes var. *deltoides*
  O. deltoides (L.) Schwantes var. *major* (Weston) Schwantes
  O. pedunculata (N.E.Br.) Schwantes

LITERATURE
  GLEN, H.F. 1978. A taxonomic monograph of *Lampranthus* and allied genera (Mesembryanthemaceae). Unpublished Ph.D. dissertation, University of Cape Town.

  GLEN, H.F. 1980. Proposal (537) to conserve the generic name 2405 *Lampranthus* N.E.Br. (1930) against *Oscularia* Schwantes (1927) (Mesembryanthemaceae). *Taxon* 29: 693–694.

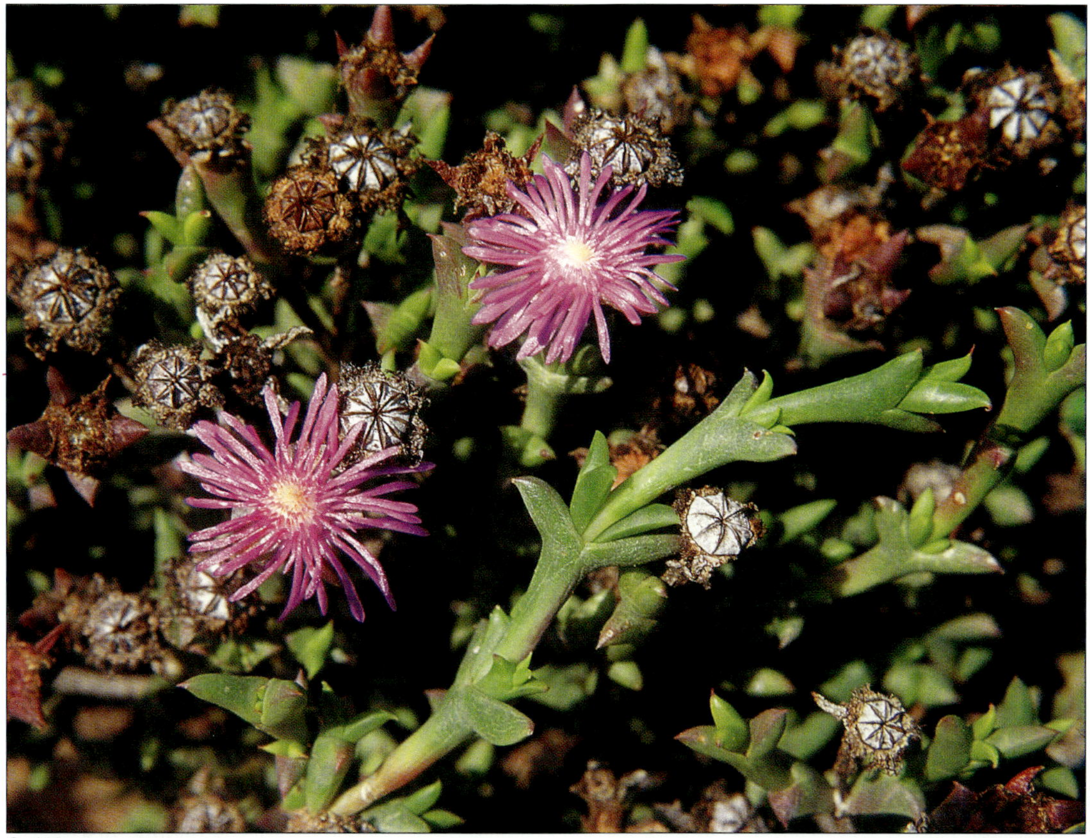
Ben-Erik van Wyk

S. viride

Ben-Erik van Wyk

Leaves of *S. viride*

Ernst van Jaarsveld

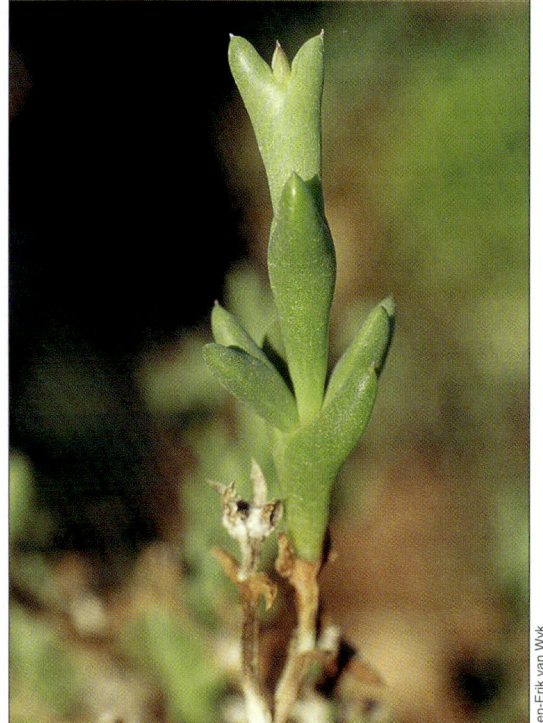

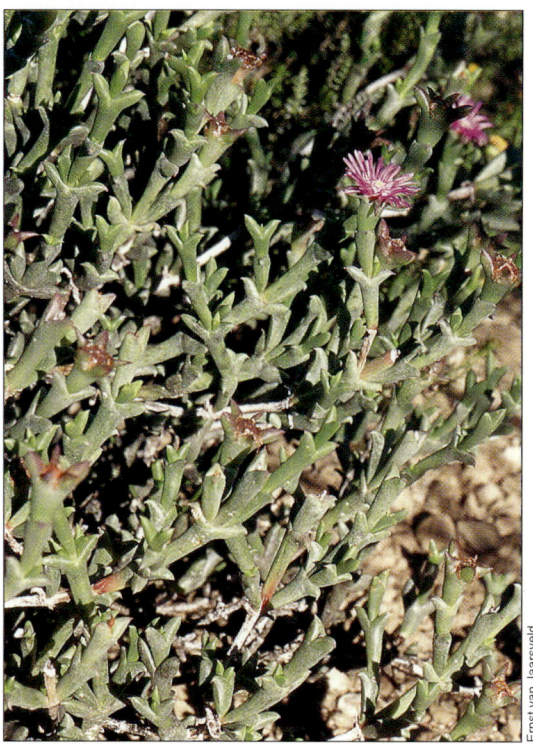

S. viride in its habitat

Wooleya

Derivation of genus name The genus is named in honour of Major C.H.F. Woolley, who for many years contributed succulents from various parts of South Africa to Kirstenbosch.

Common names The genus is known locally as *vaalvygie* (grey mesemb).

Description Plants are moderately branched, sprawling shrubs up to 200 mm in height. The young branches are ash-grey to purplish becoming grey-brown. Woody stems can grow up to 10 mm in diameter in older plants. The ash-grey to purplish leaves are obscurely three-angled to cylindrical with the upper surface slightly flattened and the lower surface convex. They are fused towards their bases, with the keel and margins reduced to reddish lines and the tip ending in a point. The white flowers are solitary, borne on a short stalk. The four sepals are unequal, with the outer ones quite rounded and the inner ones without membranous margins. There are numerous stamens with few filamentous staminodes gathered into a cone around the stamens. The nectar glands are arranged in a continuous ring. Fruit capsules have 11 to 12 locules and are woody, reddish at first, and convex on top. The valves have spreading wings. Covering membranes are present, but closing bodies are absent. Seeds are smooth.

Distinguishing characters Plants are distinguished by their leaves, which look as if they have been dusted with flour; the robust but sprawling habit; the solitary white flowers; and the multi-locular fruit capsules without closing bodies.

Flowering time Plants flower mostly in winter (March to August in South Africa). As with so many coastal mesembs, flowering is erratic. The flowers open in the morning and close in the evening.

Distribution and ecology *Wooleya* is restricted to the coastal plains of Namaqualand, Northern Cape Province, South Africa. They grow in sandy soil in the succulent karoo where rainfall is mainly in winter and ranges from 50 to 100 mm per year.

Cultivation Plants thrive in cultivation but are best grown in containers under controlled conditions. They are readily propagated from seed sown during autumn. Germination is within a month. Cuttings root well in sand. Plants are slow growers but should be regrown from cuttings or seeds when they eventually become untidy. Keep dry during the summer months.

Notes It is curious that Major Woolley's surname was correctly spelled when he discovered *Haworthia woolleyi*, yet the name of this genus was originally published in the form given here.

Wooleya L.Bolus

Number of species/subspecies/varieties (1/0/0)

Species list and conservation status
W. farinosa (L.Bolus) L.Bolus

Literature
BOLUS, H.M.L. 1961. Notes on *Mesembryanthemum* and allied genera. *Journal of South African Botany* 27: 48.

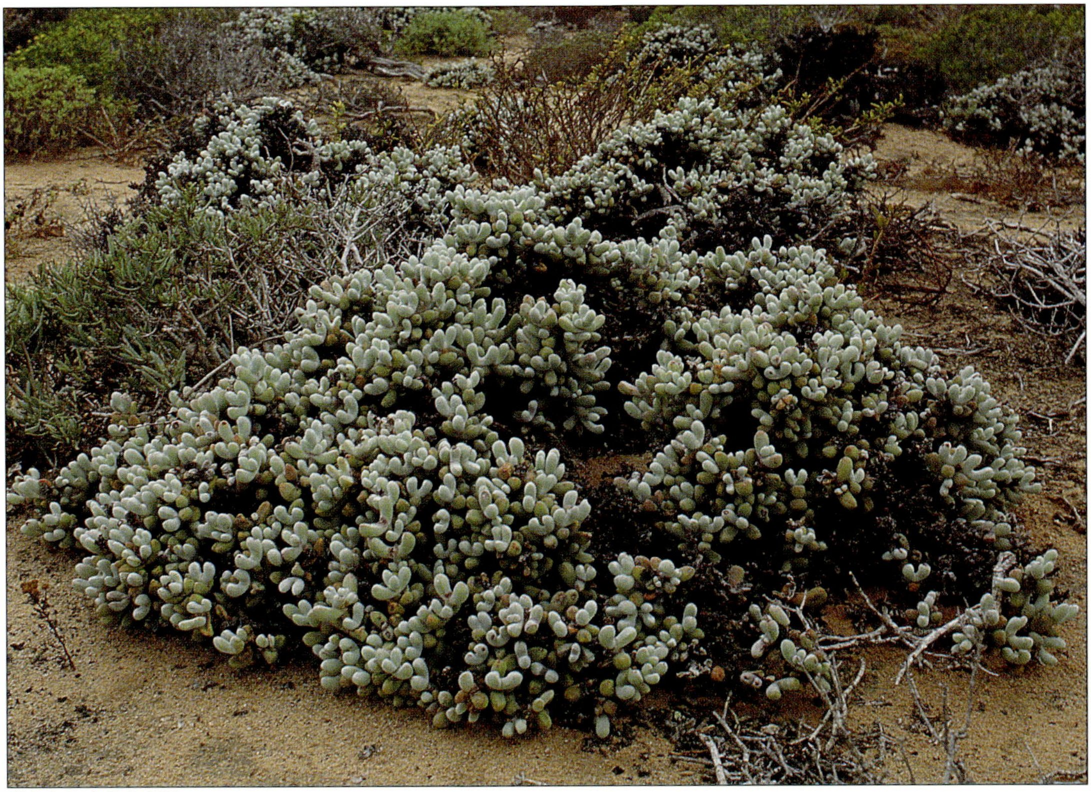
Ernst van Jaarsveld

W. farinosa in its habitat

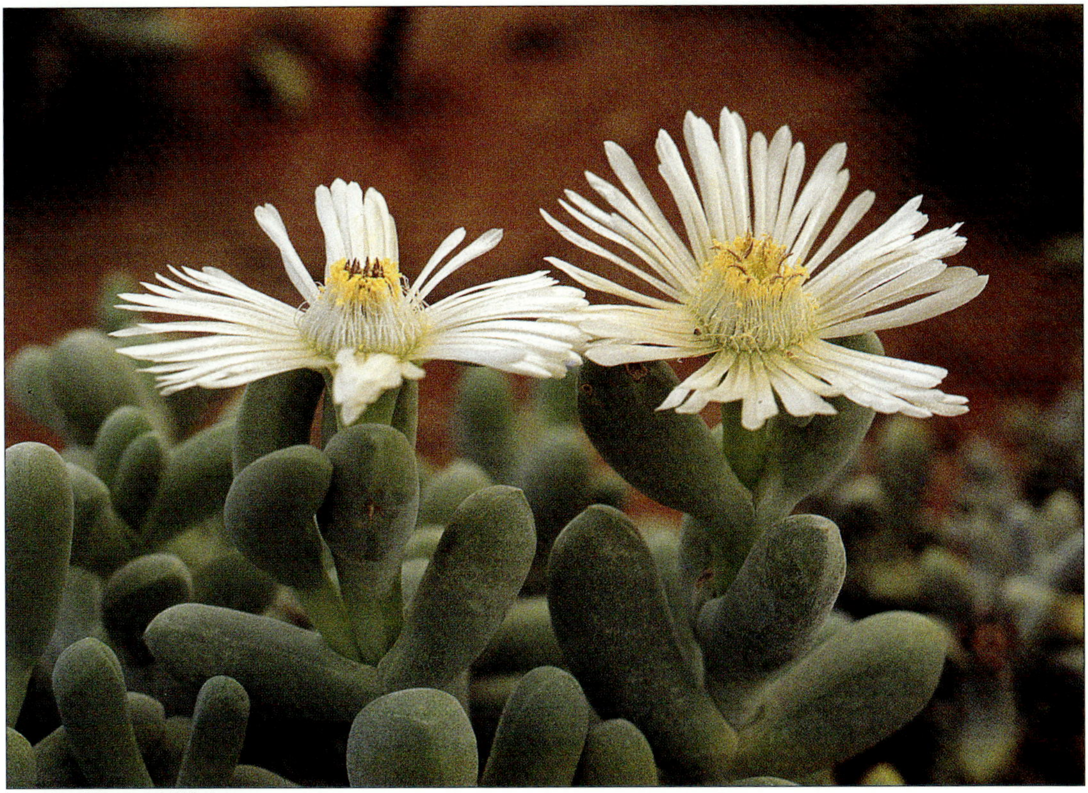
Ernst van Jaarsveld

Leaves and flowers of *W. farinosa*

Ruschia schollii

Ruschia caroli

Ruschia spinosa

Stayneria neilii

Ruschia-like shrubby Mesembs

GROUP 13

Shrubby plants with creeping or erect stems with characteristic fruit capsules of the *Ruschia*-type.

Ruschia-type fruit are easily recognised by the consistent absence of valve wings (except in *Eberlanzia*) in combination with small, rod-shaped closing bodies and stout, convex covering membranes over the locules. These usually have upright rims and closing rodlets at their distal extremities. This fruit type is usually five-locular.

RUSCHIA-LIKE SHRUBBY MESEMBS comprise 8 genera and 241 species.

Arenifera (4 species)	*Ruschia* (220 species)
Astridia (7 species)	*Ruschianthemum* (1 species)
Eberlanzia (1 species)	*Stayneria* (1 species)
Polymita (2 species)	*Stoeberia* (5 species)

Arenifera

Derivation of genus name The name is derived from the Latin words *arena* (sand) and *ferre* (to carry), pertaining to the sticky leaves carrying sand.

Common names The names s*andvygie* and *plakkertjie* (little sticker) have been recorded.

Description The plants are small, succulent shrublets of up to 150 mm in height, with reddish brown, densely branched stems. The leaves are three-sided, slightly fused at their bases and the two of each pair tend to be unequal in size and shape. Young leaves are green and sticky, and become increasingly covered by adhering dust and sand with age. Older leaves are grey and waxy, with rough, warty surfaces. Flowers occur in threes, but the side flowers sometimes develop slower than the middle one. There are only four sepals, of which the inner pair has broad membranous margins. Numerous pale rose to light violet petals occur in two whorls. The stamens are surrounded by numerous filamentous staminodes. The ovary is slightly bulging on top, with six to eight stigmas and dark green nectar glands fused into a ring. The fruit capsules have six to eight locules and are of the *Ruschia*-type, but the valve wings are woody and cup-like. The dry remains of the stigmas persist, forming small peaks in the centre of the fruits. The valves are transversely ribbed, with two conspicuous, recurved teeth on the tip. Covering membranes half cover the locules, with a prominent distal closing rodlet. The closing bodies are rodlet-shaped and conspicuous. The light yellowish brown seeds are pear-shaped and each of them is finely ribbed along its length.

Distinguishing characters The plants are many-branched shrublets bearing reddish brown stems with small, sticky, sand-carrying, three-sided leaves. The woody fruit capsules have six to eight locules and the valve margins are prominent, forming a central column.

Flowering time Flowering occurs in autumn and winter. The flowers open at midday.

Distribution and ecology *Arenifera* species occur in the Northern and Western Cape Provinces of South Africa, from the Richtersveld to Calvinia in the Northern Cape, and the Knersvlakte to Matjiesfontein in the Western Cape. They grow mainly in succulent karoo in the winter rainfall region and also on the margin of the winter and summer rainfall areas. The habitat consists of sandy flats and rocky hillsides in full sunlight. Rainfall varies from 75 to 200 mm per year.

Cultivation *Arenifera* plants are easily cultivated in a greenhouse and require full sun to light shade. They thrive in containers, growing from autumn to spring and then entering a rest phase in summer, when they should be kept dry. Propagation is by seed or cuttings in autumn.

Notes The genus was originally considered to comprise only one species, *Arenifera pillansii*, which has a localised distribution. Three more species have recently been added to *Arenifera*.

Arenifera A.G.J.Herre

Number of species/subspecies/varieties (4/0/0)

Species list and conservation status
 A. *pillansii* (L.Bolus) A.G.J.Herre
 A. *pungens* H.E.K.Hartmann
 A. *spinescens* (L. Bolus) H.E.K. Hartmann
 A. *stylosa* (L. Bolus) H.E. K. Hartmann

Literature
HARTMANN, H.E.K. 1996. Miscellaneous taxonomic notes on Aizoaceae. *Bradleya* 14: 29–56.

A. pillansii

Leaves and flowers of *A. pillansii*

A. stylosa

Flowers of *A. pillansii*

ASTRIDIA

DERIVATION OF GENUS NAME The genus is named after Mrs Astrid Schwantes, wife of Professor G. Schwantes.

COMMON NAMES No vernacular names have been recorded.

DESCRIPTION The plants are robust, woody, erect shrublets, 300 to 600 mm in height, with the erect or spreading leaves widely spaced so that the stems are visible. The large greyish green or blue-grey leaves are slightly three-sided, smooth or velvety, with the leaf bases extending down along the stem, forming a short sheath around it. When old leaves break off, a portion of the leaf base always adheres to the stem. The large flowers of 40 to 75 mm in diameter occur solitarily or in clusters of three, but the lateral ones develop only after a while. The flower stalk is short, with boat-shaped bracts directly below the flowers. The six sepals are unequal, the inner ones narrower, with membranous margins. Petals vary in colour from white, lemon, orange, pink to various shades of purple and red; the centres of the flower are often differently coloured to the rest. The outer petals are longer and occur in one or two whorls; the inner ones are narrower and form two or three whorls. Staminodes may be present or absent. The fruit capsules have six locules with distinctive raised ridges on top. They are of the *Ruschia*-type, with expanded valves widely spreading and wingless and with well-developed closing bodies and covering membranes. There may be two additional cone-shaped closing devices on the lower distal surfaces of the covering membranes. The seeds' surfaces may be rough, with minute, sharp-pointed tubercles, or may be covered with long, brown hairs.

DISTINGUISHING CHARACTERS Plants are easily recognised by their distinctive appearance: the large, single, bird-pollinated flowers with visible nectar, and the greyish green, often velvety, half-moon-shaped leaves. Also distinctive are the exceptionally large seeds, the sharp-pointed hairs on the seed surfaces (not unique to *Astridia*), the boat-shaped bracts directly below the flowers and the robust fruit capsule with its six locules.

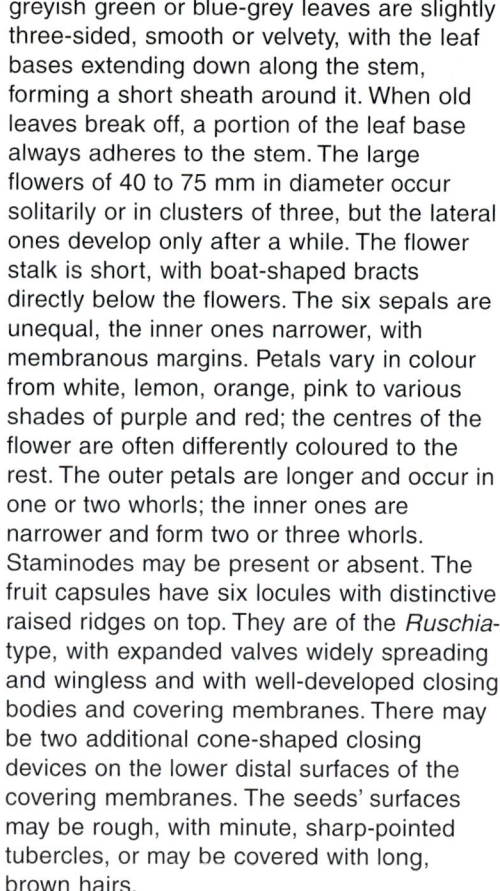

FLOWERING TIME Plants flower mainly in late summer (from January to March). The flowers open in the daytime.

DISTRIBUTION AND ECOLOGY The geographical distribution of the genus extends across the border between South Africa and Namibia, along the lower reaches of the Orange River. This is an extremely arid area.

CULTIVATION *Astridia* plants rarely thrive, except under specialised greenhouse conditions. Plants grow easily from cuttings and seeds germinate readily.

NOTES The ridges on top of the fruit capsules are not always visible but they may help to distinguish *Astridia* species from other mesembs with similar robust leaves. Flower colour varies considerably, for example *A. longifolia* varies from white to yellow to brick red, or even magenta within a few kilometres at the famous Helskloof in the Richtersveld.

Astridia Dinter

NUMBER OF SPECIES/SUBSPECIES/VARIETIES (7/0/0)

SPECIES LIST AND CONSERVATION STATUS
A. *citrina* (L.Bolus) L.Bolus [R]
A. *hallii* L.Bolus
A. *herrei* L.Bolus [K]
A. *longifolia* (L.Bolus) L.Bolus
A. *speciosa* L.Bolus [I]
A. *vanheerdei* L.Bolus [K]
A. *velutina* Dinter & Schwantes

LITERATURE
GLEN, H.F. 1986. Numerical taxonomic studies in the subtribe Ruschiinae (Mesembryanthemaceae) — *Astridia, Acrodon* and *Ebracteola. Bothalia* 16: 203–226.

A. longifolia

A. speciosa

White form of *A. longifolia*

A. citrina

Leaves and fruit of *A. hallii*

Flower of *A. hallii*

EBERLANZIA

DERIVATION OF GENUS NAME The genus was named in honour of the late Mr F. Eberlanz of Lüderitz, Namibia.

COMMON NAMES No vernacular names seem to be recorded. The common name *doringvygie* applies only to the well-known thorny species which are now included in the genus *Ruschia* (see below).

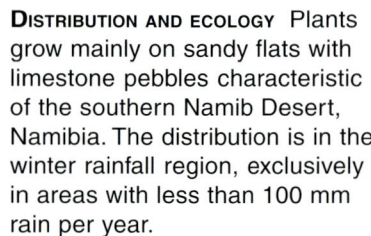

FLOWERING TIME Flowering occurs throughout the year with a peak season in winter.

DISTRIBUTION AND ECOLOGY Plants grow mainly on sandy flats with limestone pebbles characteristic of the southern Namib Desert, Namibia. The distribution is in the winter rainfall region, exclusively in areas with less than 100 mm rain per year.

DESCRIPTION The plants are erect to flat-growing thorny shrubs of up to 300 mm in height (they are often broader than they are tall). The stems are almost white or light yellow. The greyish green leaves are short and fat with rounded sides, tipped with an abrupt point. The flower clusters are richly branched and form blunt spines, with relatively small flowers which may or may not have petals. Small bracts are found on the stalks of the flowers. The stamens are grouped together into a central cone. The fruit capsules have five locules, are bell-shaped at the bottom and convex at the top, with high valve rims and broad, rectangular valve wings which taper to the base and the tip. The closing bodies are small and the covering membranes are complete, with distal closing rodlets. The amber-coloured seeds are less than 1 mm long, with a smooth surface.

DISTINGUISHING CHARACTERS The inflated appearance of the leaves with their nipple-like tips are useful characters to distinguish *Eberlanzia* species. The branched flower clusters are very spiny, but the arrangement of spines is irregular and their number increases during flowering and fruit ripening. The spines are noticeably blunt to the touch, a feature easily recognised in the field. Other distinguishing features are the glandular leaf surfaces, tapering valve wings, and the lack of petals (in *E. sedoides*).

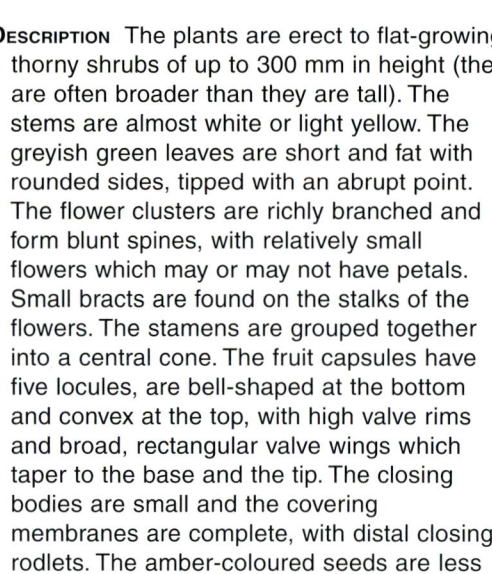

CULTIVATION Plants of this genus are not cultivated.

NOTES *Eberlanzia* is a distinct genus, close to *Ruschia* in that it has a deep capsule base, small closing bodies and closing rodlets, but the broad, rectangular valve wings exclude it from *Ruschia*. The genus name was used for many spiny mesembs (26 names were used). Presently only one spiny species remains in *Eberlanzia* and most of the species known in the past as *Eberlanzia* are now members of *Ruschia* subgenus *Spinosae*. Some new species of *Eberlanzia* have been discovered, but these have not yet been formally described. An example is illustrated here.

Eberlanzia Schwantes

NUMBER OF SPECIES/SUBSPECIES/VARIETIES (1/0/0 + some undescribed species)

SPECIES LIST AND CONSERVATION STATUS
E. *sedoides* (Dinter & A.Berger) Schwantes

LITERATURE
HARTMANN, H.E.K. & STÜBER, D. 1993. On the spiny Mesembryanthema and the genus *Eberlanzia* (Aizoaceae). *Contributions from the Bolus Herbarium* 15: 1–75.

HARTMANN, H.E.K. 1996. Miscellaneous taxonomic notes on Aizoaceae. *Bradleya* 14: 29–56.

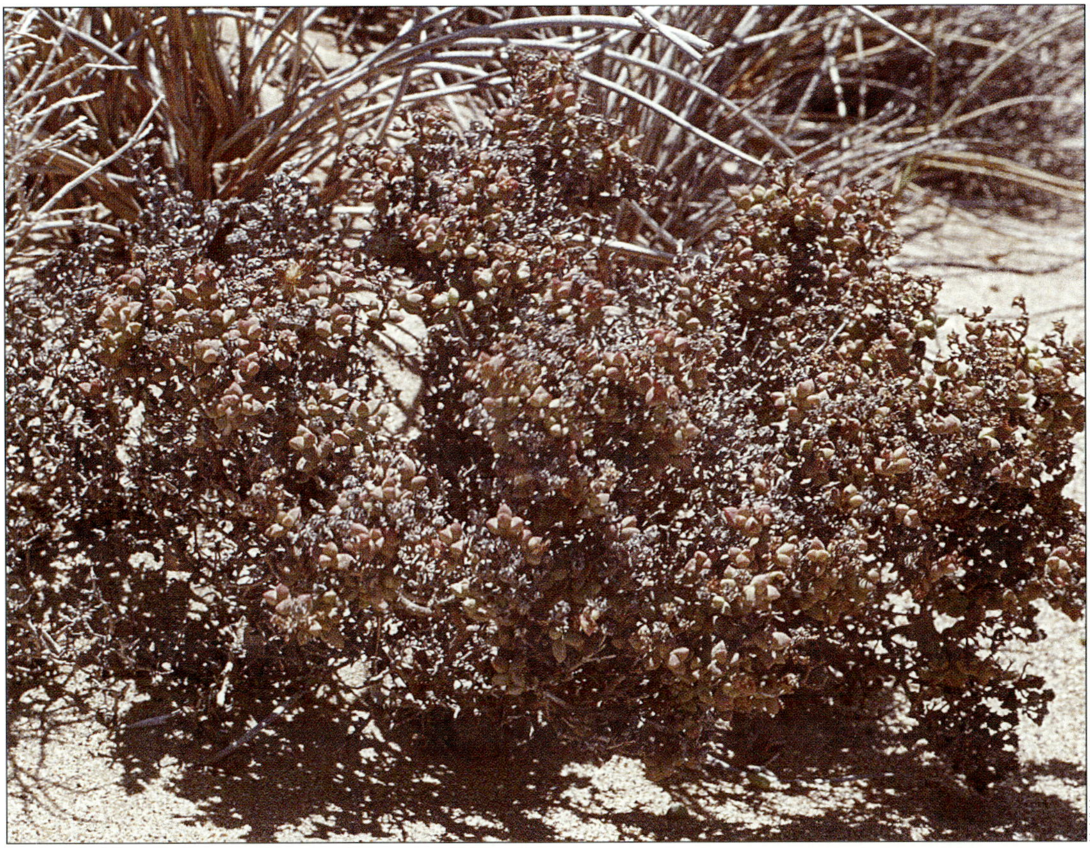

Heidi Hartmann

E. sedoides

Antje Burke

Eberlanzia sp.

Antje Burke

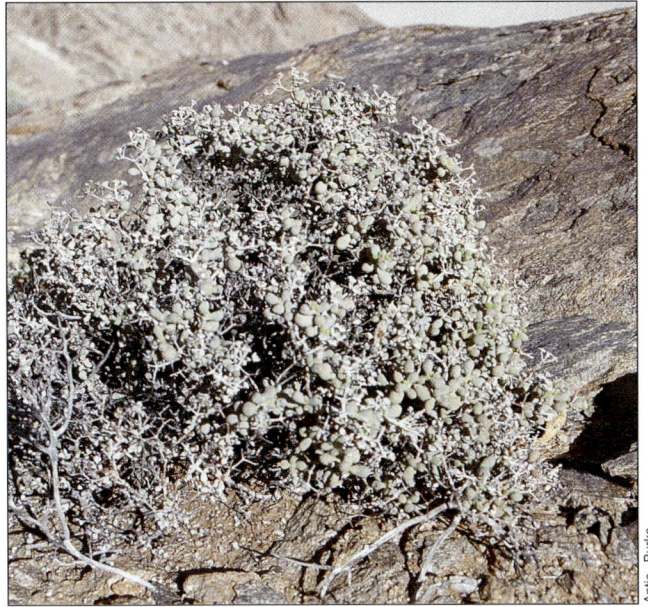

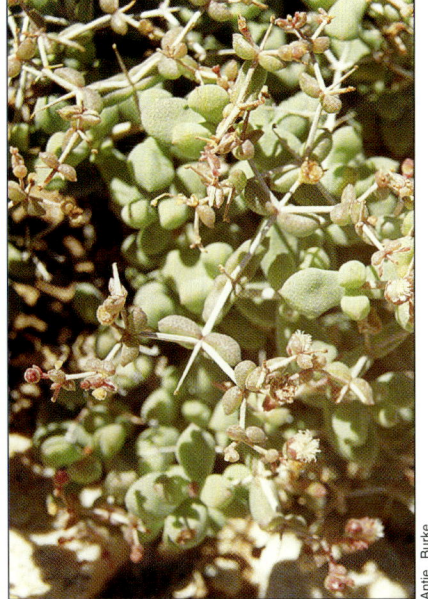

E. sedoides

POLYMITA

DERIVATION OF GENUS NAME The name is derived from the Greek words *poly* (many) and *mita* (very small petals).

COMMON NAMES No vernacular names seem to have been recorded.

DESCRIPTION These mesembs are hairless shrubs of up to 400 mm in height, with peculiar leaves which are fused for about half their length. The three-sided leaves have longitudinal lines at their edges, the tips are widely divergent, ending in sharp tips and the margins and lower edge (keel) are somewhat horny, often toothed. The flowers occur singly or in clusters, with bracts enclosing the lower part of the flower. The white petals are long and thin, often occurring in five groups. The stamens are curved inwards and are concealed by the longer and more numerous white filamentous staminodes which occur in three to four whorls. There are eight to 10 (or up to 18) awl-shaped stigmas. The fruit capsules are hard and stiffly erect, with the top elevated and white, and the bases more or less funnel-shaped. They are close to the *Ruschia*-type, but with valve wings and rather shallow seed cavities. The expanding keels diverge abruptly towards their tips, and have broad wings. Covering membranes are straight, without a rim at the top, and the closing bodies are large and white (*P. albiflora*) or smaller and dark (*P. steenbokensis*). Seeds are smooth.

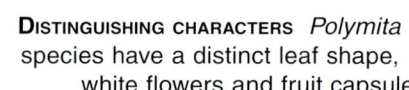

DISTINGUISHING CHARACTERS *Polymita* species have a distinct leaf shape, white flowers and fruit capsules with up to 18 locules.

FLOWERING TIME Flowering occurs in summer (November in South Africa). The scented flowers open during the day and close at night.

DISTRIBUTION AND ECOLOGY The plants occur in northern Namaqualand in the Northern Cape Province, South Africa, where they grow on quartzite slopes or in shallow pockets on granites.

CULTIVATION *Polymita* thrives in cultivation and can be grown in containers or in rockeries. However, out of its range, plants should be grown in a greenhouse.

NOTES The recently described *P. steenbokensis* was for a long time confused with *P. albiflora*, but it is a more modest shrublet.

Polymita N.E.Br.

NUMBER OF SPECIES/SUBSPECIES/VARIETIES (2/0/0)

SPECIES LIST AND CONSERVATION STATUS
P. albiflora (L.Bolus) L.Bolus
P. steenbokensis H.E.K.Hartmann

LITERATURE
HARTMANN, H.E.K. 1996. Miscellaneous taxonomic notes on Aizoaceae. *Bradleya* 14: 29–56.

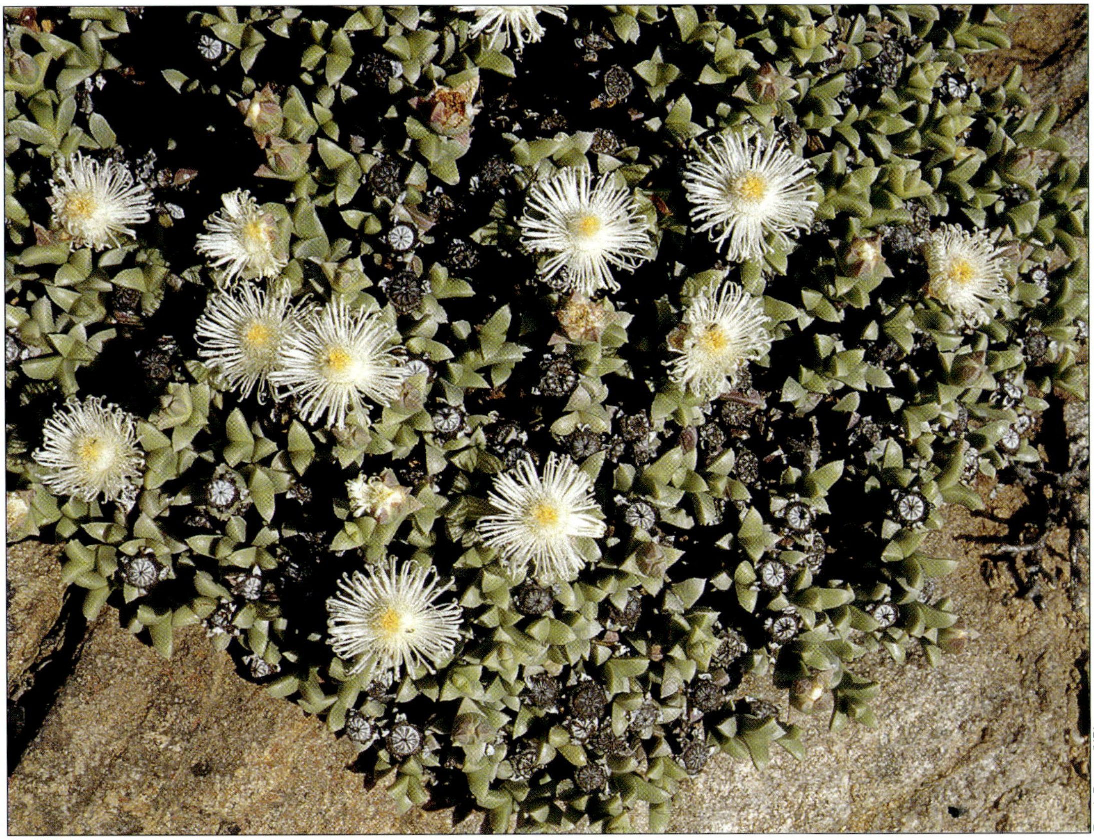

Frank Brusse/NBI

P. albiflora

Ernst van Jaarsveld

P. albiflora

Ernst van Jaarsveld

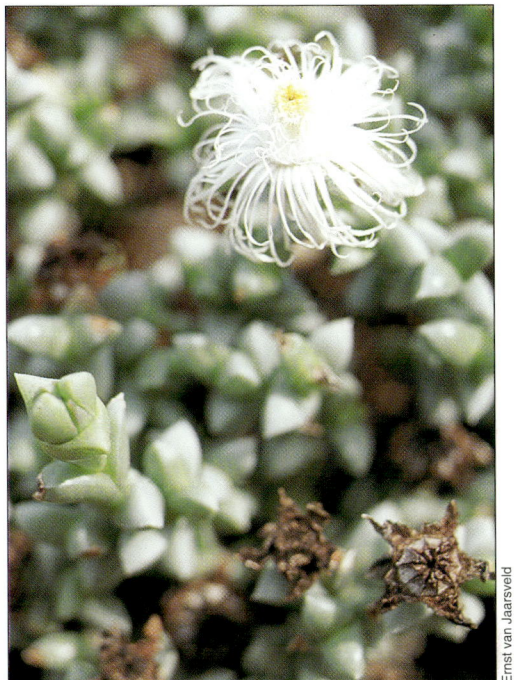

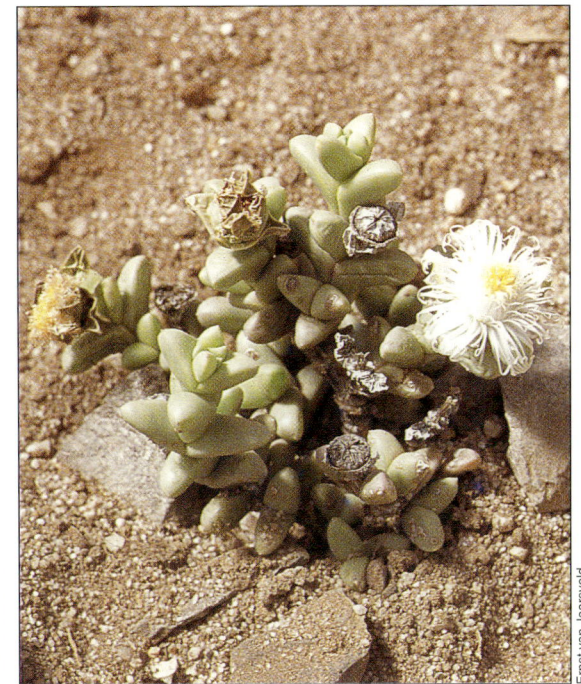

P. steenbokensis

Ruschia

DERIVATION OF THE GENUS NAME The genus was named after the Namibian farmer Ernst Rusch.

COMMON NAMES The spiny members of this genus are called *doringvygie*, *steekdoring* or similar names, all referring to the spines. *Beesvygie* (cattle mesemb) has been used for *R. canonotata* and *R. hamata*, while *R. unidens* is known as *rooibergvygie* (red mountain mesemb) or *swaelstert(tjie)vygie* (swallow tail mesemb). *Muisvygie* (mouse mesemb) is used for tufted, creeping forms.

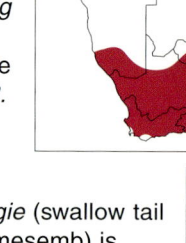

DESCRIPTION Plants are shrubby with mostly erect, rarely curving or creeping branches. Five species possess spines. The leaves are mostly three-sided, sometimes with teeth along the edges. The flowers occur solitarily or in clusters and the flower stalks have small bracts below the flower. The petals are pink to purple, rarely white and are sometimes arranged in five bundles. Filamentous staminodes surround the central cone of stamens. Fruit capsules have five or rarely six locules, a deep capsule base and no valve wings. They mostly have small, hook-shaped closing bodies, rarely larger ones or sometimes none. The expanding keels are rather short and spread sideways, opening the valves into an upright position in most cases. The covering membranes usually have distinct closing rodlets at their top ends.

DISTINGUISHING CHARACTERS The small, hook-shaped closing bodies and the short, tangentially arranged expanding keels in most fruit capsules are the main differences between *Ruschia* and *Antimima*. Almost 100 species of *Ruschia* have been transferred to *Antimima*. *Ruschia* species usually have fairly small, purple flowers, in contrast to the multi-coloured species of *Lampranthus*, a genus with which they may be confused.

FLOWERING TIME Flowering takes place virtually throughout the year, but a peak can be expected in spring and autumn. Flowers are diurnal and sweet-smelling.

DISTRIBUTION AND ECOLOGY The genus has a very wide distribution from near Helmeringhausen in Namibia and eastward in South Africa to Vereeniging in Gauteng and to Harrismith in the Free State, down to the western and southern coasts, also including the mountains of Lesotho. The highest frequency of species is found in the drier southwestern regions near the sea with less than 100 mm rainfall per year, mainly in winter. In the high lying localities in the east, high summer rainfall of up to 800 mm per year occurs. The plants may tolerate frost and fires.

CULTIVATION Species of *Ruschia* thrive in cultivation and are frequently used in gardens, but less so than the spectacular species of *Lampranthus*, because they are less showy. Since some *Ruschia* species are hardy, they are well suited for rockeries and embankments. Some species are very floriferous.

NOTES The following species are frequently encountered in gardens: *Ruschia viridifolia* with its very green leaves, *R. caroli* native to the environs of Cape Town, the creeping *R. sarmentosa* with its bicoloured petals, *R. intrusa* with its flaky wax layer, and *R. maxima* with its enormous half-moon-shaped leaves. After the removal of numerous species to *Acrodon*, *Amphibolia*, *Antimima*, *Eberlanzia* and *Ebracteola*, only 220 of the more than 400 described species remain in *Ruschia* at present, some of which will have to be transferred to other genera once the revisions are completed.

Ruschia Schwantes

NUMBER OF SPECIES/SUBSPECIES/VARIETIES (220/1/0)

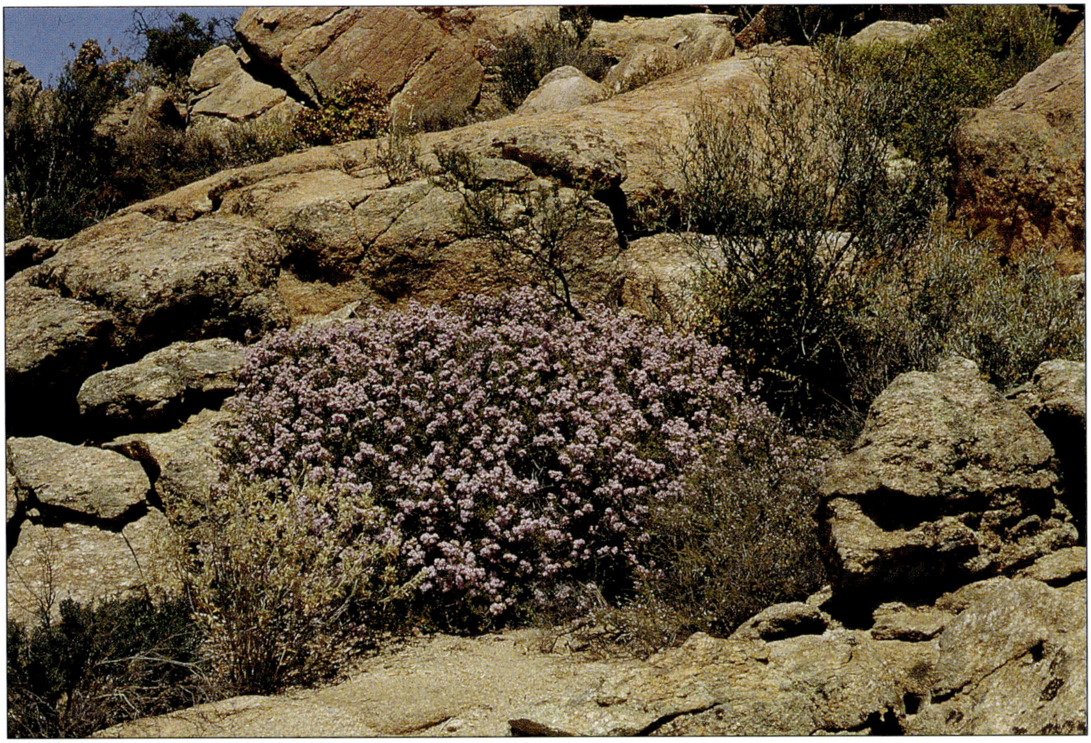

Frank Brusse/NBI

R. laxiflora

Frank Brusse/NBI

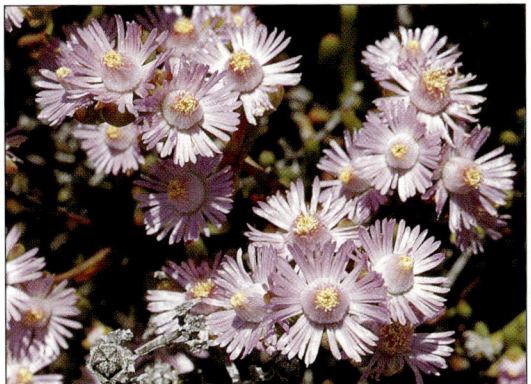

Flowers of *R. laxiflora*

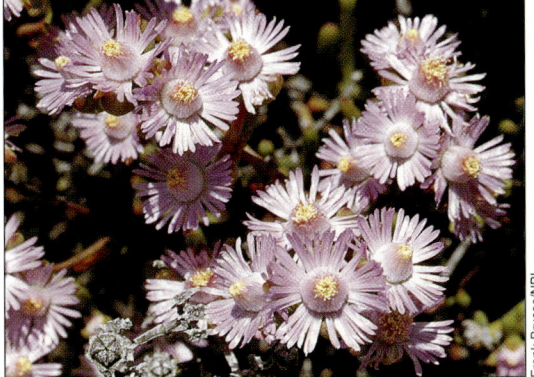

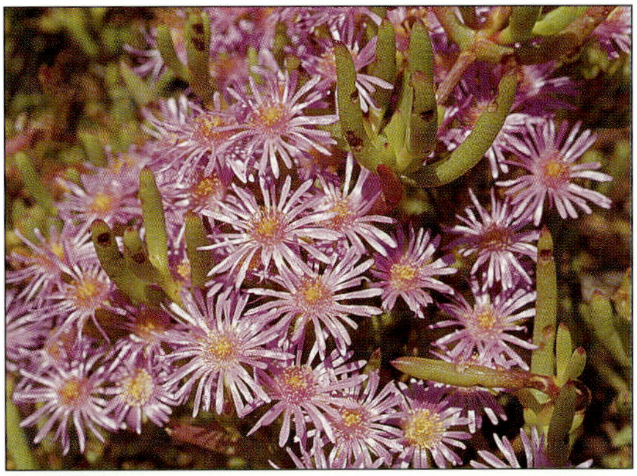

NBI

R. viridifolia

NBI

R. sarmentosa

NBI

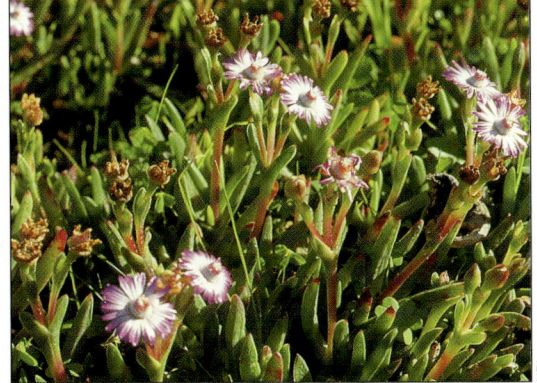

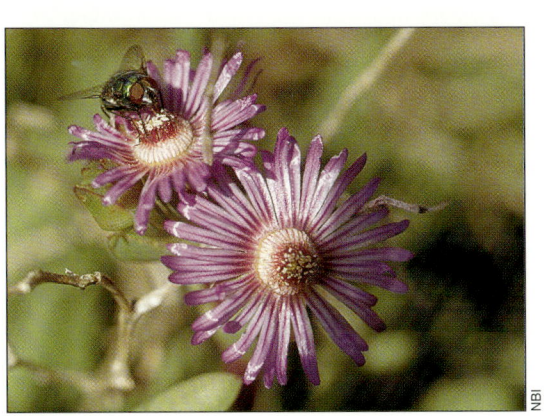

R. caroli

Species list and conservation status

R. abbreviata L.Bolus
R. acocksii L.Bolus
R. acuminata L.Bolus
R. acutangula (Haw.) Schwantes
R. aggregata L.Bolus
R. alata L.Bolus
R. altigena (L.Bolus) L.Bolus
R. amicorum (L.Bolus) Schwantes [R]
R. ampliata L.Bolus
R. approximata (L.Bolus) Schwantes
R. archeri L.Bolus
R. aristulata (Sond.) Schwantes
R. aspera L.Bolus
R. atrata L.Bolus
R. axthelmiana (Dinter) Schwantes
R. barnardii L.Bolus
R. beaufortensis L.Bolus
R. bijliae L.Bolus
R. bipapillata L.Bolus
R. bolusiae Schwantes
R. brakdamensis (L.Bolus) L.Bolus
R. breekpoortensis L.Bolus
R. brevibracteata L.Bolus
R. brevicyma L.Bolus
R. brevifolia L.Bolus
R. brevipes L.Bolus
R. britteniae L.Bolus
R. burtoniae L.Bolus
R. calcarea L.Bolus
R. calcicola (L.Bolus) L.Bolus
R. callifera L.Bolus
R. campestris (Burch.) Schwantes
R. canonotata (L.Bolus) Schwantes
R. capornii (L.Bolus) L.Bolus
R. caroli (L.Bolus) Schwantes
R. caudata L.Bolus
R. cedarbergensis L.Bolus
R. centrocapsula H.E.K.Hartmann & Stüber
R. ceresiana L.Bolus
R. cincta (L.Bolus) L.Bolus
R. clavata L.Bolus
R. complanata L.Bolus
R. congesta (Salm-Dyck) L.Bolus
R. copiosa L.Bolus
R. costata L.Bolus
R. cradockensis (O.Kuntze) H.E.K.Hartmann & Stüber subsp. cradockensis
R. cradockensis (O.Kuntze) H.E.K.Hartmann & Stüber subsp. triticiformis (L.Bolus) H.E.K.Hartmann & Stüber
R. crassa (L.Bolus) Schwantes
R. crassisepala L.Bolus
R. cupulata (L.Bolus) Schwantes
R. curta (Haw.) Schwantes
R. cymbifolia (Haw.) L.Bolus
R. cymosa (L.Bolus) Schwantes
R. decumbens L.Bolus
R. decurrens L.Bolus
R. decurvans L.Bolus
R. dejagerae L.Bolus
R. deminuta L.Bolus
R. densiflora L.Bolus
R. depressa L.Bolus
R. dichroa (Rolfe) L.Bolus
R. dilatata L.Bolus
R. divaricata L.Bolus
R. diversifolia L.Bolus
R. duthiae (L.Bolus) Schwantes
R. edentula (Haw.) L.Bolus
R. elineata L.Bolus
R. erecta (L.Bolus) Schwantes
R. esterhuyseniae L.Bolus
R. excedens L.Bolus
R. exigua L.Bolus
R. extensa L.Bolus
R. festiva (N.E.Br.) Schwantes
R. filamentosa (L.) L.Bolus [I]
R. filipetala L.Bolus
R. firma L.Bolus [R]
R. floribunda L.Bolus
R. foliosa (Haw.) Schwantes
R. fourcadei L.Bolus
R. framesii L.Bolus
R. fredericii (L.Bolus) L.Bolus
R. fugitans L.Bolus
R. geminiflora (Haw.) Schwantes
R. glauca L.Bolus
R. goodiae L.Bolus
R. gracilipes L.Bolus
R. gracilis L.Bolus
R. griquensis (L.Bolus) Schwantes
R. grisea (L.Bolus) Schwantes
R. hamata (L.Bolus) Schwantes
R. haworthii H.Jacobsen & G.D.Rowley
R. heteropetala L.Bolus
R. hexamera L.Bolus
R. holensis L.Bolus
R. hutchinsonii L.Bolus
R. imbricata (Haw.) Schwantes
R. impressa L.Bolus
R. inclusa L.Bolus
R. inconspicua L.Bolus
R. incumbens L.Bolus
R. incurvata L.Bolus
R. indecora (L.Bolus) Schwantes

Leaves of *R. indurata*

Flowers of *R. indurata*

R. spinosa

R. spinosa showing stem thorns

R. intrusa

R. senaria

R. indurata (L.Bolus) Schwantes
R. insidens L.Bolus
R. intermedia L.Bolus
R. intricata (N.E.Br.) H.E.K.Hartmann & Stüber
R. intrusa (Kensit) L.Bolus
R. jacobseniana L.Bolus
R. karrachabensis L.Bolus
R. karrooica (L.Bolus) L.Bolus
R. kenhardtensis L.Bolus
R. klipbergensis L.Bolus
R. knysnana (L.Bolus) L.Bolus
R. kuboosana L.Bolus
R. langebaanensis L.Bolus
R. lapidicola L.Bolus
R. lavisii L.Bolus
R. laxa (Willd.) Schwantes
R. laxiflora L.Bolus
R. laxipetala L.Bolus
R. leptocalyx L.Bolus
R. leucosperma L.Bolus
R. lerouxiae (L.Bolus) L.Bolus
R. lineolata (Haw.) Schwantes
R. lisabeliae L.Bolus
R. littlewoodii L.Bolus
R. macowanii (L.Bolus) Schwantes
R. mariae L.Bolus
R. marianae (L.Bolus) Schwantes
R. maxima (Haw.) L.Bolus
R. middlemostii L.Bolus
R. misera (L.Bolus) L.Bolus
R. mollis (A.Berger) Schwantes
R. montaguensis L.Bolus
R. muelleri (L.Bolus) Schwantes
R. muiriana (L.Bolus) Schwantes
R. multiflora (Haw.) Schwantes
R. muricata L.Bolus
R. namusmontana Friedrich
R. nana L.Bolus
R. nelii Schwantes
R. nieuwerustensis L.Bolus
R. nivea L.Bolus
R. nonimpressa L.Bolus
R. odontocalyx (Schltr. & Diels) Schwantes
R. orientalis L.Bolus
R. pallens L.Bolus
R. paripetala (L.Bolus) L.Bolus
R. parviflora (Haw.) Schwantes
R. parvifolia L.Bolus
R. patens L.Bolus
R. patulifolia L.Bolus
R. pauciflora L.Bolus
R. paucipetala L.Bolus

R. perfoliata (Mill.) Schwantes
R. phylicoides L.Bolus
R. pillansii L.Bolus
R. pinguis L.Bolus
R. polita L.Bolus
R. pollardii Friedrich
R. primosii L.Bolus
R. pulchella (Haw.) Schwantes
R. pulvinaris L.Bolus
R. pungens (A.Berger) H.Jacobsen
R. purpureostyla (L.Bolus) P.V.Bruyns
R. putterillii (L.Bolus) L.Bolus
R. radicans L.Bolus
R. rariflora L.Bolus
R. rigens L.Bolus
R. rigida (Haw.) Schwantes
R. rigidicaulis (Haw.) Schwantes
R. robusta L.Bolus
R. rostella (Haw.) Schwantes
R. rubricaulis (Haw.) L.Bolus [I]
R. rupicola (Engler) Schwantes
R. ruralis (N.E.Br.) Schwantes
R. ruschiana (Dinter) Dinter & Schwantes
R. sabulicola Dinter
R. sandbergensis L.Bolus
R. sarmentosa (Haw.) Schwantes
R. scabra H.E.K.Hartmann
R. schollii (Salm-Dyck) Schwantes
R. semidentata (Haw.) Schwantes
R. semiglobosa L.Bolus
R. senaria L.Bolus
R. serrulata (Haw.) Schwantes
R. singula L.Bolus
R. solitaria L.Bolus
R. spinosa (L.) Dehn
R. staminodiosa L.Bolus
R. stricta L.Bolus
R. strubeniae (L.Bolus) Schwantes
R. suaveolens L.Bolus
R. subpaniculata L.Bolus
R. subsphaerica L.Bolus
R. subteres L.Bolus
R. tardissima L.Bolus
R. tecta L.Bolus
R. tenella (Haw.) Schwantes
R. testacea L.Bolus
R. tribracteata L.Bolus
R. triflora L.Bolus
R. truteri L.Bolus
R. tumidula (Haw.) Schwantes
R. uitenhagensis (L.Bolus) Schwantes
R. umbellata (L.) Schwantes
R. uncinata (L.) Schwantes

R. radicans

R. diversifolia

R. robusta

R. geminiflora

Leaves and flowers of *R. maxima*

R. maxima

R. unidens (Haw.) Schwantes
R. vaginata (Haw.) Schwantes
R. valida Schwantes
R. vanbredai L.Bolus
R. vanderbergiae L.Bolus
R. vanheerdei L.Bolus
R. vanniekerkiae L.Bolus
R. versicolor L.Bolus
R. victoris (L.Bolus) L.Bolus
R. virens L.Bolus
R. virgata (Haw.) L.Bolus

R. viridifolia L.Bolus
R. vulvaria (Dinter) Schwantes
R. willdenowii Schwantes

LITERATURE
DEHN, M. 1993. Untersuchungen zum Verwandtschaftskreis der Ruschiinae (Mesembryanthemaceae). *Mitteilungen aus dem Institut für Allgemeine Botanik, Hamburg* 24: 91–198.

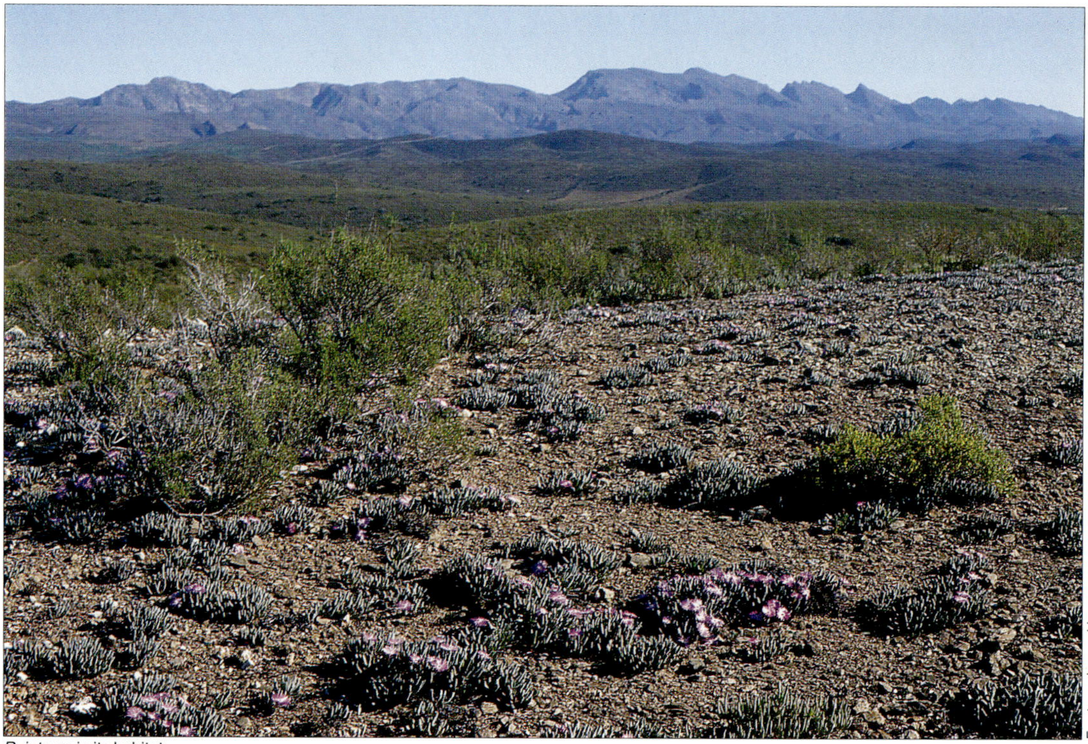

R. intrusa in its habitat

R. crassa

Ruschianthemum

Derivation of genus name The genus was named in honour of Mr Ernst Rusch; the generic name is a combination of the name Rusch and the Greek word *anthemis* (flowers).

Common names *Mandjievygie* or basket mesemb has been used for this plant.

Description The plant is an erect succulent shrub of up to 1,2 m in height, with whitish grey branches. The leaves are slightly fused at their bases and are more or less flat above and three-sided towards the tips. The small flowers occur in a much-branched cluster and each one has small bracts on the flower stalk. There are five fleshy sepals, of which the outer two are the largest. The short petals are purple to white. They occur in a single whorl and are about as long as the sepals. Stamens and filamentous staminodes are present, grouped together into a central cone. The fruit capsules have five locules and are top-shaped, but break into nutlets at maturity. Each nutlet consists of parts of the capsule which enclose a single seed. The vascular bundles in the fruit capsules form a fibrous skeleton which remains behind when the capsules have broken apart. Expanding keels have ornate, rudimentary wings, and the closing bodies are large, filling most of the seed cavities, which only have one or two seeds each.

Distinguishing characters The shrubs may be recognised by their club-shaped leaves and small whitish flowers but they are at once distinguished from all other mesembs by the mature fruit capsule which splits into nutlets. These break away, leaving behind the basket-like capsule remains. There are only one or two seeds in each seed cavity.

Flowering time The plants flower in winter and early spring (July to September in southern Africa).

Distribution and ecology *Ruschianthemum* is confined to a small area of the winter rainfall region of the lower Orange River Valley of the Northern Cape Province, South Africa, and Namibia. In this area the rainfall is only 25 to 75 mm per year. Plants are locally abundant in succulent karoo, occurring in windy, sandy spots on flats and hills, often among rocks.

Cultivation Plants thrive in cultivation. Seeds germinate readily and plants are fast growing. For the best results, seed should be sown during autumn. Provide sufficient space and sandy mineral-rich soil. Keep dry during the summer months. *Ruschianthemum* is mainly grown as a curiosity plant as the flowers are small and unattractive.

Notes The characteristic basket-like old fruit bases distinguish this mesemb at once from all other vygies.

Ruschianthemum Friedrich

Number of species/subspecies/varieties (1/0/0)

Species list and conservation status
R. gigas (Dinter) Friedrich

Literature
FRIEDRICH, H.-Chr. 1960. Mesembryanthemenstudien 1. Beitrag zur Kenntnis der Gattungen *Stoeberia* Dtr. & Schw. und *Ruschianthemum* Friedr. gen nov. *Mitteilungen der Botanischen Staatssammlung, München* 3: 554–567.

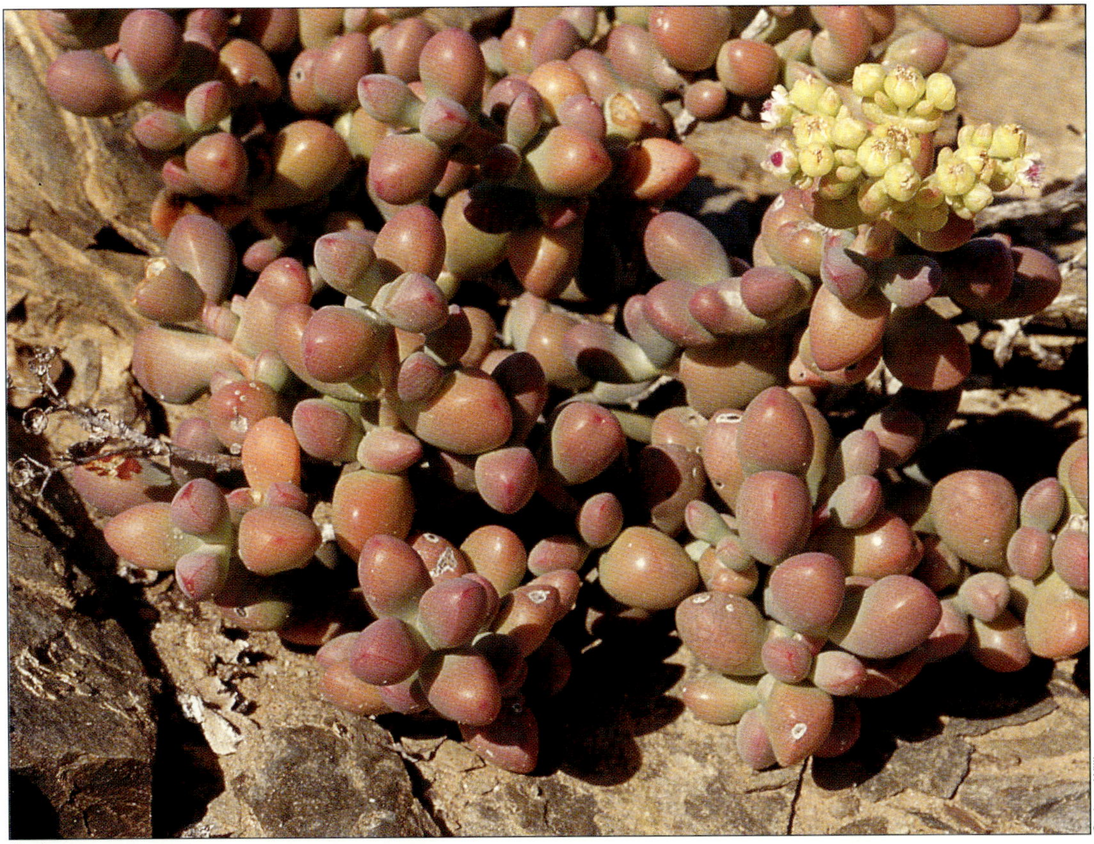

Graham Williamson

R. gigas

Ben-Erik van Wyk

Flowers of *R. gigas*

Ernst van Jaarsveld

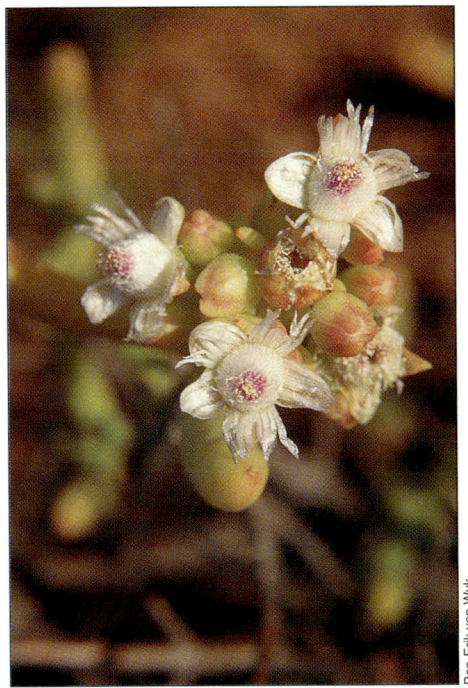

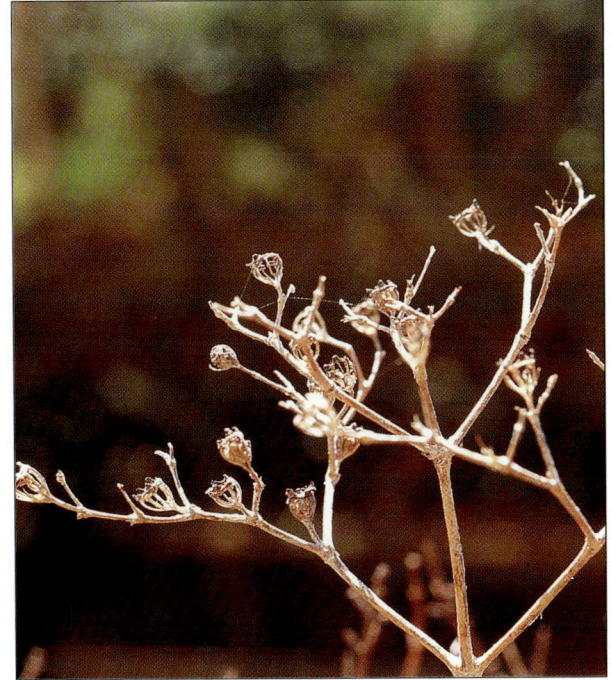

Fruit capsules of *R. gigas*

STAYNERIA

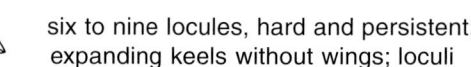

DERIVATION OF GENUS NAME This plant was named in honour of Mr Frank Stayner (1907 to 1981), a previous curator of the Karoo National Botanical Garden in Worcester, South Africa.

COMMON NAMES The name *witblomvygie* (white-flowered mesemb) has been used.

DESCRIPTION This plant is an erect, woody shrub of up to 800 mm in height, with persistent hardened remains of old leaves separated by long internodes of 30 to 40 mm in length. The stem-sheathing leaves (about 70 mm long and 8 mm wide) are three-sided, with a sharp bottom edge. Flowers usually occur in groups of three to seven on a short stalk, with bracts below the flowers. There are five or six more or less equal sepals. The white petals occur in one or two whorls. The stamens are surrounded by numerous filamentous staminodes. The number of stigmas vary from six to nine. Fruit capsules usually have eight locules, but the number varies from six to nine. They are close to the *Ruschia*-type, but the valves are fused at their bases so that they open only partially and remain open. The expanding keels are contiguous at the base and widely diverging upwards, without wings. Covering membranes are well developed. Closing bodies are absent. The seeds are egg-shaped, brown in colour and about 1 mm long.

DISTINGUISHING CHARACTERS The leaves of *Stayneria* smell like berries and flowers are small in relation to the leaf size. Plants are recognised by the following combination of characters: rigid and erect shrubs; leaves oblong and cylindrical but with a ridge along the bottom edge, light green in colour; flowers white to pink and often untidy, sometimes in groups of three; stamens grouped together into a cone; capsules with six to nine locules, hard and persistent; expanding keels without wings; loculi roofs present; closing bodies absent; once opened, the capsule remains open.

FLOWERING TIME Flowering occurs during the winter and early spring (July to September in South Africa). The flowers open by day.

DISTRIBUTION AND ECOLOGY The plant is confined to the Breede River Valley, Western Cape Province, South Africa, from Worcester in the west to McGregor in the east. It grows in succulent karoo on acid, quartzitic sandstone soil. Rainfall in this area is mainly in winter and ranges from 200 to 300 mm per year.

CULTIVATION Plants are easily propagated by seed or cuttings. They can be grown in well-drained rockeries in the winter rainfall region. Outside of its habitat *Stayneria* is best grown in containers in well-drained, slightly acid soil in a greenhouse. Keep dry in summer.

NOTES Plants are locally abundant in their natural habitat. Two species were previously recognised, namely the usual white-flowered *Stayneria neilii*, and the more showy, pink-flowered *S. littlewoodii*. The latter may well prove to be a distinct species.

Stayneria L.Bolus

NUMBER OF SPECIES/SUBSPECIES/VARIETIES (1/0/0)

SPECIES LIST AND CONSERVATION STATUS
S. *neilii* (L.Bolus) L.Bolus [K]

LITERATURE
BOLUS, H.M.L. 1961. Notes on *Mesembryanthemum* and allied genera. *Journal of South African Botany* 27: 47.

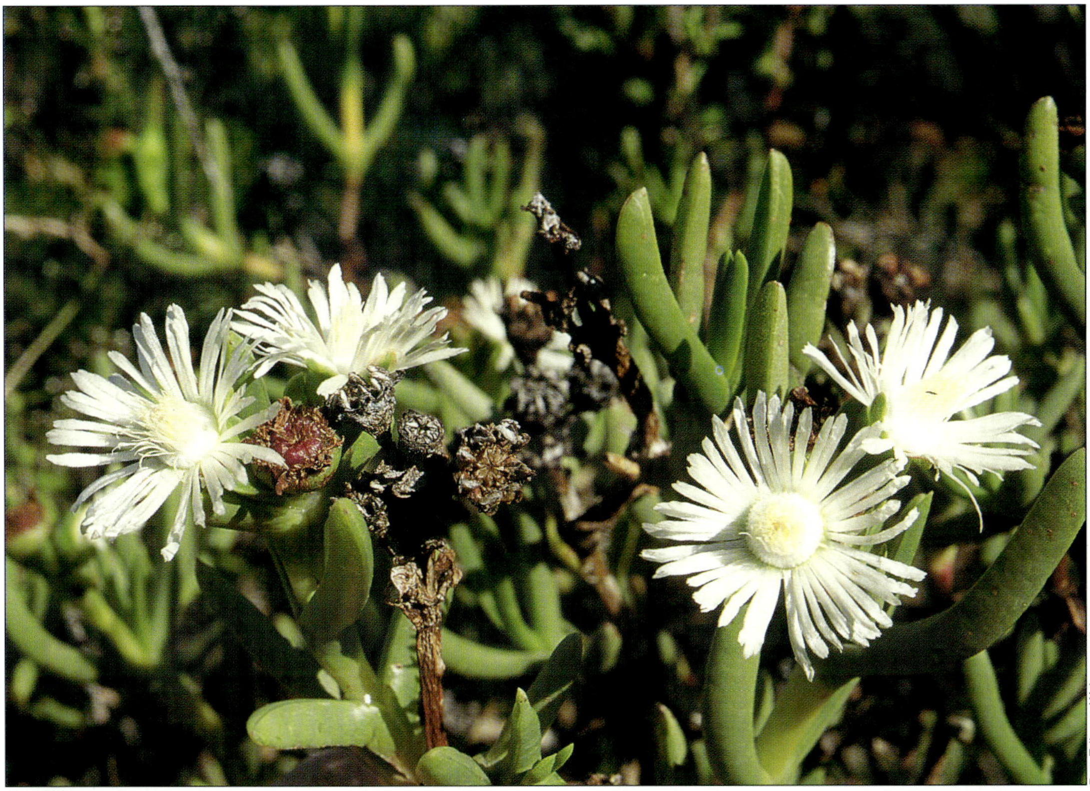

Ben-Erik van Wyk

S. neilii

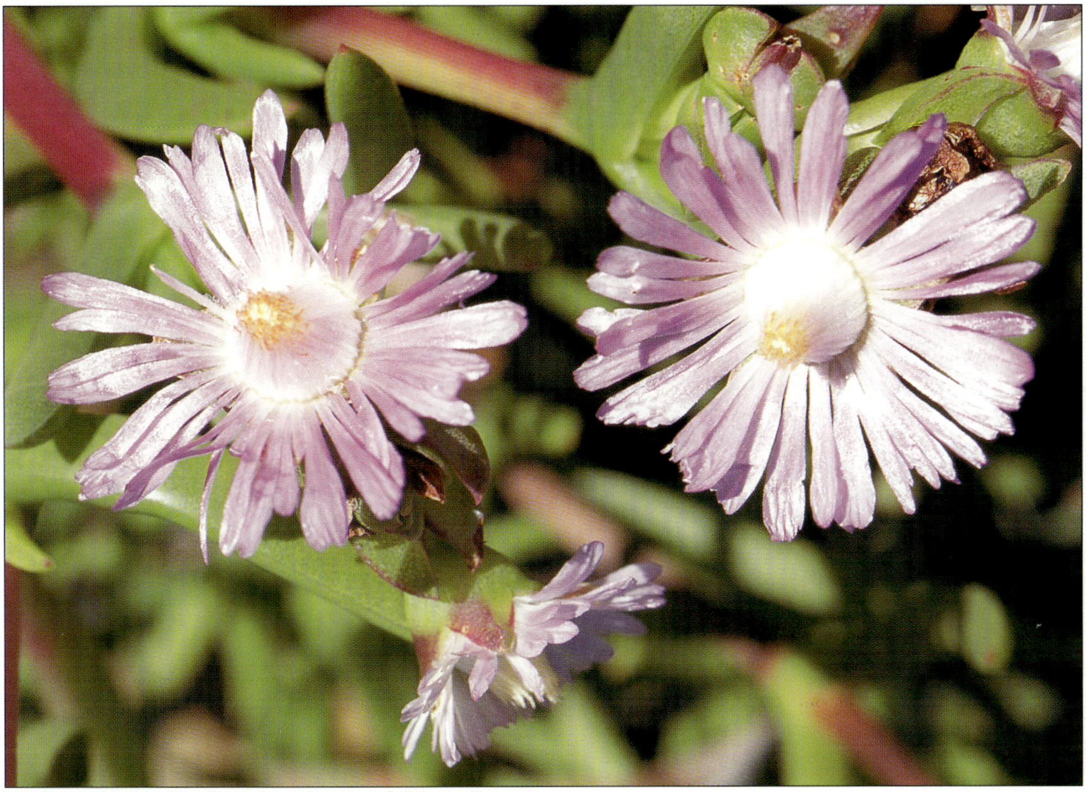

Peter Bruyns

Pink-flowered form of *S. neilii*, previously known as *S. littlewoodii*

Stoeberia

Derivation of genus name The genus was named after Mr E. Stoeber of Lüderitz, Namibia.

Common names There are several vernacular names such as *rooivye* (*S. arborea*), *donkievye* (*S. arborescens*), *vaalvye* (*S. beetzii*), *slaplootvye* (*S. carpii*), and *rooivyge* (*S. utilis*).

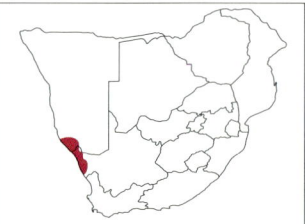

Description The plants are shrubby with succulent leaves, low and spreading to tall and erect (up to 2,5 m in height). Leaves are boat-shaped to somewhat club-shaped, with slightly flattened upper surfaces and with a ridge along the bottom, but only towards the tip. Flowers occur in much-branched clusters. There are no bracts on the flower stalks. Five or six triangular sepals are present, with short petals in a more or less single whorl and numerous filamentous staminodes and stamens. The fruit capsules have five or six locules, with valves remaining open. The valves are broad and bony in texture, valve wings are present and large closing bodies occur deep inside the seed cavities. The wind-dispersed seeds are pear-shaped with rough surfaces.

Distinguishing characters The genus may be recognised by the shrubby growth form, the capsules which remain open and the flattish, wind-dispersed seeds.

Flowering time Flowering occurs in winter and spring in southern Africa.

Distribution and ecology *Stoeberia* species are confined to Namaqualand, Northern Cape Province, South Africa, and Namibia. Rainfall is mainly in winter and ranges from 50 to 300 mm per year. Plants occur from the sandy coastal plains to mountainous regions. They are often locally abundant. *S. frutescens* is a pioneer on disturbed sites. Capsules remain open and the seeds are dispersed by wind.

Cultivation *Stoeberia* plants are easily grown from seed but are difficult to grow from cuttings. Sow seeds during autumn. Plants are rapid growers as long as plenty of light and well-drained soil are available. Keep dry in summer.

Notes *Stoeberia utilis* and *S. arborea* are the tallest of all mesembs and the only species sometimes considered to be tree-like in their growth form. They have brittle wood which is utilised locally as firewood.

Stoeberia Dinter & Schwantes emend. Friedrich; emend. Dehn

Number of species/subspecies/varieties (5/0/0)

Species list and conservation status
  S. arborea Van Jaarsv.
  S. beetzii (Dinter) Dinter & Schwantes
  S. carpii Friedrich
  S. frutescens (L.Bolus) Van Jaarsv.
  S. utilis (L.Bolus) Van Jaarsv.

Literature
FRIEDRICH, H.-Chr. 1960. Mesembryanthemenstudien 1. Beitrag zur Kenntnis der Gattungen *Stoeberia* Dtr. & Schw. und *Ruschianthemum* Friedr. gen nov. *Mitteilungen der Botanischen Staatssammlung, München* 3: 554–567.

VAN JAARSVELD, E.J. 1994. A synopsis of *Stoeberia*. *Aloe* 31: 68–76.

HARTMANN, H.E.K. 1996. Miscellaneous taxonomic notes on Aizoaceae. *Bradleya* 114: 29–56.

Flowers of *S. utilis*

S. utilis

S. beetzii

S. arborea

Leaves of *S. carpii*

Flowers of *S. carpii*

Leipoldtia schultzei

Hallianthus planus

Leipoldtia-like shrubby Mesembs

GROUP 14

More or less shrubby plants with creeping or erect stems with characteristic fruit capsules of the *Leipoldtia*-type.

Leipoldtia-type fruits are easily recognised by the large, round, often whitish, stalked closing bodies at the exit of the locules. Valve wings are broad and covering membranes concave, unlike the previous types, and quite stiff. They are distinctly recurved at their distal margins. Fruit capsules in this group have five to many locules and rounded tops.

LEIPOLDTIA-LIKE SHRUBBY MESEMBS comprise 5 genera and 23 species.

Hallianthus (1 species)	*Ottosonderia* (2 species)
Leipoldtia (11 species)	*Vanzijlia* (1 species)
Octopoma (8 species)	

HALLIANTHUS

DERIVATION OF GENUS NAME The genus was named in honour of Mr H. Hall, past curator of Kirstenbosch National Botanical Garden, who made significant contributions to the study of mesembs.

COMMON NAMES No vernacular names seem to have been recorded.

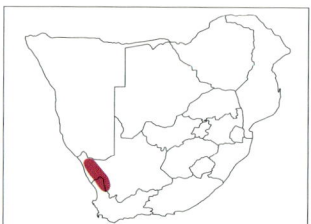

DESCRIPTION The plants are flat-growing shrubs with three-sided and often laterally compressed leaves. The white, pink or bicoloured flowers are borne singly or in groups of three and have numerous filamentous staminodes which are brightly coloured towards their tips and encircle the central cone of stamens. The fruit capsules normally have ten locules and very much resemble those of *Leipoldtia* except that they are turret-like with prominent elevations on the covering membranes.

DISTINGUISHING CHARACTERS Plants are distinguished by long thin internodes, the flattened leaves and the flowers, which are always borne close to the ground. The turret-like capsules are remarkable, but they vary considerably.

FLOWERING TIME Plants flower in autumn and winter (in South Africa), and in cultivation also flower in early spring. The flowers open during the sunny hours of the day.

DISTRIBUTION AND ECOLOGY *Hallianthus* is found in Namaqualand in the Northern Cape Province of South Africa, on granite, sandstone or quartzite hills and also further south in the Western Cape Province in the Tanqua Karoo and the Namaqualand coastal belt (both are winter rainfall areas), in regions with less than 200 mm of annual precipitation. Plants grow in rock crevices and sandy depressions.

CULTIVATION Plants grow readily in cultivation.

NOTES Their occurrence in rock crevices and in depressions can be seen as an ecological adaptation, as these specialised habitats retain more moisture than others in the same area. The sole species of this genus was originally included in the genus *Leipoldtia*. Based on a detailed analysis of growth form, stems, roots, leaves, flowers and capsules, the species was recently transferred to the monotypic genus *Hallianthus*, as it does not quite fit into any other genus.

Hallianthus H.E.K. Hartmann

NUMBER OF SPECIES/SUBSPECIES/VARIETIES (1/0/0)

SPECIES LIST AND CONSERVATION STATUS
H. planus (L.Bolus) H.E.K. Hartmann

LITERATURE
HARTMANN, H.E.K. 1983. Monographien der Subtribus Leipoldtiinae. V. Monographie der Gattung *Hallianthus* (Mesembryanthemaceae). *Botanische Jahrbücher* 104: 143–169.

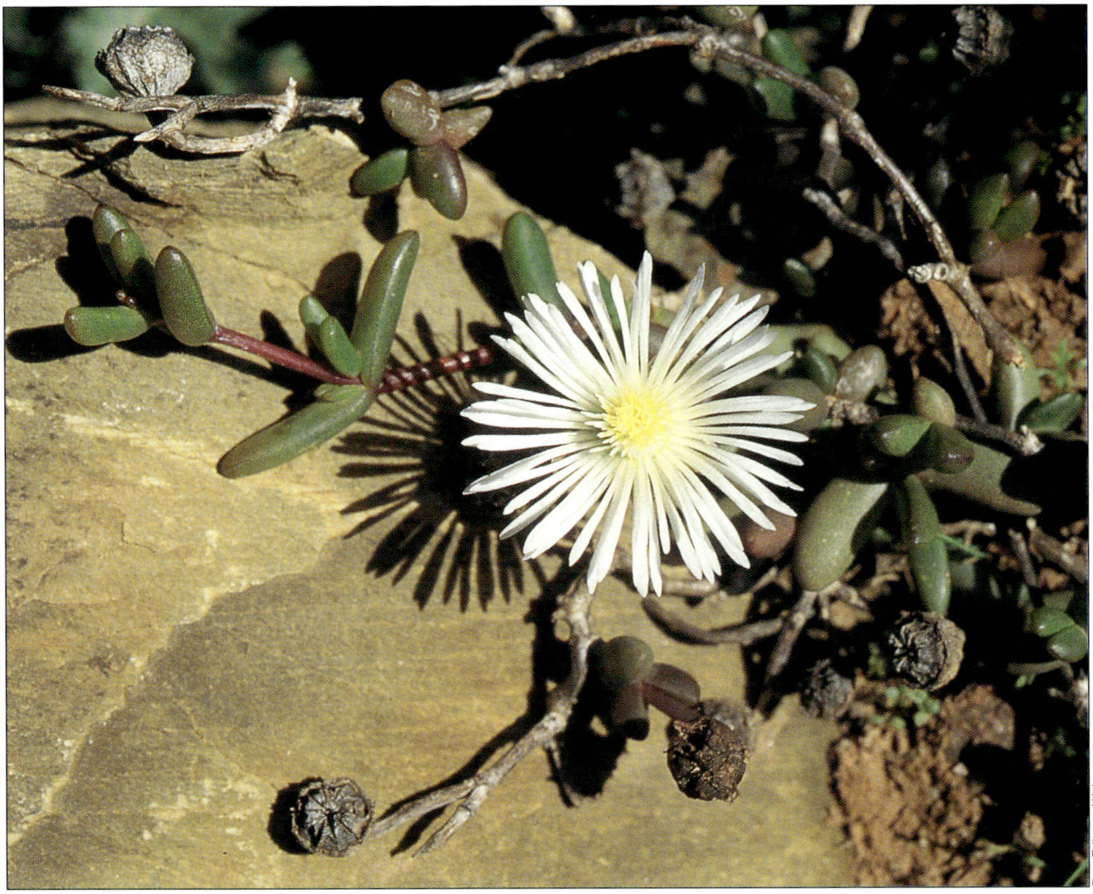
Ben-Erik van Wyk

White form of *H. planus*

Graham Williamson

H. planus

Ben-Erik van Wyk

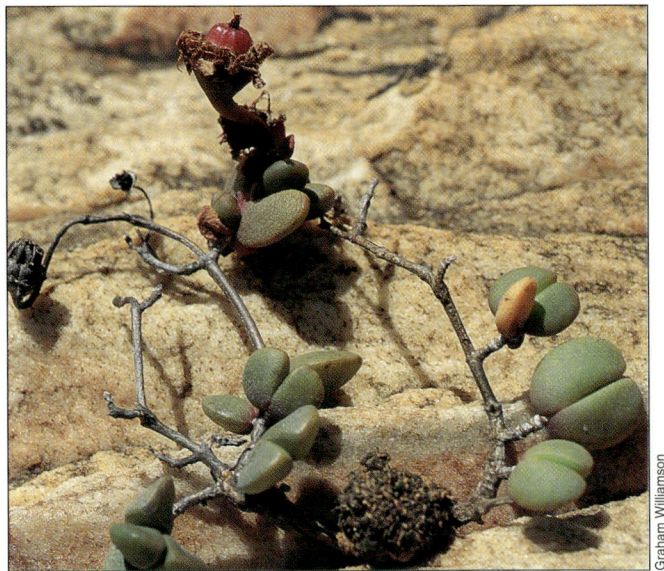

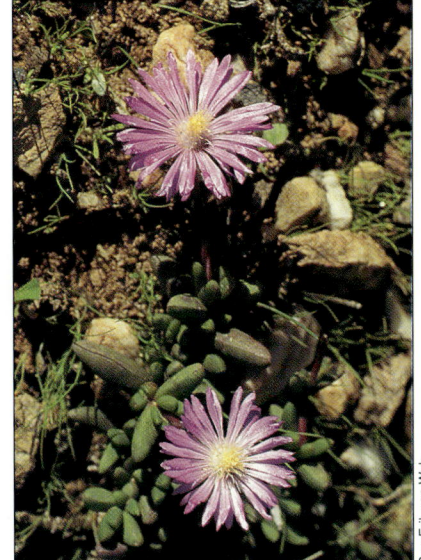

Pink form of *H. planus*

Leipoldtia

Derivation of genus name The genus is named after the Afrikaans poet and plant collector C.L. Leipoldt (1880 to 1947).

Common names *Kussingvygie* (cushion mesemb) is used for some species.

Description The woody shrubs grow erect or flat and spreading, sometimes with branches rooting again some distance from the mother plant. The leaves are borne upright and are three-sided, greyish and smooth. Flowers are mostly developed in many-flowered clusters and in two species some flower stalks turn into spines. Pink, rarely white or yellow petals surround the filamentous staminodes which are arranged in a cone. The fruit capsules, which are borne on erect or bent stalks, mostly have 10 locules but sometimes up to 16. They have wavy covering membranes with closing rodlets at their distal ends, broad valve wings, radially spreading expanding keels, and large white closing bodies.

Distinguishing characters Among the shrubs with many-locular fruit capsules, *Leipoldtia* can best be recognised by the *Leipoldtia* fruit type, as described above. The fruit capsules are never brown when viewed from above but are whitish to grey. The sparse growth form is also characteristic to some extent.

Flowering time Flowering occurs mainly in winter but odd flowers can also be found at other seasons. Several flowers per cluster may open at the same time. They open around midday and remain open well into the afternoon.

Distribution and ecology The distribution area extends from near Aus, Namibia, down to South Africa in the region of Worcester, into the Little Karoo in the Western Cape Province, and eastwards to the Eastern Cape Province. Most species occur in the vicinity of the Richtersveld. Rainfall is expected mainly in winter, but also in March and November. In the gravelly flats inland, the plants form a considerable portion of the succulent vegetation.

Cultivation Plants thrive in cultivation but are seldom grown. Some species are suited for gardens, rockeries and embankments. Plants are propagated from cuttings or seeds sown during the warmer months. A wonderful display can be expected.

Notes The genus has been divided into four groups, based on flower colour, habit and the presence of spines.

Leipoldtia L.Bolus

Number of species/subspecies/varieties (11/2/0)

Species list and conservation status
L. *alborosea* (L.Bolus) H.E.K.Hartmann & Stüber
L. *calandra* (L.Bolus) L.Bolus
L. *compacta* L.Bolus
L. *frutescens* (L.Bolus) H.E.K.Hartmann
L. *klaverensis* L.Bolus
L. *laxa* L.Bolus
L. *lunata* H.E.K.Hartmann & Rust
L. *rosea* L.Bolus
L. *schultzei* (Schltr. & Diels) Friedrich
L. *uniflora* L.Bolus
L. *weigangiana* L.Bolus subsp. *grandifolia* (L.Bolus) H.E.K.Hartmann & Rust
L. *weigangiana* L.Bolus subsp. *littlewoodii* (L.Bolus) H.E.K.Hartmann & Rust
L. *weigangiana* L.Bolus subsp. *weigangiana*

Literature
HARTMANN, H.E.K. & RUST, S. 1994. Monographien der Leipoldtiinae IX. Monographie der Gattung *Leipoldtia* L.Bolus s.lat. (Aizoaceae). *Verhandlungen des Naturwissenschaftlichen Vereins in Hamburg* 34: 275–351.

HARTMANN, H.E.K. & STÜBER, D. 1994. Erweiterung der Gattung *Leipoldtia* um zwei dornentragende Arten (Mesembryanthema, Aizoaceae). *Verhandlungen des Naturwissenschaftlichen Vereins in Hamburg* 34: 353–372.

L. weigangiana subsp. *grandifolia*

L. uniflora

L. schultzei

L. frutescens

OCTOPOMA

DERIVATION OF THE GENUS NAME The name is derived from the Greek words *okto* (eight) and *poma* (lid), describing the eight locules of the fruit capsule.

COMMON NAMES No vernacular names seem to have been recorded.

DESCRIPTION The thick leaves are the most conspicuous feature in most of these shrubs; branches can be erect to spreading, and the stems are hard. The erect to spreading leaves are only a little fused at their bases in most species. Their surfaces are rough to almost smooth. The mostly solitary flowers are often embraced by the bracts on the flower stalks. There are mostly four sepals, the petals are white to purple and filamentous staminodes surround the stamens, which are arranged in a cone in the centre of the flower. The firm capsules have six to eight locules and are brown and markedly convex (bulging) on top in most cases. The covering membranes rise to the centre. The distal closing ledge is inconspicuous and the closing body is large and white. The seeds have a rough surface.

DISTINGUISHING CHARACTERS Species placed in *Octopoma* are not a natural group since they have been included merely on the basis of the number of locules in the fruit capsule. Eight locules are however not at all rare in the mesembs, so that a re-arrangement of the species can be expected. Therefore, at present, the only distinguishing features which can be given for this genus are the firm brown fruit capsules, mostly borne well above the plant.

FLOWERING TIME Flowering occurs mainly in winter in South Africa, and the flowers open from midday onwards.

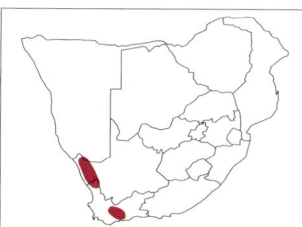

DISTRIBUTION AND ECOLOGY As circumscribed at present, the genus occurs mainly in two separate areas in South Africa, one in northern Namaqualand, Northern Cape Province; the other in the Little Karoo, Western Cape Province. Both these regions are in or at the edge of the winter rainfall area.

CULTIVATION The plants can be grown from seed, and less readily from cuttings. Water sparingly during winter and summer. Outside of their habitat plants are best grown in a greenhouse.

NOTES The genus is an assembly of species which has capsules with eight locules. There are obvious differences in other important characters, so that some species will have to be transferred to other genera and the rest will probably fall into two different genera. This problem is currently being studied.

Octopoma N.E.Br.

NUMBER OF SPECIES/SUBSPECIES/VARIETIES (8/0/0)

SPECIES LIST AND CONSERVATION STATUS
O. abruptum L.Bolus
O. calycinum (L.Bolus) L.Bolus
O. conjunctum (L.Bolus) L.Bolus
O. connatum (L.Bolus) L.Bolus
O. inclusum (L.Bolus) N.E.Br.
O. octojuge (L.Bolus) N.E.Br.
O. rupigenum (L.Bolus) L.Bolus
O. subglobosum (L.Bolus) L.Bolus

LITERATURE
BROWN, N.E. 1930. *Mesembryanthemum* and some new genera separated from it. *Gardener's Chronicle* 87: 72.

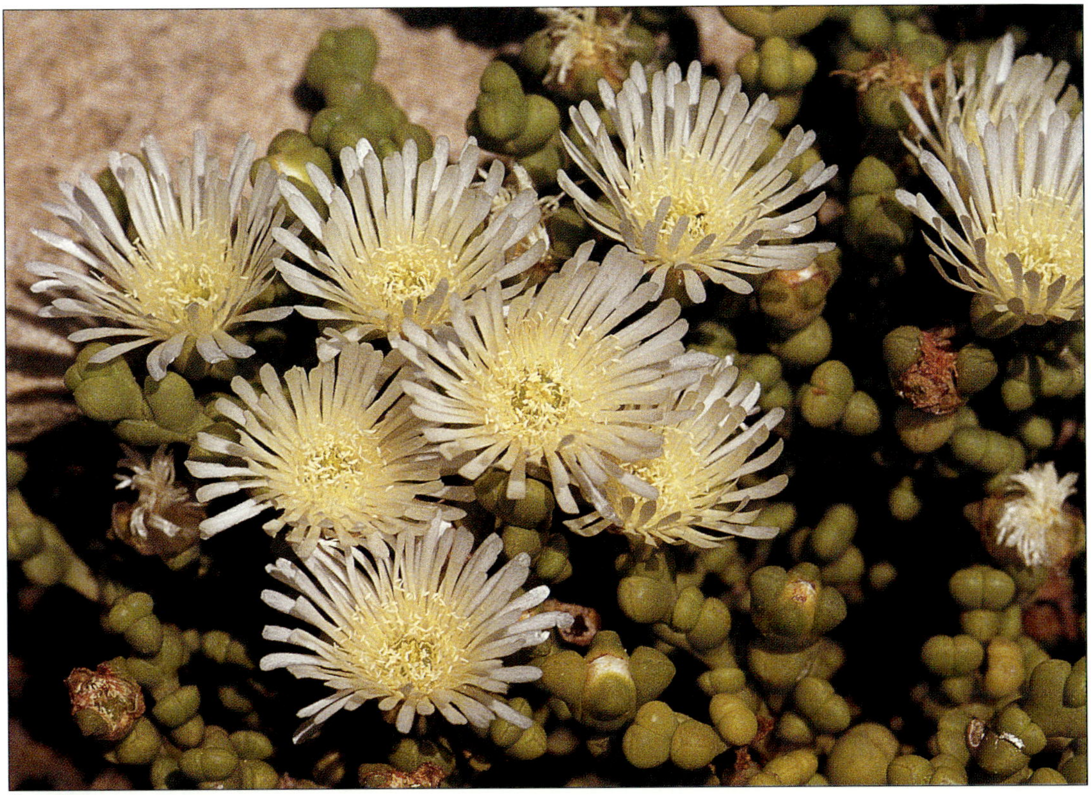

Ernst van Jaarsveld

O. octojuge

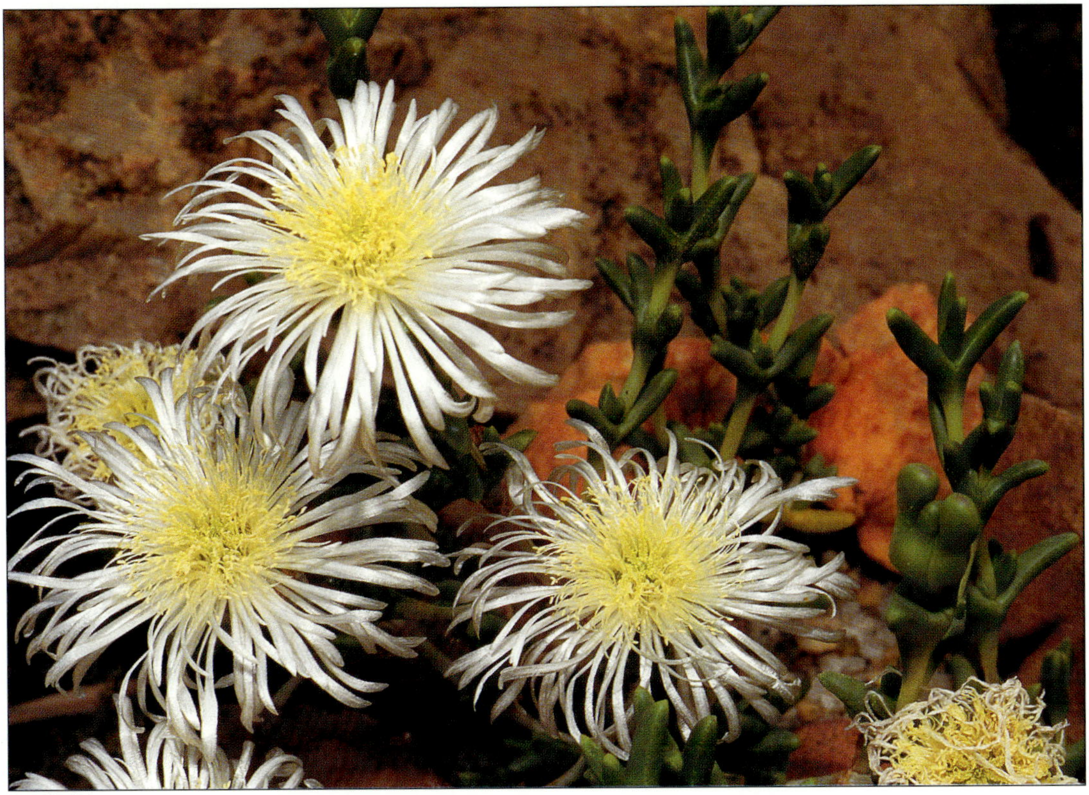

Bob Chinnock

O. calycinum

Ottosonderia

Derivation of genus name The genus was named in honour of Dr Otto Sonder of Hamburg, Germany, who was one of the editors of the *Flora Capensis* volumes.

Common names No vernacular names seem to have been recorded.

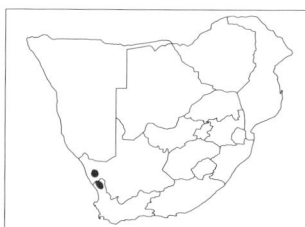

Description The plants are compact shrublets with short vegetative growth, elongated floral branches and prominent persistent clusters of flowers. The lower leaves are markedly different from those within the flower clusters. The flower clusters are unique in that their development extends over a period of years; in the first year, a solitary, terminal flower is formed, with two fused bracts below it; in the second year, a new flower cluster is formed from each bract; this process of subdivision is repeated annually until a broad flower cluster is produced, with scars of the old bracts in its lower part, hardened remains of bracts in upper parts and remains of old fruit capsules throughout. Flowers occur singly on either side of each terminal fruit. There are four sharp-tipped sepals – the outer pair with a ridge along the lower end, and the inner ones with brown, membranous margins. The petals are purplish pink and occur in two whorls. Numerous filamentous staminodes are also present. Fruit capsules have six to eight locules and are close to the *Ruschia*-type but the valves have low rims, yet form a rounded dome. The valves are erect when open. Wings are absent, expanding keels are thick and erect, covering membranes are well developed, and closing bodies are large. The seeds are more or less egg-shaped and brown in colour.

Distinguishing characters Species of *Ottosonderia* are easily identified by the persistent, succulent flower clusters.

Flowering time Flowering occurs in winter (June and July in South Africa).

Distribution and ecology The plants are restricted to South Africa and have a very localised distribution in Namaqualand, with two separate distributions, the one to the north and the other just south of the border between the Northern and Western Cape Provinces.

Cultivation *Ottosonderia* species are easily cultivated but they are not particularly attractive and are mostly grown by collectors as curiosity plants. They are best grown in a greenhouse outside of their habitat in sandy, well-drained soil. Keep the plants dry in summer.

Notes *Ottosonderia monticola* differs from *O. obtusa* in the more southern distribution and in its larger number of locules in the fruit capsules. Vegetatively the two are identical. Although fruit capsules are similar to the *Ruschia*-type in having no value wings, the presence of large closing bodies and other features justify its inclusion in the *LEIPOLDTIA*-LIKE SHRUBBY MESEMBS.

Ottosonderia L.Bolus

Number of species/subspecies/varieties (2/0/0)

Species list and conservation status
O. *monticola* (Sond.) L.Bolus
O. *obtusa* L.Bolus

Literature
BOLUS, H.M.L. 1954. *Notes on* Mesembryanthemum *and allied genera*. University of Cape Town, Cape Town. Vol. 3: 292.

O. monticola

Vanzijlia

Derivation of the genus name The genus was named after Dorothy van Zijl, a plant collector and grower from Clanwilliam in the Western Cape Province, South Africa.

Common names No vernacular names seem to have been recorded.

Description The plants have numerous short shoots forming compact clumps at the centre and also several long shoots which may be trailing or climbing into other shrubs. The leaves on short shoots and the first pair of the long shoots consist of long sheaths with only short free tips, forming a body during the dry season. Leaf pairs on long shoots have basal sheaths less than half their lengths. Flowers are solitary at the end of long shoots, the white petals thin and spreading. The apically pink or purple filamentous staminodes elongate while the flower opens and closes over several days, eventually drooping with age. Fruit capsules have 10 locules and are long-stalked; they possess valve wings, covering membranes with distal closing bulges, and large, white closing bodies.

Distinguishing characters The different leaf shapes and the climbing growth of the long shoots are most typical of this genus. A remarkable character is the distinctly elongating filamentous staminodes in the flowers.

Flowering time Flowering occurs in winter and early spring. The flowers open mainly over midday and close in the late afternoon.

Distribution and ecology The single species of Vanzijlia occurs along the western coast of the Northern and Western Cape Provinces, South Africa, with an extension inland into the Knersvlakte, where it grows in gravelly soils receiving winter rainfall of about 125 mm per year.

Cultivation Vanzijlia thrives in strandveld gardens where plants can be used as a ground cover. Propagation is easy by cuttings which should be planted directly into the desired spot. Outside of their habitat plants should be grown in containers and kept dry in summer. Prune to maintain a compact growth.

Notes The original museum specimen of Vanzijlia at Berlin consisted of a mixture of two different plants belonging to two different genera (the true Vanzijlia and a species of Dicrocaulon). As a consequence, much discussion arose between H.M.L. Bolus describing Vanzijlia as a new genus, and N.E. Brown studying a member of the genus Dicrocaulon. The problem could only be cleared up when both pieces were compared.

Vanzijlia L.Bolus

Number of species/subspecies/varieties (1/0/0)

Species list and conservation status
V. annulata (A.Berger) L.Bolus

Literature
BOLUS, L. 1927. Vanzijlia annulata. *Flowering Plants of South Africa* 7: t. 262.

BROWN, N.E. 1929. Vanzijlia L.Bol. *Kew Bulletin* 1929: 61-62.

HARTMANN, H.E.K. 1983. Monographien der Subtribus Leipoldtiinae. IV. Monographie der Gattung Vanzijlia (Mesembryanthemaceae). *Botanische Jahrbücher* 103: 499–538.

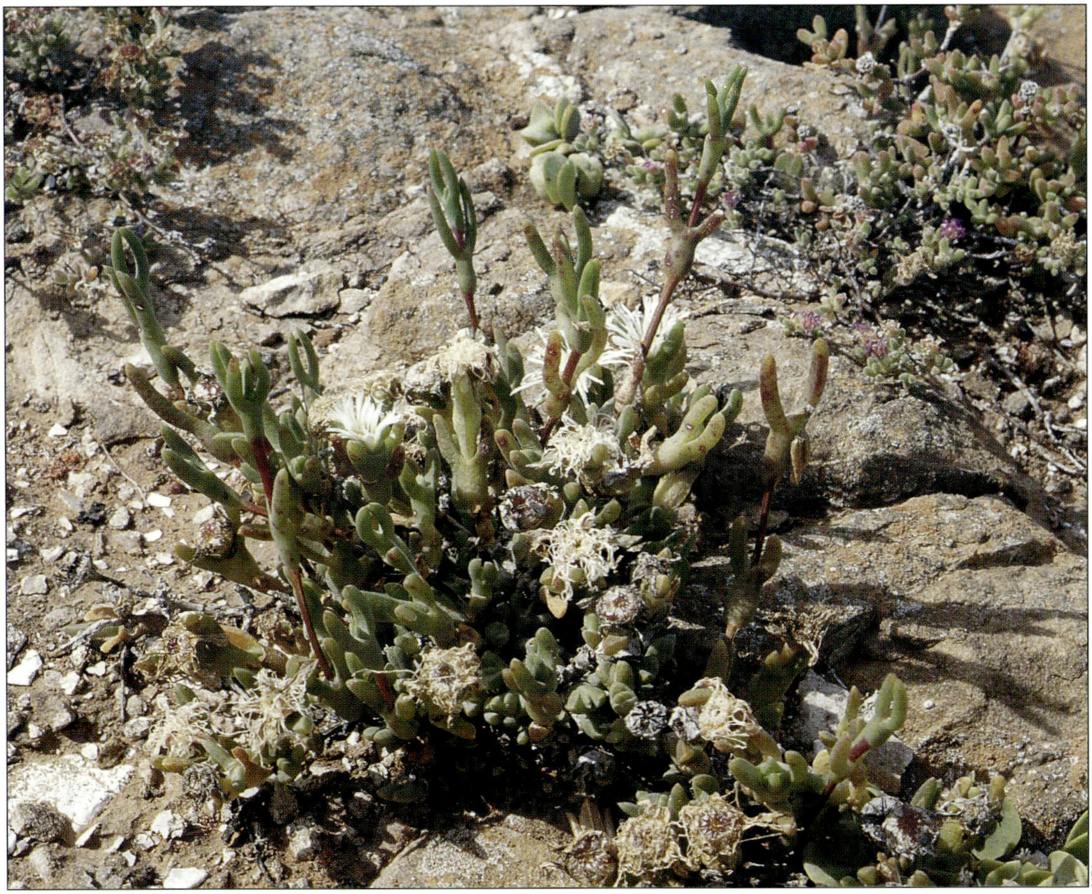

NBI

V. annulata in its habitat

Heidi Hartmann

Fruit capsules of *V. annulata*

Ben-Erik van Wyk

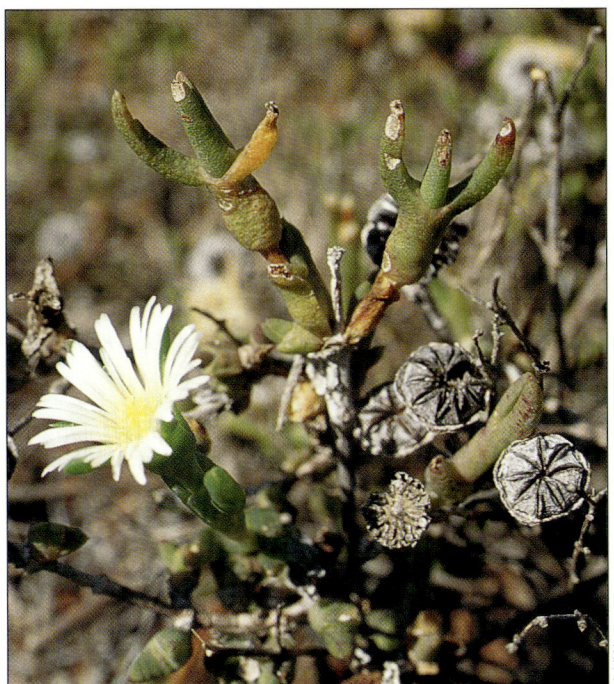

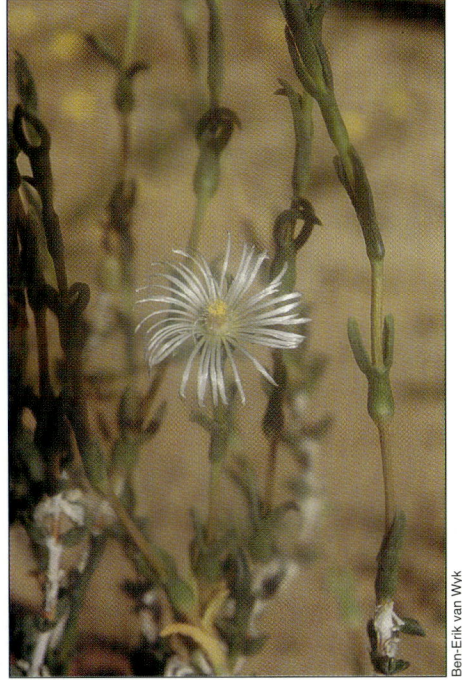

V. annulata

FURTHER READING

BITTRICH, V. & HARTMANN, H. 1988. Aizoaceae – a new approach. *Botanical Journal of the Linnean Society* 97: 239–254.

BITTRICH, V. & STRUCK, M. 1989. What is primitive in Mesembryanthemaceae? An analysis of evolutionary polarity of character states. *South African Journal of Botany* 55: 321–331.

BOLUS, H.M.L. 1928–1958. *Notes on Mesembri(y)anthemum and (some) allied genera*. Specialty Press (Part 1) and Bolus Herbarium (Parts 2 & 3), Cape Town.

BROWN, N.E., TISCHER, A. & KARSTEN, M.C. 1931. *Mesembryanthema*. L. Reeve, Kent.

CHESSELET, P., MÖSSMER, M. & SMITH, G.F. 1995. Research priorities in the succulent plant family Mesembryanthemaceae Fenzl. *South African Journal of Science* 91: 197–209.

CHESSELET, P. & DE WET, B.C. 1997. Mesembryanthemaceae. In G.F. Smith, E.J. van Jaarsveld, T.H. Arnold, F.E. Steffens, R.D. Dixon & J.A. Retief (eds), *List of southern African succulent plants*. Pp. 89–122. Umdaus Press, Pretoria.

CROIZAT, L. 1993. On the structural and developmental history of the capsule of *Mesembryanthemum* s.l. *Kirkia* 14: 145–169.

HAMMER, S.A. 1995. Mastering the art of growing mesembs. *Cactus & Succulent Journal (US)*: 67: 195–247.

HARTMANN, H.E.K. 1983a. Interaction of ecology, taxonomy and distribution in some Mesembryanthemaceae. *Bothalia* 14: 653–659.

HARTMANN, H.E.K. 1983b. Untersuchungen zum Merkmalsbestand und zur Taxonomie der Subtribus Leipoldtiinae (Mesembryanthemaceae). *Bibliotheca Botanica* 136: 1–67

HARTMANN, H.E.K. 1988. Fruit types in Mesembryanthema. *Beiträge zur Biologie der Pflanzen* 63: 313–349.

HARTMANN, H.E.K. 1989. Biological adaptation in Mesembryanthema. *Excelsa* 14: 39–44.

HARTMANN, H.E.K. 1991a. Mesembryanthema. *Contributions from the Bolus Herbarium* 13: 75–157.

HARTMANN, H.E.K. 1991b. Keys to the genera of Mesembryanthema. *South African Journal of Botany* 57: 95–106.

HARTMANN, H.E.K. 1993. Aizoaceae. In K. Kubitzki, J. G. Rohwer & V. Bittrich (eds), *The families and genera of vascular plants* 2: 37-69. Springer Verlag, Berlin.

HARTMANN, H.E.K. 1994. On the phytogeography and evolution of Mesembryanthema (Aizoaceae). Proceedings of the XIII Plenary Meeting of AETFAT 1990, Malawi 2: 1165–1180. Zomba, Malawi.

HARTZER, P. 1997. Mesembryanthemaceae. In N. Meyer, M. Mössmer & G.F. Smith (eds), Taxonomic literature of southern African plants. *Strelitzia* 5: 89–95. National Botanical Institute, Pretoria.

HERRE, H. 1971. *The genera of the Mesembryanthemaceae*. Tafelberg, Cape Town.

HILTON-TAYLOR, C. 1996. Red Data List of southern African plants. *Strelitzia* 4: 1–117. National Botanical Institute, Pretoria.

IHLENFELDT, H.-D. 1994. Diversification in an arid world: the Mesembryanthemaceae. *Annual review of ecology and systematics* 25: 521–546.

JACOBSEN, H. 1977. *Lexicon of succulent plants*. Part II. Family Mesembryanthemaceae. Pp. 395–583. Blandford, Poole.

JACOBSEN, H. 1986. *A handbook of succulent plants*. Volume III. Mesembryanthemums (Ficoidaceae). Pp. 865-1380. Blandford, Poole.

SCHWANTES, G. 1957. *Flowering stones and midday flowers*. Ernst Benn, London.

Key to mesemb genera

There are few keys available to identify mesemb genera. These are either outdated or highly technical, prove difficult to use and identifications are not always successful. Here we have attempted to provide a key that follows the structure and groupings found in this book to facilitate identification. Our broad groupings are in some respects problematic due to numerous exceptions to the general growth form into which a genus has been grouped. A number of difficulties also arise from unclear generic delimitation, particularly in the large, heterogeneous genera *Lampranthus*, *Ruschia* and *Antimima*. It is recommended that for any shrubby mesemb, the fruit capsule should be opened and looked at with a magnifying glass before using the key. Our descriptions of basic capsule types should prove useful in identifying the correct group to place the shrubby mesembs i.e. GLITTERING SHRUBBY MESEMBS, *LAMPRANTHUS*-LIKE SHRUBBY MESEMBS, *RUSCHIA*-LIKE SHRUBBY MESEMBS or *LEIPOLDTIA*-LIKE SHRUBBY MESEMBS. Other, more obvious characteristics should be used for the other groups. If an identification proves unsuccessful go back to the couplet where there was uncertainty and take the other option. Good luck!

1. Plants often annual, often found as weeds in disturbed places such as roadsides or ploughed lands; ovules/ young seeds attached to central axis in ovary or fruit; usually with only slightly succulent leaves or succulent stems; stems/ branches always visible; leaves usually glistening or greyish green and soft in texture; open fruit capsules with expanding keels reaching from the central column to the tip of the valves, fruit capsules either opening and closing repeatedly when wet and dry, or very rarely a dry nut	WEEDY MESEMBS ...14
– Plants mostly perennial, found in various habitats; ovules/ young seeds attached to walls/ floor of ovary/ fruit, rarely to central axis (*Hymenogyne*, *Caryotophora*, *Faucaria*, *Neohenricia* and *Cylindrophyllum*); leaves slightly or highly succulent, or a combination of succulent leaves and succulent stems; stems/ branches visible or plants stemless; leaves glistening, smooth, rough or pubescent, of various shades of green, soft or hard in texture; open fruit capsules with expanding keels never reaching the centre of the fruit; fruit opening and closing repeatedly when wetted and dried, or splitting up when dry, or rarely fleshy or nut-like	ALL OTHER GROUPS ...2
2. Plants often with a leafy appearance, leaves flat, lance to spoon-shaped, elongated or sometimes slightly cylindrical (*Conicosia*); plants mostly annual, sometimes perennial; leaves smooth, hairy or covered with glistening surface water cells; flowers mostly yellow or white, or multicoloured as in *Dorotheanthus*. (Some species of *Delosperma* have flat leaves but are not included in this group.)	FLAT-LEAVED MESEMBS ...24
– Plants succulent to various degrees with cylindrical, three-sided, rounded sometimes ± fused or sometimes slightly flattened leaves, mostly perennial, sometimes with annual shoots on a perennial plant; leaves smooth, or hairy, or glistening, or variously textured; flowers mostly pink, magenta, yellow, white or variously coloured (e.g. *Cephalophyllum*, *Drosanthemum*)	...3
3. Plants highly succulent with leaf pairs ± fused forming plant bodies, or free and rounded, or forming small clumps (*Antegibbaeum*); internodes of stems not visible and leaves in single pairs or in small groups; leaf tips sometimes windowed, variously patterned or not specialised, sometimes highly textured (species of *Conophytum* previously included in *Berrisfordia*); plants often mimic the colours of their habitat and may resemble stones; during the dry season plants may disappear underground or may be covered with papery leaf remains	FLOWERING STONES ...33

– Plants succulent with leaf pairs free or partially fused; internodes of stems visible to ± hidden; leaf tips rounded, pointed, chin-like, or with teeth or bristles; plants may or may not mimic their habitat with colouring and texture; during the dry season papery sheaths may cover the buds of next year's growth	...4
4. Plants mostly clump-forming with stems not clearly visible; leaves highly succulent, mostly oblong and slightly flattened, sometimes grape-shaped, rarely toothed along the leaf margins (*Juttadinteria* and *Schwantesia*); surface may be covered by a waxy layer	TONGUE-LEAVED MESEMBS ...**49**
– Plants clump-forming, or creeping or shrubby; stems clearly visible or not visible; leaves mostly cylindrical, three-angled and sometimes toothed	...5
5. Leaves mostly arranged in rosettes, with rough, finely or coarsely textured surfaces; plants rarely creeping (*Neohenricia*); stems usually not clearly visible; flowers in various shades of yellow, occasionally pink, often with a darker, longitudinal stripe along the center of the petals; stamens usually grouped together into a cone-like structure in the centre of the flower (this group does not include genera with leaves roughened by adhering sand i.e. *Psammophora* and *Arenifera*)	ROUGH-LEAVED MESEMBS ...**57**
– Leaves variously arranged, with smooth, hairy or dotted leaf surfaces; stems clearly or not clearly visible; flowers variously coloured, rarely with a darker longitudinal stripe along the center of the petals; stamens loosely arranged or gathered together into a central cone	...6
6. Leaves usually with one or more teeth along the leaf margins or near the leaf tips; plants usually short-stemmed, sometimes with leaves arranged in rosettes, rarely partially fused to form plant bodies; plants sometimes with creeping stems which root at the nodes	TOOTH-LEAVED MESEMBS ...**62**
– Leaves rarely toothed (some species of *Oscularia* and *Ruschia*), sometimes with cartilaginous or serrated leaf margins (*Erepsia* and *Ruschia*); plants short-stemmed and clump-forming, or creeping, or shrubby	...7
7. Plants with short stems, internodes visible or not, often with thickened fleshy or woody roots branching underground; finger-like leaves erect and crowded together just above the ground; leaves may be cylindrical or three-sided with pointed tips	TUFTED MESEMBS ...**70**
– Plants with long or short, upright or creeping stems, internodes mostly visible, plants rarely with thickened roots, tubers (*Delosperma* and *Ruschia*) or a caudex (*Mestoklema*); leaves variously arranged along the stems, cylindrical, three-sided with pointed or rounded tips	...8
8. Plants with consecutive leaf pairs differing from each other, one pair often forming plant bodies, so that the new (elongate) leaf pair differs markedly from the older (round) leaf pair from which it has emerged; the older globular leaf pairs dry out and papery sheaths cover the new leaf pairs during the dry season	BEAD-LEAVED MESEMBS ...**81**
– Plants with consecutive leaf pairs of the same shape, without papery sheaths; when differently-shaped consecutive leaf pairs are present (*Cheiridopsis* and *Antimima*), they may be finely textured (*Antimima* subgenus *Microphylla*)	...9
9. Plants usually creeping along the ground, forming dense or sparse mats or cushions with leaves variable in size and shape; plants sometimes root at the nodes or originate from a thick tap root; some species are upright (some species of *Antimima*) or with finely textured, sharp-tipped leaf pairs that differ from each other. (Mat-forming species of *Lampranthus*, *Drosanthemum*, *Delosperma* and *Ruschia* are not included in this group)	MAT-FORMING MESEMBS ...**85**

– Plants mostly upright shrubs or subshrubs, or creeping; consecutive leaf pairs rarely differing from each other (*Vanzijlia*); leaves variable in size and shape	...10
10. Plants usually upright dwarf shrubs; leaves sometimes finely pubescent, or smooth, greyish or with a reddish tinge at the leaf tips	DWARF SHRUBBY MESEMBS...**92**
– Plants usually upright shrubs, sometimes creeping; leaves smooth or textured by fine water cells or dotted, of various shades of green	...11
11. Plants shrubby with visible internodes and mostly erect, but sometimes creeping stems; leaves with tiny surface cells storing water, giving the leaf surface a glittering appearance and rough texture; stems smooth, sometimes reddish or pale, often with long or short hairs, particularly on the pale-stemmed plants; fruit capsules of the *Drosanthemum*-type or *Delosperma*-type	GLITTERING SHRUBBY MESEMBS ...**98**
– Plants shrubby with visible internodes and mostly erect, but sometimes creeping stems; leaves dotted, waxy or smooth; stems smooth, reddish, whitish or brown, without hairs; fruit capsules of the *Lampranthus*-type, the *Ruschia*-type or the *Leipoldtia*-type	...12
12. Plants shrubby, with creeping or erect stems; fruit capsules of the *Lampranthus*-type with sterile funicular hairs at the seed exits, leaves sometimes toothed (some species of *Oscularia*)	*LAMPRANTHUS*-LIKE SHRUBBY MESEMBS ...**102**
– Plants shrubby, with creeping or erect stems; fruit capsules of the *Ruschia*-type or the *Leipoldtia*-type; leaves sometimes toothed (some species of *Ruschia*)	13
13. Shrubby plants with creeping or erect stems with fruit capsules of the *Ruschia*-type; valve wings absent, except in the spiny species (*Ruschia* subgenus *Spinosae*, previously *Eberlanzia*) and closing bodies rod-shaped (one species of *Polymita* has large white closing bodies)	*RUSCHIA*-LIKE SHRUBBY MESEMBS ...**111**
– Shrubby plants with creeping or erect stems with fruit capsules of the *Leipoldtia*-type; closing bodies large, whitish and rounded, valve wings present, except in *Ottosonderia*	*LEIPOLDTIA*-LIKE SHRUBBY MESEMBS ...**118**

WEEDY MESEMBS

14. Plants with articulated stems with persistent succulent green internodes	...15
– Plants with continuous stems, only the youngest stems are green, soon becoming woody	...18
15. Water cells similar on stems and leaves, ± reduced; filamentous staminodes gathered into a cone	...***Psilocaulon***
– Stems with densely arranged, tall, cylindrical, xeromorphic water cells, i.e. firm to the touch, visible in cross section as a conspicuous white layer; leaves with distant, mesomorphic water cells, i.e. soft to the touch; filamentous staminodes not gathered into a cone	...16
16. Plants lying flat on the ground or low with tips ascending; leaves flat or semi–cylindrical in the ± erect climbers; seeds dark brown, rough	...***Aptenia***
– Plants erect with ± cylindrical leaves; seeds ochre to light brown	...17

17. Leaves free to the base, with tubular leaf sheaths, partly enclosing one another towards base; flowers solitary, up to 45 mm in diameter; seeds up to 2,2 mm long	...*Aspazoma*
– Leaves fused towards the base or free, but never forming a tubular sheath; flowers solitary or in a few many-flowered inflorescences, 5 to 25 mm in diameter; seeds less than 1 mm long	...*Brownanthus*
18. Plants annual or biennial, with conspicuous, prominent water cells or rarely with a smooth epidermis and small water cells along the leaf margins	...19
– Plants perennial, rarely short-lived, with the epidermis ± smooth to papillate with water cells less prominent	...20
19. Leaves opposite, fused towards the base into a disc or a cone, upper leaves smaller and ± free	...*Synaptophyllum*
– Leaves alternate or opposite, but never fused into a disc or a cone	...*Mesembryanthemum*
20. Fruit capsules with deep locules (lower part of capsule funnel-shaped); water cells flattened (epidermis smooth)	...21
– Fruit capsules with lower part of capsule as long as upper part; epidermis smooth or with water cells	...22
21. Plant a geophyte with stems lying flat on the ground and a thick subterranean rootstock	..."*Caulipsolon*"
– Plant upright or lying flat on the ground without a thickened rootstock	...*Prenia*
22. Plants with a smooth epidermis, water cells reduced	...*Aridaria*
– Plants with a papillate epidermis, water cells rarely reduced	...23
23. Old leaves withering to a skeleton, persisting and enclosing the young leaves	...*Sceletium*
– Old leaves never withering to a skeleton	...*Phyllobolus*

FLAT-LEAVED MESEMBS

24. Leaves and stems entirely covered with glistening water cells	...25
– Leaves and stems smooth, or hairy, sometimes with surface water cells	...27
25. Leaf margins irregularly lobed; flowers yellow; fruit capsules flat on top	...*Aethephyllum*
– Leaf margins straight or lobed; flowers white or multicoloured; fruit capsules ribbed on top	...26
26. Flower stalk curved (S-shaped); valve wings broad and rectangular	...*Cleretum*
– Flower stalk more or less straight, or curved; valve wings tapering into awns, remains of stigmas persist on the ripe fruit capsules	...*Dorotheanthus*

27. Flowers white	...28
– Flowers ivory to yellow	...30
28. Fruit with 3 or 4 very hard, nut-like parts, each with two chambers containing a single seed	...*Caryotophora*
– Fruit with 5 to 7 locules, with seeds entirely or partially embedded in seed pockets	...29
29. Leaves alternating with each other, limp, lance-shaped, with surface water cells; fruit capsules open spontaneously upon drying out and remain open—the ends of the valves split apart for about half the height of the capsule; a circular convolution surrounds the fruit which contains several seeds; receptacle without ridges	...*Saphesia*
– Leaves opposite or alternating with each other, spoon-shaped, smooth and without water cells; leaf margins undulate at early developmental stages; fruit capsules open spontaneously upon drying out and remain open, the valves then separate and spread, with two seeds per locule enclosed in chambers of the outer wall of the capsule; receptacle with prominent ridges	...*Skiatophytum*
30. Petals without hairs along their margins or bases; stigmas fused into a single column	...*Hymenogyne*
– Petals with minute hairs along their margins or bases; stigmas free	...31
31. Plants covered with visible hairs, with flattened leaves and a prominent mid-vein	...*Carpanthea*
– Plants smooth, leaves flat or sub-cylindrical	...32
32. Leaves flat	...*Apatesia*
– Leaves long and narrow, three-angled or sub-cylindrical	...*Conicosia*

FLOWERING STONES

33. Plants usually velvety or hairy	...34
– Plants without hairs or very minutely hairy, surfaces waxy, textured or smooth	...35
34. Leaves unequal, the leaf pair often resembling a shark's mouth	...*Gibbaeum*
– Leaves fused to form a single plant body with a small fissure just below the apex	...*Muiria*
35. Plants with highly textured, bumpy or dotted leaf surfaces	...36
– Plants with a relatively smooth leaf surface, sometimes vaguely dotted, rarely textured (*Conophytum* species previously in *Berrisfordia*)	...39
36. Fruit capsules with 9 to 15 locules	...*Pleiospilos*
– Fruit capsules with 5 to 7 locules	...37
37. Plants with persistent trailing stems	...*Vlokia*
– Plants with indistinct stems	...38

38. Leaves fused, each pair with an apical fissure; flowers arising on a bracteate stalk	...*Oophytum*
– Leaves shortly fused at the base with free parts diverging; flowers sessile	...*Diplosoma*
39. Leaf pairs conically united, with an apical aperture; petals united to form a tube	...*Conophytum*
– Leaf pairs partially fused or free; petals free or united into a tube	...40
40. Plant bodies comprising two distinct leaves, joined at base and fused to various degrees	...41
– Plants in clumps and leaves quite separate	...45
41. Leaves keeled along the lower surfaces	...42
– Leaves not keeled	...44
42. Plants producing two flowers simultaneously, one on each side of the plant body, leaves covered in a thick coating of wax	...*Didymaotus*
– Plants producing flowers centrally in the plant body; leaves smooth or rough with indentations or ridges	...43
43. Leaves smooth and silvery	...*Argyroderma*
– Leaves windowed or variously marked	...*Dinteranthus*
44. Leaves windowed, capsules 5 or 6-locular	...*Lithops*
– Leaves not windowed, capsules 10-locular	...*Tanquana*
45. Leaves with windowed tips	...46
– Leaves with unspecialised tips	...47
46. Fruit capsules of *Leipoldtia*-type, 8 to 16-locular; leaves smooth	...*Fenestraria*
– Fruit capsules 5 or 6-locular, fragile; leaves textured with fine water-filled cells	...*Frithia*
47. Fruit capsules 6 to 8-locular	...*Lapidaria*
– Fruit capsules with more than 8 locules	...48
48. Flowers yellow, tinged on outside with purple; fruit with 10 to 15 locules	...*Ihlenfeldtia*
Flowers pale or dark pink, square-looking; fruit with 9 to 28 locules	...*Namibia*

TONGUE-LEAVED MESEMBS

49. Calyx with 4 sepals; bracts absent	...50
– Calyx with more than 4 sepals; bracts present or absent	...51

50. Flowers yellow, nectaries absent; covering membranes well developed; leaves never toothed	...*Glottiphyllum*
– Flowers white, nectaries in a ring; covering membranes poorly developed; leaves sometimes toothed	...*Juttadinteria*
51. Calyx with 6 to 8 sepals; bracts present	...52
– Calyx with 5 sepals; bracts present or absent	...53
52. Flowers pink to magenta; fruit capsules with 6 to 8 locules	...*Antegibbaeum*
– Flowers white or cream; capsules 8 to 10-locular; leaves with brackish taste	...*Drosanthemopsis*
53. Bracts absent; fruit capsules 5-locular, without covering membranes	...*Schwantesia*
– Bracts present; fruit capsules 5, or 8 to 14-locular, with or without covering membranes	...54
54. Fruit capsules without covering membranes; petals stiff, flowers white or pale yellow, nectar visible	...*Nelia*
– Fruit capsules with covering membranes; petals purple, white or yellow	...55
55. Fruit capsules 8 to 14-locular; nectaries in a ring	...*Dracophilus*
– Fruit capsules 5-locular; nectaries separate	...56
56. Leaves purplish green with a wax coating; flowers purple, white or yellow	...*Cerochlamys*
– Leaves yellowish white, without wax coating; flowers yellow	...*Bijlia*

ROUGH-LEAVED MESEMBS

57. Plants minute, less than 20 mm high, creeping, rooting at the nodes; leaves highly textured with minute warts or spines	...*Neohenricia*
– Plants in an aloe-like rosette, internodes not easily visible; leaves highly textured with warts or covered with dots or bumps, or inconspicuously dotted	...58
58. Leaves velvety and dust-covered	...*Deilanthe*
– Leaves dotted, inconspicuously dotted or minutely bumpy, warty or pustulate on the upper surface	...59
59. Leaves conspicuously warty and highly textured with pinkish or whitish bumps, spoon-shaped; with shortly stalked or sessile flowers, petals not striped; locules usually 6 (5 to 10)	...*Titanopsis*
– Leaves pointed, club-shaped or spoon-shaped (*Aloinopsis*), not conspicuously warty; petals striped or not striped; locules 5 or 6 to 14	...60

60. Flowers white to yellow, without central stripe on petals, leaf surface minutely covered with hard white protuberances	...*Rhinephyllum*
– Flowers yellowish, salmon, pink or rose, usually with reddish stripe, leaves with dotted surface	...61
61. Seeds flat	...*Nananthus*
– Seeds egg-shaped	...*Aloinopsis*

TOOTH-LEAVED MESEMBS

62. Plants creeping, rooting at the nodes	...63
– Plants in clumps or rosettes, not creeping or rooting at the nodes	...64
63. Plants with small erect leaves and soft small teeth at the leaf tips; fruit capsules with partial covering membranes	...*Chasmatophyllum*
– Plants with two kinds of leaves, fused at their bases forming sheaths, with small teeth at their tips and a flaky wax layer on the surface; fruit capsules with ample covering membranes	...*Hammeria*
64. Flowers white to pink, or orange-pink with a longitudinal stripe along the apically notched petals; pedicels long, with bracts	...*Acrodon*
– Flowers in shades of yellow (rarely white or pink), petals not striped, pedicels long or short	...65
65. Plants with bright yellowish green leaves; flowers with very long stalks (up to 100 mm), with one or two pairs of bracts	...*Carruanthus*
– Plants with variously coloured and textured leaves; long or short pedicels, with or without bracts	...66
66. Fruit capsules with 7 to 15 locules; flowers with long or short stalks	...67
– Fruit capsules with 5 or 6 locules; flowers more or less stalkless	...68
67. Leaves dull greyish green, keeled; flowers pale yellow to deep yellowish orange, up to 30 mm in diameter; fruit capsules close completely after opening, often with a hairy base, 8 to 11-locular	...*Odontophorus*
– Leaves yellowish green, felty or smooth, fused for half their length or more; flowers a rich yellow with an orange tinge, large, up to 50 mm in diameter; fruit capsules 7 to 15-locular, valves, once open, do not close again completely	...*Vanheerdea*
68. Leaf surfaces minutely pimpled; small white soft teeth on leaf margins; shallow fruit capsules without covering membranes	...*Stomatium*
– Leaf surface smooth, tubercled, marbled or ribbed; deep fruit capsules with more or less perpendicular covering membranes	...69

106. Shrubs with smooth stems, often branching from the base of the plant; leaves with sharp tips or lacerated edges; flowers with distinct hypanthium (calyx tube), filamentous staminodes conceal the stamens	...*Erepsia*
– Flat-growing to erect shrubs, or cushion-forming soft shrubs, leaves finely toothed or smooth, with sharp or blunt tips; flowers with filamentous staminodes present or absent	...107
107. Sprawling shrubs with pale stems; fruit capsules with valve wings and small closing bodies; flowers with star-like colour pattern, stamens arranged in a cone at the centre of the flower	...*Amphibolia*
– Shrubs or shrublets with reddish, or bluish green, smooth stems; flowers not with star colour pattern, stamens collected into a cone or loosely arranged	...108
108. Bluish green plants that form large cushions; leaves with blunt tips and smooth surfaces; flowers small and yellow in many-flowered clusters; fruit capsules without valve wings, covering membranes partially covering locules	...*Scopelogena*
–Leaves cylindrical, 3-angled, club-shaped to sickle-shaped, smooth or finely dentate; flowers solitary or in clusters; fruit capsules with valve wings	...109
109. Flowers solitary, large, yellow, filamentous staminodes absent, filaments broadened and connate at the base	...*Circandra*
– Flowers solitary or in clusters, variable in colour, filamentous staminodes present or absent, filaments not broadened at base	...110
110. Leaves always grey-green from a waxy layer, sometimes toothed, stamens and staminodes always arranged in a cone; flowers white to purple	...*Oscularia*
– Leaves variable; filamentous staminodes often absent; flowers variable in colour, usually solitary and without bracts	...*Lampranthus*

RUSCHIA-LIKE SHRUBBY MESEMBS

111. Young leaves green and sticky, increasingly becoming covered by adhering dust and sand with age	...*Arenifera*
– Young leaves not sticky	...112
112. Fruit capsules once open do not close again completely, or close only partially, with or without closing bodies and valve wings	...113
– Fruit capsules open and close readily when hydrated and dried, or split into nutlets leaving basket-like capsule remains	...114
113. Shrubs low and spreading or tall and erect; leaves boat-shaped to club-shaped with slightly flattened upper surfaces; flowers in much-branched clusters, pedicels without bracts; fruit capsules with 5 or 6 locules	...*Stoeberia*
– Shrubs erect and stems woody with long internodes and persistent remains of old leaves; leaves 3-angled with a sharp bottom edge, stem-sheathing, long (up to 70 mm); flowers in groups of 3 to 7 on short pedicels with bracts; fruit capsules with 6 to 9, mostly 8, locules	...*Stayneria*

114. Fruit an indehiscent capsule with 1 or 2 seeds per locule that has a basket-like appearance as walls disintegrate and a network of veins remains	...*Ruschianthemum*
– Fruit capsules with more than 2 seeds per locule that does not break up leaving a network of veins	...115
115. Plants robust, woody and erect, grey-green leaves often velvety; sepals 6; fruit capsules 6-locular; flowers solitary or in three's, large (up to 70 mm in diameter), white, lemon, orange, pink, purple or red; seeds coarsely hairy (echinate)	...*Astridia*
– Plants erect or creeping; leaves in various shades of green, usually smooth, often dotted; sepals 4 but mostly 5; fruit capsules with up to 18 locules; flowers solitary or in clusters, white or various shades of pink or purple; seeds smooth or textured	...116
116. Fruit capsules mostly 8 to 10-locular, with up to 18 locules, with valve wings; flowers white, petals grouped into 5 bundles, bracts enclose lower part of the flower; leaves fused, with diverging pointed tips and longitudinal markings	...*Polymita*
– Fruit capsules 5 or 6-locular, without or rarely with valve wings; flowers pink to purple, rarely white; petals sometimes grouped into 5 bundles; bracts and sometimes bracteoles present on flower stalks; leaves 3-angled or short and thick, sometimes finely toothed	...117
117. Stems whitish or pale yellow; leaves grey-green, short and thick with an abrupt point; fruit capsules 5-locular with broad, tapering valve wings; flowers with or without petals, in richly branched spiny clusters, tips of spines blunt to the touch	...*Eberlanzia*
– Stems reddish, brownish or greenish and succulent; leaves variable in colour, mostly 3-sided; fruit capsules 5-locular (rarely 6-locular), mostly without, rarely with valve wings; flowers with pink to purple petals, rarely white, petals often arranged in 5 bundles; flowers solitary or in clusters, inflorescences sometimes spiny, tips of spines not blunt	...*Ruschia*

LEIPOLDTIA-LIKE SHRUBBY MESEMBS

118. Fruit capsules with 6 to 8 locules; sepals mostly 4	...119
– Fruit capsules with 10 (up to 16) locules; sepals 5	...120
119. Plants with low, vegetative growth and taller, highly succulent floral branches, with the lower leaves markedly different from those within the flower clusters; fruit capsules close to the *Ruschia*-type, without valve wings but with large closing bodies	...*Ottosonderia*
– Branches succulent when young, becoming woody with age, vegetative and reproductive branches not differentiated; fruit capsules firm and brown, close to the *Leipoldtia*-type with large, white closing bodies	...*Octopoma*
120. Plants with numerous short shoots forming compact shrubs and with several long shoots which may be trailing or climbing onto other shrubs; leaf pairs fused with free tips, forming bodies during the dry season	...*Vanzijlia*
– Plants not differentiated into long and short shoots; leaf pairs fused at the base, free for most of their length	...121
121. Plants with long thin internodes, more or less creeping; flowers borne close to the ground; fruit capsules turret-like with prominent elevations on the covering membranes	...*Hallianthus*
– Plants erect to low-reclining with upright tips, sometimes spiny, fruit capsules of the *Leipoldtia*-type, whitish grey	...*Leipoldtia*

Index

INDEX INFORMATION

bold italics
genera & species described or illustrated

italics
all other genera & species mentioned

normal
all common names

CAPITAL LETTERS
GROUP NAMES

bold page numbers
description and/or illustration page

Acrodon 24, 177, **178**, 354
Acrodon bellidiflorus 178, **179**
Aethephyllum 24, 57, **58**, 70
Aethephyllum pinnatifidum 58, **59**
Aizoaceae 7
Aloinopsis 24, 155, **156**, 162, 198
Aloinopsis hilmarii 160, **161**
Aloinopsis lodewykii **159**
Aloinopsis loganii **158**
Aloinopsis luckhoffii 156, **157**
Aloinopsis malherbei 154, **157**
Aloinopsis orpenii 156, **158**
Aloinopsis rosulata 156, **159**
Aloinopsis rubrolineata 156, **157**
Aloinopsis schooneesii 156, **159**
Aloinopsis setifera 13, **154**
Aloinopsis spathulata 156, **157**
Aloinopsis villetii 156, **159**
altydvygie 322
Amoebophyllum 42
Amphibolia 25, 315, **316**, 354
Amphibolia maritima **317**
angular pigface 252
Anisocalyx 142
Anisocalyx salarius 142
Antegibbaeum 24, 133, **134**, 204, 334
Antegibbaeum fissoides **135**
Antimima 25, 188, 245, **246**, 278, 354
Antimima evoluta **250**
Antimima fenestrata **249**
Antimima lawsonii 246, **247**
Antimima leipoldtii **249**
Antimima meyerae **251**
Antimima piscodora 246
Antimima pygmaea 246, **249**
Antimima saxicola 246
Antimima subtruncata 246
Antimima ventricosa 246, **247**
Apatesia-group 57
Apatesia 24, 57, **60**, 68
Apatesia helianthoides **61**
Apatesia sabulosa 60, **61**
Aptenia 14, 16, 24, 27, **28**, 34, 48
Aptenia cordifolia 8, **26**, 28, **29**
Aptenia lancifolia 28, **29**
Aptenia sp. **29**
Arenifera 25, 224, 345, **346**
Arenifera pillansii 346, **347**
Arenifera stylosa **347**
Argyroderma 24, 81, **82**, 104, 246
Argyroderma crateriforme **85**
Argyroderma delaetii 82, **83**, 85
Argyroderma fissum 82, **85**
Argyroderma framesii subsp. *framesii* 83

Argyroderma framesii subsp. *hallii* **85**
Argyroderma pearsonii 82, **85**
Argyroderma testiculare **83**, 84
Aridaria 24, 27, **30**, 42
Aridaria brevicarpa 30, **31**
Aridaria noctiflora 30, **31**
Aridaria serotina **31**
asbos 14, 48
asbosskerm 14, 48
Aspazoma 24, 27, 28, **32**, 34, 48
Aspazoma amplectens **33**
Astridia 25, 334, 345, **348**
Astridia citrina **349**
Astridia hallii **349**
Astridia longifolia 348, **349**
Astridia speciosa **349**
axile placentation 20, 27
bababoudjies 82
baby toes 100
basket mesemb 362
BEAD-LEAVED MESEMBS 16, 25, 188, **233**, 236
beeskloutjie 112
beesvygie 354
Bergeranthus 24, 201, **202**, 214
Bergeranthus artus **203**
Bergeranthus glenensis 202
Bergeranthus jamesii **203**
Bergeranthus scapiger **203**
Bergeranthus sp. **203**
bergvygie 300
Berrisfordia 88
Bijlia 24, 133, **136**, 138
Bijlia dilatata **132**, 136, **137**
Bijlia tugwelliae **137**
bishop's cap 240
Bok Bay vygie 8, 70
bokbaaivygie 70
brack mesemb 28, 162
brakslaai 38
brakveldvygie 42, 162
brakveldwitvygie 30
brakvy 38
brakvygie 28, 162
Braunsia 21, 25, 275, **276**, 334
Braunsia apiculata **11**, 276, **277**
Braunsia geminata **277**
Braunsia stayneri **277**
Braunsia vanrensburgii **274**, 276
Brownanthus 24, 27, 28, 32, **34**, 48
Brownanthus ciliatus 34, **35**
Brownanthus corallinus 34, **35**
Brownanthus marlothii **35**
Brownanthus nucifer **35**

Brownanthus pubescens **35**
Buckbay-vygie 70
button plant 86
Calamophyllum 24, 201, **204**
Calamophyllum sp. **205**
calendar plant 240
Callistigma 38
Capsules (see fruit capsules)
Carpanthea 16, 24, 57, 60, **62**, 68
Carpanthea pomeridiana 62, **63**
Carpobrotus 8, 10, 14, 16, 25, 245, **252**, 260, 284, 324
Carpobrotus acinaciformis 14, **244**, 252, **255**
Carpobrotus aequilaterus 252, **253**, 254
Carpobrotus chilensis **253**, 254
Carpobrotus deliciosus 14, 252, **255**
Carpobrotus dimidiatus 252, **255**
Carpobrotus edulis 14, **15**, 254
Carpobrotus edulis subsp. ***edulis*** 252, **255**
Carpobrotus edulis subsp. *parviflorus* 252
Carpobrotus glaucescens **253**, 254
Carpobrotus mellei 252, **255**
Carpobrotus modestus 252, 254
Carpobrotus muirii 254
Carpobrotus quadrifidus **253**, 254
Carpobrotus rossii **253**, 254
Carpobrotus virescens **253**, 254
Carruanthus 24, 177, **180**
Carruanthus peersii **176**, 180, **181**
Carruanthus ringens 180, **181**
Caryotophora 24, 57, **64**, 78
Caryotophora skiatophytoides **56**, 64, **65**
"***Caulipsolon***" 24, 27, **36**
"***Caulipsolon rapaceum***" 36, **37**
Cephalophyllum 10, 25, 206, 245, **256**, 334
Cephalophyllum subgenus *Cephalophyllum* 256
Cephalophyllum subgenus *Homophyllum* 256
Cephalophyllum alstonii 256, **257**
Cephalophyllum caespitosum 257
Cephalophyllum diversiphyllum 257
Cephalophyllum ebracteatum 244
Cephalophyllum fulleri 257
Cephalophyllum loreum 259
Cephalophyllum numeesense 259
Cephalophyllum parvulum 12
Cephalophyllum pillansii 256, **259**
Cephalophyllum rigidum 257
Cephalophyllum spissum 257
Cephalophyllum spongiosum 259
Cephalophyllum tricolorum 259
Cerochlamys 24, 133, **138**
Cerochlamys pachyphylla 138, **139**

Chasmatophyllum 21, 24, 168, 177, **182**
Chasmatophyllum braunsii **183**
Chasmatophyllum musculinum 182, **183**
Cheiridopsis 24, 108, 188, 198, 201, **206**
Cheiridopsis acuminata **207**
Cheiridopsis aurea 206
Cheiridopsis cigarettifera 206, **207**
Cheiridopsis cigarettifera
Cheiridopsis glomerata 206, **209**
Cheiridopsis imitans **207**
Cheiridopsis meyeri **200**
Cheiridopsis peculiaris 206, **209**
Cheiridopsis pillansii **207**
Cheiridopsis purpurea 206
Cheiridopsis robusta 206, **209**
Cheiridopsis rostrata **209**
Cheiridopsis speciosa 206
Cheiridopsis umdausensis **208**, **209**
Circandra 25, 315, **318**
Circandra serrata 12, 318, **319**
CITES 12
Cleretum 24, 57, 58, **66**, 70, 72
Cleretum herrei **56**, 66, **67**
Cleretum lyratifolium 66
Cleretum papulosum 66
Cleretum papulosum subsp. *papulosum* 66
Cleretum papulosum subsp. ***schlechteri*** 66, **67**
clock plant 206, 214, 240
closing bodies 20
closing rodlets 20
cone plant 86
Conicosia 24, 57, 60, **68**
Conicosia elongata 68, **69**
Conicosia pugioniformis 68, **69**
Conicosia pugioniformis subsp. *muirii* 68
cono 86
Conophyllum 240
Conophytum 16, 21, 24, 81, **86**, 104, 112
Conophytum angelicae subsp. ***angelicae*** 91
Conophytum burgeri 13, 86, **87**, 88
Conophytum calculus 80
Conophytum ectypum subsp. ***ectypum*** 91
Conophytum flavum 19
Conophytum hians 87
Conophytum limpidum 89
Conophytum lithopsoides 80
Conophytum longum 91
Conophytum luckhoffii 87
Conophytum maughanii subsp. ***armeniacum*** 93
Conophytum meyeri 87
Conophytum minimum 19
Conophytum obcordellum 88, **93**

Conophytum pellucidum var. **neohallii** 93
Conophytum rugosum 86, **91**
Conophytum schlechteri 89
Conophytum smorenskaduense 86, **89**
Conophytum sp. **9**
Conophytum tantillum subsp. *lindenianum* **91**
Conophytum verrucosum 93
conservation 12
coral plant 166
Corpuscularia 25, 275, **278**, 294
Corpuscularia lehmannii 278, **279**
Corpuscularia taylorii 278, **279**
covering membranes 20
Crocanthus 268
Cryophytum 38
cultivation 16
cuttings 16
Cylindrophyllum 24, 201, **210**
Cylindrophyllum comptonii 211
Cylindrophyllum dyeri **200**, 210, **211**
Cylindrophyllum hallii 210
Cylindrophyllum tugwelliae 210, **211**
Dactylopsis digitata 42
Dactylopsis 42
dagger plants 220
dehiscence 20
Deilanthe 24, 155, **160**
Deilanthe peersii 160, **161**
Deilanthe thudichumii 160, **161**
Deilanthe sp. **161**
Delosperma 10, 16, 20, 21, 25, 278, 282, 291, **292**, 300, 306, 308, 310
Delosperma-type capsule 21, **22**, 25, 245, 278, 291
Delosperma ashtonii 293
Delosperma asperulum 293
Delosperma carolinense 292
Delosperma cooperi 295
Delosperma echinatum 293
Delosperma esterhuyseniae 294, **297**
Delosperma floribundum 292, **295**
Delosperma herbeum 292, **295**
Delosperma leendertziae 297
Delosperma litorale 292, 294, **295**
Delosperma macellum 292
Delosperma nubigenum 8
Delosperma obtusum 297
Delosperma ornatulum 292
Delosperma pottsii 299
Delosperma prasinum 299
Delosperma repens **293**, 294
Delosperma rileyi 293
Delosperma rogersii 293

Delosperma sutherlandii 297
Delosperma vinaceum 297
Delosperma vogtsii 292
Derenbergia 88
Derenbergiella 38
Dicrocaulon 25, 233, **234**, 238, 242, 378
Dicrocaulon brevifolium 234, **235**
Dicrocaulon grandiflorum 234, **235**
Dicrocaulon humile 234
Dicrocaulon sp. **232**
Didymaotus 24, 81, **94**
Didymaotus lapidiformis 95
Dinteranthus 18, 24, 81, **96**, 110
Dinteranthus subgenus *Lapidaria* 110
Dinteranthus microspermus subsp. **puberulus** 97
Dinteranthus pole-evansii 96, **97**
Dinteranthus pole-evansii
Dinteranthus vanzylii **80, 97**
Diplosoma 24, 81, **98**
Diplosoma luckhoffii 98, **99**
Diplosoma retroversum 98, **99**
dispersal 10, 20
Disphyma 10, 16, 25, 245, **260**, 268
Disphyma australe 260, **261**
Disphyma crassifolium **244**, 260, 268
Disphyma crassifolium subsp. *clavellatum* 260, **261**
Disphyma crassifolium subsp. *crassifolium* 260, **263**
Disphyma dunsdonii 260, **263**
Disphyma papillatum 260
dissemination 20
distribution 10
donkievye 366
donkievygie 308, 310
doringvygie 350, 354
Dorotheanthus-group 57
Dorotheanthus 16, 24, 57, 58, 66, **70**
Dorotheanthus apetalus 70, 72
Dorotheanthus bellidiformis **17**, 70
Dorotheanthus bellidiformis subsp. **bellidiformis** 70, **71**, 72, **73**
Dorotheanthus bellidiformis subsp. *hestermalensis* 70, 72
Dorotheanthus booysenii 70, 72, **73**
Dorotheanthus criniflorus 70
Dorotheanthus gramineus 70, 72, **73**
Dorotheanthus maughanii 70, **71**, 72
Dorotheanthus rourkei **56**, 70, **71**, 72
Dorotheanthus tricolor 70
Dorotheanthus ululars 72, **73**
Dracophilus 24, 120, 133, **140**
Dracophilus dealbatus 141

Dracophilus delaetianus 141
Dracophilus montis-draconis 141
Dracophilus proximus 141
Drosanthemopsis 24, 133, **142**
Drosanthemopsis vaginata 143
Drosanthemum 16, 21, 25, 142, 268, 291, 294, **300**, 308, 326
Drosanthemum-type capsule 21, **22**, 25, 291
Drosanthemum albiflorum 303
Drosanthemum bellum 305
Drosanthemum bicolor 17, 302, **305**
Drosanthemum calycinum 303
Drosanthemum candens 300
Drosanthemum collinum 301, 302
Drosanthemum diversifolium 305
Drosanthemum floribundum 302, **305**
Drosanthemum hispidum 290, 302, **303**
Drosanthemum latipetalum 303
Drosanthemum micans 17, 302, **305**
Drosanthemum speciosum 9, **11**, 300, **301**, **305**
Drosanthemum splendens 301
Drosanthemum striatum 290, 300, 302
Drosanthemum sp. 17, **303**
druiwetrosvygie 264
duimpie-snuif 104
duinevygie 310
dumplings 86
DWARF SHRUBBY MESEMBS 21, 25, 254, **275**, 276
Eberlanzia 21, 25, 345, **350**, 354
Eberlanzia sedoides 350, **351**
Eberlanzia sp. **351**
Ebracteola 10, 24, 201, **212**, 354
Ebracteola derenbergiana 212, **213**
Ebracteola fulleri 212, **213**
Ebracteola montis-moltkei 212, **213**
Ebracteola wilmaniae 200, 212, **213**
Echinus 276
ecology 10
Ectotropis 25, 291, **306**
Ectotropis alpina 307
edible fruits 14
eendvoetvygie 98
elephant's feet 102
Enarganthe 25, 315, **320**
Enarganthe octonaria 320, **321**
Endangered 12
Erepsia 25, 315, **322**
Erepsia aspera 323
Erepsia inclaudens 322, **323**
Erepsia lacera 322, **325**
Erepsia forficata 323
Erepsia pillansii 322, 324, **325**

Erepsia pillansii
Erepsia promontorii 12, 322
Erepsia saturata 325
ertjievygie 242
eseloor 206
Esterhuysenia 21, 25, 275, **280**
Esterhuysenia alpina 280, **281**, 307
Esterhuysenia alpina 280
Esterhuysenia sp. **274**, **281**
Eurystigma 40
Eurystigma clavatum **11**, **41**
expanding keels 10, 20
Extinct 12
Faucaria 24, 177, 180, **184**, 192, 194, 210, 220
Faucaria bosscheana 228
Faucaria britteniae 185
Faucaria candida 185
Faucaria felina **176**, 184
Faucaria gratiae 187
Faucaria longidens 176
Faucaria tigrina 184, **187**
Faucaria tuberculosa **185**, **187**
feeding 16
Fenestraria 24, 81, **100**, 102
Fenestraria rhopalophylla 101
Fenestraria rhopalophylla subsp. **aurantiaca** 100, **101**
Fenestraria rhopalophylla subsp. **rhopalophylla** 100, **101**
fertilizers 16
fig-marigolds 7
FLAT-LEAVED MESEMBS 24, **57**
FLOWERING STONES 7, 24, **81**
Frithia 24, 81, **102**
Frithia pulchra 102, **103**
Frithia pulchra var. *minor* 102
fruit capsules 20
fruit capsule types 20, 21, **22**, 23
fujiyama plant 88
fungicide 18
gansies 68
gaukum 252
geelbergvygie 182
geelswaelstertvygie 182
geelvingerkanna 268
geelvingervygie 268
geophytes 27
germination 10, 20
Gibbaeum 21, 24, 81, **104**, 106, 134, 150
Gibbaeum album 104, **105**, 106, 118
Gibbaeum dispar 104, **105**
Gibbaeum esterhuyseniae 105
Gibbaeum gibbosum 104, **105**

Gibbaeum haagei var. ***parviflorum*** 107
Gibbaeum heathii 104, **105**, 106
Gibbaeum johnstonii 107
Gibbaeum pachypodium 104, 106, **107**
Gibbaeum petrense 105, **106**
Gibbaeum pilosulum 107
Gibbaeum pubescens subsp. ***pubescens*** 107
Gibbaeum velutinum 104, **107**
GLITTERING SHRUBBY MESEMBS 21, 25, 291
Glottiphyllum 24, 133, **144**
Glottiphyllum depressum 144
Glottiphyllum grandiflorum 146
Glottiphyllum latum var. *cultratum* 144
Glottiphyllum linguiforme 144, **145**
Glottiphyllum longum 132, 144, **145**
Glottiphyllum muirii 144
Glottiphyllum neilii 147
Glottiphyllum nelii 132, **147**
Glottiphyllum ochraceum 146, **147**
Glottiphyllum peersii 144, **147**
Glottiphyllum regium 147
Glottiphyllum sp. **145**
glue plant 224
gomvy 224
greenhouses 18
grootlidjies 48
habitat destruction 12
habitat transformation 12
Halenbergia 40
Hallianthus 25, 369, **370**
Hallianthus planus 368, **371**
Hammeria 24, 177, **188**
Hammeria salteri 188, **189**
Hartmanthus 25, 275, **282**, 294
Hartmanthus hallii 282, **283**
Hartmanthus pergamentaceus 282, **283**
harvesting seeds 18
Henricia 166
Hereroa 24, 201, 202, 212, **214**, 226
Hereroa angustifolia 217
Hereroa brevifolia 214
Hereroa calycina 217
Hereroa dyeri 214
Hereroa fimbriata 214
Hereroa glenensis 214
Hereroa pallens 215
Hereroa puttkameriana 217
Hereroa tenutifolia 215
Hereroa tugwelliae 136
Hereroa wilmaniae 212
Herrea 68
Herreanthus 88

hitchhiker plant 42
hongerdoring 308
horticulture 8, 14
Hydrodea 40
hygrochastic 20
Hymenocyclus 268
Hymenogyne 24, 57, 60, 68, **74**
Hymenogyne conica 74
Hymenogyne glabra 74, **75**
ice plant 7, 38
identification 20, 381
Ihlenfeldtia 24, 81, **108**, 198
Ihlenfeldtia excavata 108, **109**
Ihlenfeldtia vanzylii 108, **109**
Imitaria 106
Indeterminate 12
indoor cultivation 18
inland pigface 252
Insufficiently Known (conservation status) 12
invasion of alien plant species 12
IUCN Red Data categories 12
Jacobsenia 25, 233, **236**
Jacobsenia hallii 236, **237**
Jacobsenia kolbei 236, **237**, 240
jakkalsniertjie 82
jam 15
Jensenobotrya 25, 245, **264**, 275
Jensenobotrya lossowiana 265
Jordaaniella 16, 25, 245, **266**, 334
Jordaaniella clavifolia 267
Jordaaniella cuprea 267
Jordaaniella dubia 244, **267**
Juttadinteria 24, 120, 133, 140, **148**
Juttadinteria subgenus *Namibia* 120
Juttadinteria albata 148, 282
Juttadinteria attenuata 148
Juttadinteria ausensis 148, **149**
Juttadinteria deserticola 148, **149**
Juttadinteria elizae 148
Juttadinteria insolita 148
Juttadinteria kovisimontana 148, **149**
Juttadinteria simpsonii 148, **149**
Juttadinteria suavissima 148, **149**
Juttadinteria tetrasepala 148
kalk(veld)vygie 172
kalkklipvygie 292
kama 38
kanna 52
kareemoervygie 310
Kensitia 324
key to groups and genera 381
khadi root 218
khadi 218
Khadia 10, 14, 24, 201, 214, **218**

Khadia acutipetala 218, **219**
Khadia alticola 218, **219**
Khadia borealis 218, **219**
Khadia carolinensis 218, **219**
khadiwortel 218
kierievygie 310
Kirstenbosch 17
klein-s'keng-keng 214
klipplant 124
klipvygie 110, 128, 246, 292
knopies 86
kookskerm 14, 48
kougoed 14, **15**, 52
kussingvygie 194, 246, 372
kwaggavy 124
LAMPRANTHUS-LIKE SHRUBBY MESEMBS 21, 25, **315**
Lampranthus 10, 16, 18, 25, 276, 280, 315, 316, **326**, 338, 354
Lampranthus-**type capsule** 21, **23**, 25, 275, 315
Lampranthus amoenus 314
Lampranthus arbuthnotiae 326
Lampranthus aurantiacus 331
Lampranthus aureus 326, **331**
Lampranthus austricolus 329
Lampranthus bicolor 333
Lampranthus blandus 326, **329**
Lampranthus coralliflorus 327
Lampranthus emarginatus 329
Lampranthus explanatus 331
Lampranthus falciformis 329
Lampranthus filicaulis 333
Lampranthus haworthii 326, **329**, **333**
Lampranthus lunatus 326
Lampranthus multiradiatus 327
Lampranthus ornatus 326
Lampranthus piquetbergensis 337
Lampranthus reptans 333
Lampranthus roseus 326, **327**
Lampranthus saturatus 329
Lampranthus sauerae 331
Lampranthus schlechteri 12
Lampranthus steenbergensis 337
Lampranthus tegens 327
Lampranthus vanzijliae 12
Lampranthus variabilis 331
Lampranthus vernalis 327
Lampranthus sp. 333
Lapidaria 24, 81, **110**, 152
Lapidaria margaretae 111
legislation 12
LEIPOLDTIA-LIKE SHRUBBY MESEMBS 21, 25, **369**

Leipoldtia 10, 25, 266, 369, 370, **372**
Leipoldtia-**type capsule** 21, **23**, 25, 245, 275, 286, 369, 372
Leipoldtia frutescens 373
Leipoldtia weigangiana subsp. **grandifolia 373**
Leipoldtia schultzei 368, **373**
Leipoldtia uniflora 373
lewerplant 124
lewervygie 124
lidjiesganna 308
Lithops 16, 18, 20, 24, 81, 86, 88, 96, 110, **112**, 150
***Lithops divergens* 113**
***Lithops herrei* 117**
***Lithops julii* 19**, **115**
Lithops karasmontana subsp. ***eberlanzii*** 117
Lithops karamontana var. ***tischeri* 115**
Lithops lesliei subsp. ***burchellii* 115**
Lithops lesliei var. ***lesliei* 113**
***Lithops localis* 113**
***Lithops marmorata* 117**
***Lithops meyeri* 117**
***Lithops naureeniae* 113**
***Lithops optica* 113**
***Lithops otzeniana* 115**
Lithops ruschiorum 80, **117**
***Lithops schwantesii* 117**
Lithops schwantesii var. ***marthae* 115**
***Lithops steineckeana* 9** ,114
***Lithops terricolor* 113**
liver plant 124
Livingstone daisy 70
locules 20
loogasbossie 48
Machairophyllum 24, 201, **220**
***Machairophyllum acuminatum* 221**
Machairophyllum albidum 220, **221**
Machairophyllum brevifolium 220, **221**
***Machairophyllum latifolium* 221**
Malephora 10, 25, 245, **268**, 278
Malephora crassa 268, **271**
Malephora crocea 268
Malephora crocea var. ***crocea* 269**
Malephora crocea var. ***purpureo-crocea* 269**
Malephora herrei 268
Malephora lutea 268, **270**
Malephora lutea 268
Malephora uitenhagensis 268
mandjievygie 362
Marlothistella 24, 201, **222**
Marlothistella stenophylla 200, **223**
MAT-FORMING MESEMBS 25, **245**, 284

Maughania 98
Maughaniella 98
medicinal uses 14
Mesembrianthemum 38
Mesembryanthemaceae 7
Mesembryanthemoideae 20, 27
Mesembryanthemum 10, 14, 24, 27, **38**, 42, 54
***Mesembryanthemum*-type capsule** 21, **22**
***Mesembryanthemum alatum* 39**
***Mesembryanthemum barklyi* 39**
Mesembryanthemum cryptanthum 40, **41**
Mesembryanthemum crystallinum 27, **39**, 40
Mesembryanthemum fissoides 204
Mesembryanthemum guerichianum 26, **39**
***Mesembryanthemum hypertrophicum* 41**
***Mesembryanthemum longistylum* 41**
***Mesembryanthemum* sp.** 9, **11**, **26**, **41**
Mestoklema 10, 14, 21, 25, 291, 294, **308**
***Mestoklema arboriforme* 309**
***Mestoklema tuberosum* 309**
Meyerophytum 25, 233, **238**
Meyerophytum meyeri var. ***holgatense*** 238, **239**
Meyerophytum meyeri var. ***meyeri*** 239
Micropterum 66
midday flowers 7
Mimetophytum 240
mimicry plants 94, 104, 134, 156, 168
Mitrophyllum 10, 25, 233, 236, **240**, 242
Mitrophyllum dissitum 240, **241**
Mitrophyllum grande 240, **241**
***Mitrophyllum mitratum* 241**
Mitrophyllum roseum 240
***Mitrophyllum abbreviatum* x *M. roseum* 241**
moervygie 162
moerwortel 14
moerwortelvygie 14
Monilaria 25, 233, 234, 238, **242**
Monilaria chrysoleuca 232, **243**
Monilaria chrysoleuca var. *chrysoleuca* 242
Monilaria moniliformis 242, **243**
Monilaria obconica 242, **243**
***Monilaria scutata* 243**
Mossia 25, 245, **272**
Mossia intervallaris 272, **273**
Mossia intervallaris 272
Muiria 24, 81, **118**
Muiria hortenseae 118, **119**
Muirio-gibbaeum 118
muiskopvygie 118
muisvygie 354
namakwavygie 334

Namaquanthus 25, 206, 315, 320, **334**
***Namaquanthus vanheerdei* 335**
Namibia 24, 81, **120**, 140
Namibia cinerea 120, **121**
Namibia ponderosa 120, **121**
Nananthus 10, 24, 155, 156, **162**
Nananthus aloides 162, **163**
***Nananthus broomii* 163**
***Nananthus transvaalensis* 163**
Nananthus vittatus 162
Nananthus wilmaniae 162
***Nananthus* sp. 164**, **165**
Nelia 24, 133, **150**, 278, 334
Nelia meyeri 150
Nelia meyeri var. *longipetala* 150
Nelia pillansii 150, **151**
Nelia robusta 150
Nelia schlechteri 150, **151**
Neohenricia 24, 155, **166**, 210
Neohenricia sibbettii 154, 166, **167**
Neohenricia spiculata 166, **167**
not threatened 12
nuta 38
Octopoma 25, 198, 369, **374**
***Octopoma calycinum* 375**
***Octopoma octojuge* 375**
Odontophorus 24, 177, **190**
Odontophorus angustifolius subsp. ***angustifolius*** 190, **191**
Odontophorus angustifolius subsp. *protoparcoides* 190
Odontophorus angustifolius subsp. ***protoparcoides* 176**
Odontophorus angustifolius 190
Odontophorus herrei 190
Odontophorus marlothii 190, **191**
Odontophorus nanus 190, **191**
ogies 86
olifantslaai 38
Oophytum 24, 81, **122**
Oophytum nanum 122, **123**
Oophytum nordenstamii 122, **123**
Oophytum oviforme 122, **123**
Ophthalmophyllum 88
Opophytum 40
Opophytum aquosum 26, **41**
opstandingsvygie 234
Orthopterum 24, 177, **192**
Orthopterum coegana 192, **193**
Orthopterum waltoniae 192, **193**
Oscularia 25, 280, 315, 328, **336**
Oscularia deltoides 336, **337**
***Oscularia* sp. 337**
Ottosonderia 25, 369, **376**

Ottosonderia monticola 376, **377**
Ottosonderia monticola 376
Ottosonderia obtusa 376
oumasepram 42
outdoor cultivation 16
over-collection 12
Paarlvygie 322
papegaaibek 104
parietal placentation 20
Peersia 168, 170
perdeklou 112
perdevy 252
perdevygie 310
Pherelobus 72
Phyllobolus 16, 24, 27, **42**
Phyllobolus subgenus *Aridaria* 30
Phyllobolus subgenus *Prenia* 46
Phyllobolus subgenus *Sceletium* 52
***Phyllobolus abbreviatus* 43**
***Phyllobolus canaliculatus* 45**
Phyllobolus digitatus 42
***Phyllobolus digitatus* subsp. *digitatus* 45**
***Phyllobolus digitatus* subsp. *littlewoodii* 45**
***Phyllobolus rabiei* 42, 43**
***Phyllobolus roseus* 45**
Phyllobolus splendens 42
***Phyllobolus tenuiflorus* 43**
***Phyllobolus tetragonus* 43**
***Phyllobolus* sp. 43**, **45**
Piketbergvygie 322
pioneers 10
plakkertjie 346
platblaarvygie 78
Platythyra haeckeliana 28, **29**
Pleiospilos 24, 81, **124**, 128
Pleiospilos bolusii 80, 124, **126**, **127**
Pleiospilos borealis 124
Pleiospilos compactus 9
Pleiospilos compactus subsp. *canus* 124
***Pleiospilos compactus* subsp. *minor* 13**, 124
***Pleiospilos compactus* subsp. *sororius* 125**
Pleiospilos nelii 80, 124, **125**
Pleiospilos simulans 124, **127**
pollination 10
Polymita 25, 345, **352**, 340
Polymita albiflora 352, **353**
Polymita albiflora 352
Polymita steenbokensis 352, **353**
porseleinbos 300
Prenia 24, 27, 30, **46**
Prenia pallens 46
***Prenia pallens* subsp. *lutea* 47**
Prenia sladeniana 46, **47**
Prenia tetragona 46, **47**

Prenia vanrensburgii 46, **47**
Prepodesma orpenii 156
Prince Albert vygie 136
pronkvingertjies 138
propagation from seeds 18
Psammophora 24, 168, 201, **224**
Psammophora longifolia 224, **225**
Psammophora modesta 224, **225**
Psammophora nissenii 224, **225**
Psammophora saxicola 224, **225**
Pseudobrownanthus 34
Psilocaulon 10, 14, 24, 27, 34, 36, 46, **48**
***Psilocaulon articulatum* 49**
Psilocaulon coriarium 14, 48
***Psilocaulon dinteri* 48, 51**
***Psilocaulon granulicaule* 48, 51**
Psilocaulon junceum 14, 15, 48, **49**
***Psilocaulon leptarthron* 49**
Psilocaulon parviflorum 48, **50**
Psilocaulon rapaceum 36
***Psilocaulon* sp. 26**
Rabiea 24, 201, 202, **226**
Rabiea albinota 226, **227**
***Rabiea albipuncta* 227**
***Rabiea difformis* 227**
rankvygie 256, 266, 326
Rare 12
Rare/Vulnerable 12
Red Data conservation status 12
Red Data List of southern African Plants 12
Rhinephyllum 24, 155, **168**, 210
Rhinephyllum broomii 168, **171**
Rhinephyllum comptonii 168
Rhinephyllum frithii 168, **171**
Rhinephyllum inaequale 168
Rhinephyllum macradenium 168, **169**
Rhinephyllum muirii 168, **169**
***Rhinephyllum parvifolium* 169**
Rhinephyllum parvifolium 168
Rhinephyllum pillansii 168, **171**
Rhinephyllum rouxii 168
Rhombophyllum 24, 201, **228**
Rhombophyllum dolabriforme 214, 228, **229**
Rhombophyllum rhomboideum 228, **229**
rooibergvygie 292, 354
rooivye 366
rooivyge 366
rooivygie 326
rosa-de-jericho 38
rotsvygie 292
ROUGH-LEAVED MESEMBS 24, **155**
RUSCHIA-LIKE SHRUBBY MESEMBS 21, 25, 345
Ruschia 10, 25, 188, 222, 246, 280, 316, 326,

340, 345, 350, **354**
***Ruschia*-type capsule** 21, **23**, 25, 345, 346, 348, 352, 364, 376
Ruschia subgenus *Spinosae* 350
Ruschia canonotata 354
Ruschia caroli 344, 354, **355**
Ruschia crassa 340, **361**
***Ruschia diversifolia* 359**
***Ruschia geminiflora* 359**
Ruschia hamata 354
***Ruschia indurata* 357**
Ruschia intrusa 354, **357**, **360**
***Ruschia laxiflora* 355**
Ruschia maxima 354, **359**
***Ruschia radicans* 359**
***Ruschia robusta* 359**
Ruschia sarmentosa 354, **355**
***Ruschia schollii* 344**
***Ruschia senaria* 357**
Ruschia spinosa 344, **357**
Ruschia unidens 354
Ruschia viridifolia 354, **355**
***Ruschia* sp. 11**
Ruschianthemum 25, 345, **362**
***Ruschianthemum gigas* 363**
Ruschianthus 24, 201, **230**
***Ruschianthus falcatus* 231**
Ruschioideae 20
sandsteenvygie 336
sandvygie 70, 346
Saphesia 24, 57, **76**
***Saphesia flaccida* 56**, 76, **77**
Saphesia flaccida
Sarcozona 10, 25, 254, 275, **284**
Sarcozona bicarinata 284, **285**
Sarcozona praecox 284, **285**
Sceletium 14, 21, 24, 27, **52**
Sceletium emarcidum 52, **53**
***Sceletium rigidum* 53**
Sceletium tortuosum 52, **53**
Sceletium tortuosum 52
Sceletium varians 52
***Sceletium* sp. 15, 26**
Schlechteranthus 25, 275, **286**
Schlechteranthus hallii 286, **287**
Schlechteranthus maximiliani 286, **287**
schmoo plant 118
Schonlandia 278
Schwantesia, 24, 133, **152**
Schwantesia borcherdsii 152, **153**
Schwantesia herrei 152
***Schwantesia herrei* var. *herrei* 153**
Schwantesia pillansii 152, **153**
Schwantesia ruedebuschii 152, **153**

Schwantesia triebneri 152
Scopelogena 25, 315, **338**
Scopelogena gracilis 338, **339**
Scopelogena veruculata 338, **339**
seeds 18, 20
seepampoen 46
seepbossie 48
Semnanthe 322, 324
sheep's tongue 172
Sineoperculum 72
skaapvygie 292
s'keng-keng 226
skerpioenvygie 48
Skiatophytum 24, 57, **78**
Skiatophytum tripolium 64, **79**
skilpadkos 70, 144
skilpadvoetjies 172
skotteloor 46
slaaibos 38
slaapvygie 214
slaplootvye 366
slypvygie 292
Smicrostigma 25, 315, **340**
***Smicrostigma viride* 341**
snotwortel 68
soap-leaved mesemb 150
soap-making 14
soetaartappel 310
soil 10, 16, 18
sour fig 252
soutslaai 38
soutvygie 52
sphaeroids 86
Sphalmanthus 42
split rock 124
spookvygie 288
Stayneria 25, 345, **364**
Stayneria littlewoodii 364, **365**
Stayneria neilii 344, 364, **365**
steekdoring 354
Sterropetalum 150
Stoeberia 10, 20, 25, 316, 345, **366**
Stoeberia arborea 366, **367**
Stoeberia beetzii 366, **367**
Stoeberia carpii 366, **367**
Stoeberia frutescens 366
Stoeberia utilis 366, **367**
Stomatium 24, 168, 177, **194**
***Stomatium agninum* 195**
Stomatium alboroseum 194, **197**
Stomatium ermininum 194
***Stomatium lesliei* 197**
Stomatium mustellinum 194, **197**
***Stomatium niveum* 195**

Stomatium suaveolens 195
Stomatium suricatinum 197
Stomatium villetii 195
stone face 112
stone plant 112
strandvygie 266
streepvygie 156
stryvygie 162
subfamilies 20
suurvy 252
swaelstert(tjie)vygie 354
Synaptophyllum 24, 27, **54**
Synaptophyllum juttae **26**, **55**
tankwa-beesklou 128
Tanquana 24, 81, **128**, 198
Tanquana archeri 128, **129**
Tanquana hilmarii 128, **129**
Tanquana prismatica 128, **129**
tierbekvygie 180, 184, 194
tiger-jaw 184
Tischleria 180
Titanopsis 18, 24, 108, 155, 156, 168, **172**, 198
Titanopsis calcarea 9, 155, 172, **173**, **174**, 175
Titanopsis fulleri 172, **173**, 175
Titanopsis hugo-schlechteri 172
Titanopsis hugo-schlechteri var. *alboviridis* 172, **173**, **174**
Titanopsis hugo-schlechteri var. *hugo-schlechteri* **173**, **175**
Titanopsis luederitzii 172
Titanopsis primosii **154**, 172
Titanopsis schwantesii 172, **175**
tongblaarvygie 144
TONGUE-LEAVED MESEMBS 24, **133**
toontjies 86
TOOTH-LEAVED MESEMBS 24, 152, **177**
Trichodiadema 10, 14, 21, 25, 291, 294, 300, 310
Trichodiadema barbatum 313
Trichodiadema bulbosum 313
Trichodiadema decorum 311
Trichodiadema densum 310, **311**
Trichodiadema fergusoniae 302, **303**
Trichodiadema mirabile 313
Trichodiadema pygmaeum **311**, 312
TUFTED MESEMBS 24, **201**
tweelingvygie 94
uses of mesembs 14
vaalvye 366
vaalvygie 342
Vanheerdea 24, 108, 177, **198**
Vanheerdea primosii 198, **199**
Vanheerdea roodiae 198, **199**
Vanzijlia 25, 369, **378**
Vanzijlia annulata **379**
varkslaai 68
varkswortel 68
vegetable golf ball 96
vensterplant 100
vetkousie 62
vinger-en-duim 104
vingerkanna 42, 268
vingertjie-en-duimpie 42
vingervygie 82
visbekvygie 104
vlaktevygie 162
vleisbos 286, 300, 302, 326
Vlokia 24, 81, **130**
Vlokia ater 130, **131**
volstruiskos 144
volstruisslaai 38
volstruistone 104, 134
volstruiswater 104
Vulnerable 12
vybossie 308
vyerank 252
waterblasies 86
watering 16, 18
WEEDY MESEMBS 14, 20, 21, 24, **27**, 54, 166, 210
window plant 100
wishbone plant 260
witbergvygie 292
witblomvygie 364
Wooleya 25, 315, **342**
Wooleya farinosa **343**
ysplant 70
Zeuktophyllum 21, 25, 275, **288**
Zeuktophyllum suppositum **289**